Channel-Adaptive Technologies and Cross-Layer Designs for Wireless Systems with Multiple Antennas

Channel-Adaptive Technologies and Cross-Layer Designs for Wireless Systems with Multiple Antennas

Theory and Applications

Vincent K. N. Lau
The Hong Kong University of Science and Technology

Yu Kwong Ricky Kwok
The University of Hong Kong

Published by John Wiley & Sons, Inc., Hoboken, New Jersey.
Published simultaneously in Canada.

Library of Congress Cataloging-in-Publication Data:

Lau, Vincent K. N.
Channel adaptive technologies and cross layer designs for wireless systems with multiple antennas : theory and applications / Vincent K. N. Lau, Yu-Kwong Ricky Kwok.
p. cm.
ISBN-13: 978-0-471-64865-9
ISBN-10: 0-471-64865-5
1. Wireless communication systems—Design and construction. I. Kwok, Yu-Kwong Ricky. II. Title.

TK5103.4.L56 2006 2005043673

Printed in the United States of America.

10 9 8 7 6 5 4 3 2 1

To our wives: Elvina and Fion

CONTENTS

LIST OF FIGURES

LIST OF TABLES

PREFACE

Wireless communications is one of the most active areas of research over the past and the current decades. In fact, the demand for wireless services has been changing from the regular voice telephony services to mixed voice, data, and multimedia services over the wireless media. Since the mid-1990s, the wireless industry has been advancing at an incredible speed. For example, the 2G cellular systems (such as GSM, D-AMPS, and IS-95) have significantly improved the spectral efficiency and network capacity to support wireless telephony services. Fueled by the explosion of demands for applications over the fixed-line and broadband Internet access, there is a parallel development in the wireless domains to support high-quality and high-speed data and multimedia services. For example, we have the development of 3G systems (CDMA2000, UMTS), 3.5G systems (HSDPA, EV-DO, EV-DV), B3G systems (Beyond 3G), wireless LAN (IEEE 802.11a/b/g), ultrawideband (UWB) systems, and Wi-MAX (IEEE 802.16) as well as Wi-MAN (IEEE 802.20) systems. These technologies have spurred a lot of research in the signal processing and cross-layer design for wireless communications.

Realizing reliable and efficient communications over the wireless channel has been a very challenging topic for over 50 years. This is attributed to the hostile nature of the wireless channel in the form of rapid time variation, extreme fading, and multipath. For instance, the transmission of signals over the wireless channels is affected by time-varying channel attenuation, called *fading*. The received signal strength can fluctuate over a wide range of 80 dB in the order of milliseconds. On one occasion, the transmission may experience good fading and the transmission error probability will be low. On the other hand, the transmission may experience bad fading on other occasions and the error probability will be high. Hence, in general, the fading effects of wireless channels impose additional challenges for signal transmissions besides the regular channel noise. Two very promising more recent approaches address the challenging problems of wireless transmissions: the *multiple-antenna technologies* and the *cross-layer transmitter adaptation designs*. In addition, these two techniques can be combined to achieve significant performance advantages. In the subsequent chapters, we shall follow a bottom–up

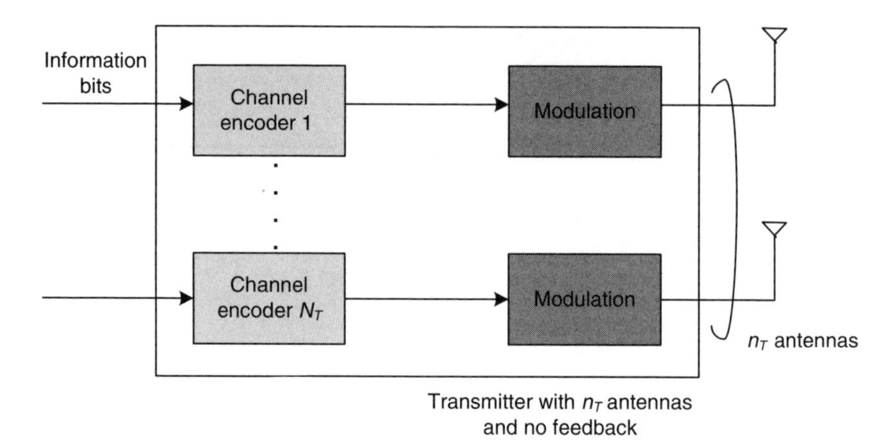

Figure P.1. Architecture of multiantenna transmitter without feedback.

approach to elaborate on the adaptive physical layer design and adaptive MAC layer design as well as adaptive routing layer design combined with multiple antenna technologies.

Multiple-Antenna Technologies

In point-to-point wireless links, the primary design objective is to increase the data rate, and a promising way to do so (without increasing the bandwidth and power budget) is through multiple-antenna technologies. Specifically, the transmitter is equipped with n_T transmit antennas and the receiver is equipped with n_R receive antennas as illustrated in Figure P.1. The antennas are assumed to be sufficiently separated so that they are spatially uncorrelated.

The advantage of having multiple antennas at the transmitter and the receiver is to transform the original wireless fading channels into *multiple-input multiple-output* (MIMO) wireless fading channels. It has been shown [126] that the link capacity can be increased by $m = \min(n_T, n_R)$ times relative to single-antenna wireless links. This is because there are $m$spatial channels created as a result of the multiple antennas and the scattering environment surrounding the transmitter and the receiver. Hence, independent information streams can be delivered on the$m$parallel spatial channels to realize the increased transmission bit rate; this is called *spatial multiplexing*. On the other hand, one can deliver the same information bits over multiple spatial channels to exploit the *spatial diversity* so as to enhance the reliability of the transmission. These important concepts of spatial diversity and spatial multiplexing have been employed in the framework of *spacetime coding* design. We will discuss the advantage of multiple antennas and spacetime coding design in Chapters 2–5.

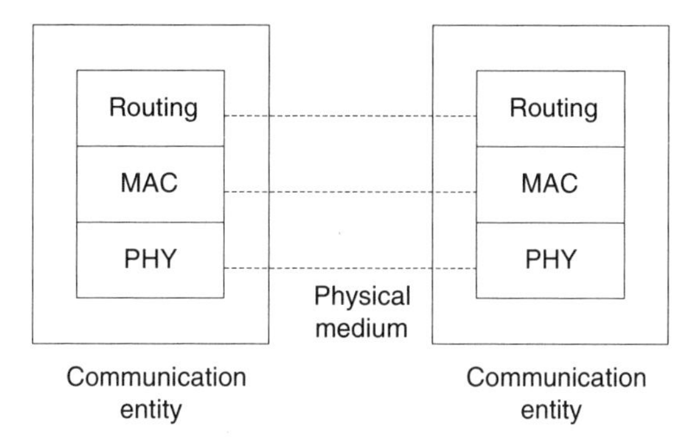

Figure P.2. Layered architecture of communication systems.

Cross-Layer Adaptive Transmission

Atypical communication system can be conveniently modeled by a layered approach (physical layer, MAC layer, routing layer) where each layer has a specific role and performance measure (as illustrated in Figure P.2). For example, the *physical layer* is responsible for the reliable and efficient delivery of information bits. The *MAC layer* is responsible for the resource management among multiple users in the system. The *routing layer* is responsible for the delivery of packets to the destination.

The traditional approach of communication system design is based on an isolated procedure while optimization is isolated within layers. In other words, there is no cross-optimization between layers. This isolated approach is reasonable for fixed-line network because the fixed-line channel is essentially time invariant. However, such an isolated approach usually results in a suboptimal design for wireless systems because the wireless channel is a time-varying channel. Hence, adaptation techniques are needed at various layers to enhance the wireless communication system design to exploit the time-varying nature of the channel. In other words, a jointly adaptive design or a *cross-layer design* is needed. The roles and challenges of the physical layer, MAC layer, and the routing layer are elaborated below.

Physical Layer. The role of the physical layer is to deliver information bits across a wireless channel in an efficient and reliable manner given a limited resource. *Resource* in this context refers to the bandwidth and transmit power; *performance* refers to the bit rate (bits per second) and the frame error rate. Information bits (source) are first protected by adding redundancy in the "channel encoder" so that error recovery is possible at the receiver. Following channel encoding is the modulation process where coded bits are mapped into physical channel symbols. Two research directions are followed to tackle the physical layer design: the information-theoretic approach and the practical coding design approach. For information-theoretic design, Shannon's coding theory has shown that error-free transmission is possible when the bit rate is less than the channel capacity. In other words, the Shannon's capacity represents the best data rate achievable given a particular channel model. Unfortunately, the information-theoretic approach does not reveal how to achieve the channel capacity.

On the other hand, the coding design approach focuses on finding practical encoding and decoding algorithms that could approach the Shannon capacity. The design objective is to increase the bit rate at a given target frame error rate with fixed bandwidth and power budget.

Various approaches have been investigated to improve the performance of the physical layer. One promising approach is to utilize channel feedback information at the transmitter. We call it the "channel—adaptive approach." Another promising approach is to utilize multiple antennas at the transmitter and the receiver. We call this the "MIMO approach" (multiple-input multiple-output). Depending on the level of feedback information available at the transmitter, we have different transmission strategies for the MIMO systems. For FDD systems with perfect feedback of channel state information (CSI), channel adaptation can be done at the transmitter as illustrated in Figure P.3. This will introduce significant capacity gain on top of the linear capacity gain of open-loop MIMO systems.

With the availability of perfect channel state information (channel matrix), power adaptation (in both the temporal and spatial domains) and rate adap-

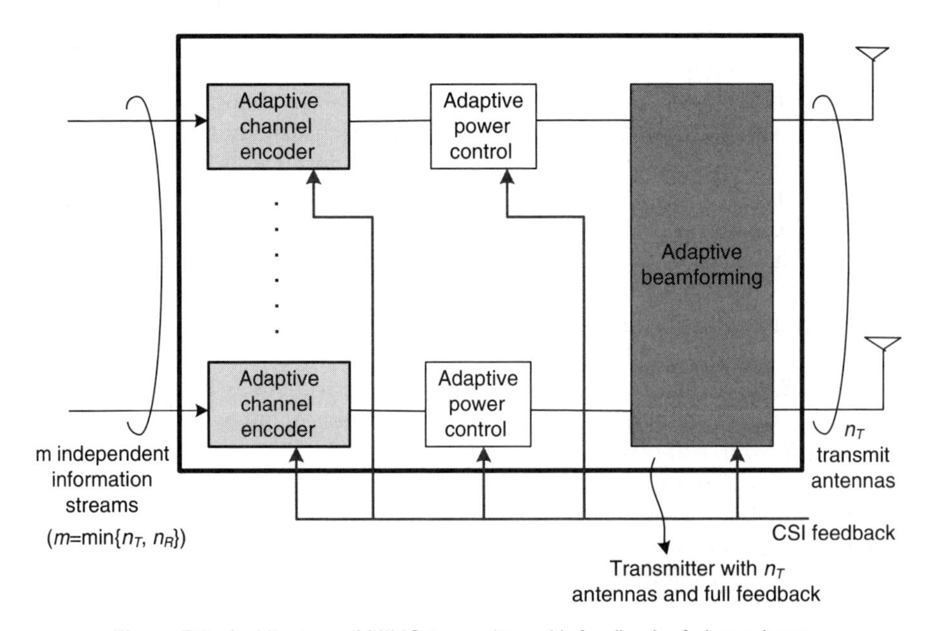

Figure P.3. Architecture of MIMO transmitter with feedback of channel state.

tation can be performed at the MIMO transmitter. In fact, the adaptation across the spatial domain contributes to significant performance gains over open-loop performance and therefore has received tremendous attention in the research community more recently.

MAC Layer. In a wireless system with a centralized access point (base station) and multiple mobile users, the physical layer design is just part of the big picture because it focuses on the point-to-point link performance only. On the other hand, the MAC layer is a very crucial component in multiuser communication systems because it is responsible for resource allocation (scheduling) among multiple competing users.

A MAC layer usually consists of a request collection sublayer and a scheduling sublayer as illustrated in Figure P.4. The request collection sublayer is responsible for the collection of payload transmission requests from the active users. On the other hand, the scheduling sublayer is responsible for the prioritization and the allocation of resource among the competing users.

Conventional MAC layer designs for wireless systems follow the isolated approach where there is no cross-optimization across the physical layer and the MAC layer. For instance, a lot of research effort has been devoted to designing efficient request collection sublayer. Examples are slotted ALOHA, dynamic TDMA, and PRMA. The scheduling sublayer is essentially very simple in the sense that the "first come–first serve" scheduling is done. The focus is to integrate realtime (voice, video) and background (email) sources nicely into the MAC layer.

More recently, because of the wide acceptance of adaptive physical layer, plenty of research effort has been devoted to considering a jointly adaptive

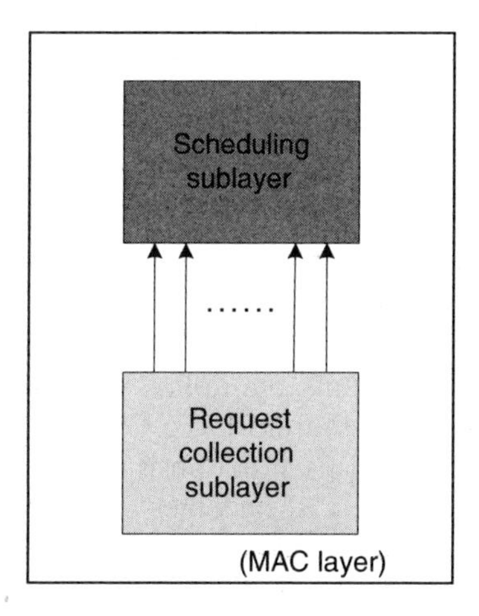

Figure P.4. Architecture of MAC layer with channel state feedback.

physical and MAC layer design, with focus on improving system performance by taking advantage of the time-varying throughput delivered by the adaptive physical layer.

We characterize "system performance" as a utility function of the throughputs achieved by all the active users. Depending on the forms of the utility function, we can have different emphasis on system performance. For example, we may focus on system capacity (where the utility function is the sum of average user throughput) for better resource utilization. We may also wish to strike a compromise between resource utilization and fairness among users, and we may set the utility function to be "proportionally fair." In any case, priority should be given to users with good channel conditions because such users could utilize the limited radio resource more effectively (enjoying a highly throughput due to the adaptive physical layer). Therefore, we could always find "good users" at any time to transmit at high-throughput modes, and as a result, the overall multiuser system capacity is greatly enhanced. This is called *multiuser selection diversity*.

Routing Layer. Ad hoc network is important for mobile devices due to its robustness with respect to hostile propagation environment. While traditional table-based or on-demand routing protocols can be used, it is much more efficient to use a routing protocol that is channel-adaptive—judiciously selecting links that can transmit at higher data rates to form a route. Devising channel-adaptive routing protocols is a very hot field. We shall provide a detailed survey of existing techniques. We also describe a reactive ad hoc routing algorithm, called *RICA* (receiver-initiated channel-adaptive) protocol, to intelligently utilize the multirate services (based on different modulation schemes). NS-2 simulation results show that the RICA protocol is highly effective.

In conclusion, channel adaptation technologies can benefit all the three layers described above. However, one fundamental requirement for channel adaptation is the knowledge of channel state information (CSI) at the transmitter. In frequency-division duplexing (FDD) systems, the CSI is estimated at the receiver and fed back to the transmitter. In time-division duplexing (TDD) systems, the CSI of the downlink can be estimated on the basis of the uplink pilots. In either case, obtaining perfect knowledge of CSI at the transmitter is not feasible, and in practice, there is always some imperfection associated with the CSI knowledge at the transmitter. For example, in FDD systems, the feedback link usually has a limited capacity and this results in *limited feedback*. In TDD systems, the uplink pilot power transmitted by the mobiles is usually limited and therefore the CSI estimation noise at the base station side is not negligible. This situation is aggregated by the presence of multiple antennas because the CSI becomes a matrix of $n_R \times n_T$ dimensions. Hence, a significant portion of the book is devoted to addressing this practical concern—channel adaptation and cross layer design in the presence of imperfect CSI.

Audience

This is a graduate-level book intended for readers who are graduate-level research students and would like to know more about the theory and practice of channel-adaptive wireless communication technologies. This book assumes that the reader has a solid background in basic communication theory and information theory as well as queueing theory.

Organization of the Book

This book is organized into three parts. In Part 1, we shall focus on the theoretical aspects of channel adaptation in wireless communications for point-to-point and multiuser systems with multiple antennas. In Part 2, we shall focus on the applications of the channel-adaptive technologies in practical systems such asUMTS. In Part 3, we shall focus on some advanced topics such as multiuser scheduling for wideband systems, combined queueing theory, and information theory as well as ad hoc routing.

Part 1. Chapter 1 discusses the basic and fundamental theories behind SISO/MIMO communications. It addresses the channel capacity as well as spacetime code design for SISO/MIMO physical layer. Readers with a strong background in communication theory and information theory can skip Chapter 1. Chapter 2 discusses the optimal transmission adaptation strategy and feedback strategy for a point-to-point multiantenna communication link with perfect CSI knowledge based on the information-theoretic approach. The notion of ergodic capacity and outage capacity will be elaborated and the design insights from the results will be discussed. Chapter 3 extends the discussion in Chapter 2 to consider the optimal transmission and feedback strategies in the presence of imperfect CSI.

Chapter 4 gives the practical design considerations of adaptive physical layers based on the theories developed in Chapter 2. Specifically, we shall elaborate on spacetime coding and decoding techniques that could achieve spatial diversity and spatial multiplexing, respectively. The optimal MIMO transmitter and receiver architectures in fast fading and slow fading channels will be discussed. The fundamental tradeoff between *spatial diversity* and *spatial multiplexing* is elaborated. Finally, the design of adaptation thresholds, modulation levels, and encoding rates in a MIMO link is discussed.

Chapter 5 extends the discussions in Chapter 4 to consider the case with imperfect CSI. Specifically, we shall focus on the constellation and coding design for MIMO link with imperfect CSIR.

Chapter 6 addresses the optimal adaptive MIMO multiuser scheduling design. This is an extension of Chapters 2 and 3 (which address the adaptive MIMO link) to a multiuser scenario. Specifically, we shall focus on the capacity and coverage performance gains as a result of cross-layer adaptation. Cross-layer scheduling is formulated as an optimization problem, and the optimal and heuristic scheduling algorithms will be introduced. One commonly employed heuristic algorithm—the "greedy"-based algorithm—was shown to be optimal for single-antenna systems but suboptimal for multiple antenna systems. This motivates the genetic-based scheduling algorithm, which achieves near-optimal performance at significant computational savings. Finally, we consider the cross-layer design and system performance in the presence of imperfect CSI.

Part 2. Chapter 7 provides a high level review of MAC layer design for contemporary wireless systems with design examples for TDMA-based and CDMA-based systems. Chapter 8 contains overviews of various practical fairness notions and scheduling algorithms. Chapter 9 gives a detailed description of how cross-layer scheduling is applied in UMTS (W-CDMA) systems for packet-switched data users.

Part 3. Chapter 10 discusses an advanced topic of cross-layer adaptive scheduling design for wideband systems. Both DS-CDMA and OFDM are promising physical layer technologies for dealing with wideband multipath channels. We shall compare and contrast the multiuser performance of the cross-layer designs based on DS-CDMA and OFDMA systems. Chapter 11 gives the advanced application of information theory and queueing theory for the cross layer design. The concepts of stability region, throughput optimal scheduler as well as delay optimal scheduler will be discussed. Finally, Chapter 12 provides a detailed account of the design of a channel-adaptive ad hoc routing algorithm.

VINCENT AND RICKY

Clear Water Bay, Hong Kong, and South Pasadena, California, USA

ACKNOWLEDGMENTS

The authors would like to thank all the people who have helped in this book in one way or another. In particular, the authors would like to thank Professor Ross Murch, soon-to-be doctor Mr. Edward Au Miss Zaleta Ho, Ph. Peter Chan for the detail review and comments on the book. Furthermore, we would like to thank Mr. Ray Wang, David Hui, Yi Wu, and Tianyu Wu for helping us to contribute some of the materials and figures in the book.

Indeed, we owe a deep sense of gratitude to our friends and families, especially now that this project has been completed.

V. R.

PART 1

THEORY

1

BASIC CONCEPTS IN WIRELESS COMMUNICATIONS

1.1 OVERVIEW

In this chapter, we review the important and basic concepts in wireless communications. In Section 1.2, we first review different types of wireless channel models, namely, *time dispersion*, *multipath dispersion*, and *spatial dispersion* in microscopic fading. Concepts of frequency-selective fading, frequency flat fading, fast fading, slow fading, *coherence bandwidth*, *coherence time*, and *coherence distance* will be introduced. In Section 1.3, we establish the equivalence of discrete-time and continuous-time models in wireless communications for both the frequency flat fading and frequency-selective fading channels. In Section 1.4, we review the important and fundamental concepts of entropy, mutual information, and channel capacity, which are critical to the understanding of the materials and approaches in the subsequent chapters. Finally, in Section 1.5, we conclude with a brief summary of main points.

1.2 WIRELESS CHANNEL MODELS

A typical communication system consists of a transmitter, a receiver, and a channel. The *channel* is defined as the physical medium linking the transmitter output and the receiver input. For instance, telephone wire, optical fiber, and the atmosphere are different examples of communication channels. In fact, communication channel plays a very important role in communication system design because the transmitter and receiver designs have to be optimized with respect to the target channel.

Channel-Adaptive Technologies and Cross-Layer Designs for Wireless Systems with Multiple Antennas: Theory and Applications. By V. K. N. Lau and Y.-K. R. Kwok
ISBN 0-471-64865-5

In this book, we focus on the wireless communication channels involving radiofrequencies. In other words, the atmosphere is the medium carrying radiowaves. Please refer to References 15 and 114 for a more detailed introduction to wireless communication channels. Specifically, we briefly review the statistical models of wireless communication channels for single-antenna and multiple-antenna systems, which are frequently used in the analysis and the design of wireless communication systems.

1.2.1 AWGN Channel Model

We consider the simplest wireless channel, the *additive white Gaussian noise* (AWGN) channel. Without loss of generality, we consider single-antenna systems as illustrative in this section. The received signal ($y(t)$) is given by the transmitted signal ($x(t)$) plus a white Gaussian noise ($z(t)$)

$$y(t) = \sqrt{L}x(t) + z(t) \tag{1.1}$$

where L is the power attenuation from the transmitter to the receiver. In free space, L obeys the inverse square law.[1]

The AWGN channel is in fact quite accurate in deep-space communications and the communication links between satellite and Earth station. However, it is far from accurate in most terrestrial wireless communications, due to multipath, reflection, and diffraction. Yet, AWGN channel serves as an important reference on the performance evaluation of communication systems.

In terrestrial wireless communications, signals travel to the receiver via multiple paths, and this creates additional distortion to the transmitted signal on top of the channel noise. In general, the effect of multipath and reflections could be modeled as wireless fading channels or *microscopic fading*. Factors affecting the microscopic fading include multipath propagation, speed of the mobile (unit), speed of the surrounding objects, the transmission symbol duration, and the transmission bandwidth of the signal.

1.2.2 Linear Time-Varying Deterministic Spatial Channel

Consider a general linear channel that can be characterized by a *lowpass equivalent* time-domain impulse response denoted by $h(t; \tau, r)$ (where t is the time-varying parameter, τ is the path delay parameter, and r is the spatial position parameter). The general linear channel is therefore characterized by three independent dimensions: the *time dimension* (characterized by the time parameter t), the *delay dimension* (characterized by the delay parameter τ), and the *spatial dimension* (characterized by the position parameter r). Given a lowpass equivalent input signal $x(t)$, the lowpass equivalent received signal

[1]This is the received power level reduced by 4 times whenever the distance between the transmitter and the receiver increases by 2 times.

$y(t, r)$ through the general linear deterministic channel at time t and position r is given by

$$y(t,r) = \int_{-\infty}^{\infty} h(t;\tau,r)x(t-\tau)d\tau + z(t,r) \tag{1.2}$$

where the input signal (in time domain) is mapped into output signal (in time domain and spatial domain) through the impulse response $h(t; \tau, r)$. For simplicity, we shall discuss the channel characterization based on single-antenna systems. Extension to the MIMO systems will be straightforward. For example, to extend the model to MIMO systems, the transmitted signal $x(t)$ is replaced by the $n_T \times 1$ vector $\mathbf{x}(t)$:

$$\mathbf{x}(t) = \begin{bmatrix} x_1(t) \\ \vdots \\ x_{n_T}(t) \end{bmatrix}$$

The received signal $y(t, r)$ and the noise signal $z(t, r)$ are replaced by the $n_R \times 1$ vector $\mathbf{y}(t)$ and $\mathbf{z}(t)$, respectively:

$$\mathbf{y}(t) = \begin{bmatrix} y_1(t, r_1) \\ \vdots \\ y_{n_R}(t, r_{n_R}) \end{bmatrix}$$

$$\mathbf{z}(t) = \begin{bmatrix} z_1(t) \\ \vdots \\ z_{n_R}(t) \end{bmatrix}$$

The time-varying channel impulse response is replaced by the $n_R \times n_T$ matrix $\mathbf{h}(t; \tau, \mathbf{r})$, given by

$$\mathbf{h}(t;\tau,\mathbf{r}) = \begin{bmatrix} h_{1,1}(t;\tau,r_{1,1}) & \cdots & h_{1,n_T}(t;\tau,r_{1,n_T}) \\ \vdots & \ddots & \vdots \\ h_{n_R,1}(t;\tau,r_{n_R,1}) & \cdots & h_{n_R,n_T}(t;\tau,r_{n_R,n_T}) \end{bmatrix}$$

where $\mathbf{h}[i, j]$ is the channel response corresponding to the jth transmit antenna and the ith receive antenna and $\mathbf{r}$ is the corresponding position parameter.

1.2.2.1 Spectral Domain Representations. While Equation (1.2) gives the fundamental input–output relationship of the linear deterministic channels, Fourier transforms are sometimes useful for gaining additional insights in channel analysis. Since the channel impulse responses $h(t; \tau, r)$ are defined over the time, delay, and position domains, Fourier transforms may be defined for each of these domains, and they are elaborated as follows:

Frequency Domain. The spectral domain of the delay parameter τ is called the *frequency domain* ν. They are related by the Fourier transform relationship $h(t;\tau,r) \leftrightarrow H(t;\nu,r)$. For example, $H(t;\nu,r)$ is given by

$$H(t;\nu,r)=\int_{-\infty}^{\infty}h(t;\tau,r)\exp(-j2\pi\tau\nu)d\tau \tag{1.3}$$

Since $x(t-\tau)=\int_{-\infty}^{\infty}X(\nu)\exp(j2\pi\nu(t-\tau))d\nu$, substituting into Equation (1.2), we have

$$\begin{aligned}y(t,r)&=\int_{-\infty}^{\infty}h(t;\tau,r)\int_{-\infty}^{\infty}X(\nu)\exp(j2\pi\nu(t-\tau))d\nu\,d\tau\\&=\int_{-\infty}^{\infty}X(\nu)\exp(j2\pi\nu t)\left(\int_{-\infty}^{\infty}h(t;\tau,r)\exp(-j2\pi\nu\tau)d\tau\right)d\nu\\&=\int_{-\infty}^{\infty}H(t;\nu,r)X(\nu)\exp(j2\pi\nu t)d\nu\end{aligned} \tag{1.4}$$

Hence, the channel response can also be specified by the *time-varying transfer function* $H(t;\nu,r)$. In addition, it can be found from Equation (1.4) that the output signal (in time domain) $y(t,r)$ is mapped from the input signal (in frequency domain) $X(\nu)$ through the time-varying transfer function $H(t;\nu,r)$.

Doppler Domain. The spectral domain of the time parameter t is called the *Doppler domain* f. They are related by the Fourier transform relationship $h(t;\tau,r) \leftrightarrow H(f;\tau,r)$. For example, $H(f;\tau,r)$ is given by

$$H(f;\tau,r)=\int_{-\infty}^{\infty}h(t;\tau,r)\exp(-j2\pi tf)dt \tag{1.5}$$

Similarly, the input signal (in delay domain) $x(\tau)$ can be mapped into the output signal (in frequency domain) $Y(f,r)$ through the transfer function $H(f;\tau,r)$:

$$Y(f,r)=\int_{-\infty}^{\infty}H(f;\tau,r)x(t-\tau)d\tau \tag{1.6}$$

Wavenumber Domain. The spectral domain of the position parameter r is called the *wavenumber domain* k. The wavenumber in three-dimensional space has a physical interpretation of the plane-wave propagation direction. The position and wavenumber domains are related by the Fourier transform relationship $h(t;\tau,r) \leftrightarrow H(t;\tau,k)$. For example, $H(t;\tau,k)$ is given by

$$H(t;\tau,k)=\int_{-\infty}^{\infty}h(t;\tau,r)\exp(-j2\pi rk)dr \tag{1.7}$$

In general, two important concepts are applied to describe these linear deterministic channels: spreading and coherence. The *spreading* concept deals with the physical spreading of the received signal over the parameter space (τ, f, k) when a narrow pulse is transmitted in the corresponding domain. The coherence *concept* deals with the variation of the channel response with respect to another parameter space (ν, t, r). These concepts are elaborated in the text below.

1.2.2.2 Channel Spreading. The channel spreading concepts of describing the general linear deterministic channels focus on the spreading of the received signals over the parameter space (τ, f, k) when a narrow pulse in the corresponding parameter is transmitted. We therefore have three types of channel spreading:

Delay Spread. If we transmit a test pulse that is narrow in time, the received signal will have a spread in propagation delay τ due to the sum of different propagation delays of multipaths at the receiver. From Equation (1.2), when the transmit signal is narrow in time, we have $x(\tau) = \delta(\tau)$. Hence, the received signal is given by $y(t, r) = h(t; \tau = t, r)$. The plot of $|h(t; t, r)|^2$ versus time is called the *power-delay profile* as illustrated in Figure 1.1a. The range of delays where we find significant power is called the *delay spread* σ_τ.

Doppler Spread. If we transmit a test pulse narrow in frequency $X(\nu) = \delta(\nu)$, the received signal in general will experience a spread in the received spectrum. The range of spectrum spread in the frequency domain of the received signal $Y(f, r)$ refers to Doppler spread. The Doppler spread is given by $f_d = \frac{v}{\lambda}$, where v is the maximum speed between the transmitter and the receiver and λ is the wavelength of the carrier. This is illustrated in Figure 1.2a.

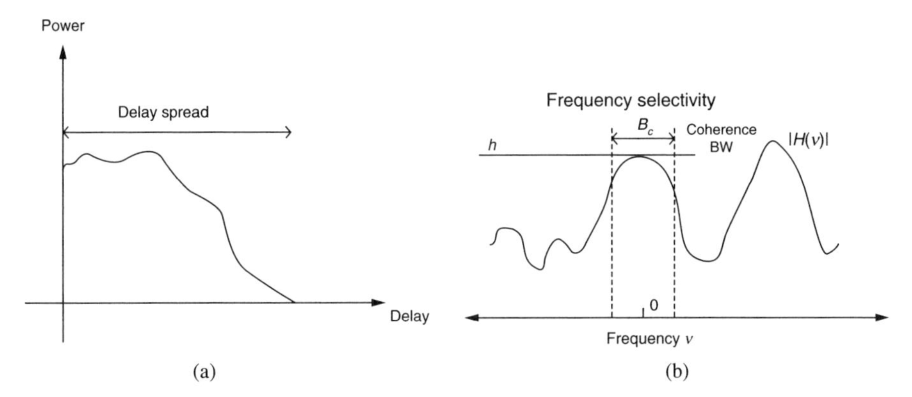

Figure 1.1. Delay spread (a) and coherence bandwidth (b).

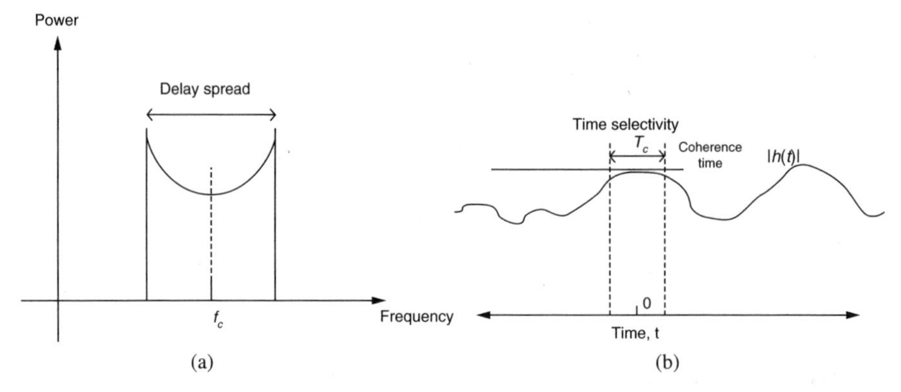

Figure 1.2. Doppler spread (a) and coherence time (b).

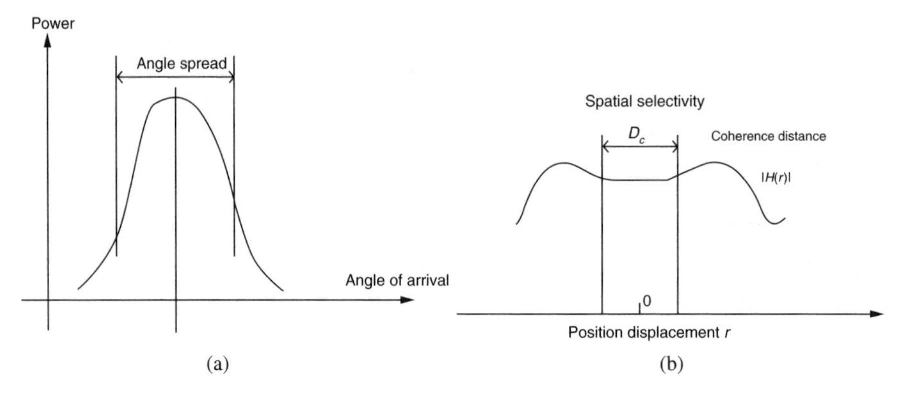

Figure 1.3. Illustration of angle spread and coherence distance.

Angle Spread. Finally, the scattering environment introduces variation in the spatial parameter r, which is equivalent to the spreading in the wavenumber domain k, this is called the *angle spread.* For example, if a test pulse narrow in direction is transmitted, the received signal will experience a spread in the wavenumber domain (angle of arrivals) due to the scattering surroundings; this is called the *angle spread* as illustrated in Figure 1.3a.

1.2.2.3 Channel Coherence.

On the other hand, we can describe the linear deterministic channels by looking at the channel coherence or channel selectivity properties over the parameter space (ν, t, r). A channel is said to be selective in the corresponding dimension if the channel response varies as a function of that parameter. The opposite of selectivity is *coherence.* A channel has coherence in the corresponding dimension if it does not change significantly as a function of that parameter. The channel coherence properties with respect to the frequency, time, and position dimensions are elaborated below.

Frequency Coherence or Frequency Selectivity. A wireless channel has frequency coherence if the magnitude of the carrier wave does not change over a frequency window of interest. This window of interest is usually the bandwidth of the transmitted signal. Hence, mathematically, we can quantify the frequency coherence of the channels by a parameter called the *coherence bandwidth* B_c

$$|H(t;\nu,r)| \approx H_0(t,r) \quad \text{for} \quad |\nu| \leq \frac{B_c}{2} \tag{1.8}$$

where $H_0(t, r)$ is a constant in the frequency domain ν and B_c is the size of the frequency window where we have constant channel response. The largest value of B_c for which Equation (1.8) holds is called the *coherence bandwidth* and can be interpreted as the range of frequencies over which the channel appears static. Figure 1.1b illustrates the concept of coherence bandwidth. In fact, if the bandwidth of the transmitted signal is larger than the coherence bandwidth of the channel, the signal will experience frequency distortion according to Equation (1.4). Such a channel is classified as a *frequency-selective fading channel.* On the other hand, if the transmitted signal has bandwidth smaller than the coherence bandwidth of the channel, frequency distortion will be no introduced to the signal and therefore, the channel will be classified as a *frequency flat fading channel.* Frequency selectivity introduces *intersymbol interference*, and this results in irreducible error floor in the BER (bit error rate) curve. Hence, this is highly undesirable. Whether a signal will experience frequency-selective fading or flat fading depends on both the environment (coherence bandwidth) and the transmitted signal (transmitted bandwidth).

Time Coherence or Time Selectivity. A wireless channel has temporal coherence if the envelope of the unmodulated carrier does not change over a time window of interest. The time coherence of channels can be specified by a parameter called the *coherence time* T_c

$$|H(t;\nu,r)| \approx H_0(\nu,r) \quad \text{for} \quad |t| \leq \frac{T_c}{2} \tag{1.9}$$

where $|H(t; \nu, r)|$ is the envelope of the response at the receiver (at a fixed position r) when a single-tone signal (at a fixed frequency ν) is transmitted, $H_0(\nu, r)$ is a constant in the time domain t and T_c is the size of the time window where we have constant channel response. The largest value of T_c for which Equation (1.9) holds is called the *coherence time* and can be interpreted as the range of time over which the channel appears static as illustrated in Figure 1.2b. In wireless fading channels, temporal incoherence (or time selectivity) is caused by the motion of the transmitter, the receiver or the scattering objects in the environment. Time selectivity can degrade the performance of wireless communication systems. If the transmit data rate is comparable to the coherence time, it becomes extremely difficult for the receiver to demodulate the transmitted signal reliably because the time selectivity within a symbol duration causes catastrophic distortion on the received pulseshape. Hence, when the transmit symbol duration T_s is longer than the coherence time T_c, we have *fast fading channels*. On the other hand, when the transmit symbol duration is shorter than the coherence time, we have *slow fading channels*. In the extreme case of slow fading the channel remains static for the entire transmit frame.

Spatial Coherence or Spatial Selectivity. A wireless channel has spatial coherence if the magnitude of the carrier wave does not change over a spatial displacement of the receiver. Mathematically, the spatial coherence can be parameterized by the *coherence distance*, D_c

$$|H(t;\nu,r)| \approx H_0(t,\nu) \quad \text{for} \quad |r| \leq \frac{D_c}{2} \tag{1.10}$$

where $|H(t;\nu,r)|$ is the envelope of the response at the receiver when a single-tone signal (at a fixed frequency ν) is transmitted (at a fixed time t), $H_0(t;\nu)$ is a constant with respect to the spatial domain r, and D_c is the size of the spatial displacement where we have constant channel response. The largest value of D_c for which Equation (1.10) holds is called the *coherence distance* and can be interpreted as the range of displacement over which the channel appears static as illustrated in Figure 1.3b. Note that for a wireless receiver moving in three-dimensional space, the coherence distance is a function of the direction that the receiver travels; that is, the position displacement $\mathbf{r}$ is a vector instead of a scalar. Hence, the study of spatial coherence is much more difficult than the study of the scalar quantities of temporal or frequency coherence. While frequency selectivity is a result of multipath propagation arriving with many different time delays τ, spatial selectivity is caused by the multipath propagation arriving from different directions in space. These multipath waves are superimposed on each other, creating pockets of constructive and destructive interference in the three-dimensional spatial domain so that the received signal power does not appear to be constant over small displacements of receiver position. Hence, if the distance traversed by a receiver is greater than the coherence distance, the channel is said to be *spatially selective* or *small-scale fading*. On the other hand, if the distance traversed by a receiver is smaller than the coherence distance, the channel is said to be *spatially flat*. Spatially selective or spatial flat fading is important when we have to apply spatial diversity (or spatial multiplexing) and beamforming. For instance, in order to produce a beam of energy along the designated direction through antenna array, the dimension of the antenna array must be within the coherence distance of the channels. On the other hand, to effectively exploit the spatial multiplexing or spatial diversity of MIMO systems, the spacing of the antenna array must be larger than coherence distance of the channels.

Figure 1.4 summarizes the various behaviors of microscopic fading channels.

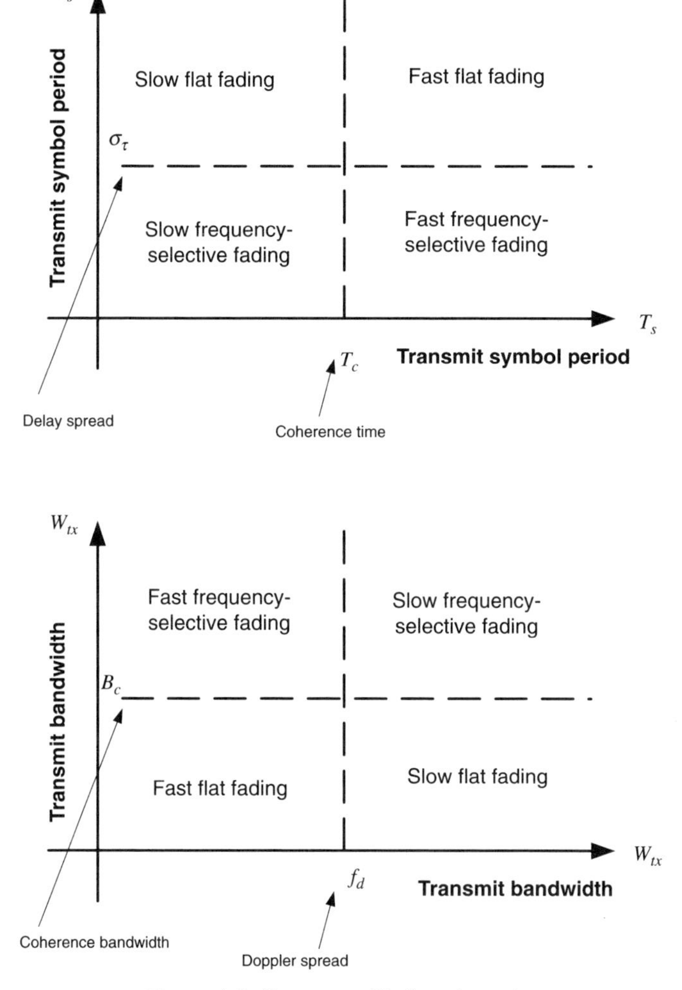

Figure 1.4. Summary of fading channels.

1.2.3 The Random Channels

In Section 1.2.2, we have introduced the general linear deterministic channel where the relationship of output given an input signal is modeled as a general time-varying system. However, in practice, the wireless fading channels we experience are random instead of deterministic; that is, $h(t; \tau, r)$ is a random process instead of a deterministic quantity. Hence, in this section, we shall extend the model of linear deterministic channels to cover the random channels.

For simplicity, let's consider a channel response on the time dimension only. The most common way to characterize the statistical behavior of the random process $h(t)$ is by means of autocorrelation:

$$R_h(t_1, t_2) = \varepsilon[h(t_1)h^*(t_2)] \tag{1.11}$$

The random process is called *wide-sense stationary* (WSS) if the autocorrelation $R_h(t_1, t_2)$ is a function of $|t_1 - t_2|$.

On the other hand, we can also consider the correlation in the spectral domain of t. Specifically, after Fourier transform on $h(t)$, we have a random frequency-varying process $H(f)$. The autocorrelation of the random process $H(f)$ is given by

$$S_H(f_1, f_2) = \varepsilon[H(f_1)H^*(f_2)] \tag{1.12}$$

Lemma 1.1 (Wide-Sense Stationary) A random process is WSS if and only if its spectral components are uncorrelated.

Proof Since $H(f)$ is the Fourier transform of $h(t)$, we have

$$R_h(t_1, t_2) = \int_{-\infty}^{\infty}\int_{-\infty}^{\infty} \varepsilon[H(f_1)H^*(f_2)]\exp(j2\pi[f_1t_1 - f_2t_2])df_1\, df_2$$

Suppose that we have uncorrelated spectral components: $S_H(f_1, f_2) = S_h(f_1)\delta(f_1 - f_2)$. Hence $\varepsilon[H(f_1)H^*(f_2)] = 0$ for $f_1 \neq f_2$. For this case, we can write the complex exponent as $\exp(j2\pi f_1[t_1 - t_2])$ and therefore, $R_h(t_1, t_2)$ is a function of $|t_1 - t_2|$.

On the other hand, if $R_h(t_1, t_2)$ is a function of $|t_1 - t_2|$, the multiplier in the integration must be zero for $f_1 \neq f_2$ because otherwise, there is no way to force $\exp(j2\pi[f_1t_1 - f_2t_2])$ to be a function of $|t_1 - t_2|$ only. Hence, we must have $\varepsilon[H(f_1)H^*(f_2)] = 0$ for $f_1 \neq f_2$.

In fact, the condition of uncorrelated spectral components is often referred to as *uncorrelated scattering* (US).

1.2.3.1 Joint Correlation and Spectrum. Now, let's consider the general random channel response $H(t; \nu, r)$ with respect to the time t, frequency ν, and position r. To accommodate all the random dependencies of such a channel, it is possible to define a joint correlation of $H(t; \nu, r)$ with respect to (t, ν, r). The joint correlation of the channel response is given by

$$R_H(t_1, \nu_1, r_1; t_2, \nu_2, r_2) = \varepsilon[H(t_1, \nu_1, r_1)H^*(t_2, \nu_2, r_2)] \tag{1.13}$$

For simplicity, we assume the random channel is a *wide-sense stationary, uncorrelated scattering* (WSS-US) random process. Hence, the joint correlation $R_H(t_1, \nu_1, r_1; t_2, \nu_2, r_2)$ is a function of $(\Delta t, \Delta\nu, \Delta r)$ only where $\Delta t = |t_1 - t_2|$, $\Delta\nu = |\nu_1 - \nu_2|$ and $\Delta r = |r_1 - r_2|$. Note that WSS refers to wide-sense stationary with respect to the time parameter t, the frequency parameter ν, and the position parameter r. On the other hand, uncorrelated scattering refers to uncorrelation of the spectral components (as a result of Lemma 1.1) in the *Doppler* parameter f, *delay* parameter τ, and *wavenumber* parameter k

$$\begin{aligned} S_H(f_1, \tau_1, k_1; f_2, \tau_2, k_2) &= \varepsilon[H(f_1, \tau_1, k_1)H^*(f_2, \tau_2, k_2)] \\ &= S_H(f_1, \tau_1, k_1)\delta(f_1 - f_2)\delta(\tau_1 - \tau_2)\delta(k_1 - k_2) \end{aligned} \tag{1.14}$$

where $S_H(f_1, \tau_1, k_1)$ is the *power spectral density* of the random process $H(t, \nu, r)$. The Wiener–Khintchine theorem for WSS-US process leads to the following Fourier transform relationship between the autocorrelation function $R_H(\Delta t, \Delta\nu, \Delta r)$ and the power spectral density $S_H(f, \tau, k)$:

$$R_H(\Delta t, \Delta\nu, \Delta r) \leftrightarrow S_H(f, \tau, k) \tag{1.15}$$

1.2.3.2 Time–Frequency Transform Mapping. Since the joint correlation function $R_H(\Delta t, \Delta\nu, \Delta r)$ is a function of three independent parameters, it is easier for illustration purposes to fix one dimension and focus on the interrelationship between the other two dimensions. For instance, consider single-antenna systems with the receiver at a fixed position r. Thus, this random channel has no dependence on r. Hence, the statistical properties of the random channels can be specified by either the *time–frequency autocorrelation* $R_H(\Delta t, \Delta\nu)$ or the *delay–Doppler spectrum* $S_H(f, \tau)$ as illustrated in Figure 1.5. In a WSS-US channel, knowledge of only one is sufficient as they are two-dimensional Fourier transform pairs.

In Section 1.2.2, we have introduced the concepts of *coherence time* and *coherence bandwidth* or equivalently, *Doppler spread* and *delay spread* for deterministic channels. We will try to extend the definition of these parameters for WSS-US random channels. From the time–frequency autocorrelation function, the correlation in time dimension is given by

$$R_H(\Delta t) = R_H(\Delta t, \Delta\nu)|_{\Delta\nu=0} \tag{1.16}$$

The coherence time T_c for the random channel is defined to be value of Δt such that $R_H(\Delta t) < 0.5$.

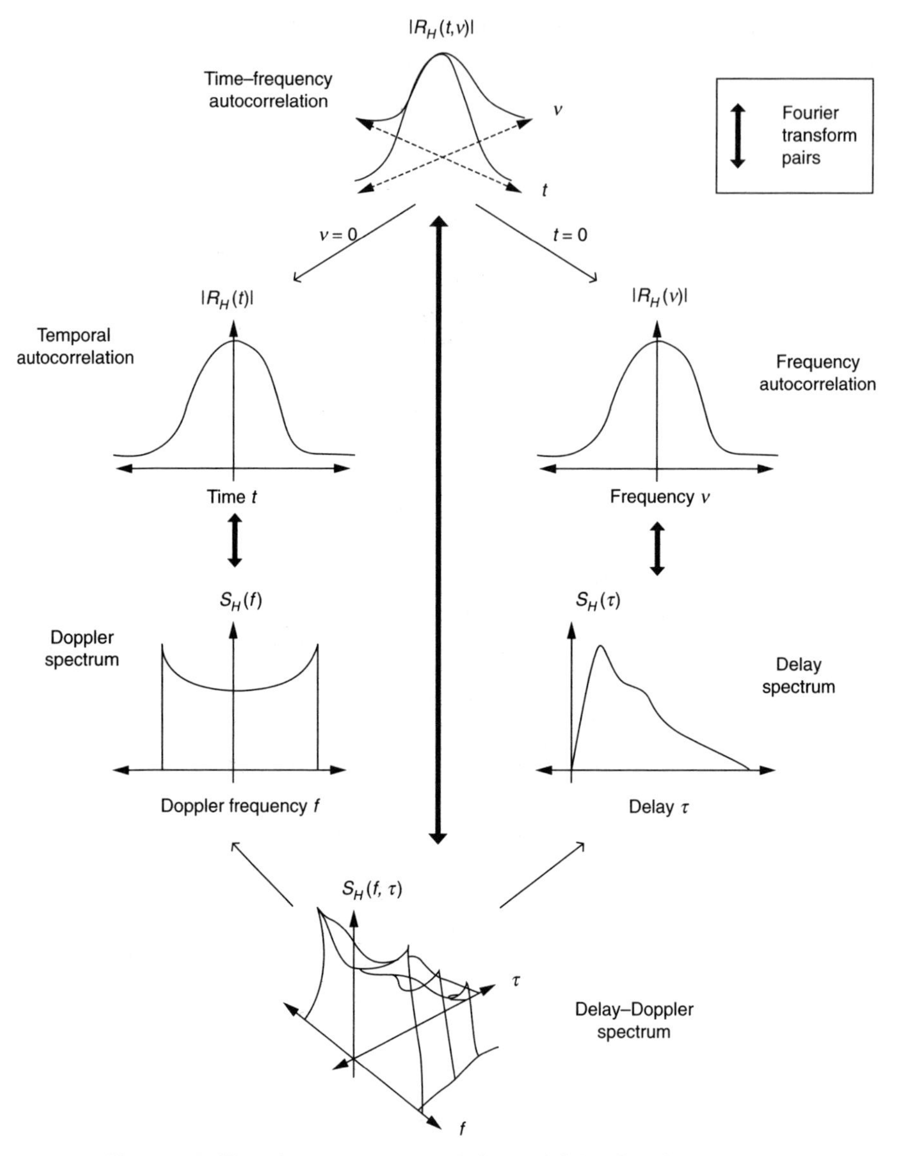

Figure 1.5. Time–frequency autocorrelation and delay–Doppler spectrum.

Similarly, the correlation in the frequency dimension is given by

$$R_H(\Delta\nu) = R_H(\Delta t, \Delta\nu)|_{\Delta t=0} \tag{1.17}$$

The coherence time B_c for the random channel is defined to be value of $\Delta\nu$ such that $R_H(\Delta\nu) < 0.5$.

On the other hand, we can characterize the random channel on the basis of the *delay–Doppler spectrum*. For instance, the *Doppler spectrum* is given by

$$S_H(f) = \int_{-\infty}^{\infty} S_H(f, \tau) d\tau \quad (1.18)$$

The Doppler spread σ_f^2 is defined as the second centered moment of the Doppler spectrum:

$$\sigma_f^2 = \frac{\int_{-\infty}^{\infty} f^2 S_H(f) df}{\int_{-\infty}^{\infty} S_H(f) df} - \left(\frac{\int_{-\infty}^{\infty} f S_H(f) df}{\int_{-\infty}^{\infty} S_H(f) df} \right)^2 \quad (1.19)$$

Similarly, the *power-delay profile* is given by

$$S_H(\tau) = \int_{-\infty}^{\infty} S_H(f, \tau) df \quad (1.20)$$

The delay spread σ_τ^2 is defined as the second centered moment of the power-delay profile:

$$\sigma_\tau^2 = \frac{\int_{-\infty}^{\infty} \tau^2 S_H(\tau) d\tau}{\int_{-\infty}^{\infty} S_H(\tau) d\tau} - \left(\frac{\int_{-\infty}^{\infty} \tau S_H(\tau) d\tau}{\int_{-\infty}^{\infty} S_H(\tau) d\tau} \right)^2 \quad (1.21)$$

Since the Doppler spectrum and the time autocorrelation function are Fourier transform pairs, a large Doppler spread σ_f^2 will result in small coherence time T_c and therefore faster temporal fading and vice versa. Similarly, the power-delay profile and the frequency autocorrelation function are Fourier transform pairs. Hence, a large delay spread σ_τ^2 will result in a small coherence bandwidth B_c and vice versa. In practice, the four parameters are related by

$$B_c \approx \frac{1}{5\sigma_\tau} \quad (1.22)$$

and

$$T_c \approx \frac{1}{5\sigma_f} \quad (1.23)$$

1.2.3.3 Frequency–Space Transform Mapping. For a static channel, we may extend the time–frequency map described in the previous section for the *frequency–space* relationship as illustrated in Figure 1.6.

In this diagram, the joint space–frequency autocorrelation $R_H(\Delta v, \Delta r)$ and the joint delay–wavenumber spectrum $S_H(\tau, k)$ are related by Fourier trans-

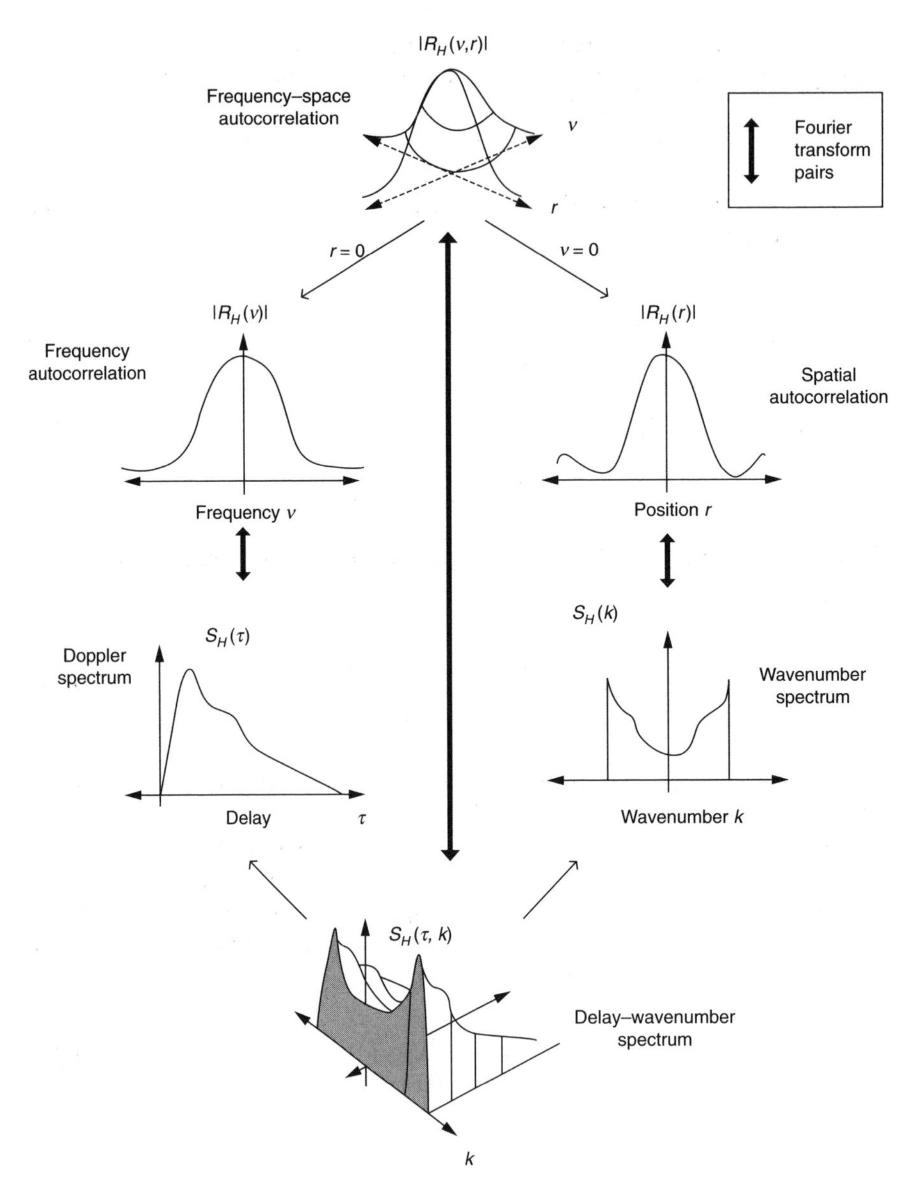

Figure 1.6. Illustration of frequency–space autocorrelation and delay–wavenumber spectrum.

form pairs. In Section 1.2.2, we have introduced the concepts of *coherence distance* and *angle spread* for deterministic channels. We shall try to extend the definition of these parameters for WSS-US random channels. From the frequency–space autocorrelation function, the single dimension spatial autocorrelation of the random channels is given by

$$R_H(\Delta r) = R_H(\Delta\nu, \Delta r)|_{\Delta\nu=0} \tag{1.24}$$

The coherence distance D_c is therefore defined as the maximum Δr such that $R_H(\Delta r) < 0.5$.

Similarly, we can characterize the statistical behavior of the random channels by the delay–wavenumber spectrum $S_H(\tau, k)$. Consider the single-dimension wavenumber spectrum $S_H(k)$:

$$S_H(k) = \int_{-\infty}^{\infty} S_H(\tau, k)d\tau \tag{1.25}$$

The *angle spread* σ_k^2 is defined to be the second centered moment of the wavenumber spectrum:

$$\sigma_k^2 = \frac{\int_{-\infty}^{\infty} k^2 S_H(k)dk}{\int_{-\infty}^{\infty} S_H(k)dk} - \left(\frac{\int_{-\infty}^{\infty} kS_H(k)dk}{\int_{-\infty}^{\infty} S_H(k)dk} \right)^2 \tag{1.26}$$

An important indication of the nature of the channel is called the *spread factor*, given by B_cT_c. If $B_cT_c < 1$, the channel is said to be *underspread*; otherwise, it is called *overspread*. In general, if $B_cT_c \ll 1$, the channel impulse response could be easily measured and the measurement could be utilized at the receiver for demodulation and detection or at the transmitter for adaptation. On the other hand, if $B_cT_c \gg 1$, channel measurement would be extremely difficult and unreliable. In this book, we deal mainly with *underspread* fading channels.

1.2.4 Frequency-Flat Fading Channels

We shall consider the effect of fading channels on a transmitted signal. We first look at a simple case called a *flat fading channel*. Let $x(\tau)$ be the lowpass equivalent signal transmitted over the channel and $X(\nu)$ be the corresponding Fourier transform. The lowpass equivalent received signal $y(t, r)$ is given by

$$y(t,r) = \int_{-\infty}^{\infty} h(t;\tau,r)x(t-\tau)d\tau + z(t) = \int_{-\infty}^{\infty} H(t;\nu,r)X(\nu)\exp(j2\pi\nu t)d\nu + z(t) \tag{1.27}$$

where $h(t; \tau, r)$ is the *time-varying impulse response* and $H(t; \nu, r)$ is the *time-varying transfer function* of the channel. Note that both $H(t; \nu, r)$ and $h(t; \tau, r)$ are random processes. Suppose that the two-sided bandwidth W of $x(t)$ is less than the coherence bandwidth B_c. According to the definition of the correlation function $R_H(0; \Delta\nu, 0)$, the random channel fading $H(t, \nu, r)$ is highly correlated within the range of the transmitted bandwidth $\nu \in [-W/2, W/2]$. Hence, all the frequency component of $X(\nu)$ undergoes the same complex fading within the range of frequencies $\nu \in [-W/2, W/2]$. This means that within the bandwidth W of $X(\nu)$, we obtain $H(t, \nu, r) = h(t; \tau, r) = h(t, r)$, where $h(t, r)$ is a complex stationary random process in t and r only. This results in both the envelope attenuation and phase rotation on the transmitted signal. The received signal simplifies to

$$y(t,r) = h(t,r)\int_{-\infty}^{\infty} X(\nu)\exp(j2\pi\nu t)d\nu + z(t) = h(t,r)x(t) + z(t) \qquad (1.28)$$

Hence, the signal $x(t)$ is said to experience the *flat fading channel*. Therefore, the flat fading channel has a time-varying multiplicative effect on the transmitted signal. In this case, the multipath components are not *resolvable* because the signal bandwidth $W \ll B_c = 1/(5T_c)$. In accordance with the central-limit theorem, the random process $h(t, r)$ (as a result of multipath aggregation) can be well approximated by complex stationary zero-mean Gaussian random process. A flat fading channel is also called a *slowly fading channel* if the time duration of a transmitted symbol T_s is much smaller than the coherence time of the channel T_c. Otherwise, it is called a *fast fading channel*. Since in general $W \geq 1/T_s$, a slowly flat fading channel is underspread because $B_c T_c \ll 1$.

1.2.5 Frequency-Selective Fading Channels

When the transmit signal bandwidth $W \gg B_c$, frequency components in $X(\nu)$ have frequency separation greater than B_c. In this case, the random fading components $H(t; \nu_1, r)$ and $H(t; \nu_2, r)$ become uncorrelated whenever $|\nu_1 - \nu_2| \geq B_c$. Hence, some frequency components in $X(\nu)$ (within the transmitted bandwidth W) may experience independent fading. In such case, the channel is said to be *frequency-selective*.

Any lowpass equivalent signal $x(t)$ with two-sided bandwidth W and time duration $t \in \{0, T\}$ may be represented geometrically by a $(N = WT)$-dimensional vector $\mathbf{x} = [x(0), x(1/W), \ldots, x(T - 1/W)]$ in the signal space spanned by the orthonormal basis $\{\psi_0(t), \ldots, \psi_{N-1}(t)\}$. Specifically, by sampling theorem, $\psi_n(t) = \text{sinc}(\pi W(t - n/W))/\pi W(t - n/W)$, and the lowpass equivalent signal is given by

$$x(t) = \sum_n \mathbf{x}[n]\psi_n(t) \qquad (1.29)$$

where $\mathbf{x}[n]$ denotes the nth component of the vector $\mathbf{x}$. The corresponding Fourier transform of $x(t)$ is given by

$$x(\nu) = \begin{cases} \frac{1}{W}\sum_n \mathbf{x}[n]\exp\frac{-j2\pi\nu n}{W} & |\nu| \leq W/2 \\ 0 & |\nu| > W/2 \end{cases} \qquad (1.30)$$

Hence, the received signal as a result of frequency-selective fading is given by

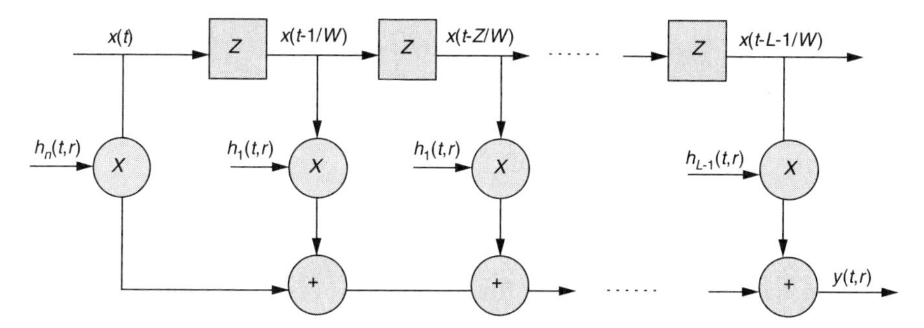

Figure 1.7. Illustration of tapped-delay-line frequency-selective channel model.

$$
\begin{aligned}
y(t,r) &= \int_{-\infty}^{\infty} H(t;v,r)X(v)\exp(j2\pi vt)dv + z(t) \\
&= \frac{1}{W}\sum_n \mathbf{x}[n]\int_{-\infty}^{\infty} H(t;v,r)\exp\left(j2\pi v\left(t-\frac{n}{W}\right)\right)dv + z(t) \\
&= \frac{1}{W}\sum_n x\left(\frac{n}{W}\right)h\left(t;t-\frac{n}{W},r\right) + z(t) \\
&= \sum_{n=0}^{L-1} x\left(t-\frac{n}{W}\right)h_n(t,r) + z(t)
\end{aligned} \tag{1.31}
$$

where $h_n(t,r) = \frac{1}{W}h\,(t;n/W,r)$ and $L = [W/B_c]$ is the number of *resolvable multipaths* as seen by the transmitted signal. Hence, the larger the transmitted signal bandwidth W (or the smaller the channel coherence bandwidth B_c), the larger the number of resolvable multipaths that the transmit signal will see. Figure 1.7 illustrates the equivalent view of the frequency-selective channel model in Equation (1.31).

According to the statistical characterization of the channel presented above, the channel taps $h_n(t, r)$ are complex stationary random processes. Specifically, due to the central-limit theorem, the random process can be well approximated by zero-mean complex Gaussian stationary random process. Furthermore, $h_n(t)$ and $h_m(t)$ are statistically independent if $n \neq m$ due to the uncorrelated scattering assumption.

1.3 EQUIVALENCE OF CONTINUOUS-TIME AND DISCRETE-TIME MODELS

While the physical transmitted signals and received signals are all in continuous time domain, it is quite difficult to gain proper design insights by working in the original forms. For example, given the received signal[2] $y(t)$, it is not clear

[2] Without loss of generality, ignore the notation about the position parameter r in the received signal $y(t)$ to simplify notation.

what should be the optimal structure to extract the information bits carried by the signal. Hence, for statistical signal detection and estimation problems, it is usually more convenient to formulate the problems in the *geometric domain* or *discrete-time domain*. In this section, we try to establish the equivalence relationship of the two domains.

1.3.1 Concepts of Signal Space

Theorem 1.1 (Orthonormal Basis for Random Process) For any wide-sense stationary (WSS) complex random process $x(t)$ for $t \in [0, T]$, there exists a set of orthonormal[3] basis functions $\{\psi_0(t), \ldots, \psi_{N-1}(t)\}$ and complex random variables $\{x_0, \ldots, x_{N-1}\}$ such that $\lim_{N\to\infty}\varepsilon[|x(t) - x_N(t)|^2] = 0$ for all t where $x_N(t)$ is given by:

$$x_N(t) = \sum_{n=0}^{N-1} x_n \psi_n(t)$$

and $x_n = \langle x(t), \psi_n(t) \rangle$. In particular, the basis functions are the solution (called eigenfunctions) of the integral equation:

$$\int_0^T R_x(t-\tau)\psi_n(\tau)d\tau = \lambda_n \psi_n(t)$$

(where $R_x(t)$ is the autocorrelation function of $x(t)$) and $\varepsilon[x_n x_m^*] = \lambda_n \delta_{nm}$ (uncorrelated).

In addition, if $x(t)$ is bandlimited to W ($x(t)$ has significant energy only for $f \in [-W, W]$) and $WT \gg 1$, we have $\lambda_n \approx 0$ for $n > 2WT$ and $\lambda_n \approx 1$ for $n < 2WT$.

Proof Please refer to [122] for the proof.

Since λ_n represents the energy of the basis function $\psi_n(t)$ over $[0, T]$, in other words only about $2WT$ coefficients in $x_N(t)$ have significant energy and we might say that the signal $x(t)$ lies in a signal space of $2WT$ dimensions.

1.3.2 Sufficient Statistics

Before we discuss the equivalence of continuous-time model and discrete-time model, we have to introduce a concept called *sufficient statistics*.

Definition 1.1 (Sufficient Statistics) Suppose that we have a probability density function on the random sample $f(x; \theta)$, where θ is an unknown parameter. Let $T(X_1, \ldots, X_N)$ (a function of N random samples $X_1, \ldots, X_N$ from the population) be a statistical estimate of the unknown parameter θ. The estimate T is called *sufficient statistic* for θ if $\mathbf{X} = (X_1, \ldots, X_N)$ is independent of θ given $T(X_1, \ldots, X_N)$:

$$\Pr[\mathbf{X}|T, \theta] = \Pr[\mathbf{X}|T]$$

[3] Orthonormal basis refers to $\int_0^T \psi_n(t)\psi_m^*(t)dt = \delta(n-m)$.

Example 1.1 (Sufficient Statistics) Given: $X_1, \ldots, X_N, X_n \in \{0, 1\}$, is an independent and identically distributed (i.i.d.) sequence of coin tosses of a coin with unknown parameter $\theta = \Pr[X_n = 1]$. Given N, the number of 1s is a sufficient statistic for θ. Here $T(X_1, \ldots, X_N) = \Sigma_n X_n$. T is a sufficient statistic for θ because

$$\begin{aligned}
\Pr\{(X_1, \ldots, X_N) = (x_1, \ldots, x_N) | T = k, \theta\} \\
&= \Pr\{(X_1, \ldots, X_N) = (x_1, \ldots, x_N) | T = k\} \\
&= \begin{cases} \dfrac{1}{\binom{N}{k}}, & \text{if } \sum_n x_n = k \\ 0, & \text{otherwise} \end{cases}
\end{aligned}$$

Example 1.2 (Sufficient Statistics) If X is Gaussian distributed with mean θ and variance 1, and $X_1, \ldots, X_N$ are drawn independently according to this distribution, then $T = \frac{1}{N}\Sigma_n X_n$ is a *sufficient statistic* for θ. This is because $\Pr[(X_1, \ldots, X_N)|T]$ is independent of θ.

In other words, the concept of sufficient statistics allows us to discard data samples that are not related to the parameter in question because the conditional pdf (probability distribution function) of X given T is independent of θ. Note that sufficient statistic is not unique. There could be many sufficient statistics for the same parameter θ. As shown by Examples 1.1 and 1.2, it is relatively easy to *verify* whether a statistic T is sufficient with respect to a parameter θ. However, it is a more difficult problem to *identify* potential sufficient statistics for a parameter θ. Below we summarize the Neyman–Fisher factorization theorem, which can be used to find sufficient statistics in several examples.

Theorem 1.2 (Necessary and Sufficient Condition for Sufficient Statistics) If we can factorize the pdf $p(\mathbf{X}; \theta)$ as $p(\mathbf{X}; \theta) = g(T(\mathbf{X}), \theta)h(\mathbf{X})$, where g is a function depending on $\mathbf{X}$ only through $T(\mathbf{X})$ and h is a function depending only on $\mathbf{X}$, then $T(\mathbf{X})$ is a sufficient statistic for θ. Conversely, if $T(\mathbf{x})$ is a sufficient statistic for θ, then the pdf $p(\mathbf{X}; \theta)$ can be factorized as shown above.

Some examples are given below to illustrate the use of this theorem to find a sufficient statistic.

Example 1.3 (Sufficient Statistics) Let X be Gaussian distributed with unknown mean θ and unit variance, with $X_1, \ldots, X_N$ drawn independently according to this distribution. The pdf of the data is given by

$$f(\mathbf{X}; \theta) = \frac{1}{(2\pi)^{N/2}} \exp\left[-\frac{1}{2}\sum_{n=1}^{N}(X_n - \theta)^2\right]$$

Since $\Sigma_{n=1}^{N}(X_n - \theta)^2 = \Sigma_{n=1}^{N}X_n^2 - 2\theta\Sigma_{n=1}^{N}X_n + N\theta^2$, the pdf could be factorized into

$$f(\mathbf{X};\theta) = \underbrace{\frac{1}{(2\pi)^{N/2}} \exp\left[-\frac{1}{2}\left(N\theta^2 - 2\theta\sum_{n=1}^{N} X_n\right)\right]}_{g(T(\mathbf{X}),\theta)} \underbrace{\exp\left[-\frac{1}{2}\left(\sum_{n=1}^{N} X_n^2\right)\right]}_{h(\mathbf{X})}$$

Hence, $T(\mathbf{X}) = \Sigma_{n=1}^{N} X_n$ is a sufficient statistic for θ because g is a function depending on $\mathbf{X}$ only through T. In fact, any one-to-one mapping of $T(\mathbf{X})$ is also a sufficient statistic for θ.

Example 1.4 (Sufficient Statistics) Let X be Gaussian distributed with zero mean and unknown variance θ, with $X_1, \ldots, X_N$ drawn independently according to this distribution. The pdf of the data is given by

$$f(\mathbf{X};\theta) = \frac{1}{(2\pi\theta)^{N/2}} \exp\left[-\frac{1}{2}\sum_{n=1}^{N} X_n^2\right] = \underbrace{\frac{1}{(2\pi\theta)^{N/2}} \exp\left[-\frac{1}{2}\sum_{n=1}^{N} X_n^2\right]}_{g(T(\mathbf{X}),\theta)} \underbrace{1}_{h(\mathbf{X})}$$

Hence, $T(\mathbf{X}) = \Sigma_{n=1}^{N} X_n^2$ is a sufficient statistic for θ.

1.3.3 Discrete-Time Signal Model—Flat Fading

A digital transmitter can be modeled as a device with input as information bits and continuous-time analog signals matched to the channel or the medium as output. A string of k information bits is passed to a channel encoder (combined with modulator) where redundancy is added to protect the raw information bits. N encoded symbols $\{x_1, \ldots, x_N\}$ (where $x_n \in \mathcal{X}$) are produced at the output of the channel encoder. The N encoded symbols are mapped to the analog signal (modulated) and transmitted out to the channel. The channel input signal $x(t)$ is given by

$$x(t) = \sum_{n} x_n g(t - nT_s) \tag{1.32}$$

where T_s is the *symbol duration* and $g(t)$ is the low pass equivalent *transmit pulse* with two-sided bandwidth W and pulse energy $\int_{-\infty}^{\infty} |g(t)|^2 dt = 1$. Equation (1.32) is a general model for digitally modulated signals, and the signal set $\mathcal{X}$ is called the *signal constellation*. For example, $\mathcal{X} = \{e^{j2\pi m/M} : m = \{0, 1, \ldots, M-1\}\}$ represents MPSK modulation. $\mathcal{X} = \{x_R + jx_I : x_R, x_I \in \{\pm\frac{1}{2}, \pm\frac{3}{2}, \ldots, \pm\frac{M}{4}\}\}$ represents MQAM modulation. The signal constellations for MPSK and MQAM are illustrated in Figure 1.8.

From Equation (1.28), the low pass equivalent received signal through flat fading channel can be expressed as

$$y(t) = h(t)x(t) + z(t) = \sum_{n} x_n h(t) g(t - nT_s) + z(t) \tag{1.33}$$

where $z(t)$ is the low pass equivalent white complex Gaussian noise, $h(t)$ is a zero-mean unit variance complex Gaussian random process, and only $x(t)$ contains information.

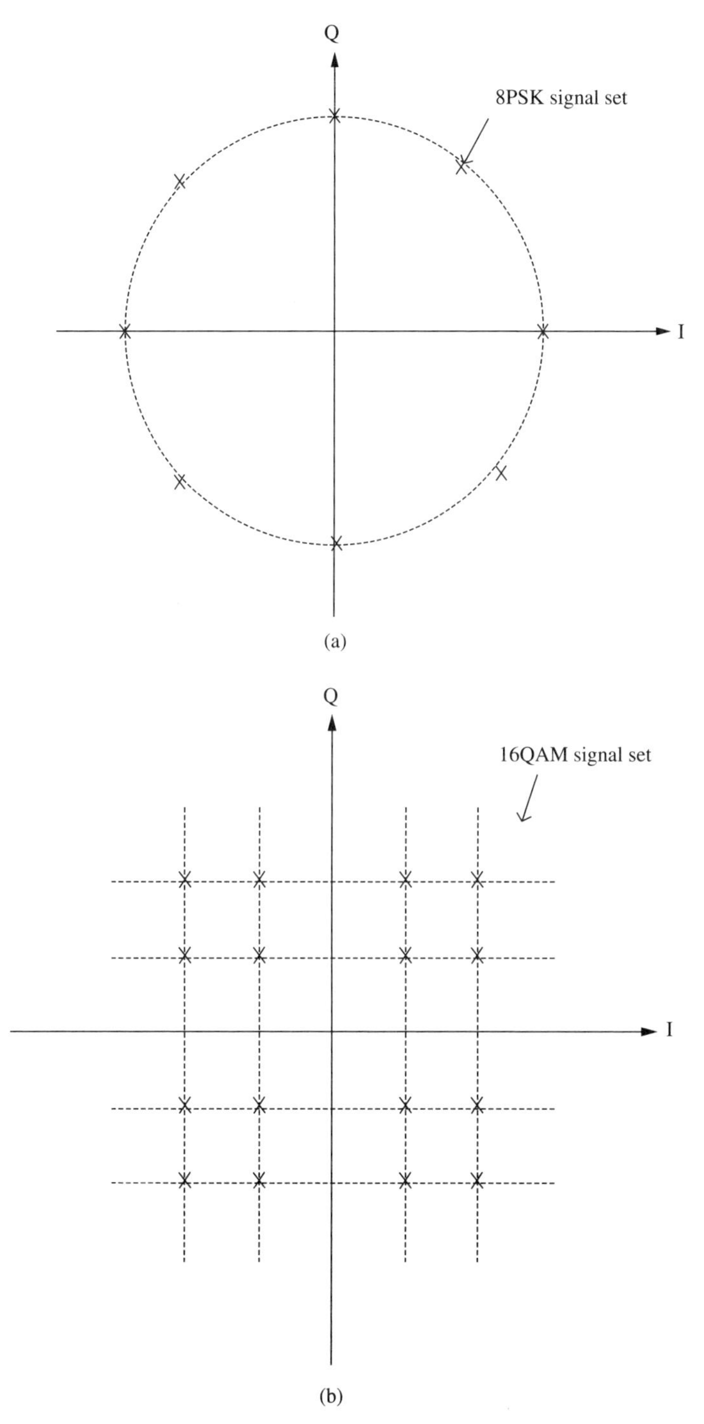

Figure 1.8. Illustration of signal constellations: (a) MPSK constellation; (b) MQAM constellation.

At the receiver side, the decoder produces the string of k decoded information bits based on the observation $y(t)$ from the channel. Hence, the receiver problem can be modeled as a *detection problem*; that is, given the observation $y(t)$, the receiver has to determine which one of the 2^k *hypothesis* is actually transmitted. It is difficult to gain very useful design insight if we look at the problem from continuous-time domain. In fact, as we have illustrated, the continuous-time signal $y(t)$ can be represent equivalently as vectors in signal space $\mathbf{y}$. Hence, the receiver detection problem [assuming knowledge of channel fading $H(t)$] is summarized below.

Problem 1.1 (Detection) Let $\omega \in \{1, 2^k\}$ be the message index at the input of the transmitter. The decoded message $\hat{\omega}$ at the receiver is given by

$$\hat{\omega} = \arg\max_{\omega} p(\mathbf{y}|\omega, h(t))$$

Such a receiver is called the *maximum-likelihood* (ML) receiver, which minimizes the probability of error if all the 2^k messages are equally probable.

Due to the white channel noise term in (1.33), we need infinite dimension signal space in general to represent the received signal $y(t)$ as vector $\mathbf{y}$. However, since only $h(t)x(t)$ contains information and $h(t)x(t)$ can be represented as vector in a finite-dimensional signal subspace, not all components in $\mathbf{y}$ will contain information. Hence, the detection problem in Problem 1.1 could be further simplified, and this is expressed mathematically below.

Let $\Theta_y = \{\Psi_1(t), \ldots, \Psi_n(t), \ldots\}$ denote the infinite-dimensional signal space that contains $y(t)$ and $\Theta_s = \{\Psi_1(t), \ldots, \Psi_{D_s}(t)\}$ denote a D_s-dimensional subspace of Θ_y that contains $s(t) = h(t)x(t)$; that is

$$y(t) = \sum_{j=1}^{\infty} y_j \Psi_j(t) \tag{1.34}$$

and

$$s(t) = \sum_{j=1}^{D_s} s_j \Psi_j(t) \tag{1.35}$$

where $y_j = \langle y(t), \Psi_j(t) \rangle = \int_{-\infty}^{\infty} y(t)\Psi_j^*(t)dt$ and $s_j = \langle s(t), \Psi_j(t) \rangle = \int_{-\infty}^{\infty} s(t)\Psi_j^*(t)dt$.

Let $\mathbf{y}$ and $\mathbf{s}$ be the vectors corresponding to $y(t)$ and $s(t) = h(t)x(t)$ with respect to the signal spaces Θ_y and Θ_s, respectively. Define $\mathbf{v}$ as the projection of $\mathbf{y}$ over the signal space Θ_s. Since $p(\mathbf{y}|\omega, h(t)) = p(\mathbf{v}|\omega, h(t))$, $\mathbf{v}$ forms a *sufficient statistics* for the unknown parameter ω. Therefore, it is sufficient to project the received signal $y(t)$ over the signal space Θ_s, and there is no loss of information with respect to the detection of the message index ω.

The vector representation of $s(t) = h(t)x(t)$ is given by $(s_1, \ldots, s_{D_s})$, where

$$s_j = \langle h(t)x(t), \Psi_j(t) \rangle = \int_{-\infty}^{\infty} h(t)x(t)\Psi_j^*(t)dt = \sum_n x_n \int_{-\infty}^{\infty} h(t)g(t - nT_s)\Psi_j^*(t)dt$$

In the special case of slow fading where $T_c \gg T_s$, we have $h(t) \approx h_n$ for $t \in [(n-1)T_s, nT_s]$. The vector representation of $s(t)$ becomes

$$\begin{aligned} s_j &= \sum_n x_n \int_{-\infty}^{\infty} h(t) g(t - nT_s) \Psi_j^*(t) dt \\ &= \sum_n x_n h_n \int_{-\infty}^{\infty} g(t - nT_s) \Psi_j^*(t) dt \\ &= \sum_n x_n h_n g_{n,j} \quad \forall j \in [1, D_s] \end{aligned}$$

where $g_{n,j} = \int_{-\infty}^{\infty} g(t - nT_s)\Psi_j^*(t)dt$, $D_s = WNT_s$ and N is the number of transmitted symbols. Therefore, we have $\mathbf{s} = \Sigma_n x_n h_n \mathbf{g}_n$, where $\mathbf{g}_n$ is the vector representation of the time-delayed transmit pulse $g(t - nT_s)$. The equivalent discrete-time flat fading model with channel input $\mathbf{s}$ and channel output $\mathbf{v}$ (the projection of $y(t)$ onto the signal space $\ominus_s$) is therefore given by

$$\mathbf{v} = \mathbf{s} + \mathbf{z} = \sum_n x_n h_n \mathbf{g}_n + \mathbf{z} \tag{1.36}$$

where the vectors are defined with respect to the NWT_s-dimensional signal space $\ominus_x$ that contains $x(t)$. If $z(t)$ is the low pass equivalent white Gaussian noise and the two-sided noise spectral power density of the real and quadrature components are both $\eta_0/2$, the noise covariance of $\mathbf{z}$ is given by

$$\varepsilon[z_j z_k^*] = \int_{-\infty}^{\infty}\int_{-\infty}^{\infty} \varepsilon[z(t) z^*(\tau)] \Psi_j^*(t) \Psi_k(\tau) dt d\tau = \eta_0 \delta(j-k)$$

In other words, we have i.i.d. noise vector components in $\mathbf{z}$, each with variance η_0.

In addition, if the channel noise is a white Gaussian random process, the likelihood function of Problem 1.1 (with knowledge of channel fading at the receiver) can be expressed as

$$p(\mathbf{v}|\omega, \mathbf{h}) = \frac{1}{(\pi\eta_0)^N} \exp\left[-\frac{1}{2\eta_0}|\mathbf{v} - \mathbf{s}(\omega)|^2\right]$$

Maximizing the likelihood function is equivalent to minimizing the *distance metric* $d(\mathbf{v}, \mathbf{s})$ between the observation $\mathbf{v}$ and the hypothesis $\mathbf{s}(\omega)$. The distance metric is given by: $d(\mathbf{v}, \mathbf{s}) = |\mathbf{v} - \mathbf{s}(\omega)|^2$. This can be further simplified as follows:

$$d(\mathbf{v}, \mathbf{s}) = |\mathbf{v}|^2 - 2\sum_{\mathbf{n}} \mathbf{x}_{\mathbf{n}}^*(\omega) \mathbf{h}_{\mathbf{n}}^* \mathbf{v} \cdot \mathbf{g}_{\mathbf{n}}^* + \left|\sum_{\mathbf{n}} \mathbf{h}_{\mathbf{n}} \mathbf{x}_{\mathbf{n}} \mathbf{g}_{\mathbf{n}}\right|^2$$

Since $|\mathbf{v}|^2$ is independent of ω, the maximum-likelihood (ML) metric of ω, $\mu(\mathbf{v}, \omega)$, is given by

$$
\begin{aligned}
\mu(\mathbf{v},\omega) &= 2\sum_n x_n^*(\omega)h_n^*\mathbf{v}\cdot g_n^* - \left|\sum_n h_n x_n g_n\right|^2 \\
&= 2\sum_n x_n^*(\omega)h_n^*\langle y(t), g(t-nT_s)\rangle - \sum_{n,m} h_n h_m^* x_n x_m^* \left\langle g_n, g_m^*\right\rangle \\
&= 2\sum_n x_n^*(\omega)h_n^* \int_{-\infty}^{\infty} y(t)g^*(t-nT_s)dt \\
&\quad - \sum_{n,m} h_n h_m^* x_n x_m^* \int_{-\infty}^{\infty} g(t-nT_s)g^*(t-mT_s)dt \\
&= 2\sum_n x_n^*(\omega)h_n^* q_n - \sum_{n,m} h_n h_m^* x_n x_m^* R_g(n-m)
\end{aligned} \tag{1.37}
$$

where $R_g(n-m) = \int_{-\infty}^{\infty} g(t-nT_s)g^*(t-mT_s)dt$ and $q_n = \int_{-\infty}^{\infty} y(t)g^*(t-nT_s)dt$ is the matched-filter output (with impulse response $g^*(-t)$) of the received signal $y(t)$ (sampled at $t = nT_s$). Observe that the ML metric $\mu(\mathbf{v}, \omega)$ of ω in Equation (1.37) depends on the received signal $y(t)$ only through $q_n = \int_{-\infty}^{\infty} y(t)g(t-nT_s)dt$. Once $\mathbf{q} = [q_1, \ldots, q_N]$ are known, the ML metric $\mu(\mathbf{v}, \omega)$ can be computed. Hence, the matched-filter outputs, $q_n = \int_{-\infty}^{\infty} y(t)g(t-nT_s)dt$, are also the *sufficient statistics* with respect to ω. Therefore, the ML metric $\mu(\mathbf{v}, \omega)$ in Equation (1.37) can be expressed in terms of the *sufficient statistics* $\mathbf{q} = [q_1, \ldots, q_N]$ directly as $\mu(\mathbf{q}, \omega)$.

If $g(t) = 0$ for all $t > T_s$ or $t < 0$ [*zero intersymbol interference* (ISI) due to the transmit pulseshape $g(t)$], we have $\int_{-\infty}^{\infty} g(t-nT_s)g^*(t-mT_s)dt = \delta(n-m)$ and the ML metric $\mu(\mathbf{q}, \omega)$ can be further simplified as

$$
\mu(\mathbf{q},\omega) = \sum_n x_n^*(\omega)h_n^* q_n - \sum_n |h_n|^2 |x_n(\omega)|^2 \tag{1.38}
$$

Figure 1.9 illustrates the optimal detector structure based on *matched filtering*.

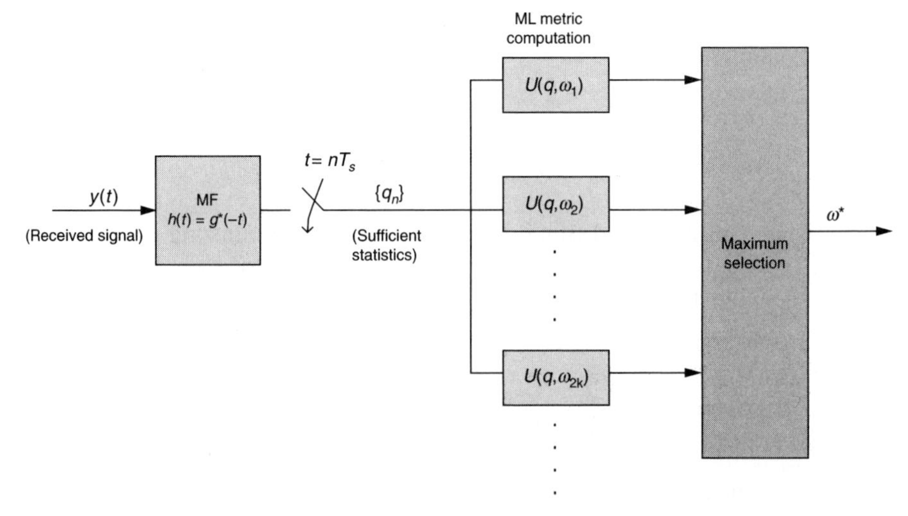

Figure 1.9. Optimal detector structure based on matched filtering for fading channels with zero ISI transmit pulse.

The sufficient statistics $\mathbf{q}$ are first generated from $y(t)$ through matched filtering and periodically sampling at $t = nT_s$. The sequence of sufficient statistics is passed to the ML metric computation, where the ML metrics $\mu(\mathbf{q}, \omega)$ for each hypothesis ω are computed. Hence, the hypothesis that maximizes the metrics is the decoded information $\hat{\omega}$. Finally, the following theorem establish the equivalence of the continuous-time flat fading channel and the discrete-time flat fading channels.

Theorem 1.3 (Equivalence of Continuous-Time and Discrete-Time Models) Given any continuous-time flat fading channels with the transmitted signal $x(t)$ given by Equation (1.32) and the received signal $y(t)$ given by Equation (1.33), there always exists an equivalent discrete-time channel with input x_n and output q_n such that

$$q_n = \sum_k h_k x_k R_g(n-k) + z_n \tag{1.39}$$

where $R_g(n-k) = \int_{-\infty}^{\infty} g(t - kT_s)g^*(t - nT_s)dt$, z_n is a zero-mean Gaussian noise with variance η_0, and η_0 is the power spectral density of $z(t)$. The term *equivalent channels* means that both channels will have the same ML detection results with respect to the message index ω. Furthermore, if $g(t) = 0$ for all $t > T_s$ or $t < 0$ such that

$$q_n = h_n x_n + z_n \tag{1.40}$$

Proof Since q_n is a sufficient statistic with respect to the detection of ω, there is no loss of information if the receiver can only observe q_n instead of the complete received signal $y(t)$. Hence, substituting $y(t)$ in Equation (1.33) into q_n, we have

$$\begin{aligned} q_n &= \sum_k h_k x_k \int_{-\infty}^{\infty} g(t-kT_s)g^*(t-nT_s)dt + \int_{-\infty}^{\infty} z(t)g^*(t-nT_s)dt \\ &= \sum_k h_k x_k R_g(n-k) + z_n \end{aligned}$$

where $R_g(n-k) = \int_{-\infty}^{\infty} g(t - kT_s)g^*(t - nT_s)dt$ and $z_n = \int_{-\infty}^{\infty} z(t)g^*(t - nT_s)dt$. Since the message index ω is contained in the symbols $[x_1, \ldots, x_N]$ only, there is no loss of information if the receiver only observes q_n due to sufficient statistic properties. Hence, the two channels produce the same ML detection results. If the pulse duration of $g(t)$ is T_s, then $g_{n-k} = \delta(n - k)$ and Equation (1.40) follows immediately.

Extending Equation (1.40) to MIMO channels, let $\mathbf{X} \in \mathcal{X}$ be the $(n_T \times 1)$-dimensional input vector, $\mathbf{Y} \in \mathcal{Y}$ be the $n_R \times 1$ output vector, and $\mathbf{H} \in \mathcal{H}$ be the $(n_R \times n_T)$-dimensional channel fading of the discrete-time MIMO channel. The received symbol $\mathbf{y}_n$ is given by

$$\mathbf{y}_n = \mathbf{h}_n \mathbf{x}_n + \mathbf{z}_n \tag{1.41}$$

1.3.4 Discrete-Time Channel Model—Frequency-Selective Fading

The transmitted signal $x(t)$ is given by

$$x(t) = \sum_n x_n g(t - nT_s) \tag{1.42}$$

where T_s is the *symbol duration* and $g(t)$ is the low pass equivalent *transmit pulse* with two-sided bandwidth W and pulse energy $\int_{-\infty}^{\infty} |g(t)|^2 dt = 1$. From Equation (1.31), the received signal after passing through a frequency-selective fading channel is given by

$$y(t) = \sum_{l=0}^{L-1} h_l(t) x(t - lT_s) + z(t) \tag{1.43}$$

where $L = \lfloor \frac{W}{B_c} \rfloor$ is the number of resolvable paths. Letting $s(t) = \Sigma_{l=0}^{L-1} h_l(t) x(t - lT_s)$, using similar arguments, the vector projection of $\mathbf{y}$ onto the signal space containing $s(t)$, $\mathbf{v}$, is a sufficient statistic for the message index ω.

The vector representation of $s(t)$ is given by $(s_1, \ldots, s_{D_s})$, where

$$\begin{aligned} s_j &= \langle s(t), \Psi_j(t) \rangle = \sum_{l=0}^{L-1} \int_{-\infty}^{\infty} h_l(t) x(t - lT_s) \Psi_j^*(t) dt \\ &= \sum_{l=0}^{L-1} \sum_n x_n \int_{-\infty}^{\infty} h_l(t) g(t - (n+l)T_s) \Psi_j^*(t) dt \end{aligned}$$

In the special case of slow fading where $T_c \gg T_s$, we have $h_l(t) \approx h_{m,l}$ for $t \in [(m-1)T_s, mT_s]$. Hence, the vector representation of $s(t)$ becomes

$$\begin{aligned} s_j &= \sum_{l=0}^{L-1} \sum_n x_n \int_{-\infty}^{\infty} h_l(t) g(t - (n+l)T_s) \Psi_j^*(t) dt \\ &= \sum_{l=0}^{L-1} \sum_n x_n h_{n+l,l} \int_{-\infty}^{\infty} g(t - (n+l)T_s) \Psi_j^*(t) dt \\ &= \sum_{l=0}^{L-1} \sum_n x_n h_{n+l,l} g_{n+l,j} \end{aligned}$$

where $g_{m,j}$ is the component of the projection of $g(t - mT_s)$ onto $\Psi_j(t)$ given by

$$g_{m,j} = \int_{-\infty}^{\infty} g(t - mT_s) \Psi_j^*(t) dt \tag{1.44}$$

Therefore, we have $\mathbf{s} = \Sigma_{l=0}^{L-1} \Sigma_n x_n h_{n+l,l} \mathbf{g}_{n+l}$. The equivalent discrete-time frequency-selective fading model with input vector $\mathbf{s}$ and output vector $\mathbf{v}$ is therefore given by

$$\mathbf{v} = \mathbf{s} + \mathbf{z} = \sum_{l=0}^{L-1} \sum_{n} x_n h_{n+l,l} g_{n+l} + \mathbf{z} \tag{1.45}$$

where the vectors are defined with respect to the signal space that contains $s(t)$. Similarly, since $z(t)$ is the low pass equivalent white Gaussian noise with two-sided power spectral density of the real and quadrature components both given by $\eta_0/2$, we have i.i.d. noise vector components in $\mathbf{z}$ with noise variance η_0.

When the channel noise is white Gaussian, the likelihood function is given by

$$p(\mathbf{v}|\omega, \mathbf{h}) = \frac{1}{(\pi\eta_0)^N} \exp\left[-\frac{1}{2\eta_0}|\mathbf{v} - \mathbf{s}|^2\right]$$

Factorizing the likelihood function according to Theorem 1.2, we have the ML metric $\mu(\mathbf{v}, \omega)$ given by

$$\begin{aligned}
\mu(\mathbf{r}, \omega) &= 2\sum_n \sum_{l=0}^{L-1} x_n^*(\omega) h_{n+l,l}^* \mathbf{v} \cdot \mathbf{g}_{n+l}^* - \left|\sum_{l=0}^{L-1} \sum_n x_n h_{n+l,l} \mathbf{g}_{n+l}\right|^2 \\
&= 2\sum_n \sum_{l=0}^{L-1} h_{n+l,l}^* x_n^* \langle y(t), g(t-(n+l)T_s)\rangle \\
&\quad - \sum_{n,n'} \sum_{l,l'} h_{n+l,l} h_{n'+l',l'}^* x_n x_{n'}^* \langle g(t-(n+l)T_s), g(t-(n'+l')T_s)\rangle \\
&= 2\sum_n \sum_{l=0}^{L-1} h_{n+l,l}^* x_n^* \int_{-\infty}^{\infty} y(t) g^*(t-(n+l)T_s)dt \\
&\quad - \sum_{n,n'} \sum_{l,l'} h_{n,l} h_{n',l'}^* x_n x_{n'}^* \int_{-\infty}^{\infty} g(t-(n+l)T_s) g^*(t-(n'+l')T_s)dt \\
&= 2\sum_n \sum_{l=0}^{L-1} h_{n+l,l}^* x_n^* q_{n+l} - \sum_{n,n'} \sum_{l,l'} h_{n,l} h_{n',l'}^* x_n x_{n'}^* R_g(n+l-n'-l')
\end{aligned}$$

where

$$q_n = \int_{-\infty}^{\infty} y(t) g^*(t - nT_s)dt \tag{1.46}$$

is the matched-filter output [with impulse response $g^*(-t)$] of the received signal $y(t)$ sampled at $t = nT_s$. Observe that the ML metric depends on the received signal $y(t)$ only through the matched-filter outputs $\mathbf{q} = [q_1, \ldots, q_N]$. Once $\mathbf{q}$ is known, the ML metric $\mu(\mathbf{v}, \omega)$ can be computed and hence, the ML detection of $\hat{\omega}$ can be obtained directly from $\mathbf{q}$. In other words, there is no loss of information if the receiver can only observe $\mathbf{q}$ instead of the complete received signal $y(t)$. $\mathbf{q}$ is therefore another sufficient statistics with respect to the detection of the message ω. If $g(t) = 0$ for all $t > T_s$ or $t < 0$ [*zero ISI* due to the transmit pulse $g(t)$], the ML metric can be further simplified as

$$\mu(\mathbf{v}, \omega) = \sum_n \sum_{l=0}^{L-1} x_n^*(\omega) h_{n+l,l}^* q_{n+l} - \sum_{n,n'} \sum_{l,l':n+l=n'+l'} h_{n+l,l} h_{n'+l',l'}^* x_n x_{n'}^*. \tag{1.47}$$

Figure 1.9 illustrates the optimal detector structure for frequency-selective fading channel with zero ISI pulse based on matched filtering.

Similar to the flat fading case, we can establish a discrete-time equivalent channel with any given continuous-time frequency-selective fading channels and summarize the results by the following theorem.

Theorem 1.4 (Equivalence of Continuous-Time and Discrete-Time Models) For any continuous-time frequency-selective fading channels with input $x(t)$ given by Equation (1.42) and received signal $y(t)$ given by Equation (1.43), there always exists an equivalent discrete-time frequency-selective fading channel with discrete-time inputs $\{x_n\}$ and discrete-time outputs $\{q_n\}$ such that

$$q_n = \sum_k \sum_{l=0}^{L-1} h_{k+l,l} x_k R_g(n-k-l) + z_n \tag{1.48}$$

where $R_g(m) = \int_{-\infty}^{\infty} g(t)g^*(t - mT_s)dt$, $h_{m,l} = h_l(t)$ for $t \in [mT_s, (m+1)T_s]$, z_n is a zero-mean Gaussian noise with variance η_0, and η_0 is the power spectral density of $z(t)$. If $g(t) = 0$ for all $t > T_s$ or $t < 0$ (zero ISI), we have $R_g(m) = \delta(mT_s)$. Hence

$$q_n = \sum_{l=0}^{L-1} h_{n,n-l} x_{n-l} + z_n \tag{1.49}$$

Proof Similar to the flat fading case in Theorem 1.3, $\mathbf{q}$ is a sufficient statistic with respect to the detection of ω and there is no loss of information if the receiver can observe $\mathbf{q}$ only. Substituting $y(t)$ from Equation (1.43) into q_n, the results follow.

Extending Equation (1.49) to MIMO channels, the received symbol $\mathbf{y}_n$ is given by

$$\mathbf{y}_n = \sum_{l=0}^{L-1} \mathbf{h}_{n,n-l} \mathbf{x}_{n-1} + \mathbf{z}_n \tag{1.50}$$

1.4 FUNDAMENTALS OF INFORMATION THEORY

In this section, we give an overview on the background and mathematical concepts in information theory in order to establish Shannon's channel coding theorem. Concepts of ergodic and outage capacities will be followed. Finally, we give several examples of channel capacity in various channels. This will form the basis for the remaining chapters in the book.

1.4.1 Entropy and Mutual Information

The first important concept in information theory is *entropy*, which is a measure of uncertainty in a random variable.

Definition 1.2 (Entropy of Discrete Random Variable) The entropy of a discrete random variable X with probability mass function $p(X)$ is given by

$$H(X) = -\sum_{x} p(x)\log_2(p(x)) = -\varepsilon[\log_2(p(X))] \tag{1.51}$$

where the expectation is taken over X.

Definition 1.3 (Entropy of Continuous Random Variable) On the other hand, the entropy of a continuous random variable X with probability density function $f(X)$ is given by

$$H(X) = -\int_{x} f(x)\log_2(f(x))dx \tag{1.52}$$

For simplicity, we shall assume discrete random variable unless otherwise specified.

Definition 1.4 (Joint Entropy) The joint entropy of two random variables X_1, X_2 is defined as

$$H(X_1, X_2) = -\sum_{x_1,x_2} p(x_1, x_2)\log_2(p(x_1, x_2)) = -\varepsilon[\log_2(p(X_1, X_2))] \tag{1.53}$$

where the expectation is taken over (X_1, X_2).

Definition 1.5 (Conditional Entropy) The conditional entropy of a random variable X_2 given X_1 is defined as

$$H(X_2|X_1) = -\sum_{x_1,x_2} p(x_1, x_2)\log_2(p(x_2|x_1)) = -\varepsilon[\log_2(p(X_2|X_1))] \tag{1.54}$$

where the expectation is taken over (X_1, X_2).

After introducing the definitions of entropy, let's look at various properties of entropy. They are summarized as lemmas below. Please refer to the text by Cover and Thomas [30] for the proof.

The first lemma gives a lower bound on entropy.

Lemma 1.2 (Lower Bound of Entropy)

$$H(X) \geq 0 \tag{1.55}$$

Equality holds if and only if there exists $x_0 \in X$ such that $p(x_0) = 1$.

Proof Directly obtained from Equation (1.51).

The following lemma gives an upper bound on entropy for discrete and continuous random variable X.

Lemma 1.3 (Upper Bound of Entropy) If X is a discrete random variable, then

$$H(X) \leq \log_2(|X|) \tag{1.56}$$

Equality holds if and only if $p(X) = 1/|X|$. On the other hand, if X is a continuous random variable, then

$$H(X) \leq \frac{1}{2}\log_2(2\pi e\sigma_X^2) \tag{1.57}$$

where $\sigma_X^2 = \varepsilon[|X|^2]$ and equality holds if and only if X is Gaussian distributed with an arbitrary mean μ and variance σ_X^2.

From the lemmas presented-above, entropy can be interpreted as a measure of *information* because $H(X) = 0$ if there is no uncertainty about X. On the other hand, $H(X)$ is maximized if X is equiprobable or X is Gaussian distributed.

The chain rule of entropy is given by the following lemma.

Lemma 1.4 (Chain Rule of Entropy)

$$H(X_1, X_2) = H(X_1) + H(X_2|X_1) \tag{1.58}$$

On the other hand, as summarized in the lemma below, conditioning reduces entropy.

Lemma 1.5 (Conditioning Reduces Entropy)

$$H(X|Y) \leq H(X) \tag{1.59}$$

Equality holds if and only if $p(XY) = p(X)p(Y)$ (X and Y are independent).

Lemma 1.6 (Concavity of Entropy) $H(X)$ is a concave function of $p(X)$.

Lemma 1.7 If X and Y are independent, then $H(X + Y) \geq H(X)$.

Lemma 1.8 (Fano's Inequality) Given two random variables X and Y, let $\hat{X} = g(Y)$ be an estimate of X given Y. Define the probability of error as

$$P_e = \Pr[\hat{X} \neq X]$$

we have

$$H_2(P_e) + P_e \log_2(|X| - 1) \geq H(X|Y) \tag{1.60}$$

where $H_2(p) = -p\log_2(p) - (1 - p)\log_2(1 - p)$.

After we have reviewed the definitions and properties of entropy, we shall discuss the concept of *mutual information*.

Definition 1.6 (Mutual Information) The mutual information of two random variables X and Y is given by

$$I(X;Y) = \sum_{x,y} p(x, y)\log_2\left(\frac{p(x, y)}{p(x)p(y)}\right) = \varepsilon\left[\log_2\left(\frac{p(x, y)}{p(x)p(y)}\right)\right] \tag{1.61}$$

From Equation (1.53), we have

$$I(X;Y) = H(X) - H(X|Y) = H(Y) - H(Y|X) = H(X) + H(Y) - H(X,Y) \tag{1.62}$$

Figure 1.10 illustrates the mutual information.

Hence, mutual information is the reduction in uncertainty of X, due to knowledge of Y. Therefore, if X is the transmitted symbol and Y is the received

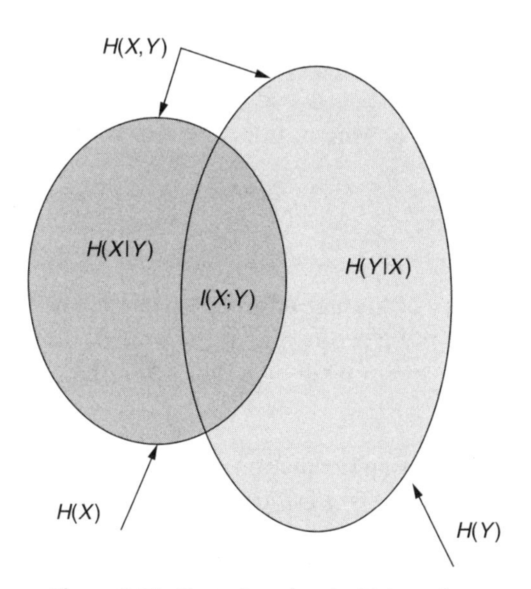

Figure 1.10. Illustration of mutual information.

symbol, mutual information could be interpreted as a measure of the amount of *information* communicated to the receiver. We shall formally establish this interpretation in the next section when we discuss Shannon's channel coding theorem.

Definition 1.7 (Conditional Mutual Information) The mutual information of 2 random variables X and Y conditioned on Z is given by

$$I(X;Y|Z) = H(X|Z) - H(X|Y,Z) = H(Y|Z) - H(Y|X,Z) \tag{1.63}$$

We shall summarize the properties of mutual information below.

Lemma 1.9 (Symmetry)

$$I(X;Y) = I(Y;X)$$

The proof can be obtained directly from definition.

Lemma 1.10 (Self-Information)

$$I(X;X) = H(X)$$

The proof can be obtained directly from the definition.

Lemma 1.11 (Lower Bound)

$$I(X;Y) \geq 0$$

Equality holds if and only if X and Y are independent.

Lemma 1.12 (Chain Rule of Mutual Information)

$$I(X_1,\ldots,X_N;Y) = \sum_{n=1}^{N} I(X_n;Y|X_1,\ldots,X_{n-1})$$

Lemma 1.13 (Convexity of Mutual Information) Let (X, Y) be the random variable with the joint pdf given by $p(X)p(Y|X)$. $I(X; Y)$ is a concave function of $p(X)$ for a given $p(Y|X)$. On the other hand, $I(X; Y)$ is a convex function of $p(Y|X)$ for a given $p(X)$.

Lemma 1.14 (Data Processing Inequality) X, Y, Z are said to form a Markov chain $X \to Y \to Z$ if $p(x, y, z) = p(x)p(y|x)p(z|y)$. If $X \to Y \to Z$, we have

$$I(X;Y) \geq I(X;Z)$$

Equality holds if and only if $X \to Z \to Y$.

Corollary 1.1 (Preprocessing at the Receiver) If $Z = g(Y)$, then $X \to Y \to Z$ and we have $I(X; Y|Z) \leq I(X; Y)$ and $I(X; Z) \leq I(X; Y)$. Equality holds in both cases if and only if Y, X are independently conditioned on $g(Y)$.

If we assume that X is the channel input and Y is the channel output, then the function of received data Y cannot increase the mutual information about X. On the other hand, the dependence of X and Y is decreased by the observation of a downstream variable $Z = g(Y)$. Since $I(X; Y)$ gives the channel capacity as we shall introduced in the next section, Corollary 1.1 states that the channel capacity will not be increased by any *preprocessing* at the receiver input.

Corollary 1.2 (Postprocessing at the Transmitter) If $Y = g(X)$, then $X \to Y \to Z$ and we have $I(X; Z) \leq I(Y; Z)$. Equality holds in both cases if and only if X, Z are independently conditioned on $g(X)$.

Similarly, any *postprocessing* at the transmitter cannot increase the channel capacity as illustrated by Corollary 1.2. These points are illustrated in Figure 1.11.

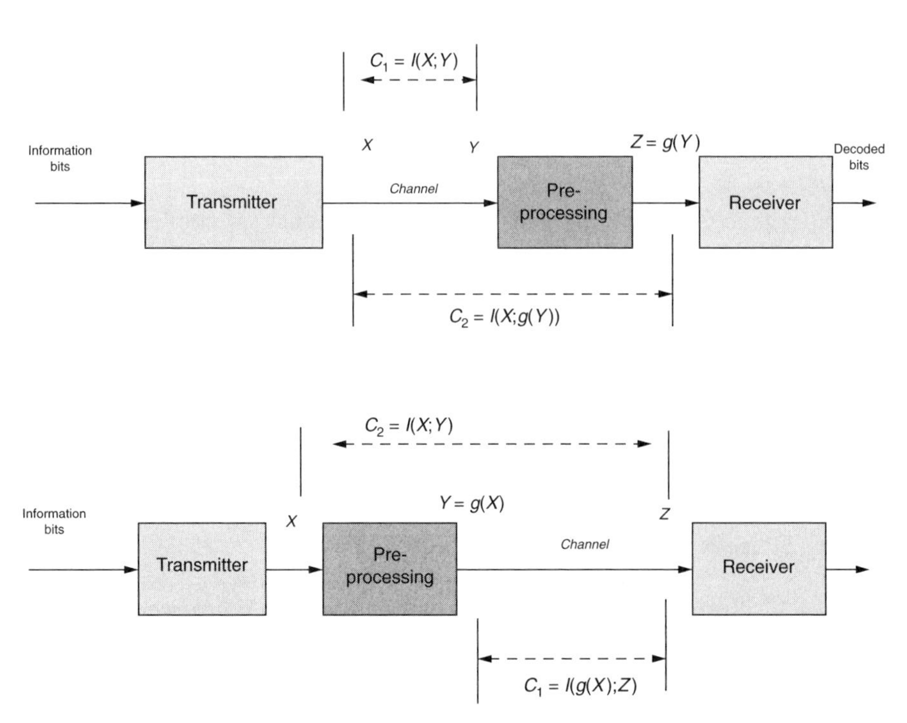

Figure 1.11. Illustration of data processing inequality. Preprocessing at the receiver and post-processing at the transmitter cannot increase mutual information; $C_1 \geq C_2$.

1.4.2 Shannon's Channel Coding Theorem

The Shannon's channel coding theorem is in fact based on the *law of large numbers*. Before we proceed to establish the coding theorem, we shall discuss several important mathematical concepts and tools.

Convergence of Random Sequence

Definition 1.8 (Random Sequence) A *random sequence* or a *discrete-time random process* is a sequence of random variables $X_1(\xi), \ldots, X_n(\xi), \ldots$, where ξ denotes an outcome of the underlying sample space of the random experiment.

For a specific event ξ, $X_n(\xi)$ is a sequence of of numbers that might or might not converge. Hence, the notion of convergence of a random sequence might have several interpretations:

Convergence Everywhere. A random sequence $\{X_n\}$ converges everywhere if the sequence of numbers $X_n(\xi)$ converges in the traditional sense[4] for every event ξ. In other words, the limit of the random sequence is a random variable $X(\xi)$, which is $X_n(\xi) \to X(\xi)$ as $n \to \infty$.

Convergence Almost Everywhere. Define $\mathcal{V} = \{\xi : \lim_{n\to\infty} x_n(\xi) = x(\xi)\}$ be the set of events such that the number sequence $x_n(\xi)$ converges to $x(\xi)$ as $n \to \infty$. The random sequence x_n is said to converge almost everywhere (or with probability 1) to x if there exists such a set $\mathcal{V}$ such that $\Pr[\mathcal{V}] = 1$ as $n \to \infty$.

Convergence in the Mean-Square Sense. The random sequence x_n tends to the random variable x in the mean-square sense if $\varepsilon[|x_n - x|^2] \to 0$ as $n \to \infty$.

Convergence in Probability. The random sequence x_n converges to the random variable x in probability if for any $\varepsilon > 0$, the probability $\Pr[|x_n - x| > \varepsilon] \to 0$ as $n \to \infty$.

Convergence in Distribution. The random sequence x_n is said to converge to a random variable x in distribution if the cumulative distribution function (cdf) of $x_n(F_n(\alpha))$ converges to the cdf of $x(F(\alpha))$ as $n \to \infty$

$$F_n(\alpha) \to F(\alpha) \quad n \to \infty$$

for every point α of the range of x. Note that in this case, the sequence $x_n(\xi)$ needs not converge for any event ξ.

Figure 1.12 illustrates the relationship between various convergence modes. For instance, "almost-everywhere convergence" implies convergence in prob-

[4] A sequence of number x_n tends to a limit x if, given any $\varepsilon > 0$, there exists a number n_0 such that $|x_n - x| < \varepsilon$ for every $n > n_0$.

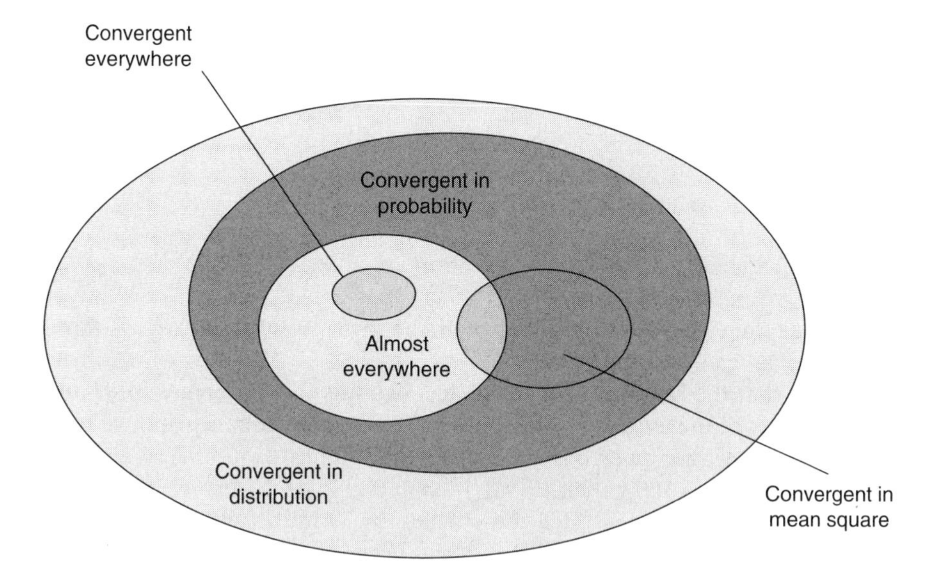

Figure 1.12. Various modes of random sequence convergence.

ability but not the converse. Hence, almost-everywhere convergence is a stronger condition than convergence in probability. Obviously, convergence everywhere implies convergence almost everywhere but not the converse. Hence, convergence everywhere is an even stronger condition compared with almost everywhere convergence. Another example is convergence in mean-square sense implies convergence in probability because

$$\Pr[|x_n - x| > \varepsilon] \leq \frac{\varepsilon\left[|x_n - x|^2\right]}{\varepsilon^2}$$

as a result of Chebyshev's inequality.[5] Hence, if $x_n \to x$ in mean-square sense, then for any fixed $\varepsilon > 0$, the right side tends to 0 as $n \to \infty$. Again, the converse is not always true. Note that convergence in mean-square sense does not always imply convergence almost everywhere and vice versa.

Law of Large Numbers. The law of large numbers is the fundamental magic behind information theory.

Weak Law of Large Numbers. If $X_1, \ldots, X_n$ are i.i.d. random variables, then

$$\lim_{n\to\infty} \frac{1}{n} \sum_n X_n = \varepsilon|X| \text{ in probability}$$

[5] For a random variable X with mean μ_X and variance σ_X^2, the probability $\Pr[|X - \mu_X| > x_0]$ is upper-bounded by $\frac{\sigma_X^2}{\sigma_0^2}$ for all $x_0 > 0$.

The expression on the left side is also called the *sample mean.*

Strong Law of Large Numbers. If $X_1, \ldots, X_n$ are i.i.d. random variables, then

$$\lim_{n\to\infty} \frac{1}{n} \sum_n X_n = \varepsilon|X|$$

with probability 1 (almost everywhere).

We shall explain the difference between the two versions of law of large numbers by an example below.

Suppose that the random variable X_n has two possible outcomes {0, 1} and the probability of the event $X_n = 1$ in a given experiment equals p and we wish to estimate p to within an error $\varepsilon = 0.1$ using the sample mean $\bar{x}(n) = (1/n)\Sigma_n x_n$ as its estimate. If $n \geq 1000$, then $\Pr[|\bar{x}(n) - p| < 0.1] \geq 1 - 1/(4n\varepsilon^2) \geq \frac{39}{40}$. Hence, if we repeat the experiment 1000 times, then in 39 out of 40 such runs, our error $|\bar{x}(n) - p|$ will be less than 0.1.

Suppose now that we perform the experiment 2000 times and we determine the sample mean $\bar{x}(n)$ not for one n but for every n between 1000 and 2000. The weak law of large numbers leads to the following conclusion. If our experiment is repeated a large number of times, then for a specific n larger than 1000, the error $|\bar{x}(n) - p|$ will exceed 0.1 only in one run out of 40. Hence, 97.5% of the runs will be "good." We cannot draw the conclusion that in the good runs, the error will be less than 0.1 for every n between 1000 and 2000. This conclusion, however, is correct but can be deduced only from the strong law of large numbers.

Typical Sequence. As a result of the weak law of large numbers, we have the following corollary.

Corollary 1.3 (Typical Sequence) If $X_1, \ldots, X_n$ are i.i.d. random variables with distribution $p(X)$, then

$$\frac{1}{n}\log_2(p(X_1, \ldots, X_n)) \to H(X)$$

in probability.

The proof follows directly by recognizing $(1/n)\log_2(p(X_1, \ldots, X_n)) = (1/n)\Sigma_n \log_2 p(X_n)$, which converges to $\varepsilon[\log_2(p(X))]$ in probability. This is called *asymptotic equipartition* (AEP) *theorem*. As a result of the AEP theorem, we define a typical sequence as follows.

Definition 1.9 (Typical Sequence) A random sequence $(x_1, \ldots, x_N)$ is called a *typical sequence* if $2^{-N(H(X)+\varepsilon)} \leq p(x_1, \ldots, x_N) \leq 2^{-N(H(X)-\varepsilon)}$. The set of typical sequences is called the typical set and is denoted by

$$\mathcal{A}_\varepsilon = \{(x_1,\ldots,x_N): 2^{-N(H(X)+\varepsilon)} \le p(x_1,\ldots,x_N) \le 2^{-N(H(X)-\varepsilon)}\}$$

Together with the AEP theorem, we have the following properties concerning typical set $\mathcal{A}_\epsilon$. Please refer to the text by Cover and Thomas [30] for the proof of the lemmas below.

Lemma 1.15 If $(x_1, \ldots, x_N) \in \mathcal{A}_\varepsilon$, then

$$H(X) - \varepsilon \le -\frac{1}{N}\log_2 p(x_1,\ldots,x_N) \le H(X) + \varepsilon$$

The proof follows directly from the definition of typical set $\mathcal{A}_\varepsilon$.

Lemma 1.16

$$\Pr[(x_1,\ldots,x_N) \in \mathcal{A}_\varepsilon] > 1 - \varepsilon$$

for sufficiently large N.

Lemma 1.17

$$|\mathcal{A}_\varepsilon| \le 2^{N(H(X)+\varepsilon)}$$

for sufficiently large N.

Lemma 1.18

$$|\mathcal{A}_\varepsilon| \ge (1-\varepsilon)2^{N(H(X)-\varepsilon)}$$

for sufficiently large N.

Figure 1.13 illustrates the concepts of typical set. As a result of AEP and Lemma 1.16, the probability of a sequence $(x_1, \ldots, x_N)$ being a member of the typical set $\mathcal{A}_\varepsilon$ is arbitrarily high for sufficiently large N. In other words, the typical set contains all *highly probable sequences*. Suppose that X_n is a binary random variable; then the total number of possible combinations in a sequence $(x_1, \ldots, x_N)$ is 2^N. However, as a result of Lemmas 1.17 and 1.18, the cardinality of $\mathcal{A}_\varepsilon$ is around $2^{NH(X)}$. Since $H(X) \le 1$, we have $|\mathcal{A}_\varepsilon| \le 2^N$. This means that the typical set is in general smaller than the set of all possible sequences, resulting in possible *data compression*. The key to achieve effective compression is to group the symbols into a long sequence to exploit the law of large numbers.

Jointly Typical Sequence. We can extend the concept of typical sequence to *jointly typical sequence*.

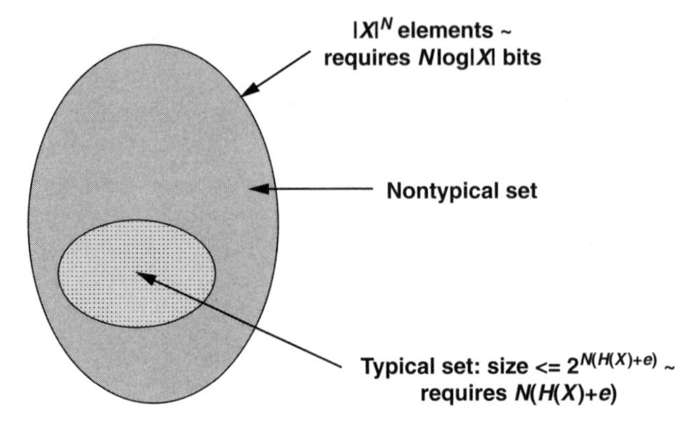

Figure 1.13. Illustration of a typical set.

Definition 1.10 (Jointly Typical Sequence) Let $\mathbf{x}^N$ and $\mathbf{y}^N$ be two sequences of length N generated according to $p(\mathbf{x}^N, \mathbf{y}^N)$. $(\mathbf{x}^N, \mathbf{y}^N)$ is called a *jointly typical sequence* if

$$\left|-\frac{1}{N}\log_2(p(\mathbf{x}^N)) - H(X)\right| < \varepsilon$$

$$\left|-\frac{1}{N}\log_2(p(\mathbf{y}^N)) - H(Y)\right| < \varepsilon$$

and

$$\left|-\frac{1}{N}\log_2(p(\mathbf{x}^N, \mathbf{y}^N)) - H(X, Y)\right| < \varepsilon$$

The jointly typical set $\mathcal{A}_\varepsilon$ is defined as

$$\mathcal{A}_\varepsilon = \{(\mathbf{x}^N, \mathbf{y}^N)\text{: the preceding three conditions are satisfied.}\}$$

Similarly, we have the following interesting properties about jointly typical sequences.

Lemma 1.19 Let $\mathbf{x}^N$ and $\mathbf{y}^N$ be two sequences of length N drawn i.i.d. according to $p(\mathbf{x}^N, \mathbf{y}^N) = \Pi_{n=1}^N p(x_n, y_n)$. For any $\varepsilon > 0$, we have

$$\Pr\{(\mathbf{x}^N, \mathbf{y}^N) \in \mathcal{A}_\varepsilon\} \to 1 \text{ as } N \to \infty \tag{1.64}$$

and

$$|\mathcal{A}_\varepsilon| \leq 2^{NH(X,Y)+\varepsilon} \text{ for some sufficiently large } N \tag{1.65}$$

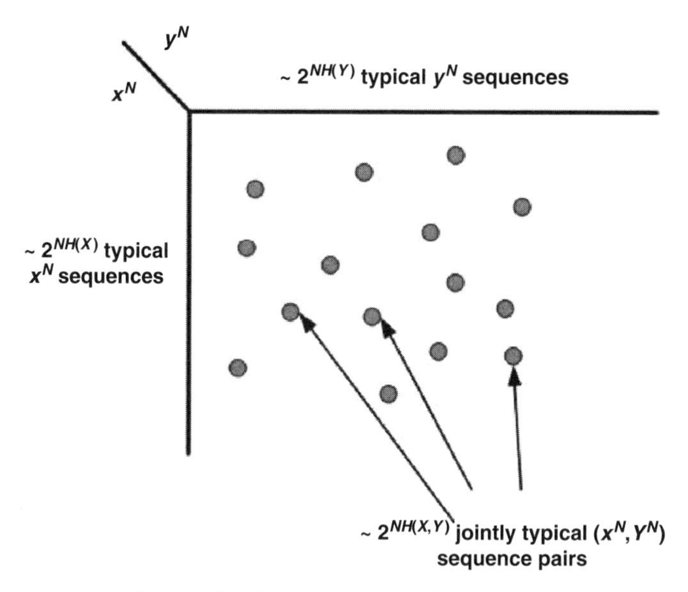

Figure 1.14. Illustration of a jointly typical set.

Furthermore, if $(\tilde{\mathbf{x}}^N, \tilde{\mathbf{y}}^N)$ are two sequences drawn i.i.d. according to $p(\tilde{\mathbf{x}}^N, \tilde{\mathbf{y}}^N) = p(\tilde{\mathbf{x}}^N)p(\tilde{\mathbf{y}}^N)$ (i.e., the distributions of $\tilde{\mathbf{X}}^N$ and $\tilde{\mathbf{Y}}^N$ have the same marginal distribution as $\mathbf{X}^N$ and $\mathbf{Y}^N$ but $\tilde{\mathbf{X}}^N$ and $\tilde{\mathbf{Y}}^N$ are independent), then

$$(1-\varepsilon)2^{-N(I(X;Y)+3\varepsilon)} \le \Pr\{(\tilde{\mathbf{x}}^N, \tilde{\mathbf{y}}^N) \in \mathcal{A}_\varepsilon\} \le 2^{-N(I(X;Y)-3\varepsilon)} \tag{1.66}$$

for sufficiently large N.

Hence, from these properties regarding typical sequences, we can see that when N is sufficiently large, there are about $2^{NH(X)}$ typical X sequences and about $2^{NH(Y)}$ typical Y sequences in the typical sets. However, not all pairs of typical X sequence and typical Y sequence are *jointly typical* as illustrated in Figure 1.14. There are about $2^{NH(X,Y)}$ such *jointly typical sequences* between X and Y. If $\mathbf{x}^N$ and $\mathbf{y}^N$ are naturally related according to their joint density function $p(X, Y)$, then there is a high chance that $(\mathbf{x}^N, \mathbf{y}^N)$ is in the *jointly typical set* for sufficiently large N. However, if $\mathbf{x}^N$ and $\mathbf{y}^N$ are randomly picked from their respective typical sets, the chance for $(\mathbf{x}^N, \mathbf{y}^N)$ in the *jointly typical set* is very small. This serves as an important property that we shall utilize to prove the channel coding theorem in the next section.

Channel Coding Theorem. We have now introduced the essential mathematical tools to establish the channel coding theorem. Before we proceed, here is a summary of the definitions of *channel encoder*, *generic channel*, *channel decoder*, and *error probability* in discrete-time domain.

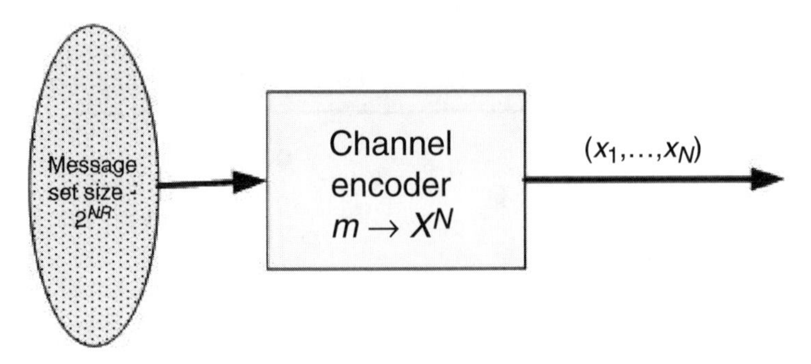

Figure 1.15. Channel encoder.

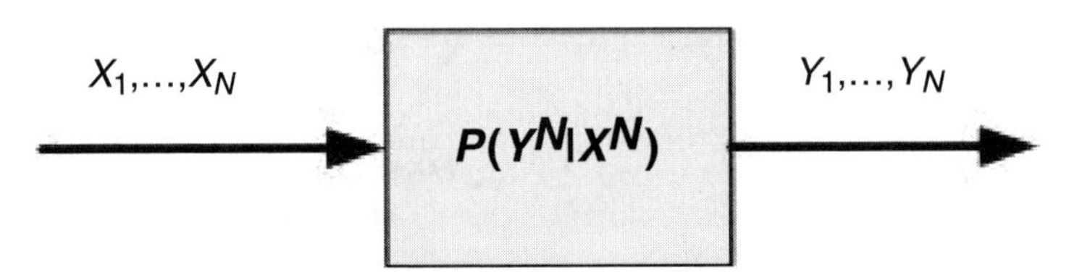

Figure 1.16. Generic channel.

Definition 1.11 (Channel Encoder) The *channel encoder* is a device that maps a message index $m \in \{1, 2, \ldots, M = 2^{NR}\}$ to the channel input $\mathbf{X}^N = [X_1, \ldots, X_N]$ *as* illustrated in Figure 1.15.

Hence, the transmitter has to use the channel N times to deliver the message m through the transmit symbols $X_1, \ldots, X_N$ and is specified by a function mapping f:f:$m \in \{1, 2, \ldots, M = 2^{NR}\} \to \mathcal{X}^N$. Alternatively, the channel encoder could be characterized by a codebook $\{\mathbf{X}^N(1), \mathbf{X}^N(2), \ldots, \mathbf{X}^N(M)\}$ with M elements. When the current message index is m, the channel encoder output will be $f(m) = \mathbf{X}^N(m)$. The encoding rate of the channel encoder is the number of information bits delivered per channel use and is given by

$$R = \frac{\log_2(M)}{N}$$

Definition 1.12 (Generic Channel) A *generic channel* in discrete time is a probabilistic mapping between the channel input $\mathbf{X}^N$ and the channel output $\mathbf{Y}^N$ as illustrated in Figure 1.16, namely, $p(\mathbf{Y}^N|\mathbf{X}^N)$. The channel is called *memoryless* if $p(\mathbf{Y}^N|\mathbf{X}^N) = \Pi_{n=1}^N p(Y_n|X_n)$, meaning that the current output symbol Y_n is independent of the past transmitted symbols.[6]

[6] Strictly speaking, this definition of memoryless channel is valid only if the encoder has no knowledge of the past channel outputs. i.e. $p(X_n|\mathbf{X}^{n-1}, \mathbf{Y}^{n-1}) = p(X_n|\mathbf{X}^{n-1})$. This condition holds in this book as the feedback channel is used to carry channel state only. Otherwise, a more general definition of memoryless channel should be: $p(Y_n|\mathbf{X}^n\mathbf{Y}^{n-1}) = p(Y_n|X_n)$ for all $n \geq 1$[100]. This more general definition reduces to the original definition if the encoder has no knowledge of past channel outputs. This is because $p(\mathbf{Y}^N|\mathbf{X}^N) = \Pi_n p(Y_n|\mathbf{X}^N\mathbf{Y}^{n-1}) = \Pi_n p(Y_n|\mathbf{X}^n\mathbf{Y}^{n-1}) = \Pi_n p(Y_n|X_n)$.

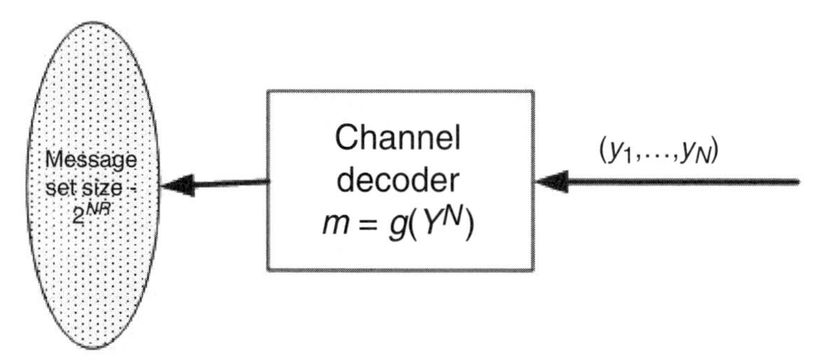

Figure 1.17. Illustration of channel decoder.

Definition 1.13 (Channel Decoder) A *channel decoder* is a deterministic mapping from the received symbols $\mathbf{Y}^N = [Y_1, \ldots, Y_N]$ to a decoded message index $\hat{m} \in \{1, 2, \ldots, 2^{NR}\}$ as illustrated in Figure 1.17

$$\hat{m} = g(\mathbf{Y}^N)$$

for some decoding functions g. The error probability associated with a decoding function is defined as

$$P_e = \Pr[g(Y^N) \neq m | m \text{ is transmitted}]$$

For simplicity and illustration purposes, we shall focus on the memoryless channel unless otherwise specified. Shannon's coding theorem is summarized below.

Theorem 1.5 (Shannon's Channel Coding Theorem) Letting $C = \max_{p(X)} I(X; Y)$. For rate $R < C$, there exists at least one channel encoder f and one channel decoder g for the memoryless channel $p(y|x)$ such that the error probability $P_e \to 0$ as $N \to \infty$. On the other hand, when $R > C$, the error probability is bounded away from zero for any channel encoder and channel decoder.

In other words, the channel coding theorem establishes a rigid bound on the maximal supportable data rate over a communication channel. From the definition of C above, the channel capacity [aftermaximization over $p(x)$] is a function of the channel $p(y|x)$ only. Note that $I(X; Y)$ is a function of both transmitter design ($p(x)$) and the channel ($p(y|x)$). We shall exploit the typical sequence to prove this powerful coding theorem on the basis of the *random codebook* argument. It is also important to go over the proof as it helps us understand the meaning of channel capacity and decoding mechanism.

Proof (Achievability) Let's first consider the proof for the achievability part, which is, given $R < C$, there exist a channel encoder f and a channel decoder g with arbitrarily low error probability. Consider a fixed distribution on x ($p(x)$). We generate 2^{NR} independent codewords at random according to the distribution $p(x)$. Hence, we formed a *random codebook* given by the matrix

$$\Omega = \begin{bmatrix} x_1(1) & \cdots & x_N(1) \\ \vdots & \vdots & \vdots \\ x_1(2^{NR}) & \cdots & x_N(2^{NR}) \end{bmatrix}$$

where the mth row represents the mth codeword that will be transmitted when the message index is m. The codebook Ω is then revealed to both the transmitter and the receiver. A message $m \in [1, 2^{NR}]$ is chosen according to a uniform distribution. The mth codeword $\Omega(m)$ {which has N symbols $\mathbf{x}(m) = [x_1(m), \ldots, x_N(m)]$} is sent out of the transmitter. The receiver receives a sequence $\mathbf{y} = [y_1, \ldots, y_N]$ according to the distribution $p(\mathbf{y}|\mathbf{x}) = \Pi_{n=1}^{N} p(y_n|x_n)$.

At the receiver, the decoder guesses which message was transmitted according to typical set decoding. The receiver declares that $\hat{m}$ was transmitted if there is one and only one index $\hat{m}$ such that $(\mathbf{x}(\hat{m}), \mathbf{y})$ is jointly typical. Hence, if there does not exist such index $\hat{m}$ or there is more than one such index, the receiver declares error. There will also be decoding error if $\hat{m} \neq m$.

To prove the achievability part, we shall show that the error probability of such encoder and decoder can achieve an arbitrarily low level. The error probability averaged over all possible random codebook realizations Ω is given by

$$\begin{aligned} \overline{P}_e &= \sum_{\Omega} P(\Omega) P_e^{(N)}(\Omega) = \sum_{\Omega} P(\Omega) \frac{1}{2^{NR}} \sum_{m} \lambda_m(\Omega) \\ &= \frac{1}{2^{NR}} \sum_{m} \left[\sum_{\Omega} P(\Omega) \lambda_m(\Omega) \right] \\ &= \sum_{\Omega} P(\Omega) \lambda_1(\Omega) \\ &= \Pr(E|W = 1) \end{aligned}$$

where

$$\lambda_m(\Omega) = \Pr\{\hat{m} \neq m | \mathbf{X}(m) \text{ is transmitted}\}$$

By symmetry of code construction, the average probability of error averaged over all codes does not depend on the particular index m being sent. Without loss of generality, assume that message $m = 1$ is transmitted. Define the events E_i as

$$E_i = \{(\mathbf{X}(i), \mathbf{Y}) \in \mathcal{A}_\varepsilon^{(N)}\} \; \forall i \in [1, \ldots, 2^{NR}]$$

Thus E_i is the event that the ith codeword $\mathbf{x}(i)$ and $\mathbf{y}$ are jointly typical. Error occurs when the transmitted codeword and the received sequence are not jointly typical or a wrong codeword is jointly typical with the received sequence. On the basis of the union bound, we have

$$\overline{P}_e = \Pr\{E_1^c \cup E_2 \cup \ldots \cup E_{2^{NR}}\} \le P(E_1^c) + \sum_{n=2}^{2^{NR}} P(E_n)$$

and by the jointly AEP theorem, we have $P(E_1^c) \le \varepsilon$ for sufficiently large N. Since by random code generation, $\mathbf{x}(1)$ and $\mathbf{x}(i)$ are independent, so are $\mathbf{y}$ and $\mathbf{x}(i)$. By the jointly AEP theorem, we have

$$P(E_i) \le 2^{-N(I(X;Y)-3\varepsilon)}$$

Hence, the average error probability is bounded by

$$\overline{P}_e \le \varepsilon + \sum_{i=2}^{2^{NR}} 2^{-N(I(X;Y)-3\varepsilon)} = \varepsilon + (2^{NR} - 1)2^{-N(I(X;Y)-3\varepsilon)} \le \varepsilon + 2^{3N\varepsilon}2^{-N(I(X;Y)-R)}$$

If $R < I(X;Y) - 3\varepsilon$, we can choose ε and N such that $\overline{P}_e \le \varepsilon$. Since the average error probability is averaged over all codebook realizations $\{\Omega\}$, there exists at least one codebook Ω^* with a small average probability of error $P_e^*(\Omega^*) \le \varepsilon$. This proved that there exists at least one achievable code with rate $R < I(X;Y) - 3\varepsilon$ for arbitrarily small error probability ε.

Converse In the converse, we have to prove that any $(N, 2^{NR})$ codes with $\lim_{N\to\infty} P_e^{(N)} = 0$ must have $R \le C$. From Fano's inequality, let $\mathbf{Y}$ be the channel output from a discrete memoryless channel when $\mathbf{X}$ is the channel input. We have $I(\mathbf{X};\mathbf{Y}) \le NC$ for all $p(\mathbf{X})$ because

$$\begin{aligned} I(\mathbf{X};\mathbf{Y}) &= \text{“}H(\mathbf{Y}) - H(\mathbf{Y}|\mathbf{X})\text{”} \\ &= H(\mathbf{Y}) - \sum_{n=1}^{N} H(Y_n|X_n) \\ &\le \sum_n H(Y_n) - \sum_n H(Y_n|X_n) \\ &= \sum_n I(X_n;Y_n) \le NC \end{aligned}$$

where the first equality is due to the fact that Y_n depends on X_n only and the last inequality is obtained directly from the definition of C. Let W be the message index drawn according to a uniform distribution over $[1, 2^{NR}]$. We have

$$\begin{aligned} NR = H(W) &= H(W|\mathbf{Y}) + I(W;\mathbf{Y}) \le H(W|\mathbf{Y}) + I(\mathbf{X}(W);\mathbf{Y}) \\ &\le 1 + P_e^{(N)}NR + I(\mathbf{X}(W);\mathbf{Y}) \le 1 + P_e^{(N)}NR + NC \end{aligned}$$

Thus

$$R \le P_e^{(N)} R + \frac{1}{N} + C \Rightarrow P_e^{(N)} \ge 1 - \frac{C}{R} - \frac{1}{NR}$$

Hence, if $R > C$, the error probability is bounded away from 0.

As a result of the *random coding proof*, we can see that good codes are actually very easy to find. For instance, we can randomly generate the M codewords to form a random codebook. We reveal this randomly generated codebook realization to the receiver so that the receiver can employ typical sequence decoding as illustrated in the proof. The random codebook argument presented above shows that there is a high chance that this randomly generated codebook is indeed a good code (or capacity achieving code). However, in practice, the problem lies in the receiving side. If our receiver has infinite processing power enabling us to exhaustively search in the decoding process, then there is almost no need for any code design at the transmitter. Because of limited processing power at the receiver, we have to create some *artificial structures* in the channel encoder to facilitate simple decoding at the receiver. For example, we have various types of codes such as *block codes*, *trellis codes*, and *turbo codes* in the literature. Once these artificial structures are added, good codes that can achieve the Shannon capacity are much harder to find. Yet, the channel coding theorem offers a tight upper bound on the maximum achievable rate so that we know how far we are from the best possible performance when we design practical codes.

1.4.3 Examples of Channel Capacity

In this section, we try to illustrate the concept of channel capacity by looking at several examples in wireless channels. We shall focus on some simple channels here. More sophisticated channels (such as the finite-state channels) will be discussed in detail in Chapter 2. In general, there is no closed-form expression for channel capacity except for a few special cases.

Discrete Memoryless Channels. Real-world channels are continuous-time with continuous-time input and continuous output. From Section 1.3, we established the equivalence between continuous-time and discrete-time channels. Hence, without loss of generality, we assume a discrete-time channel model with channel input $X \in \mathcal{X}$ and channel output $Y \in \mathcal{Y}$.

In general, the channel input $\mathcal{X}$ and the channel output $\mathcal{Y}$ are continuous complex numbers. However, if we consider a superchannel including a M-ary digital modulator, a memoryless physical channel, and a digital demodulator with Q quantization levels, we have a M-input, Q-output discrete memoryless channel as illustrated in Figure 1.18.

The $M \times Q$ discrete memoryless channel is specified by the *channel transition matrix* $p(y|x)$:

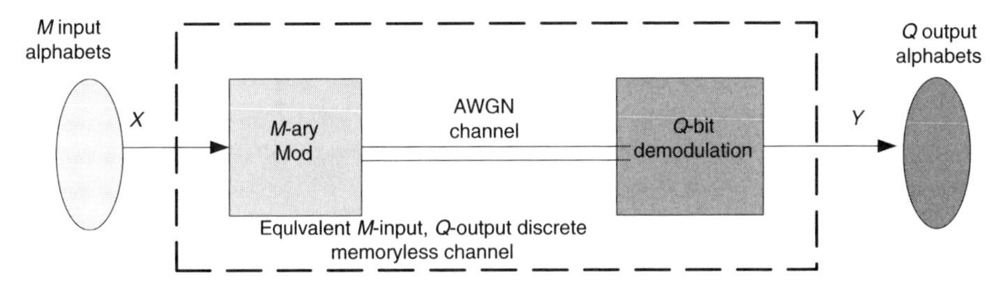

Figure 1.18. An M-input, Q-output discrete memoryless channel.

$$p(y|x) = \begin{bmatrix} p_{11} & \cdots & p_{1Q} \\ \vdots & \ddots & \vdots \\ p_{M1} & \cdots & p_{MQ} \end{bmatrix}$$

In addition, if the channel transition matrix is *symmetric* (i.e., all the rows are permutation of each other and so are the columns), the channel is called a *symmetric channel.*

Example 1.5 (Binary Symmetric Channel) A *binary symmetric channel* is a discrete channel with $\mathcal{X} = \{0, 1\}$ and $\mathcal{Y} = \{0, 1\}$. Furthermore, the probability of error $p(y = 0|x = 1) = p(y = 1|x = 0) = p$ and the probability of correct transmission is $p(y = 0|x = 0) = p(y = 1|x = 1) = 1 - p$ as illustrated in Figure 1.19.

The mutual information is given by

$$\begin{aligned} I(X;Y) &= H(Y) - H(Y|X) = H(Y) - \sum_x p(x)H(Y|X = x) \\ &= H(Y) - \sum_x p(x)H_2(p) \\ &= H(Y) - H_2(p) \\ &\leq 1 - H_2(p) \end{aligned}$$

where $H_2(p) = -p\log_2 p - (1 - p)\log_2(1 - p)$ and the last inequality follows because Y is a binary random variable. Equality holds if and only if $p(Y = 1) = p(Y = 0) = 0.5$. This is equivalent to uniform input distribution $p(X = 0) = p(X = 1) = 0.5$. Hence, we have $C = 1 - H_2(p) = 1 + p\log_2 p + (1 - p)\log_2(1 - p)$.

Example 1.6 (Binary Erasure Channel) A *binary erasure channel* is a discrete channel with $\mathcal{X} = \{0, 1\}$ and $\mathcal{Y} = \{0, 1, e\}$, where e denotes erasure. Furthermore, the probability of erasure is $p(y = e|x = 1) = p(y = e|x = 0) = p$, and the probability of correct transmission is $p(y = 0|x = 0) = p(y = 1|x = 1) = 1 - p$ as illustrated in Figure 1.20.

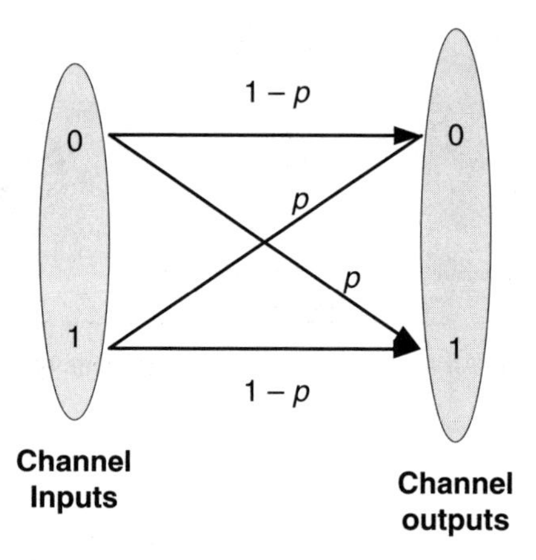

Figure 1.19. Binary symmetric channel.

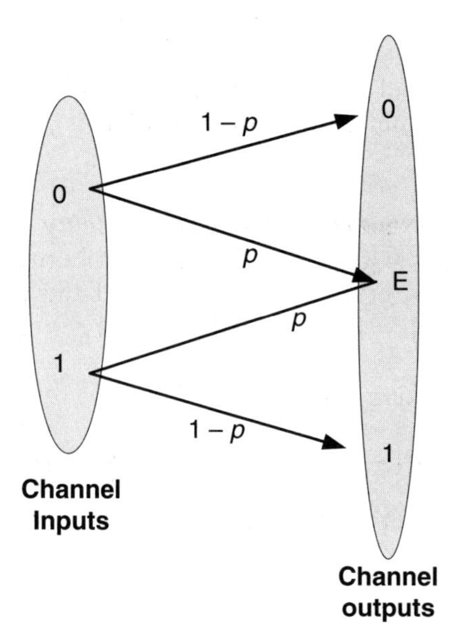

Figure 1.20. Binary erasure channel.

The channel capacity of the binary erasure channel is calculated as follows:

$$\begin{aligned} C &= \max_{p(x)} I(X;Y) \\ &= \max_{p(x)} (H(Y) - H(Y|X)) \\ &= \max_{p(x)} (H(Y) - H_2(p)) \end{aligned}$$

While the first guess for the maximum of $H(Y)$ is $\log_2(3)$, this is not achievable by any choice of input distribution $p(X)$. Let E be the event $\{Y = e\}$,

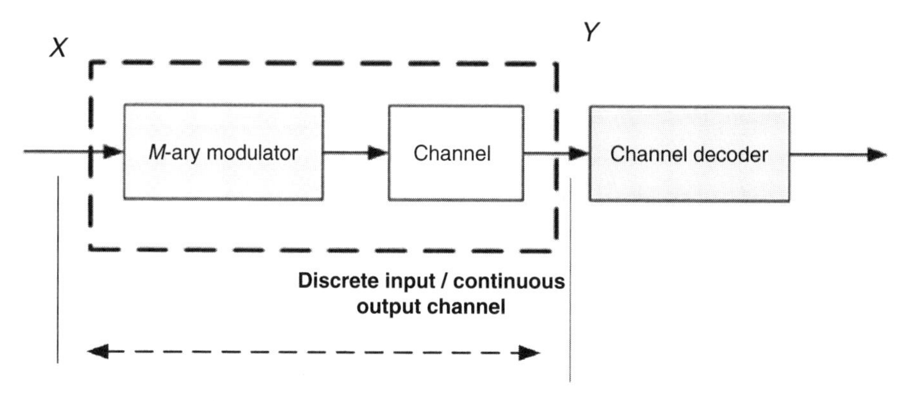

Figure 1.21. Discrete-input continuous-output channel.

(erasure). We have $H(Y) = H(Y, E) = H(E) + H(Y|E)$. Letting $\pi = p(X = 1)$, we have $H(Y) = H_2(p) + (1 - p)H_2(\pi)$. Hence

$$\begin{aligned} C &= \max_{p(x)}(H(Y) - H_2(p)) \\ &= \max_{p(x)}((1-p)H_2(\pi) + H_2(p) - H_2(p)) \\ &= \max_{\pi}((1-p)H_2(\pi)) \\ &= 1-p \end{aligned}$$

where the capacity achieving input distribution is $\pi = 0.5$. The expression for capacity has some intuitive meaning. Since a proportion p of the bits are erased in the channel, we can recover at most a proportion $(1 - p)$ of the bits. Hence, the channel capacity is at most $(1 - p)$.

Discrete-Input Continuous-Output Channels. A channel is called *discrete-input continuous-output* if the channel input alphabet $\mathcal{X}$ is discrete and the channel output alphabet $\mathcal{Y}$ is continuous. For example, the *superchannel* including a digital modulator and a physical channel as illustrated in Figure 1.21 belongs to this type of channel. The channel is characterized by a transition density $f(\mathbf{y}^N|\mathbf{x}^N)$. Similarly, the channel is memoryless if $f(\mathbf{y}^N|\mathbf{x}^N) = \Pi_n f(y_n|x_n)$.

Example 1.7 (Binary Input Continuous-output AWGN Channel) Consider a superchannel with a binary modulator and an AWGN physical channel. The discrete-time received signal Y_n is given by:

$$Y_n = X_n + Z_n$$

where X_n is the binary input and Z_n is a zero-mean white Gaussian noise with variance σ_z^2. Hence,we have $f(\mathbf{y}^N|\mathbf{x}^N) = \Pi_n f(y_n|x_n)$, where $f(y_n|x_n = -A)\mathcal{N}(-A, \sigma_z^2)$ and $f(y_n|x_n = A)\mathcal{N}(A, \sigma_z^2)$. The mutual information is given by

$$
\begin{aligned}
I(X;Y) &= H(Y) - H(Y|X) = H(Y) - (H(Y|X=A)\pi + 1(1-\pi)H(Y|X=-A)) \\
&= -\int_{-\infty}^{\infty} f(y)\log_2(f(y))dy + \pi\int_{-\infty}^{\infty} f(y|x=A)\log_2(f(y|x=A))dy \\
&\quad + (1-\pi)\int_{-\infty}^{\infty} f(y|x=-A)\log_2(f(y|x=-A))dy \\
&= \pi\int_{-\infty}^{\infty} f(y|x=A)\log_2\left(\frac{f(y|x=A)}{f(y)}\right)dy \\
&\quad + (1-\pi)\int_{-\infty}^{\infty} f(y|x=-A)\log_2\left(\frac{f(y|x=-A)}{f(y)}\right)dy
\end{aligned}
$$

where $\pi = p(X = A)$. Hence, the mutual information is maximized when $\pi = 0.5$. Note that $C \leq 1$ because the digital modulator in the "superchannel" is a binary modulator delivering at most 1 bit per modulation symbol. At finite channel noise, some of the bits transmitted will be in error. However, we observe that C approaches 1 bit per channel use as $\sigma_z^2 \to 0$. This means that at high SNR, nearly all of the transmitted bits could be decoded at the receiver and therefore, achieving a capacity of 1 bit per channel use.

In general, for M-ary modulator, the channel capacity is given by

$$
\begin{aligned}
C &= \max_{\pi}[H(Y) - H(Y|X)] \\
&= \max_{\pi}\left[\sum_{\pi} \pi(x)\int_{y} \log_2\left(\frac{f(y|x)}{\sum_x \pi(x)f(y|x)}\right)\right]
\end{aligned}
$$

where $\pi = [\pi_1, \ldots, \pi_M]$ and $\pi_q = p(X = q)$. In general, the capacity achieving distribution π has to be evaluated numerically.

Continuous-Input Continuous-Output Channels. In this case, both the channel input $\mathcal{X}$ and the channel output $\mathcal{Y}$ are continuous values (or complex values in general). The channel is specified by the *transition density* $f(y|x)$. In fact, we can consider a continuous-input channel as a limiting case of a discrete-input channel when we have infinitely dense constellation at the digital modulator. For any input distribution $p(X = x_1), \ldots, p(X = x_M)$ of an M-ary digital modulator, we could find an equivalent input distribution $f(\mathcal{X})$ in the continuous input case. In other words, we expect the channel capacity of continuous-input channels to be greater than or equal to the channel capacity of discrete input channels because the input distribution of the latter case is a subset of that in the former case.

Example 1.8 (Discrete Time AWGN Channel) Consider a discrete-time AWGN channel where $X \in \mathcal{X}$ is the channel input and $Y \in \mathcal{Y}$ is the channel output. Both the channel input and channel output are complex numbers. The channel output is given by:

$$
Y_n = X_n + Z_n
$$

where Z_n is the white Gaussian channel noise with variance σ_z^2. Therefore, the channel transition probability is given by $f(y|x) = \frac{1}{2\pi\sigma_z}\exp\left[-\frac{1}{2\sigma_z^2}|y-x|^2\right]$. The channel capacity is given by

$$
\begin{aligned}
C &= \max_{f(x)}[H(Y) - H(Y|X)] \\
&= \max_{f(x)} H(Y) - H(Z) \\
&= \max_{f(y)} H(Y) - H(Z) \\
&= \log_2\left(1 + \frac{\sigma_x^2}{\sigma_z^2}\right)
\end{aligned}
$$

where $\sigma_x^2 = \varepsilon[|X|^2]$. Note that the second equality is due to the fact that $H(Z)$ is independent of $f(x)$ and the final equality is due to the fact that $H(Y)$ is maximized if and only if Y is complex Gaussian, which is equivalent to X being complex Gaussian.

Continuous-Time Channels. In this section, we try to connect the channel capacity of discrete-time channels to continuous-time channels. The channel capacity of discrete-time channels is expressed in units of bits per channel use. On the other hand, the unit of channel capacity of continuous-time channels is bits per second. To facilitate the discussion, let's consider the following example.

Example 1.9 (Continuous-Time AWGN Channel) Consider a continuous-time AWGN channel with channel input $X(t)$ (with two-sided bandwidth W) and channel output $Y(t)$. Both the channel input and channel output are complex-valued random processes. The channel output is given by

$$
Y(t) = X(t) + Z(t)
$$

where $Z(t)$ is the white Gaussian channel noise with two-sided power spectral density η_0.

From the discussions in previous section, a bandlimited random process $X(t)$ could be represented in geometric domain as vector $\mathbf{X}$ in WT dimension signal space over the complex field. Hence, the equivalent discrete time channel is given by

$$
\mathbf{y} = \mathbf{x} + \mathbf{z}
$$

where $\mathbf{z}$ is a zero-mean Gaussian i.i.d. sequence with variance σ_z^2. The total noise power is given by

$$
P_z = \frac{\varepsilon\left[\|\mathbf{z}\|^2\right]}{T} = \frac{\sum_{i=1}^{WT}\sigma_z^2}{T} = W\eta_0
$$

Hence, we have $\sigma_z^2 = \eta_0$. The channel transition density is given by $f(\mathbf{y}|\mathbf{x}) = \Pi_{n=1}^{WT} f(y_n|x_n)$, where

$$f(y_n|x_n) = \frac{1}{2\pi\eta_0} \exp\left[-\frac{1}{\eta_0}|y_n - x_n|^2\right]$$

Treating **X** as a supersymbol, the asymptotic channel capacity [in bits per second (bps)] is given by

$$C = \lim_{T\to\infty} \max_{p(\mathbf{x})} \frac{1}{T} I(\mathbf{X}; \mathbf{Y})$$

Because of the i.i.d. nature of the noise sequence **Z**, we have

$$\max_{p(\mathbf{X})} I(\mathbf{X}; \mathbf{Y}) = \sum_{n=1}^{WT} \max_{p(X_n)} I(X_n; Y_n) = WT \log_2\left(1 + \frac{\sigma_x^2}{\eta_0}\right)$$

where $\sigma_x^2 = \varepsilon[|X_n|^2]$. On the other hand, the transmit power is given by

$$P_{tx} = \frac{1}{T}\int_0^T |x(t)|^2 dt = \frac{1}{T}\|\mathbf{x}\|^2 = W\left(\frac{1}{WT}\sum_n |x_n|^2\right) \to W\varepsilon\left[|x_n|^2\right] = W\sigma_x^2 \quad (1.67)$$

as $T \to \infty$. Substituting into the capacity equation, the channel capacity (in bps) is given by

$$C = W\log_2\left(1 + \frac{\sigma_x^2}{\eta_0}\right) = W\log_2\left(1 + \frac{P_{tx}}{W\eta_0}\right) \quad (1.68)$$

Note that $\eta_0 W$ equals to the total noise power and hence, the channel capacity could also be written as $C = \log_2(1 + SNR)$.

Define bandwidth efficiency as $\eta = R/W$, where R denotes the transmission bit rate and W denotes the bandwidth. From Shannon's capacity, the bandwidth efficiency could be expressed as

$$\eta_{\max} = \log_2\left(1 + \frac{P_{tx}}{W\eta_0}\right) = \log_2\left(1 + \frac{RE_b}{W\eta_0}\right)$$

where R is the bit rate and E_b is the bit energy. Since $R/W = \eta$, the equation above has a solution as follows:

$$\frac{E_b}{\eta_0} \geq \frac{2^{\eta_{\max}} - 1}{\eta_{\max}} \quad (1.69)$$

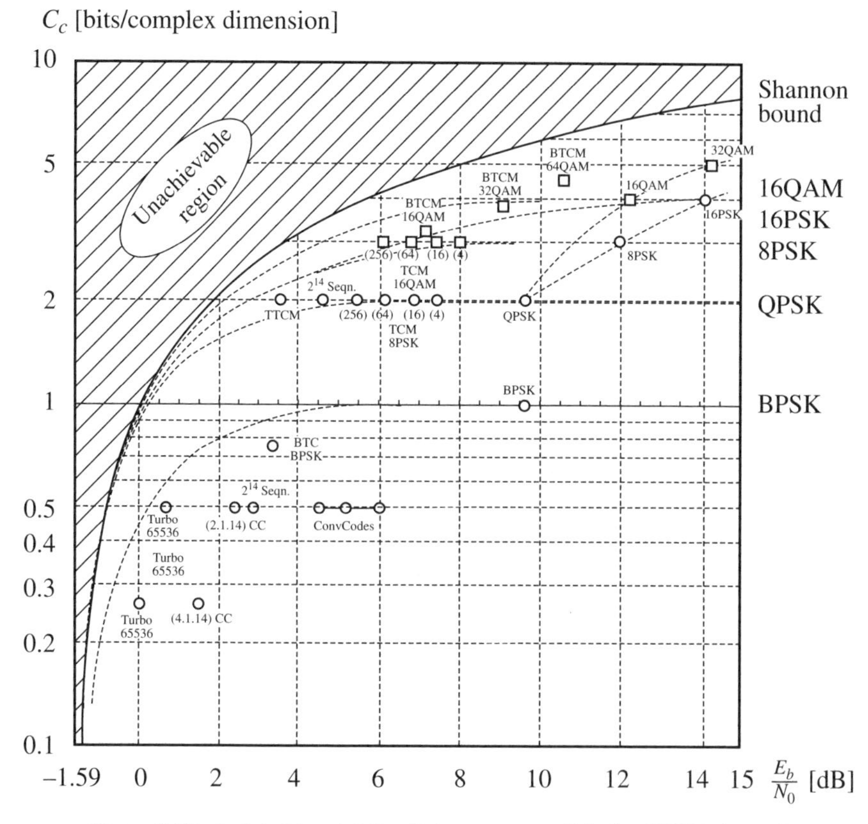

Figure 1.22. A plot of bandwidth efficiency versus E_b/η_0 for AWGN channel.

In the limiting case of very large bandwidth, we have $\eta_{\max} \to 0$ and $E_b/\eta_0 \geq \log(2) = -1.59\,\text{dB}$. This represents the absolute minimum bit energy : noise ratio that is required to reliably transmit one bit of information over a very large bandwidth.

Figure 1.22 illustrates the bandwidth efficiency versus E_b/η_0 for various modulation schemes over the AWGN channel. As mentioned, the capacity of a *continuous modulator* is an upper bound of all other discrete modulators such as BPSK, QPSK, and 16QAM. At a very low E_b/η_0 region (such as in deep-space communications), we have to sacrifice bandwidth to trade for *power efficiency* because power is the limiting factor. Hence, modulation schemes operating in the region are called *bandwidth expansion modes*. Examples are M-ary orthogonal modulations. On the other hand, when bandwidth is a limited resource (such as in terrestrial communications), we have to sacrifice power for bandwidth efficiency. Examples are 16QAM and 64QAM modulations. This region of operation is therefore referred to as consisting of *high-bandwidth-efficiency modes*.

1.5 SUMMARY

In this chapter, we have reviewed the fundamental concepts in wireless communications. We have discussed the wireless fading channels in continuous-time domain and established the equivalence between continuous-time and discrete-time models based on the concept of signal space and sufficient statistics. The wireless fading channel is modeled as a random process $H(t, \nu, r)$ in the time domain t, frequency domain ν, and the position domain r. Equivalently,we can also characterize the random fading channels from the spectral domain $H(f, \tau, k)$, which involves the *Doppler domain* f, the *delay domain* τ, and the *wavenumber domain* k. In fact, the two random processes are related by three-dimensional Fourier transform pairs ($H(t, \nu, r) \leftrightarrow H(f, \tau, k)$).

In most cases, we are concerned with the types of random channels that are wide-sense stationary (WSS) uncorrelated scattering (US). Hence, the autocorrelation of $H(t, \nu, r)$ is a function of $(\Delta t, \Delta \nu, \Delta r)$ only. To characterize the statistical properties of the WSS-US random channels, we can define *coherence time* T_c, *coherence bandwidth* B_c, and *coherence distance* D_c on the basis of the one-dimensional autocorrelation $R_H(\Delta t)$, $R_H(\Delta \nu)$ and $R_H(\Delta r)$.

On the other hand, from the spectral domains (f, τ, k), the uncorrelated scattering property implies that $\varepsilon[H(f, \tau, k)H^*(f', \tau', k')] = S_H(f, \tau, k) \delta(f - f')\delta(\tau - \tau')\delta(k - k')$. Hence, we can also define the corresponding dual parameters on the basis of the scattering function $S_H(f, \tau, k)$: the *Doppler spread* σ_f^2, the *delay spread* σ_τ^2, and the *angle spread* σ_k^2. With respect to the coherence bandwidth (or delay spread) and the transmitted bandwidth, we can deduce whether flat fading channels ($W < B_c$) or frequency-selective fading channels ($W > B_c$) will be experienced by the signal. On the other hand, on the basis of the coherence time (or Doppler spread) and the transmitted symbol duration, we can deduce whether we have fast fading channels ($T_c < T_s$) or slow fading channels ($T_c > T_s$).

Finally, we reviewed the fundamentals of information theory including entropy, mutual information, and channel capacity. We illustrate the application of information theory by evaluating the channel capacity of several simple channels.

From now on, unless specified otherwise, we shall further develop the theories and concepts based on the discrete-time model.

EXERCISES

1. *Spreading and Coherence in different environment* Recall that delay, doppler and angle spread are defined by the environment and so as the corresponding frequency, time and spatial coherence. Using the knowledge of them, determine the order of the following environment in terms of the magnitudes of spreads and coherence with 1 being the smallest and 6 being the largest.

(a) Office
(b) Elevator
(c) Residential Building
(d) Underground Trains
(e) Bullet Trains
(f) Pedestrian Pathway

2. *Wide Sense Stationary* A random processes $X(t, v, r)$ is regarded as a wide sense stationary process if the correlation with itself only depends on the difference of its parameters. Therefore, samples taking at an interval of time, frequency and space are no different from samples taken at the shifts of the interval. In other words, the randomness of the process does not depends on its parameters. $X(t, v, r)$ is the input random process of the system characterized by a system function $H(t, v, r)$ which is also random in nature and $Y(t, v, r)$ is the output random process. Suppose there is noise $N(t, v, r)$ in the system, the input-output equation is therefore

$$Y(t,v,r) = H(t,v,r)X(t,v,r) + N(t,v,r)$$

(a) Construct a random process $X(t, v, r)$ which is WSS in time, frequency and position.
(b) Under which condition(s) is output $Y(t, v, r)$ WSS?
(c) Under which condition(s) are $X(t, v, r)$ and $Y(t, v, r)$ jointly WSS?

3. *Upper Bound of differential entropy* Differential entropy is the entropy of a continuous random variable. It can be shown that the entropy is upper bounded by the following.

$$H(X) \leq \frac{1}{2}\log_2(2\pi e \sigma_x^2)$$

where σ_x^2 is the variance of the random variable X.
(a) verify that the upper bound is achieved if X is of guassian distributed.

4. *Entropy of morphism* Compare the entropy of 2 random variables X and Y if
(a) the mapping from X to Y is isomorphic
(b) the mapping from X to Y is endomorphic or subjective

5. *Entropy of markov chains* Suppose there are 2 states A and B as shown in the diagram 1. At each state, one can choose to either stay in the same state or jump to the other state. p_{AA}, p_{AB}, p_{BA}, p_{BB} are the corresponding transition probabilities. Define X_n to be the state at time instant n and X_0 is the initial state. Assume that it is equally probable for X_0 to be A or B.

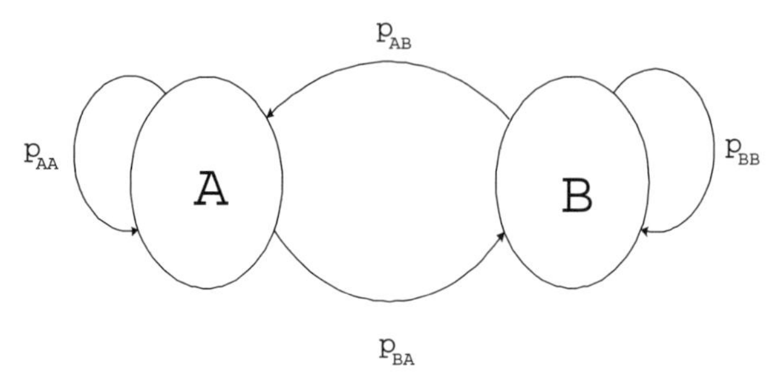

Figure 1. Markov chains with 2 states A and B.

(a) Find the probability if X_3 equals A.
(b) What is the entropy of X_n, $H(X_n)$ as $n \to \infty$?
(c) verify the result obtained in (b) if the transition probabilities are symmetric. That is $p_{AA} = p_{BB}$ and $p_{AB} = p_{BA}$.
(d) verify the result obtained in (b) if the transition probabilities are biased, say to A. That is $p_{AA} \gg p_{BA}$ and $p_{AB} \gg p_{BB}$.

6. *Fano's inequality* Suppose Y is an estimate of the random variable X. If there is no error, we can say that the conditional entropy $H(X|Y) = 0$. Thus, the estimate error is expected to be small when the conditional entropy is small. Recall the Fano's inequality,

$$H(P_e) + P_e \log(|x| - 1) \geq H(X|Y)$$

where P_e is the probability of error and $|x|$ denote the size of the support set of x. Therefore, given a target or required probability of error, it gives an upper bound of the conditional entropy. To have more insight into this inequality, we can reorganize the terms to be the following

$$P_e \geq \frac{H(X|Y) - 1}{\log(|x|)}$$

Given the conditional entropy, Fano's inequality provides a lower bound of the probability of error. Define an indicator variable E such that

$$E = \begin{cases} 1, & \text{if } \hat{X} \neq X; \\ 0, & \text{if } \hat{X} = X. \end{cases}$$

(a) Use chain rule of entropy to expand $H(E, X|Y)$
(b) Prove that $H(E|Y) \leq H(P_e)$.
(c) Prove that $H(X|E, Y) \leq P_e \log(|x| - 1)$
(d) Combining parts (a), (b) and (c) prove the Fano's inequality.

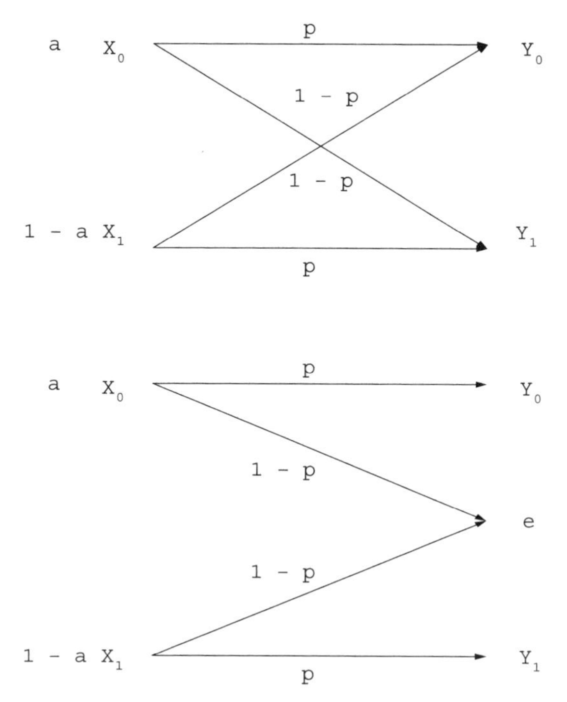

Figure 2. Above: Binary symmetric channel Below Binary erasure channel.

7. *Binary symmetric channel v.s. Binary Erasure channel* Suppose the input X of the channel is equally probable to take values X_0 and X_1, ($a = 0.5$), whereas the output can take values of Y_0 and Y_1. The probability of receiving the right symbol is denoted by p. In other words, $p(Y_0|X_0) = p(Y_1|X_1) = p$. In the binary erasure channel, the output Y can take one more value e which is called the erasure. The probability of getting the erasure is $1 - p$ and $p(Y_0|X_1) = p(Y_1|X_0) = 0$. Both channels are shown as reference in figure 2.

(a) Compute $H(X|Y)$ in the binary symmetric channel

(b) Compute $H(X|Y)$ in the binary erasure channel

(c) Compute $H(Y)$ in both cases.

(d) What does $H(Y)$ tend to if $p \to 1$? How about $p \to 0$? Explain.

8. *Venn Diagram* Use Venn Diagram to illustrate the following

(a) $I(X; Y|Z)$

(b) $I(X_1, X_2, X_3; Y)$

9. *Channel Capacity of AWGN Channels*

(a) Given an AWGN channel with channel output given by $Y_n = X_n + Z_n$, show that the capacity is

$$C = \log_2\left(1 + \frac{\sigma_x^2}{\sigma_z^2}\right)$$

where σ_x^2 is the variance of X and σ_z^2 is the variance of Z. Assume everything is complex. What is the unit of the capacity obtained.

(b) In practice, we transmit continuous waveforms $x(t)$ instead of discrete-time symbols, X_n. Using (a) and Nyquist sampling theorem, show that the channel capacity of the continuous waveform AWGN channel with bandwidth W is given by

$$C = W \log_2\left(1 + \frac{P}{W\eta_0}\right)$$

where η_0 is the single-sided noise power spectral density. What is the unit of the capacity obtained.

2

MIMO LINK WITH PERFECT CHANNEL STATE INFORMATION

2.1 OVERVIEW

In this chapter, we explore the design of channel-adaptive physical layer for wireless link with multiple antennas based on information theoretical framework. Specifically, we consider a point-to-point communication link with n_T transmit antennas and n_R receive antennas. Notationwise, we denote such multiple-input multiple-output (MIMO) link as a $n_T \times n_R$ wireless link. As we discussed in Chapter 1, wireless channels are characterized by time-varying multipath fading. For instance, the attenuation introduced by the channel fluctuates over a large dynamic range over a relatively short time interval. At one time, the channel fading may be very good, facilitating reliable communications. At other times, the channel state may be very bad and the communication link may hardly be able to deliver any information reliably. In fact, the channel capacity[1] depends on the availability of *channel state information* (CSI) at the receiver and at the transmitter. As we shall discuss, there are in general two functions of CSI. The availability of CSI at the receiver serves the purpose of detection. On the other hand, the availability of CSI at the transmitter serves the purpose of adaptation.

The $n_T \times n_R$ wireless MIMO channels, however, are characterized by time-varying fading $n_R \times n_T$ channel matrices. As a result of the multiple antennas, the MIMO channel can be viewed as a vector channel with $m^* = \min\{n_T, n_R\}$ *spatial channels*. In fact, the data rate of the MIMO link can be increased by spatial multiplexing of the m^* spatial channels and therefore, the channel

[1] *Capacity* refers to the maximum data rate that can be delivered through the wireless link with arbitrarily low error probability.

Channel-Adaptive Technologies and Cross-Layer Designs for Wireless Systems with Multiple Antennas: Theory and Applications. By V. K. N. Lau and Y.-K. R. Kwok
ISBN 0-471-64865-5

capacity increases with m^* in a linear manner asymptotically. Hence, by increasing m^*, we can increase the channel capacity at the same power and bandwidth budget. This is a very significant gain compared to increasing the capacity by increasing the transmit power. Hence, it is the focus of this chapter to study the combined benefit of MIMO and channel adaptation.

In this chapter, we shall focus on the ideal situation where the knowledge of CSI (at the transmitter and/or the receiver) is perfect. The effect of imperfect CSI knowledge on the channel capacity will be discussed in the next chapter. We begin by reviewing various discrete-time fading channels that serve as a common foundation for further discussion. Based on the channel models, we review the concept of *ergodic capacity* and *outage capacity*. The ergodic and outage capacity of the MIMO channels with various levels of CSI availability will be discussed in detail. For instance, we label the scenario as *CSIR* when the receiver has CSI and label it as *CSIT* when the transmitter has CSI. We shall discuss the channel capacities of the four cases, namely, without any CSI, with perfect CSIR, with perfect CSIT, as well as with both perfect CSIR and perfect CSIT. With perfect CSIR, channel-matched decoding can be done, resulting in enhanced channel capacity compared to the baseline case without any CSI knowledge. When the transmitter has perfect CSIT, adaptation (in both the spatial and temporal dimensions) can be performed at the transmitter, and this results in additional enhancement of channel capacity. For instance, the optimal MIMO transmission scheme with CSIR only is shown to be a bank of n_T isolated channel encoders [126]. On the other hand, the optimal MIMO transmission scheme with perfect CSIT is shown to be a cascade of channel encoder bank, adaptive power control matrix, and an *eigenbeamforming* matrix. The adaptive power control matrix is to perform *temporal and spatial* (spatiotemporal) *water filling* on each individual spatial channel. The eigenbeamforming matrix is to decompose the $n_R \times n_T$ channel matrix into $m^* = \min[n_R, n_T]$ independent spatial channels so that data carried along each spatial channel will not interfere with each other at the receiver. In other words, the $n_R \times n_T$ MIMO fading channel with CSIT and CSIR is decomposed into $m^* = \min[n_R, n_T]$ independent parallel channels.

2.2 MATHEMATICAL MODEL OF THE MIMO LINK

In Chapter 1, we have reviewed the continuous-time models for flat- and frequency-selective fading channels. We also illustrated how to discretize a continuous-time channel model to the equivalent discrete-time channel model for flat fading and frequency-selective fading channels. In this section, we shall generalize the concept of discrete-time fading channels to a probabilistic channel with states. For notation convenience, uppercase (capital) X denotes random variable, while lowercase x denotes a realization of the random variable. Bold $\mathbf{X}$ denotes a vector or matrix of random variables, while normal X denotes a scalar.

2.2.1 Probabilistic Channels with States

The MIMO transmitter produces a channel input symbol $\mathbf{X} \in \mathcal{X}$, which is an $(n_T \times 1)$-dimensional vector over the complex field, per channel use. The corresponding channel output symbol is given by $\mathbf{Y} \in \mathcal{Y}$, which is an n_R-dimensional vector over the complex field. The probabilistic MIMO channel can be characterized by the $(n_R \times n_T)$-dimensional channel state $\mathbf{H} \in \mathcal{H}$, which is again over the complex field. In general, each channel state realization $\mathbf{H} = \mathbf{h}$ specifies a *channel transition probability* $p(\mathbf{y}|\mathbf{x}, \mathbf{h})$ (if $\mathcal{X}$ and $\mathcal{Y}$ are discrete sets) or $f(\mathbf{y}|\mathbf{x}, \mathbf{h})$ (if $\mathcal{X}$ and $\mathcal{Y}$ are continuous sets). When $\mathcal{X}$ and $\mathcal{Y}$ are all finite sets, the channel is called *discrete channel with states*. For instance, given any channel state realization $\mathbf{h}$, the probability of receiving $\mathbf{y}$ given $\mathbf{x}$ is transmitted is given by $p(\mathbf{y}|\mathbf{x}, \mathbf{h})$. On the other hand, when $\mathcal{X}$ and $\mathcal{Y}$ are all continuous sets, the channel is called a *continuous channel with states*. Given any channel state realization $\mathbf{h}$, the *conditional channel transition probability* becomes $f(\mathbf{y}|\mathbf{x}, \mathbf{h})$, where the conditional probability of receiving $\mathbf{Y} \in [\mathbf{y}, \mathbf{y} + d\mathbf{y}]$ given $\mathbf{x}$ is transmitted and the current channel state $\mathbf{h}$ is given by $f(\mathbf{y}|\mathbf{x}, \mathbf{h})d\mathbf{y}$. Without loss of generality, we shall assume discrete inputs and discrete outputs for simplicity unless otherwise specified. Generalization to continuous inputs and continuous outputs follows the standard argument as in Reference 44.

The channel transition probability of MIMO channels can be completely characterized by the *conditional channel transition probability* and the *channel state sequence probability* as

$$p(\mathbf{y}_1^N|\mathbf{x}_1^N) = \sum_{\mathbf{h}_1^N} p(\mathbf{y}_1^N|\mathbf{x}_1^N, \mathbf{h}_1^N)p(\mathbf{h}_1^N)$$

where $\mathbf{x}_1^N = [\mathbf{x}_1, \ldots, \mathbf{x}_N]$ denotes a block of N transmit symbols, $\mathbf{y}_1^N = [\mathbf{y}_1, \ldots, \mathbf{y}_N]$ denotes a block of N received symbols, and $\mathbf{h}_1^N = [\mathbf{h}_1, \ldots, \mathbf{h}_N]$ denotes a block of N channel states.

In general, the output symbol $\mathbf{y}_n$ at time n depends not only on the current transmitted symbol $\mathbf{x}_n$ but also on the past and future transmitted symbols. If that's the case, the probabilistic channel is said to have *memory*. Otherwise, the probabilistic channel is said to be *memoryless*.

Definition 2.1 (Memoryless Channels) A channel is called *memoryless* if the *unconditional transition probability* can be expressed into a product form:[2]

$$p(\mathbf{y}_1^N|\mathbf{x}_1^N) = \prod_{n=1}^N p(\mathbf{y}_n|\mathbf{x}_n) \tag{2.1}$$

We look at some important examples below.

[2]Please refer to Chapter 1 for a more general definition of memoryless channe [100]. However, throughout the book, we assume the feedback channel carries channel state only and hence, the two definitions are the same.

Fast Flat Fading MIMO Channels. For example, the fast flat fading MIMO channel introduced in Chapter 1 is an example of memoryless continuous-input continuous-output MIMO channel because the channel transition probability can be decomposed into a product form. The discrete-time equivalent channel model is given by

$$\mathbf{y}_n = \mathbf{h}_n \mathbf{x}_n + \mathbf{z}_n \tag{2.2}$$

where $\mathbf{z}_n$ is an $(n_R \times 1)$-dimensional i.i.d. complex Gaussian random vector sequence with covariance $\eta_0 \mathbf{I}_{n_R}$. In the model, $\{\mathbf{x}_n\}$ are the complex-valued $n_T \times 1$ channel input, $\{\mathbf{y}_n\}$ are the complex-valued $n_R \times 1$ channel output, and $\{\mathbf{h}_n\}$ are the $n_R \times n_T$ i.i.d. complex Gaussian channel state sequence.[3] Given $\{\mathbf{h}_n\}$ and $\{\mathbf{x}_n\}$, the channel transition probability is given by

$$\begin{aligned}
p(\mathbf{y}_1, \ldots, \mathbf{y}_N | \mathbf{x}_1, \ldots, \mathbf{x}_N, \mathbf{h}_1, \ldots, \mathbf{h}_N) \\
&= \frac{1}{(\pi |\Sigma_y|)^N} \exp\left(-\sum_n (\mathbf{y}_n - \mathbf{h}_n \mathbf{x}_n)^* \Sigma_y^{-1} (\mathbf{y}_n - \mathbf{h}_n \mathbf{x}_n) \right) \\
&= \prod_n \frac{1}{(\pi |\Sigma_y|)} \exp\left(-(\mathbf{y}_n - \mathbf{h}_n \mathbf{x}_n)^* \Sigma_y^{-1} (\mathbf{y}_n - \mathbf{h}_n \mathbf{x}_n) \right) \\
&= \prod_n p(\mathbf{y}_n | \mathbf{x}_n, \mathbf{h}_n)
\end{aligned}$$

because the channel noise $\mathbf{z}_n$ is i.i.d. and

$$\Sigma_y = \varepsilon[\mathbf{y}_n \mathbf{y}_n^*] = \eta_0 \mathbf{I}_{n_R} + \underbrace{\varepsilon[\mathbf{x}_n \mathbf{x}_n^*]}_{\mathbf{Q}_x}$$

Together with the i.i.d. channel state sequence $p(\mathbf{h}_1^N) = \Pi_n p(\mathbf{h}_n)$, the fast flat fading channel is a memoryless channel because the channel transition probability can be decomposed into a product form as in Equation (2.1).

Quasistatic Flat Fading MIMO Channels. As the transmission bit rate increases and the frame duration T_f becomes shorter and shorter, we may have a *slow fading* situation across the entire symbols of a frame; that is, $T_f \ll T_c$, where T_c is the channel coherence time. In this case, the transmit symbols over a frame will share the same channel fading; that is, $\mathbf{h}_n = \mathbf{h}_m = h$. Such slow fading situation is called *quasistatic fading*.

Mathematically, a quasistatic flat fading MIMO channel has the transition probability $p(\mathbf{y}_1, \ldots, \mathbf{y}_N | \mathbf{x}_1, \ldots, \mathbf{x}_N, \mathbf{h}) = \Pi_n p(\mathbf{y}_n | \mathbf{x}_n, \mathbf{h})$. However, because of the

[3]The i.i.d. condition is with respect to the time index n; that is, $p(\mathbf{h}_n, \mathbf{h}_m) = p_H(\mathbf{h}_n) p_H(\mathbf{h}_m)$, where $p_H()$ is the joint distribution of the $n_R \times n_T$ complex channel matrix. Note that the elements of $\mathbf{h}_n$ need not be i.i.d.

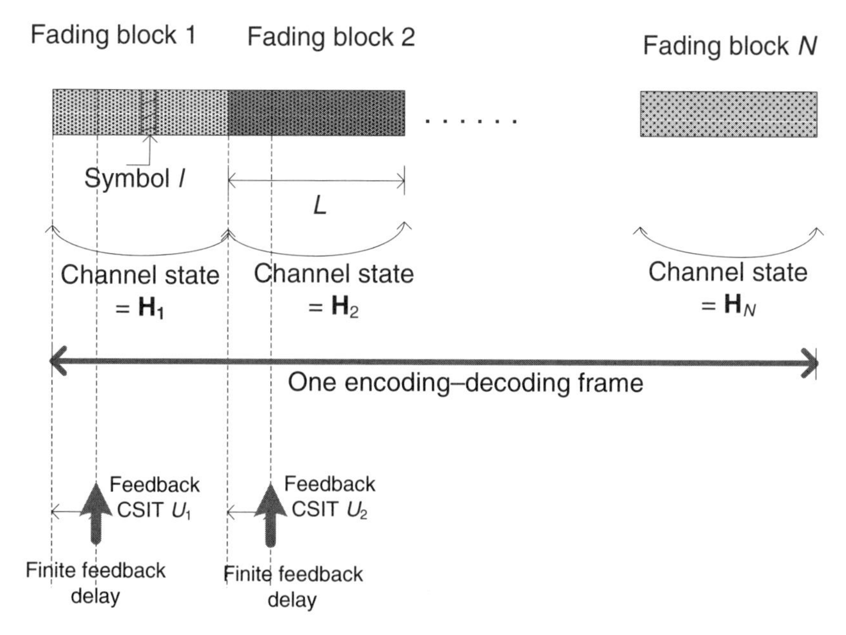

Figure 2.1. Block fading channel model.

quasistatic fading **h** over the entire coding frame, the quasistatic fading channel is not memoryless because the unconditional channel transition probability cannot be decomposed into the required product form. Moreover, within an encoding frame, there is no ergodic realization of the CSI and hence, the channel is a *nonergodic memory channel*.

Block Fading MIMO Channels. Another very important channel model is the *block fading channel*. This is a hybrid between fast fading and quasistatic fading. Specifically, an encoding frame spans over N fading blocks. Each fading block has L symbols. Channel fading remains quasistatic for symbols within a fading block and changes only between fading blocks; this is illustrated in Figure 2.1. This is quite an accurate channel model for frequency hopping with slow mobility. The channel fading remains quasistatic within a fading block and becomes i.i.d. between blocks when the frequency is hopped to another channel.

Let $\mathbf{X}_{1,1}^{L,N} = [\mathbf{X}_{1,1}, \mathbf{X}_{2,1}, \ldots, \mathbf{X}_{L,1}, \mathbf{X}_{1,2}, \ldots, \mathbf{X}_{l,n}, \ldots, \mathbf{X}_{L,N}]$ be the sequence of transmitted symbols over the encoding frame where $\mathbf{X}_{l,n} \in \mathcal{X}$ denotes the lth transmitted symbol of the nth fading block. Let $\mathbf{Y}_{1,1}^{L,N} = [\mathbf{Y}_{1,1}, \ldots, \mathbf{Y}_{l,n}, \ldots, \mathbf{Y}_{L,N}]$ be the sequence of received symbols over a decoding frame where $\mathbf{Y}_{l,n} \in \mathcal{Y}$. Similarly, $\mathbf{H}_1^N = [\mathbf{H}_1, \ldots, \mathbf{H}_N]$ denotes the sequence of channel state for the N fading blocks. The block fading channel is characterized by the conditional transition probability $p(\mathbf{y}_{1,1}^{L,N}|\mathbf{x}_{1,1}^{L,N}, \mathbf{h}_1^N)$ and the channel state sequence probability $p(\mathbf{h}_1^N)$.

The block fading channel is not memoryless, due to the fading memory over a block of L symbols in a fading slot. However, within an encoding frame, there are ergodic realizations of the CSI and therefore, the block fading channel is an ergodic memory channel. Moreover, if the fading between slots are i.i.d., the channel can be regarded as block-memoryless with respect to the *super-symbol* $\mathbf{X}_n = [\mathbf{X}_{1,n}, \ldots, \mathbf{X}_{L,n}]$.

2.2.2 General Transmission and CSI Feedback Model

Given a general probabilistic channel model with states $\{(p(\mathbf{y}_1^N|\mathbf{x}_1^N, \mathbf{h}_1^N), p(\mathbf{h}_1^N)) : \mathbf{h}_n \in \mathcal{H}\}$, Figure 2.2 illustrates a general framework for transmission and state feedback. A state sequence generator, $\mathbf{H}_1^N$, provides a state $\mathbf{H}_n \in \mathcal{H}$ to the channel once per symbol. At the same time, the receiver is fed with the corresponding CSIR sequence, $\mathbf{V}_n \in \mathcal{H}$, once per symbol n sequentially in time. The transmitter is fed with a CSIT sequence, $U_n \in \mathcal{U}$, once per symbol n sequentially in time. CSIT U_n, CSIR $\mathbf{V}_n$, and the channel state $\mathbf{H}_n$ are statistically related by $p(U_n|\mathbf{V}_n, \mathbf{V}_{n-1}, \ldots)$ and $p(\mathbf{V}_n|\mathbf{H}_n, \mathbf{H}_{n-1}, \ldots)$, respectively, meaning that U_n is derived *statistically* from $\{\mathbf{V}_n, \mathbf{V}_{n-1}, \ldots\}$ and $\mathbf{V}_n$ is derived *statistically* from $\{\mathbf{H}_n, \mathbf{H}_{n-1}, \ldots\}$. Note that if the channel state $\mathbf{H}_1^N$ is i.i.d., then $p(U_n|\mathbf{V}_n, \mathbf{V}_{n-1}, \ldots) = p(U_n|\mathbf{V}_n)$. In other words, feedback of past memory of CSIR does not provide any additional information [85] on the current channel state $\mathbf{H}_n$.

This model is general enough to cover any partial CSI feedback strategy and the corresponding adaptive transmission strategy. For example, in the first special case with no CSIT and no CSIR, we have $I(\mathbf{V}_n; \mathbf{H}_n, \mathbf{H}_{n-1}, \ldots) = 0$ and $I(U_n; \mathbf{V}_n, \mathbf{V}_{n-1}, \ldots) = 0$, meaning that the CSIT U_n is independent of the CSIR $\{\mathbf{V}_n, \mathbf{V}_{n-1}, \ldots\}$ and the CSIR $\mathbf{V}_n$ is independent of the actual channel state $\{\mathbf{H}_n, \mathbf{H}_{n-1}, \ldots\}$. In the second special case with no CSIT but perfect CSIR, we

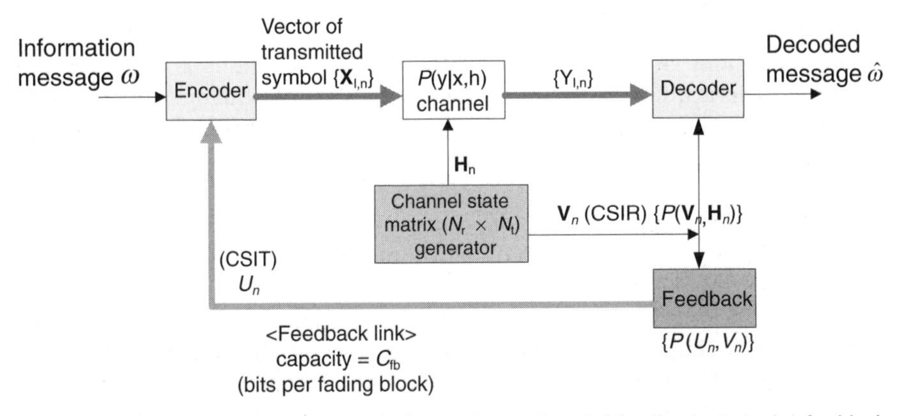

Figure 2.2. A general model of transmission and causal partial feedback strategies for block fading channels.

have $I(U_n; \mathbf{V}_n, \mathbf{V}_{n-1}, \ldots) = 0$ and $p(\mathbf{V}_n|\mathbf{H}_n = \mathbf{h}_n) = \delta(\mathbf{V}_n - \mathbf{h}_n)$, meaning that the CSIT U_n is independent of the channel states $\{\mathbf{V}_n, \mathbf{V}_{n-1}, \ldots\}$ while the CSIR $\mathbf{V}_n$ equals to the actual channel state $\mathbf{H}_n$. In the third special case with perfect CSIT and no CSIR, we have $p(U_n|\mathbf{H}_n = \mathbf{h}_n) = \delta(U_n - \mathbf{h}_n)$ and $I(\mathbf{V}_n; \mathbf{H}_n, \mathbf{H}_{n-1}, \ldots) = 0$, meaning that the CSIR is independent of the actual channel state $\mathbf{H}$ while the CSIT equals $\mathbf{H}$. In the fourth special case with perfect CSIR and perfect CSIT, we have $p(U_n|\mathbf{H}_n = \mathbf{h}_n) = \delta(U_n - \mathbf{h}_n)$ and $p(\mathbf{V}_n|\mathbf{H}_n = \mathbf{h}_n) = \delta(\mathbf{V}_n - \mathbf{h}_n)$, meaning that both CSIR and CSIT equal to $\mathbf{H}$. In the next chapter, we shall use this generic model to derive the optimal partial state feedback strategy and the optimal transmission strategy.

2.2.3 Adaptive Channel Encoding and Decoding

Figure 2.3 illustrates the general structure of adaptive channel encoder for probabilistic channels with state feedback. Information message $\omega \in \Omega$ (where Ω is the message set given by $\Omega = \{1, 2, \ldots, 2^{NR}\}$) is mapped into a frame of transmitted symbol $\mathbf{X}_1^N = [\mathbf{X}_1, \ldots, \mathbf{X}_N]$ using an *adaptive encoding function* $f_n : \Omega \times \mathcal{U}_1^n \rightarrow \mathcal{X}$. The adaptive encoding function, which is a function of the message index (ω) and causal CSIT sequence $\mathbf{U}_1^n$, is given by

$$\mathbf{X}_n = f_n(\omega, \mathbf{U}_1^n) \quad \forall n \in [1, N] \tag{2.3}$$

where $\mathbf{U}_1^n = \{U_1, \ldots, U_n\}$ is the causal CSIT sequence available to the transmitter at the nth symbol and $\mathbf{X}_n \in \mathcal{X}$ is the transmitted symbol over the complex field.

Figure 2.4 illustrates a general structure of the channel decoder with CSIR. The receiver decodes the message $\hat{\omega}$ based on the entire frame of received

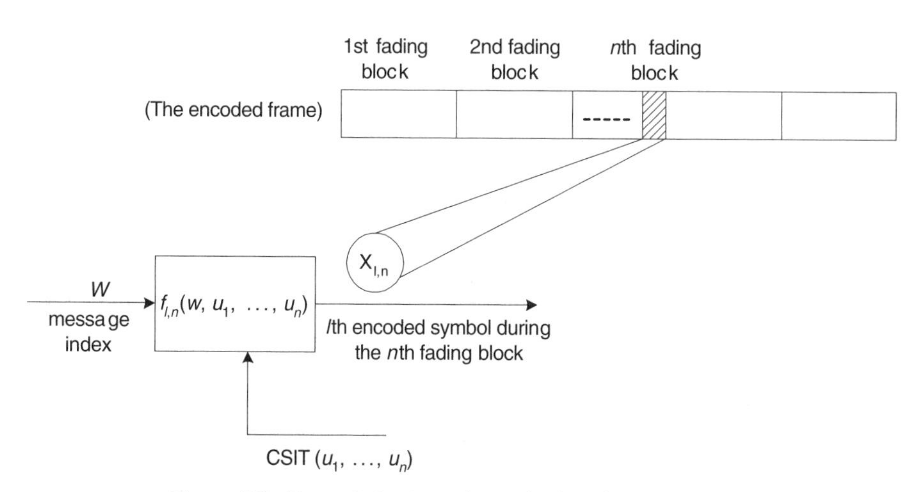

Figure 2.3. General structure of an adaptive channel encoder.

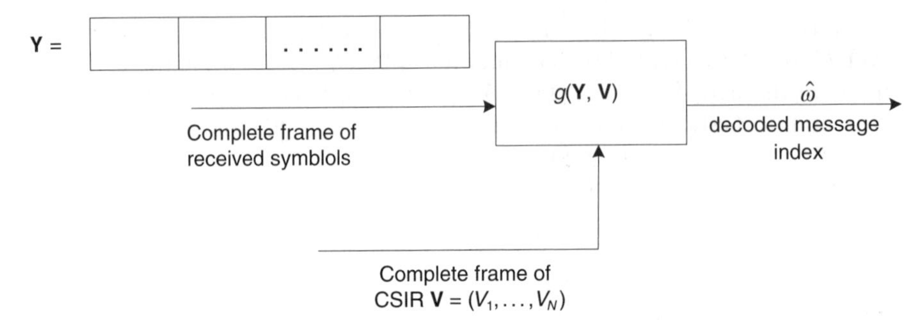

Figure 2.4. General structure of an adaptive channel decoder.

symbols, $\mathbf{Y}_1^N = [\mathbf{Y}_1, \ldots, \mathbf{Y}_N]$, and the entire sequence of CSIR, $\mathbf{V}_1^N = [\mathbf{V}_1, \ldots, \mathbf{V}_N]$. The channel decoder is given by the function $g : \mathbf{Y}^N \times \mathcal{H} \to \Omega$, given by

$$\hat{\omega} = g(\mathbf{Y}_1^N, \mathbf{V}_1^N) \tag{2.4}$$

Hence, an adaptive error correction code is characterized by (N, R), where N is the number of symbols in an encoding frame and R is the encoding rate in terms of bits per symbol. Decoding error occurs when $\hat{\omega} \neq \omega$. A code rate R is achievable if the error probability $P_e = \lim_{N\to\infty} \Pr[g(\mathbf{Y}_1^N, \mathbf{V}_1^N|\omega) \neq \omega] = 0$. The supremum of the achievable code rate is defined as the *channel capacity* [30].

2.2.4 Transmit Power Constraint

The instantaneous transmitted power at the nth transmitted symbol is given by

$$P_n = \mathrm{tr}\ \varepsilon[\mathbf{X}_n \mathbf{X}_n^*] \tag{2.5}$$

In general, for meaningful design, we must consider the capacity at a given transmitted power constraint because otherwise, we can achieve arbitrarily large capacity at the expense of arbitrarily large transmitted power. There are various possibilities of constraining the transmitted power. For example, we can have $P_n \leq P_0$ for all n where the scope of the expectation is limited to one symbol only. This refers to *peak power constraint*, which can be important in practical systems when there is nonlinearity in the power amplifier. Obviously, the peak power constraint is very strict because it applies to every transmit symbol.

On the other hand, there is a less strict constraint called the *short-term average power constraint*. The constraint is given by

$$\overline{P} = \mathrm{tr}\ \varepsilon[\mathbf{X}\mathbf{X}^*] \leq P_0 \tag{2.6}$$

where the expectation is taken for all transmit symbols within an encoding frame. This constraint is less strict because it allows the instantaneous transmit power of some symbols to exceed P_0 momentarily as long as the instantaneous transmit power of some other symbols is less than P_0 so that on average, the overall transmit power is within P_0.

A third and final possibility is called the *long-term average power constraint*, which is given by Equation (2.6), except that the expectation is taken across multiple encoding frames. This constraint is the least strict one because it allows the *short-term average transmit power* of one encoding frame to exceed P_0 as long as the average power of some other encoding frames is less than P_0 to make up the difference and at the end, the overall average power (over multiple frames) is less than P_0.

In general, the *peak power constraint* is a stronger constraint than the *short-term average power constraint*. Similarly, the *short-term average power constraint* is a stronger constraint than the *long-term average power constraint*. These three different constraints usually give different capacity results. For example, for *block fading channels with memoryless transition*, the transmitted symbols are i.i.d. and hence, the *short-term average power constraint* and the *long-term average power constraint* are equivalent while the *peak power constraint* will give different capacity result. On the other hand, for *quasistatic fading channels*,[4] the *peak power constraint* and the *short-term transmit power constraint* are equivalent while the *long-term average power constraint* will give different capacity results.

Unless otherwise specified, we shall adopt the *short-term average power constraints* for mathematical convenience.

2.2.5 Causal Feedback Constraint

As shown in Figure 2.2, the feedback channel carries the CSIT in a symbol-by-symbol manner. Hence, the CSIT has to satisfy the causality constraint. Specifically, the transmitter at the nth symbol has knowledge of the feedback CSIT $\{U_1, \ldots, U_n\}$ only. CSIT values $\{U_{n+1}, U_{n+2} \ldots\}$ correspond to future feedback values and are not yet available during the nth symbol. In other words, adaptation can be performed on the basis of current and past CSIT values only.

2.3 ERGODIC AND OUTAGE CHANNEL CAPACITY

Before we proceed to discuss the channel capacity in various CSI scenarios, we shall clarify two important concepts of Shannon's capacity for probabilistic channel with states. For ergodic channels, the interpretation of the channel

[4] *Quasistatic fading* refers to the slow fading situation where the channel fading realizations of all symbols within an encoding frame are identical.

capacity is pretty straightforward, referring to the capacity in Shannon's sense; specifically, for any rate $R < C$, there exists at least one encoder and one decoder that achieves arbitrarily small error probability. However, for non-ergodic channels, this is no longer the case because the channel capacity in Shannon's sense can be zero. In general, we may have two important notions of channel capacity, namely, the *ergodic capacity* and the *outage capacity*. These concepts are clarified below.

2.3.1 Ergodic Capacity

Ergodic capacity refers to the channel capacity in Shannon's sense; that is, for any rate $R < C$, there exists at least one encoder and one decoder that achieves arbitrarily small error probability. Conversely, if $R > C$, the error probability is always bounded away from zero with any encoder and decoder. For example, if the ergodic capacity of a fading channel is 10 bits per channel use, then we can anticipate zero error probability on the transmitted frames, provided capacity achieving codes are used. Note that not all probabilistic channels with states have finite ergodic capacity. As we shall illustrate, the ergodic capacity can be zero in some cases. That means for any finite transmission rate R, we cannot always guarantee error-free transmission no matter what encoder and decoder we use.

For example, consider the transmission of an encoding frame $\{\mathbf{X}_1, \ldots, \mathbf{X}_N\}$ with N symbols across a flat fading channel with state sequence $\{\mathbf{H}_1, \ldots, \mathbf{H}_N\}$. If the encoding frame is sufficiently long ($\gg$ coherence time of the fading channel) so that $\{\mathbf{H}_1, \ldots, \mathbf{H}_N\}$ spans across an *ergodic realization* of the underlying ergodic fading process $\mathbf{H}(t)$, the resulting ergodic capacity is nonzero and is given by

$$C = \lim_{N \to \infty} \frac{1}{N} I(\mathbf{X}_1^N ; \mathbf{Y}_1^N , \mathbf{H}_1^N) \tag{2.7}$$

In the special case of i.i.d. CSI sequence, the ergodic capacity reduces to the following well-known formula:

$$C = I(\mathbf{X}; \mathbf{Y}, \mathbf{H}) = I(\mathbf{X}; \mathbf{Y}|\mathbf{H}) \tag{2.8}$$

Hence, if the transmitted rate is lower than the ergodic capacity, the error probability is exponentially decaying with the frame length N for capacity achieving codes. More examples will be given in the next section.

2.3.2 Outage Capacity

The ergodic assumption described above is not necessarily satisfied in practical communication systems operating on fading channels. In fact, if there is a stringent delay requirement as in the case of speech transmission over wireless fading channels, the ergodicity requirement $NT_s \gg T_c$, where T_s is the modulation symbol duration and T_c is the channel coherence time, cannot be satisfied. In this case, there is no significant channel variation across the encoding frame and there may be no classical Shannon meaning attached to capacity in this typical situation. In the extreme case, when the frame is very short or the coherence time is very long, the entire encoding frame may share a single fading state realization; specifically, $h_1 = h_2 = \cdots = h_N = h$. In this case, the channel belongs to the type of *nonergodic memory channel* and the mutual information becomes *information unstable* [137]. The channel capacity in Shannon's sense is given by

$$C = \lim \inf i(\mathbf{x}_1^N; \mathbf{y}_1^N, \mathbf{h}) \tag{2.9}$$

where lim inf is in the probability sense and $i(\mathbf{x}_1^N; \mathbf{y}_1^N, \mathbf{h}) = \frac{1}{N} I(\mathbf{x}_1^N; \mathbf{y}_1^N, \mathbf{h})$ is the normalized sequence of mutual information. In fact, this mutual information can be considered as a function of the random realization $\mathbf{h}$ and therefore is a random variable itself. Hence, there may be a nonnegligible probability that the value of the transmission rate, no matter how small, exceeds the instantaneous mutual information. This situation gives rise to the error probabilities that do not decay with the increase of block length N. In these circumstances, the channel capacity is viewed as a random variable as it depends on the instantaneous channel state realization. Hence, the ergodic capacity of these channels is zero, meaning that no matter how small the transmission rate is, there is no guarantee that the transmitted frame will be error-free.

On the other hand, instead of looking at the *ergodic capacity* in Shannon's sense, it is also possible to look at the capacity from an *outage perspective*. The outage capacity C_{out} at a given *outage probability* P_{out} is defined as the maximum data rate R such that

$$\Pr\{R < \text{instantaneous mutual information}\} \leq P_{\text{out}} \tag{2.10}$$

In other words, the outage probability P_{out} is the cumulative distribution function (cdf) of the instantaneous mutual information (which is a random variable).

2.4 CHANNEL CAPACITY WITH NO CSIT AND NO CSIR

We shall first consider a special case of the model in Figure 2.2, where there is no CSIT and no CSIR. This corresponds to the case when $I(V_n; \mathbf{H}_n) = 0$ and $I(U_n; \mathbf{H}_n) = 0$. In this case, no adaptation at the encoding and decoding is possible and therefore, the channel encoding function, $f_n(\omega)$, is a function of the message index only. The channel decoding function, $g(\mathbf{Y}_1^N)$, is a function of the received symbols only.

Let $p(\mathbf{y}_1^N | \mathbf{x}_1^N, \mathbf{h}_1^N)$ be the transition probability of the probabilistic channel with states. Together with the joint distribution of the channel state sequence

$p(\mathbf{h}_1^N)$, the channel transition probability of the equivalent channel (without states) with input $\mathbf{x}_1^N$ and output $\mathbf{y}_1^N$ is given by

$$p(\mathbf{y}_1^N|\mathbf{x}_1^N) = \sum_{\mathbf{h}_1^N} p(\mathbf{y}_1^N|\mathbf{x}_1^N, \mathbf{h}_1^N)p(\mathbf{h}_1^N) \tag{2.11}$$

Treating $\mathbf{X}_1^N$ and $\mathbf{Y}_1^N$ as supersymbols and assuming *ergodic CSI sequence*, the channel capacity (bits per symbol) of the equivalent channel with transition probability in Equation (2.11) is given by

$$C_{\text{no csit, no csir}} = \lim_{N\to\infty} \max_{p(\mathbf{X}_1^N)} \frac{1}{N} I(\mathbf{X}_1^N; \mathbf{Y}_1^N) \tag{2.12}$$

We shall consider a few examples in the following subsections.

2.4.1 Fast Flat Fading MIMO Channels

We consider a fast flat fading example where the channel output $\mathbf{Y}_n$ is given by

$$\mathbf{Y}_n = \mathbf{H}_n\mathbf{X}_n + \mathbf{Z}_n$$

where $\mathbf{H}_n(i, j)$ is the i.i.d. complex Gaussian fading with unit variance and $\mathbf{Z}_n$ is the i.i.d. complex Gaussian noise with covariance $\sigma_z^2\mathbf{I}_{n_R}$. Since both the state sequence and the noise sequence are i.i.d., the channel transition probability is memoryless. The channel capacity is simplified as

$$C_{\text{no csit,no csir}} = \max_{p(\mathbf{X})} I(\mathbf{X}; \mathbf{Y}) \tag{2.13}$$

Note that although the channel noise is Gaussian, the capacity achieving distribution $p(\mathbf{X})$ is no longer complex Gaussian. As shown in reference 5, the capacity achieving distribution for a scalar channel with no CSIT and no CSIR at low SNR is binary and is given by

$$x = \begin{Bmatrix} 0 & \text{with probability } 1 - p_\alpha \\ \sqrt{\alpha} & \text{with probability } p_\alpha \end{Bmatrix} \tag{2.14}$$

such that $\alpha p_\alpha = \varepsilon[|X|^2]$. This means that the capacity achieving modulator design at low SNR is a binary constellation. In general, at higher SNR, the capacity achieving distribution is *peaky*, implying discrete constellations.

2.4.2 Block Fading Channels

Consider a MIMO block fading channel with conditional transition probability $p(\mathbf{y}_1^N|\mathbf{x}_1^N, \mathbf{h}_1^N)\Pi_n\Pi_l p(y_{l,n}|x_{l,n}, h_n)$ and channel state sequence probability $p(\mathbf{h}_1^N) = \Pi_n p(h_n)$, the channel capacity (bits per symbol) is given by

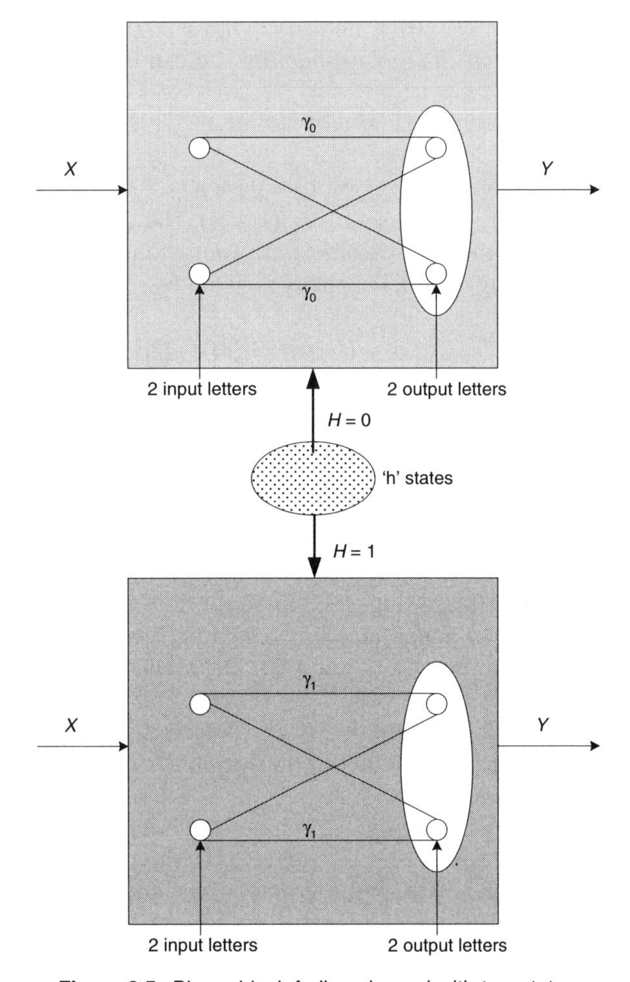

Figure 2.5. Binary block fading channel with two states.

$$C_{\text{no csir, no csit}}(L) = \frac{1}{L} \max_{p(\mathbf{X}_1^L)} I(\mathbf{X}_1^L; \mathbf{Y}_1^L) \tag{2.15}$$

where $\mathbf{X}_1^L$ is the *L-tuple transmitted supe-symbol*, $\mathbf{Y}_1^L$ is the *L-tuple received supersymbol*, $p(_1^L)$ denotes the *input distribution* of the transmit supersymbol $\mathbf{X}_1^L$, and L represents *channel memory* within a fading block.

For example, consider a discrete block fading channel where the channel output $Y \in \{0, 1\}$ and the channel input $X \in \{0, 1\}$ are related by two binary symmetric channels depending on the channel state $H \in \{0, 1\}$ as illustrated in Figure 2.5.

Assume that the state sequence is memoryless with $p(H = 0) = p_0$ and $p(H = 1) = 1 - p_0$. The channel transition probabilities are given by $p(y = 0|x = 0,$

$H = 0) = p(y = 1|x = 1, H = 0)$ γ_0 and $p(y = 0|x = 0, H = 1) = p(y = 1|x = 1, H = 1) = \gamma_1$. The *marginal channel probability* is given by

$$p(y|x) = p_0 p(y|x, H = 0) + (1 - p_0) p(y|x, H = 1) \tag{2.16}$$

Hence, we have $p(y = 0|x = 0) = p(y = 1|x = 1) = p_0\gamma_0 + (1 - p_0)\gamma_1$ and $p(y = 0|x = 1) = p(y = 1|x = 0) = p_0(1 - \gamma_0) + (1 - p_0)(1 - \gamma_1)$. The corresponding channel is therefore a binary symmetric channel with transition probability $p_0\gamma_0 + (1 - p_0)\gamma_1$. The corresponding channel capacity is given by

$$C_{\text{no csit, no csir}} = H_2(p_0\gamma_0 + (1 - p_0)\gamma_1) \tag{2.17}$$

where $H_2(p) = -p\log_2(p) - (1 - p)\log_2(1 - p)$.

2.5 CHANNEL CAPACITY WITH PERFECT CSIR

We consider the second case with perfect CSI knowledge at the receiver: $\mathbf{V}_n = \mathbf{H}_n$ and $I(U_n; \mathbf{V}_n) = 0$. In this case, no adaptation at the encoding side is possible; therefore, the channel encoding function, $f_n(\omega)$, is a function of the message index only. On the other hand,with knowledge of perfect CSI at the receiver, channel-matched decoding can be done and the channel decoding function, $g(\mathbf{Y}_1^N, \mathbf{V}_1^N)$, will be a function of the received symbols and the CSIR sequence. Hence, we anticipate better performance compared to the case without CSI information at all.

Figure 2.6 illustrates the equivalent channel model with CSIR only. The original channel model with CSIR is equivalent to the effective channel (with no CSIR and no CSIR) with input symbol ($\mathbf{X}_n$) and output *supersymbol* ($\mathbf{Y}_n$, $\mathbf{H}_n$). The equivalent channel has the transition probability given by

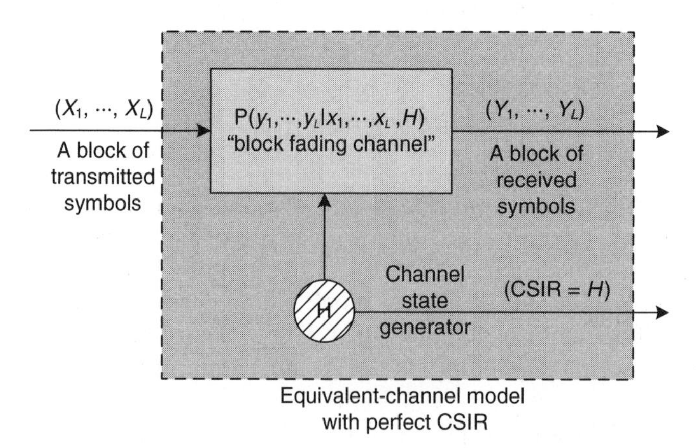

Figure 2.6. Block diagram of equivalent-channel model with CSIR only.

$$p(\mathbf{y}_1^N, \mathbf{h}_1^N | \mathbf{x}_1^N) = p(\mathbf{y}_1^N | \mathbf{x}_1^N, \mathbf{h}_1^N) p(\mathbf{h}_1^N) \tag{2.18}$$

For ergodic CSI sequence, the channel capacity (bits per channel use) is given by

$$\begin{aligned} C_{\text{csir}} &= \lim_{N\to\infty} \frac{1}{N} \max_{p(\mathbf{x}_1^N)} I(\mathbf{X}_1^N; \mathbf{Y}_1^N, \mathbf{H}_1^N) \\ &= \lim_{N\to\infty} \frac{1}{N} \max_{p(\mathbf{x}_1^N)} [I(\mathbf{X}_1^N; \mathbf{Y}_1^N | \mathbf{H}_1^N) + I(\mathbf{X}_1^N; \mathbf{H}_1^N)] \\ &= \lim_{N\to\infty} \frac{1}{N} \max_{p(\mathbf{x}_1^N)} I(\mathbf{X}_1^N; \mathbf{Y}_1^N | \mathbf{H}_1^N) \end{aligned} \tag{2.19}$$

where $I(\mathbf{X}_1^N; \mathbf{H}_1^N) = 0$ because the transmitted symbol is independent of the channel state.

Furthermore, if the conditional channel transition probability can be decomposed into product form and the channel state sequence $\mathbf{h}_1^N$ is i.i.d., the equivalent channel with input $\mathbf{X}_n$ and output $(\mathbf{Y}_n, \mathbf{H}_n)$ is memoryless. Hence, the channel capacity (*ergodic capacity*) in Equation (2.19) becomes

$$C_{\text{csir}} = \max_{p(\mathbf{X})} I(\mathbf{X}; \mathbf{Y} | \mathbf{H}) \tag{2.20}$$

We shall evaluate the channel capacity (ergodic and outage capacities) using a few important examples in the following subsections.

2.5.1 Block Fading Channels

Consider a single antenna link with block fading channels. Let $\mathbf{X}_1^N = \{X_{1,1}, \ldots, X_{L,1}, \ldots, X_{l,n}, \ldots, X_{L,N}\}$ be the channel inputs and $\mathbf{Y}_1^N = \{Y_{1,1}, \ldots, Y_{L,1}, \ldots, Y_{l,n}, \ldots, Y_{L,N}\}$ be the channel outputs, where $X_{l,n}$ denotes the lth transmit symbol of the nth fading block. Assume that the channel state remains quasi-static within a fading block but becomes i.i.d. between fading blocks. Hence, the channel states sequence is denoted by $\mathbf{H}_1^N = \{H_1, \ldots, H_N\}$ and has the following distribution:

$$p(\mathbf{h}_1^N) = \prod_{n=1}^N p(h_n)$$

The channel transition probability of the original block fading channel is given by

$$p(\mathbf{y}_1^N | \mathbf{x}_1^N, \mathbf{h}_1^N) = \prod_{n=1}^N \prod_{l=1}^L p(y_{l,n} | x_{l,n}, h_n)$$

Treating $(\mathbf{Y}_n, H_n) = (Y_{1,n}, \ldots, Y_{L,n}, H_n)$ as an output *supersymbol* and $\mathbf{X}_n = (X_{1,n}, \ldots, X_{L,n})$ as an input *supersymbol*, the original block fading channel with

CSIR is equivalent to a block memoryless channel (without states). The channel transition probability of the equiv-alent channel is given by

$$p(\mathbf{y}_1,\ldots,\mathbf{y}_N,\mathbf{h}^N|\mathbf{x}_1\ldots,\mathbf{x}_N)=p(\mathbf{y}_1,\ldots,\mathbf{y}_N|\mathbf{h}^N,\mathbf{x}_1,\ldots,\mathbf{x}_N)p(\mathbf{h}_1^N)$$
$$=\prod_{n=1}^{N}p(\mathbf{y}_n,h_n|\mathbf{x}_n) \quad (2.21)$$

As a result, the channel capacity (bits per symbol) of the equivalent channel is thus given by

$$C_{\text{csir}}(L)=\frac{1}{L}\max_{p(\mathbf{X})}I(\mathbf{X};\mathbf{Y},H)=\frac{1}{L}\max_{p(\mathbf{X})}I(\mathbf{X};\mathbf{Y}|H) \quad (2.22)$$

where the second equality is due to the fact that $I(\mathbf{X};\mathbf{Y},H)=I(\mathbf{X};H)+I(\mathbf{X};\mathbf{Y}|H)$ and $I(\mathbf{X};H)=0$ (with no CSIT).

Lemma 2.1 (Capacity of Block Fading Channels with CSIR) For "block memoryless" block fading channels, $C_{\text{csir}}(L)$ is independent of L and is given by

$$C_{\text{csir}}(L)=\overline{C}_{\text{csir}}=\max_{p(X)}I(X;Y|H)=\max_{p(X)}\varepsilon[I(X;Y|H=h)] \quad (2.23)$$

Proof We have $I(\mathbf{X};\mathbf{Y}|H=h)\leq\Sigma_{l=1}^{L}I(X_l;Y_l|H=h)$ with equality holds if $\mathbf{X}$ is i.i.d. Hence, the channel capacity $C_{\text{csir}}(L)$ is given by

$$C_{\text{csir}}(L)=\frac{1}{L}\max_{p(\mathbf{X})}\varepsilon[I(\mathbf{X};\mathbf{Y}|H=h)]=\max_{p(X)}\varepsilon I(X;Y|H=h)=\overline{C}_{\text{csir}}$$

We shall illustrate the preceding capacity formula with the following examples.

Rayleigh Block Fading Channels. For example, consider a Rayleigh block fading channel with full CSIR where the channel output $Y_{l,n}$ is related to the channel input $X_{l,n}$ by $Y_{l,n}=H_nX_{l,n}+Z_{l,n}$, where H_n is a complex Gaussian channel fading with unit variance during the nth fading block and $Z_{l,n}$ is i.i.d. Gaussian channel noise with variance σ_z^2. This corresponds to a *block fading channel with memoryless state and memoryless transitions*. The channel capacity is given by

$$C_{\text{csir}}=\int_0^{\infty}2|h|\exp\left(-|h|^2\right)\log_2\left(1+\frac{\sigma_x^2}{\sigma_z^2}|h|^2\right)d|h| \quad (2.24)$$

where the input capacity achieving distribution is complex Gaussian with variance σ_x^2.

Discrete Block Fading Channel. Consider a discrete block fading channelwhere the channel output $Y\in\{0,1\}$ and the channel input $X\in\{0,1\}$ are related by two binary symmetric channels depending on the channel state $H \in \{0, 1\}$ as illustrated in Figure 2.5. Assume that the state sequence is memoryless with $p(H = 0) = p_0$. The channel transition probabilities are given by $p(y = 0|x = 0, H = 0) = p(y = 1|x = 1, H = 0) = \gamma_0$ and $p(y = 0|x = 0, H = 1) = p(y = 1|x = 1, H = 1) = \gamma_1$. Assume that the receiver has perfect CSI knowledge H. The channel capacity with perfect CSIR is given by

$$C_{\text{csir}} = p_0 H_2(\gamma_0) + (1 - p_0) H_2(\gamma_1) \tag{2.25}$$

where $H_2(p) = -p\log_2 p - (1 - p)\log_2(1 - p)$. Obviously, since $H_2(p)$ is a convex function in p, the channel capacity with CSIR is always larger than that without CSIR in Equation (2.17)

Comparing the results with no CSIT and no CSIR, we have $\overline{C}_{\text{csir}} \geq C_{\text{no csir, no csit}}$ because no CSIR is a special case of perfect CSIR. Yet, we have an interesting observation given by the lemma below.

Lemma 2.2 (Asymptotic Block Fading Channel Capacity) In the limit of large block size, $L \to \infty$, the channel capacity of general block fading channel with memory transition $p(y_1, \ldots y_L|x_1, \ldots, x_L, h)$, no CSIT and no CSIR approach the channel capacity of the same channel with full CSIR:

$$\lim_{L\to\infty} C_{\text{no csir, no csit}}(L) = \lim_{L\to\infty} C_{\text{csir}}(L) \tag{2.26}$$

Proof Observe that

$$I(\mathbf{X}; \mathbf{Y}) \leq I(\mathbf{X}; \mathbf{Y}, H)$$

with equality holds if $\mathbf{Y}$ and H are independent. Also

$$\begin{aligned} I(\mathbf{X}; \mathbf{Y}, H) &= I(\mathbf{X}; \mathbf{Y}) + I(\mathbf{X}; H|\mathbf{Y}) \\ &= I(\mathbf{X}; \mathbf{Y}) + I(\mathbf{X}, \mathbf{Y}; H) - I(\mathbf{Y}; H) \\ &\leq I(\mathbf{X}; \mathbf{Y}) + I(\mathbf{X}, \mathbf{Y}; H) \end{aligned}$$

where equality holds if and only if $\mathbf{Y}$ and H are independent. Therefore

$$\frac{1}{L} I(\mathbf{X}; \mathbf{Y}) \leq \frac{1}{L} I(\mathbf{X}; \mathbf{Y}, H) \tag{2.27}$$

$$\leq \frac{1}{L}[I(\mathbf{X}; \mathbf{Y}) + I(\mathbf{X}, \mathbf{Y}; H)] \tag{2.28}$$

Since $\lim_{L\to\infty} \frac{1}{L} I(\mathbf{X}, \mathbf{Y}; H) = 0$, from the sandwich theorem, we have

$$\lim_{L\to\infty} \frac{1}{L} I(\mathbf{X}; \mathbf{Y}, H) = \lim_{L\to\infty} \frac{1}{L} I(\mathbf{X}; \mathbf{Y})$$

Hence, this lemma reveals that the penalty of lacking CSIR is small if the block fading memory length L is sufficiently long. Intuitively, we may probe the channel state at the beginning of every fading block with known test symbols prior to payload transmission so that CSIR could be estimated at the receiver. This consumes channel overhead but as $L \to \infty$, the overhead is asymptotically zero.

2.5.2 Fast Flat Fading MIMO Channels

For example, consider fast flat fading MIMO channels with n_T transmit antennas and n_R receive antennas. The $n_R \times 1$ channel output $\mathbf{Y}_n$ is given by

$$\mathbf{y}_n = \mathbf{h}_n \mathbf{x}_n + \mathbf{z}_n.$$

To derive the channel capacity, we consider the following background.

Definition 2.2 (Gaussian Random Vector) A complex $n \times 1$ random vector $\mathbf{x}$ is said to be Gaussian if

$$\hat{\mathbf{x}} = \begin{bmatrix} \mathrm{Re}(\mathbf{x}) \\ \mathrm{Im}(\mathbf{x}) \end{bmatrix}$$

is jointly Gaussian.

Definition 2.3 (Circularly Symmetric Gaussian Vector) A complex Gaussian $n \times 1$ random vector $\mathbf{x}$ is said to be *circularly symmetric* if

$$\varepsilon[(\hat{\mathbf{x}} - \varepsilon[\hat{\mathbf{x}}])(\hat{x} - \varepsilon[\hat{\mathbf{x}}])^*] = \frac{1}{2} \begin{bmatrix} \mathrm{Re}(\mathbf{Q}) & -\mathrm{Im}(\mathbf{Q}) \\ \mathrm{Im}(\mathbf{Q}) & \mathrm{Re}(\mathbf{Q}) \end{bmatrix}$$

where

$$\hat{\mathbf{x}} = \begin{bmatrix} \mathrm{Re}(\mathbf{x}) \\ \mathrm{Im}(\mathbf{x}) \end{bmatrix}$$

$\mathbf{Q} = \varepsilon[(\mathbf{x} - \mu)(\mathbf{x} - \mu)^*]$, and $\mu = \varepsilon[\mathbf{x}]$

For example, consider a complex Gaussian random variable $X = X_c + jX_s$ with zero mean. If $\varepsilon[X_c^2] = \varepsilon[X_s^2] = \sigma^2$ and $\varepsilon[X_c X_s] = 0$, then X is a circular symmetric complex Gaussian random variable. On the other hand, if $\varepsilon[X_c X_s] \neq 0$ or $\varepsilon[X_s^2] \neq \varepsilon[X_c^2]$, then X is not circular symmetric complex Gaussian.

Note that the probability density of a circular symmetric complex Gaussian random vector **x** can be written directly in terms of the complex vector **x** as follows:

$$f(\mathbf{x}) = \frac{1}{|\pi \mathbf{Q}|} \exp(-(\mathbf{x}-\mu)^* \mathbf{Q}^{-1}(\mathbf{x}-\mu))$$

From the definitions above, it is easy to see that if **x** and **y** are two independent circular symmetric complex Gaussian random vectors, then so are $\mathbf{w} = \mathbf{A}\mathbf{x}$ and $\mathbf{z} = \mathbf{x} + \mathbf{y}$, where **A** is a $m \times n$ constant matrix. Similar to the upper bound of entropy for scalar random variable described in Chapter 1, we have a similar upper bound on the entropy of a vector of random variables. This is summarized in the following important lemma.

Lemma 2.3 (Entropy of Random Vector) Suppose that the complex random vector **x** is zero mean and with covariance matrix $\mathbf{Q} = \varepsilon[\mathbf{x}\mathbf{x}^*]$. The entropy is upper bounded by

$$H(\mathbf{x}) \le \log_2(|\pi e \mathbf{Q}|)$$

with equality if and only if **x** is circular symmetric complex Gaussian.

The mutual information is given by

$$I(\mathbf{X}; \mathbf{Y}|\mathbf{H}) = H(\mathbf{Y}|\mathbf{H}) - H(\mathbf{Y}|X, \mathbf{H}) = H(\mathbf{Y}|\mathbf{H}) - H(\mathbf{Z})$$

From Lemma 2.3, the first entropy is upper-bounded by $\log_2(|\pi e \mathbf{Q}_y|)$, where $\mathbf{Q}_y = \varepsilon[\mathbf{Y}\mathbf{Y}^*|\mathbf{h}] = \eta_0 \mathbf{I}_{n_R} + \mathbf{h}\mathbf{Q}_x\mathbf{h}^*$ and $\mathbf{Q}_x = \varepsilon[\mathbf{X}\mathbf{X}^*]$. The second entropy is given by $\log_2(|\pi e \eta_0 \mathbf{I}_{n_R}|)$. Furthermore, we introduce a constraint on the total transmit power from the n_T antennas:

$$\mathrm{tr}[\mathbf{Q}_x] \le P_0 \tag{2.29}$$

Hence, the capacity of the fast fading MIMO channels with perfect CSIR in Equation (2.20) can be written as

$$C_{\mathrm{csir}} = \max_{\mathbf{Q}_x : \mathrm{tr}(\mathbf{Q}_x) \le P_0} \varepsilon\left[\log_2\left|\mathbf{I}_{n_R} + \frac{\mathbf{h}\mathbf{Q}_x\mathbf{h}^*}{\eta_0}\right|\right] \tag{2.30}$$

where the optimization is over all *input covariance matrix* $\mathbf{Q}_x$ satisfying the transmit power constraint in Equation (2.29). The solution of the optimization is summarized by the following theorem.

Theorem 2.1 (MIMO Capacity with CSIR) The capacity achieving input covariance matrix is given by

$$\mathbf{Q}_x = \frac{P_0}{n_T}\mathbf{I}_{n_T} \tag{2.31}$$

and the corresponding channel capacity is given by

$$C_{\text{csir}} = \varepsilon\left[\log_2\left|\mathbf{I}_{nR} + \frac{P_0\mathbf{hh}^*}{n_T\eta_0}\right|\right] \tag{2.32}$$

Proof Since $\mathbf{Q}_x$ is nonnegative definite, we can express $\mathbf{Q}_x = \mathbf{U}\tilde{\mathbf{Q}}_x\mathbf{U}^*$, where $\mathbf{U}$ is a unitary matrix and $\tilde{\mathbf{Q}}_x$ is a diagonal matrix. Hence, the capacity equation becomes

$$\begin{aligned} C_{\text{csir}}(\mathbf{Q}_x) &= \varepsilon\left[\log_2\left|\mathbf{I}_{nR} + P_0\frac{\mathbf{hU}\tilde{\mathbf{Q}}_x\mathbf{U}^*\mathbf{h}^*}{\eta_0}\right|\right] \\ &= \varepsilon\left[\log_2\left|\mathbf{I}_{nR} + P_0\frac{\tilde{\mathbf{h}}\tilde{\mathbf{Q}}_x\tilde{\mathbf{h}}^*}{\eta_0}\right|\right] \end{aligned}$$

where the second equality is due to the fact that the distribution of $\mathbf{h}$ is the same as the distribution of $\tilde{\mathbf{h}} = \mathbf{hU}$ (because $\mathbf{U}$ is unitary). Hence, without loss of generality, we can assume that the capacity achieving input covariance $\mathbf{Q}_x$ is diagonal.

Let $\Psi(\mathbf{Q}_x) = \varepsilon\left[\log_2\left|\mathbf{I}_{nR} + P_0\frac{\mathbf{hQ}_x\mathbf{h}^*}{\eta_0}\right|\right]$. For any nonnegative diagonal covariance matrix $\mathbf{Q}_x$, define $\mathbf{Q}_x^{\Pi} = \mathbf{\Pi}\mathbf{Q}_x\mathbf{\Pi}^*$, where $\mathbf{\Pi}$ is any permutation matrix. For example, if $\mathbf{Q}_x = \text{diag}[q_1, q_2, q_3]$ and

$$\mathbf{\Pi} = \begin{pmatrix} 0 & 1 & 0 \\ 1 & 0 & 0 \\ 0 & 0 & 1 \end{pmatrix}$$

then $\mathbf{Q}_x^{\Pi} = \text{diag}[q_2, q_1, q_3]$ represents a permutation between the first and the second diagonal elements. Observe that $\Psi(\mathbf{Q}_x) = \Psi(\mathbf{Q}_x^{\Pi})$ because $\mathbf{h\Pi}$ has the same distribution as $\mathbf{h}$. On the other hand, define

$$\tilde{\mathbf{Q}}_x = \frac{1}{n_T!}\sum_{\Pi}\mathbf{Q}_x^{\Pi}$$

Observe that $\Psi(\mathbf{Q}_x)$ is a concave function of $\mathbf{Q}_x$. Hence, given any diagonal $\mathbf{Q}_x$ satisfying $\text{tr}(\mathbf{Q}_x) \le P_0$, we have

$$\Psi(\tilde{\mathbf{Q}}_x) \ge \frac{1}{n_T!}\sum_{\Pi}\Psi(\mathbf{Q}_x^{\Pi}) = \Psi(\mathbf{Q}_x) \tag{2.33}$$

Equality is achieved if $\mathbf{Q}_x = \tilde{\mathbf{Q}}_x$. Therefore, the capacity achieving input covariance is given by $\tilde{\mathbf{Q}}_x$.

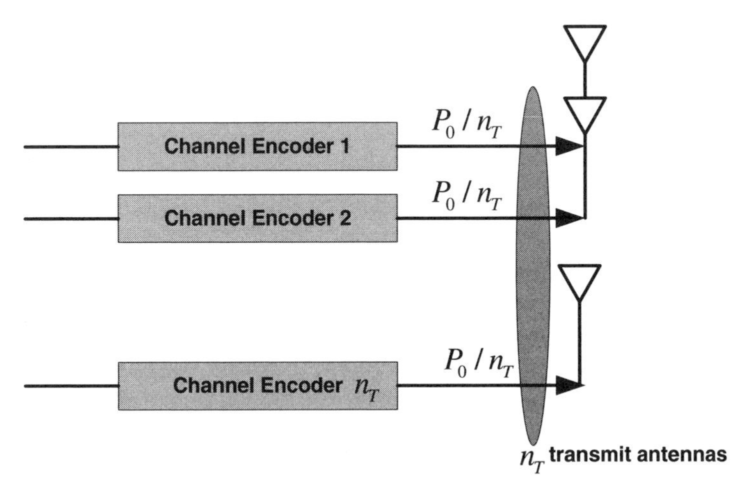

Figure 2.7. Optimal transmitter structure of fast fading MIMO channels with perfect CSIR.

On the other hand, observe that $\tilde{\mathbf{Q}}_x = \alpha \mathbf{I}_{n_T}$ because every diagonal element is an average of all possible permutations of the diagonal elements. Choosing α to satisfy the transmit power constraint, we have the capacity achieving input covariance matrix given by $\frac{P_0}{n_T}\mathbf{I}_{n_T}$.

Note that the capacity achieving the input covariance matrix implies uniform power allocation across the n_T transmit antennas. Furthermore, since $\mathbf{Q}_x$ is diagonal, one possible capacity achieving transmitter architecture[5] consists of n_T parallel and isolated channel encoders as illustrated in Figure 2.7.

Let $\mathbf{S} = (S_1 \ldots S_{n_T})$ be the output symbol vector of the n_T-independent channel encoders (with random codebooks). We have $\varepsilon[\mathbf{SS}^*] = \mathbf{I}_{n_T}$. The transmit vector symbol is given by $\mathbf{X} = \sqrt{\frac{P_0}{n_T}}\mathbf{S}$. Hence, the covariance matrix of such transmitter in Figure 2.7 is given by $\varepsilon[\mathbf{XX}^*] = \frac{P_0}{n_T}\mathbf{I}_{n_T}$. In other words, no joint encoding is needed at the transmitter to realize the MIMO channel capacity if a maximum-likelihood (ML) receiver is used.

To evaluate the MIMO capacity, we can write the capacity as

$$C_{\text{csir}} = \varepsilon\left[\log_2\left(\mathbf{I}_{n_R} + \frac{P_0}{n_T \eta_0}\mathbf{HH}^*\right)\right] = \varepsilon\left[\log_2\left(\mathbf{I}_{n_T} + \frac{P_0}{n_T \eta_0}\mathbf{H}^*\mathbf{H}\right)\right] \tag{2.34}$$

[5] Note that although one possible transmitter architecture is isolated encoding, this does not imply that isolated encoding is the only possible capacity achieving distribution. In fact, a MIMO transmitter with joint encoding can also give a diagonal input covariance matrix as well as a larger error exponent. This will be elaborated in Chapter 4.

Define

$$\mathbf{W} = \begin{cases} \mathbf{HH}^* & n_R < n_T \\ \mathbf{H}^*\mathbf{H} & n_T < n_R \end{cases}$$

This capacity can be expressed in terms of eigenvalues $\lambda_1, \ldots, \lambda_{m^*}$ of $\mathbf{W}$, where $m^* = \min[n_T, n_R]$:

$$C_{\text{csir}} = \varepsilon\left[\sum_{i=1}^{m^*} \log_2\left(1 + \frac{P_0}{n_T \eta_0}\lambda_i\right)\right] \tag{2.35}$$

where the expectation is over the distribution of the eigenvalues.

We consider several important asymptotic cases concerning the MIMO channel capacity in fast fading channels.

Transmit diversity $n_R = 1$: When $n_R = 1$, the MIMO capacity becomes

$$C_{\text{csir}}(n_R = 1) = \varepsilon\left[\log_2\left(1 + \frac{P_0\|\mathbf{h}\|^2}{n_T \eta_0}\right)\right] = \varepsilon[\log_2(1+\gamma)]$$

where $\gamma = \dfrac{P_0\|\mathbf{h}\|^2}{n_T \eta_0}$ is the effective SNR. Note that the average SNR $\varepsilon[\gamma] = \dfrac{P_0}{\eta_0}$ is independent of n_T but the diversity order of γ is n_T. Hence, we anticipate capacity increases as n_T increases. Figure 2.8a illustrates the ergodic capacity versus number of antennas.

Receive diversity $n_T = 1$: When $n_T = 1$, the channel capacity becomes

$$C_{\text{csir}}(n_T = 1) = \varepsilon\left[\log_2\left|\mathbf{I}_{nR} + \frac{P_0\mathbf{hh}^*}{\eta_0}\right|\right] = \varepsilon[\log_2(1+\gamma)]$$

where $\gamma = \dfrac{P_0\|\mathbf{h}\|^2}{\eta_0}$ is the effective SNR. In this case, γ enjoys similar n_R-order diversity gain in the statistic. In addition, the mean SNR $\varepsilon[\gamma] = \dfrac{n_R P_0}{\eta_0}$ increases as n_R increases. Hence, we anticipate the ergodic capacity to increase more effectively as n_R increases (compare with the previous case of transmit diversity). This is illustrated in Figure 2.8b. Observe that $C_{\text{csir}}(n_T = 1)$ is always $10\log_{10}(n_R)$ dB better than $C_{\text{csir}}(n_R = 1)$.

Large number of antennas $m^* = n_T = n_R \to \infty$: Since the elements of the channel matrix $\mathbf{h}$ is i.i.d., we have $\dfrac{1}{n_T}\mathbf{hh}^* \to \mathbf{I}_{nR}$, due to the law of large numbers. Hence, the MIMO ergodic capacity approaches

$$C_{\text{csir}}(m^* \to \infty) = m^* \log_2\left(1 + \frac{P_0}{\eta_0}\right)$$

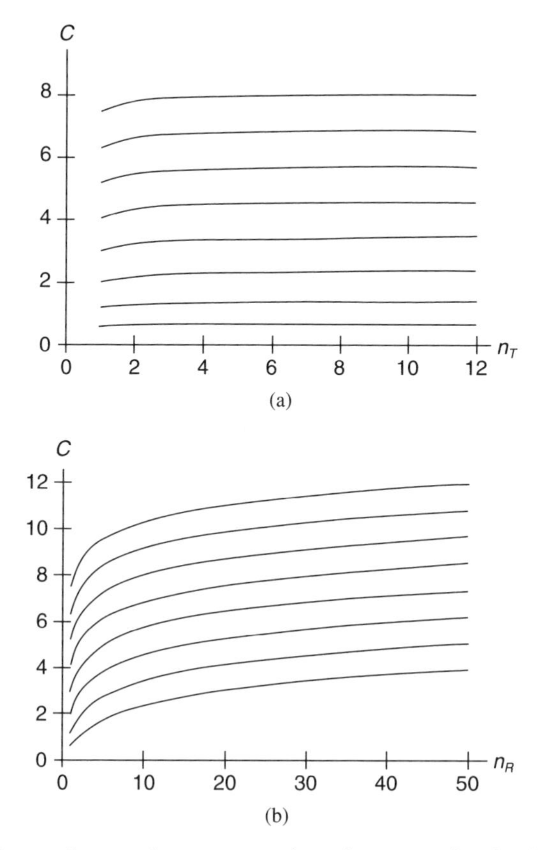

Figure 2.8. MIMO ergodic capacity versus number of antennas [126], with SNR ranges of 0–35 dB: (a) Transmit diverity—capacity versus n_T when $n_R = 1$; (b) receive diversity—capacity versus n_R when $n_T = 1$.

Hence, the channel capacity increases linearly with the number of antennas m^*. Figure 2.9 illustrates the ergodic capacity versus the number of antennas m^*.

2.5.3 Effect of Antenna Correlation on Ergodic MIMO Capacity

The evaluation of the MIMO capacity in the cases presented above assume i.i.d. channel fading in the $n_R \times n_T$ channel matrix. This is achievable when the physical separation of the transmit antennas (as well as the receive antennas) is large. However, in practice, because of the physical constraint or the scattering environment, there might be some correlation between the elements of the fading channel matrix. We shall consider the effect of correlated fading in this section.

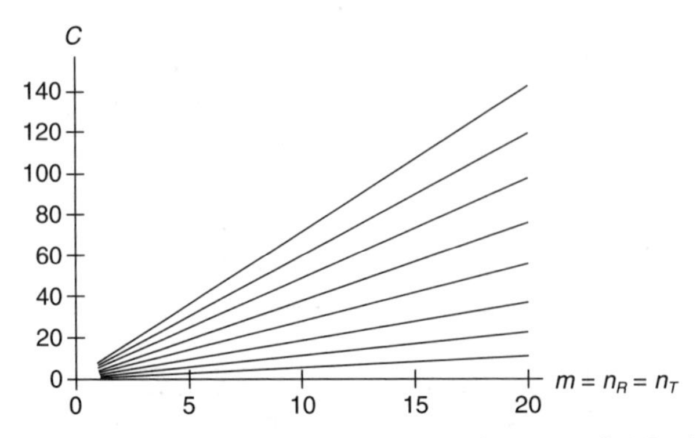

Figure 2.9. Asymptotic ergodic capacity versus number of antennas [126], with SNR ranges of 0–35 dB in 5-dB increments.

We assume that the correlation between the receive antennas does not depend on the transmit antennas and vice versa. This assumption can be justified by the fact that only intermediate antenna surroundings cause the correlation between the antenna elements and have no impact on correlation observed between the elements of the array at the other end of the link [26,101]. In such case, we can define an $n_R \times n_R$ correlation coefficient matrix, denoted by Θ_R, for the receive antennas and an $n_T \times n_T$ correlation matrix, denoted by Θ_T, for the transmit antennas. Since the correlation matrices, Θ_T and Θ_R, are symmetric and nonnegative definite, we can express them into the product form using Cholesky decomposition [59]. Hence, we have

$$\Theta_R = \mathbf{K}_R \mathbf{K}_R^*$$

where $\mathbf{K}_R$ is an $n_R \times n_R$ lower triangular matrix with positive diagonal elements and

$$\Theta_T = \mathbf{K}_T \mathbf{K}_T^*$$

where $\mathbf{K}_T$ is an $n_T \times n_T$ lower triangular matrix with positive diagonal elements. A correlated MIMO channel matrix, denoted by $\mathbf{H}_c$, can be represented as

$$\mathbf{H}_c = \mathbf{K}_R \mathbf{H} \mathbf{K}_T \tag{2.36}$$

where $\mathbf{H}$ is a $n_R \times n_T$ channel matrix with uncorrelated complex Gaussian entries.

The channel capacity of a fast fading MIMO channel with perfect CSIR and antenna correlation is thus given by

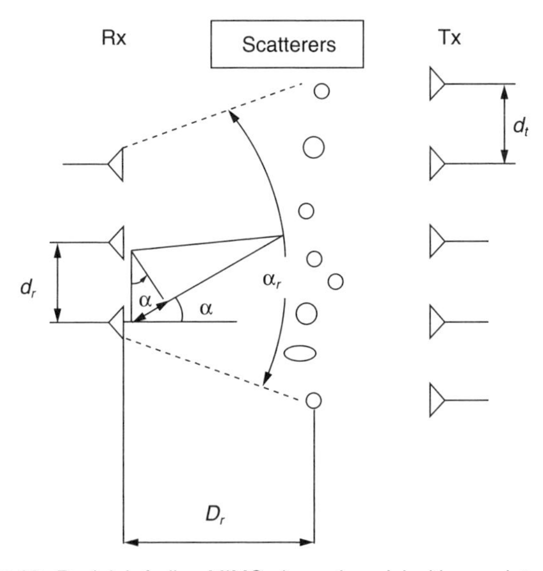

Figure 2.10. Rayleigh fading MIMO channel model with correlated fading.

$$
\begin{aligned}
C_{\text{csir}} &= \varepsilon\left[\log_2\left|\mathbf{I}_{n_R} + \frac{P_0}{n_T\eta_0}\mathbf{H}_c\mathbf{H}_c^*\right|\right] \\
&= \varepsilon\left[\log_2\left|\mathbf{I}_{n_R} + \frac{P_0}{n_T\eta_0}\mathbf{K}_R\mathbf{H}\mathbf{K}_T\mathbf{K}_T^*\mathbf{H}*\mathbf{K}_R^*\right|\right] \\
&= \varepsilon\left[\log_2\left|\mathbf{I}_{n_T} + \frac{P_0}{n_T\eta_0}\Theta_R^*\mathbf{H}\Theta_T\mathbf{H}*\right|\right]
\end{aligned} \tag{2.37}
$$

where the first expectation is with respect to $\mathbf{H}_c$ and the second and third expectations are with respect to $\mathbf{H}$. The second equality is due to $|\mathbf{I} + \mathbf{AB}| = |\mathbf{I} + \mathbf{BA}|$.

Example 2.1 (Correlation in Rayleigh Scattering Channels) Consider a linear array of n_R receive antennnas, spaced at distance d_r and surrounded by some scatterers as illustrated in Figure 2.10. There are n_T transmit antennas radiating signals, which are reflected by the scatterers surrounding the receiver. We further assume that the angle of arrival from the scatterers to the receive antennas is α. For a uniformly distributed arrival angle between $[-\pi, \pi]$, the correlation coefficient of fading between the mth and the nth receive antennas is given by

$$\theta_{m,n} = J_0\left(2\pi\frac{d_{m,n}}{\lambda}\right) \quad m \neq n$$

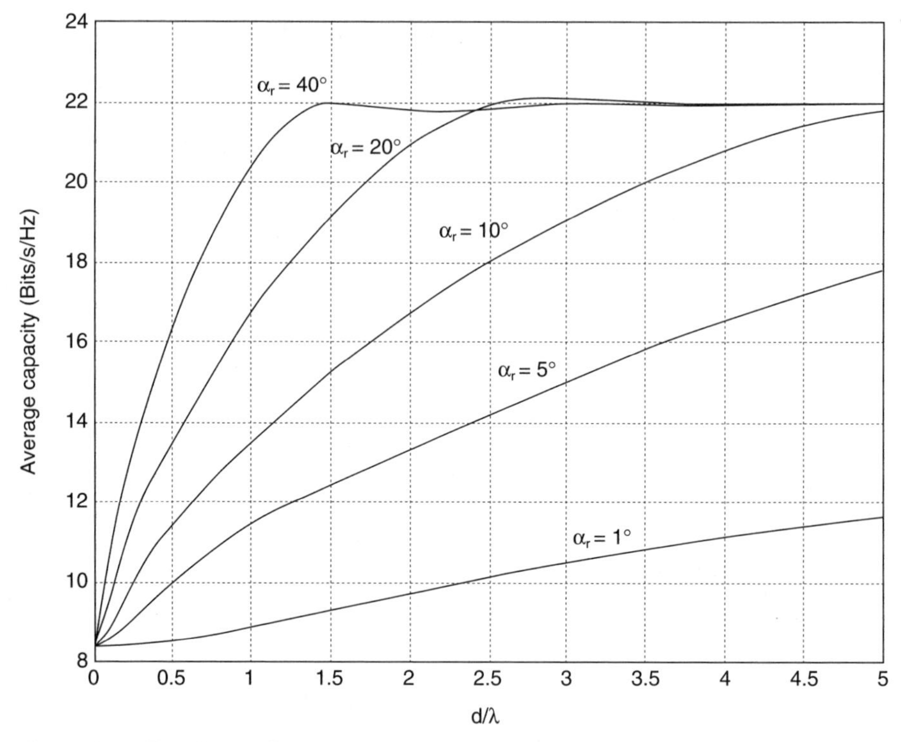

Figure 2.11. Ergodic MIMO capacity with perfect CSIR at various antenna separations and angle spreads [140]; SNR = 20 dB and $n_T = n_R = 4$.

where $J_0()$ is the zeroth-order Bessel function and λ is the wavelength of the radio signal. Hence, for small correlation, the antenna spacing should be at least $\lambda/2$. On the other hand, base stations are typically positioned high above the ground level and have a narrower angle spread. In that case, the required antenna separation is much higher (e.g., 10λ) to achieve small correlation.

Figure 2.11 illustrates the MIMO channel capacity in Equation (2.37) at various antenna separations and angle spreads with SNR = 20 dB and $n_T = n_R = 4$, based on the correlation expression in above. As illustrated, correlation in the channel matrix causes capacity degradation.

Example 2.2 (Correlation in Line-of-Sight Propagation) Consider another scenario where the transmitter and receiver are in a free-space propagation with no scattering. Figure 2.12 illustrates such a channel model with a linear array of n_T transmit and $2n_R$ receive antennas. Furthermore, we assume that the separation between the transmitter and the receiver is much larger than the antenna spacing. The wave equation of the plane-wave propagation is given by

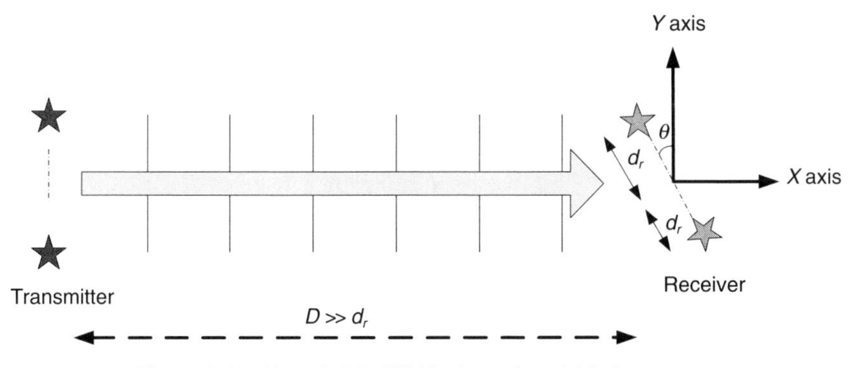

Figure 2.12. Line-of-sight MIMO channel model in free space.

$$y(x,t) = \sqrt{\frac{1}{D^2}} S_n(t) \exp(j\omega_c t - j\beta x)$$

where $\beta = \frac{2\pi}{\lambda}$, $S_t(t)$ is the lowpass equivalent transmitted signal from antenna n and x is the distance from the transmitter (in the x-axis)

$$y_r(t) = \begin{cases} \frac{S_n(t)}{\sqrt{D^2}} e^{j\omega_c t + j\beta d_r \cos\theta} & k \in [1, n_R] \\ \frac{S_n(t)}{\sqrt{D^2}} e^{j\omega_c t - j\beta d_r \cos\theta} & k \in [n_R + 1, 2n_R] \end{cases}$$

The channel matrix entry between the tth transmit and the rth receive antennas is given by

$$h_{r,t} = \begin{cases} \frac{1}{\sqrt{D^2}} e^{j\beta d_r \cos\theta} & k \in [1, n_R] \\ \frac{1}{\sqrt{D^2}} e^{-j\beta d_r \cos\theta} & k \in [n_R + 1, 2n_R] \end{cases}$$

The channel matrix is given by

$$\mathbf{H} = \frac{1}{D} \begin{bmatrix} e^{j\beta d_1 \cos\theta} & \dots & e^{j\beta d_1 \cos\theta} \\ \vdots & \dots & \vdots \\ e^{j\beta d_{nR} \cos\theta} & \dots & e^{j\beta d_{nR} \cos\theta} \\ e^{-j\beta d_{nR} \cos\theta} & \dots & e^{-j\beta d_{nR} \cos\theta} \\ \vdots & \dots & \vdots \\ e^{-j\beta d_1 \cos\theta} & \dots & e^{-j\beta d_1 \cos\theta} \end{bmatrix}$$

Hence, the channel matrix always has rank 1 because all the columns are identical. Assuming $n_T = n_R = m^*$, there is only one nonzero eigenvalue, which is equal to $\frac{1}{D^2} \cdot m^*$. The capacity of the line-of-sight channel is given by

$$C_{\text{csir}} = \log_2\left(1 + \frac{m^* P_0}{\eta_0 D^2}\right) \tag{2.38}$$

where P_0 is the total transmitted power

Observe that this is the same expression as when we have $n_T = 1$ and $n_R = m^*$. In this case, the transmit antenna does not contribute to any capacity gain. The spatial multiplexing gain of MIMO channel is lost in free-space LOS propagation.

Example 2.3 (Asymptotic Analysis) Recall that the MIMO capacity is given by Equation (2.35) in terms of the eigenvalues of **HH***. Since $\log_2(1 + (P_0/n_T\eta_0)\lambda)$ is a concave function in λ, the upper bound of the MIMO channel capacity with perfect CSIR is given by (Jensen's inequality)

$$C_{\text{csir}} \le \sum_t \log_2\left(1 + \frac{P_0}{n_T \eta_0} \varepsilon[\lambda_t]\right) \tag{2.39}$$

When the elements of the channel matrix **H** are uncorrelated, there are $m^* = \min(n_T, n_R)$ nonzero and i.i.d. eigenvalues of [**HH***] with mean equal to γ. Hence, the upper bound becomes

$$C_{\text{csir}} \le m^* \log_2\left(1 + \frac{P_0 \gamma}{n_T \eta_0}\right) \tag{2.40}$$

On the other hand, if the elements of **H** are completely correlated, there is only one nonzero eigenvalue (λ) with mean $n_R\gamma$. Hence, the capacity upper bound becomes

$$C_{\text{csir}} \le \log_2\left(1 + n_R \frac{P_0 \gamma}{n_T \eta_0}\right) \tag{2.41}$$

From the preceding upper bounds, we can see that with correlated channel fading, the spatial multiplexing advantage of MIMO is lost. However, in the special case of low SNR $P_0\gamma/\eta_0$, we have $\log_2(1 + x) \approx x$ for small x. Both capacity upper bounds become

$$C_{\text{csir}} \le \min(n_R, n_T)\left(\frac{P_0 \gamma}{n_T \eta_0}\right) \tag{2.42}$$

Hence, channel correlation has little effect on MIMO capacity in the low-SNR region but has pronounced effect in the high-SNR region.

2.5.4 Slow Flat Fading MIMO Channels

With slow flat fading MIMO channels, the encoding frame spans across a quasistatic realization of channel fading $\mathbf{H}$. Hence, the nth received symbol $\mathbf{Y}_n$ is given by

$$\mathbf{Y}_n = \mathbf{H}\mathbf{X}_n + \mathbf{Z}_n$$

Unlike the case of fast fading channels, evaluation of capacity in the slow fading channels is more tricky. Specifically, they belong to a class of special nonergodic memory channels, namely, the *compound channels* [80]. The instantaneous channel capacity conditioned on the channel realization $\mathbf{h}$ is given by

$$C_{\text{csir}}(\mathbf{h}) \leq \max_{p(\mathbf{X})} I(\mathbf{X}; \mathbf{Y}|\mathbf{h}) = \log_2 \left| \mathbf{I}_{n_R} + \frac{P_0 \mathbf{h}\mathbf{h}^*}{\eta_0 n_T} \right| \tag{2.43}$$

Hence, the Shannon's capacity of the slow fading channel is given by

$$C_{\text{csir}} = \max_{h \in \mathcal{H}} C_{\text{csir}}(\mathbf{h}) = 0.$$

In other words, the instantaneous channel capacity $C_{\text{csir}}(\mathbf{h})$ is a random variable and there is no guarantee that a frame transmission will always be error-free because the channel capacity depends on the realization of channel fading $\mathbf{h}$. While the notion of ergodic capacity becomes zero, we can still discuss the channel capacity in the *outage sense*. Letting R be the *outage capacity*, the corresponding the outage probability P_{out} is given by $P_{\text{out}}(R) = \Pr[C(\mathbf{h}) < R]$.

Figure 2.13a illustrates the cdf of the instantaneous channel capacity $C_{\text{csir}}(\mathbf{h})$ at $n_T = n_R = 1$ and $n_T = n_R = 4$ and various SNR levels. At a given outage probability, the outage capacity increases with the number of antennas.

Figure 2.13b illustrates the cdf of the instantaneous channel capacity per spatial dimension ($C_{\text{csir}}(\mathbf{h})/m^*$), where $m^* = n_T = n_R$ is the number of antennas. As shown in the figure, the outage capacity per spatial dimension approaches a constant as $m^* \to \infty$, and this illustrates the linear increase in capacity with respect to the number of antennas. On the other hand, observe that $\frac{\mathbf{h}\mathbf{h}^*}{n_T} \to \mathbf{I}_{n_R}$ as $n_T \to \infty$, due to the law of large numbers. Hence, we have

$$C_{\text{csir}}(\mathbf{h}) \to m^* \log_2\left(1 + \frac{P_0}{\eta_0}\right)$$

which is the ergodic MIMO capacity of fast fading MIMO channels with perfect CSIR. In other words, as m^* increases, the outage capacity approaches a deterministic value, which is the ergodic capacity. This implies that as m^* increases, the probability of frame transmission error [at a transmission rate R less than $m^* \log_2\left(1 + \frac{P_0}{\eta_0}\right)$] tends to zero; this is called *channel hardening*.

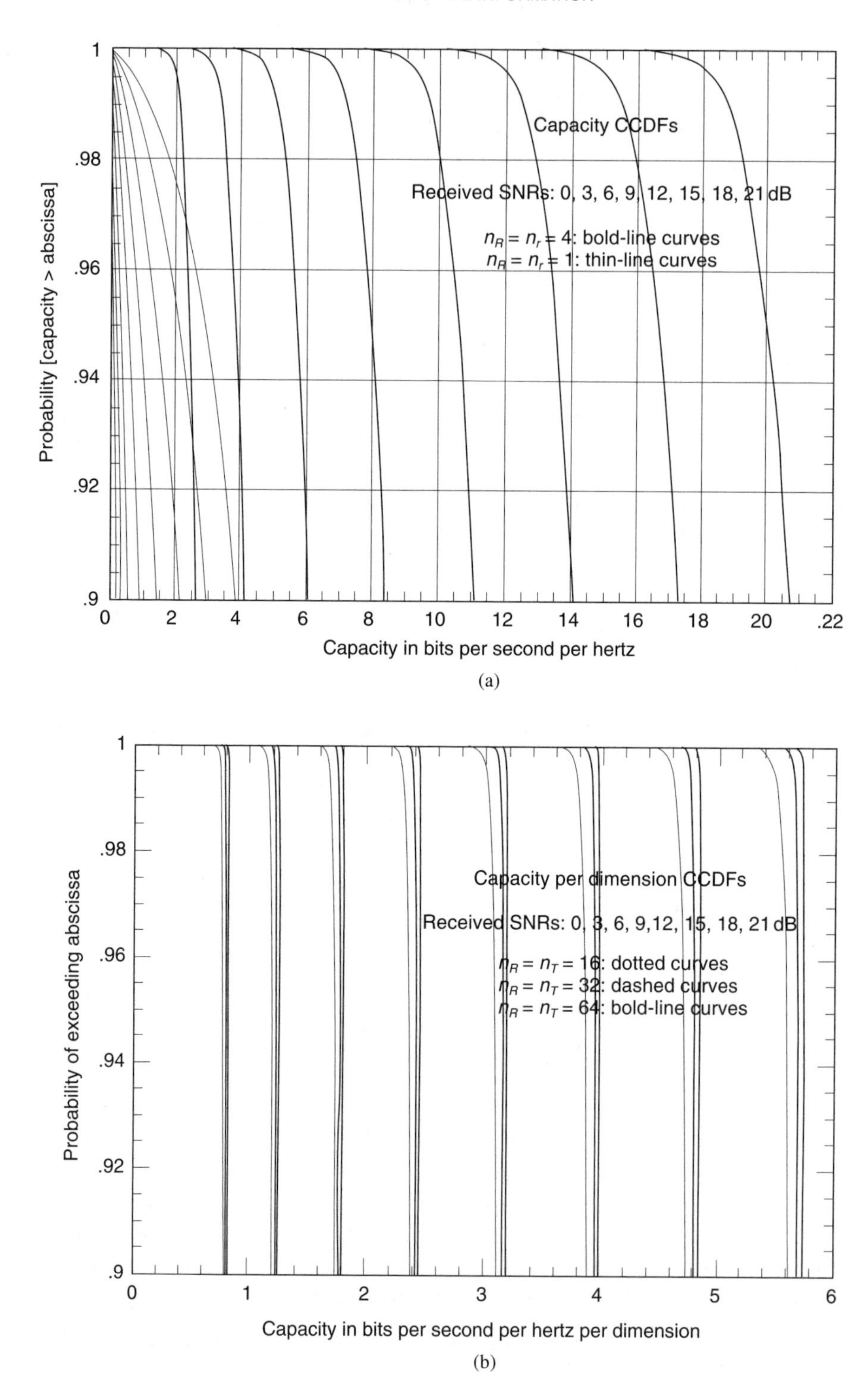

Figure 2.13. Outage capacity for MIMO channels with CSIR [126]: (a) outage capacity at $n_T = n_R = \{1, 4\}$; (b) outage capacity with a large number of antennas.

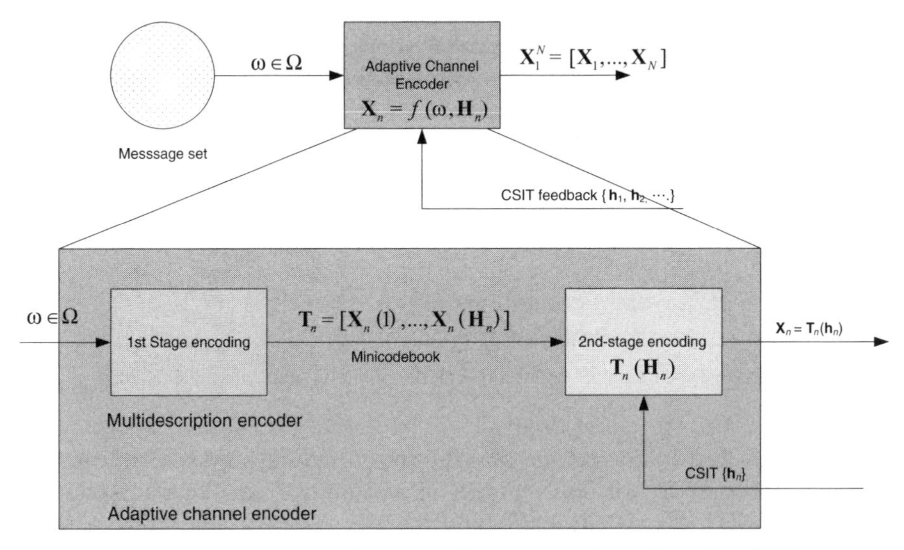

Figure 2.14. Illustration of channel-adaptive transmitter with perfect CSIT.

The channel hardening effect is also illustrated in Figure 2.13b by observing that the cdf curves become steeper as m^* increases.

2.6 CHANNEL CAPACITY WITH PERFECT CSIT ONLY

In this section, we consider the third special case of the communication system shown in Figure 2.2 with perfect CSIT only. This corresponds to $I(V; \mathbf{H}) = 0$ and $U = \mathbf{H}$. Communication with perfect state feedback at the transmitter (perfect CSIT) was first investigated in 1958 [120]. Complete and causal feedback was assumed, and the results illustrated that the feedback capacity could be expressed as that of a memoryless channel without side information at both the transmitter and the receiver.

Figure 2.6 illustrates an adaptive encoder structure with perfect CSIT (see also Figure 2.14). A message index ω is drawn from the message set Ω with cardinality 2^{NR} according to a uniform distribution. During the nth symbol duration, the channel-adaptive encoder maps the message index into the nth transmitted symbol $\mathbf{X}_n \in \mathcal{X}$, based on Equation (2.3) $\mathbf{X}_n = f_n(\omega, \mathbf{H}_1^n)$. We assume the channel state sequence $\mathbf{H}_1^N = [\mathbf{H}_1, \ldots, \mathbf{H}_N]$ is i.i.d. Hence, the knowledge of $\mathbf{H}_{n-1}, \ldots, \mathbf{H}_{-\infty}$ does not offer additional information on $\mathbf{H}_n$. Therefore, the general adaptive encoding function is a function of $\mathbf{H}_n$ only and can be written as

$$\mathbf{X}_n = f_n(\omega, \mathbf{H}_n) \tag{2.44}$$

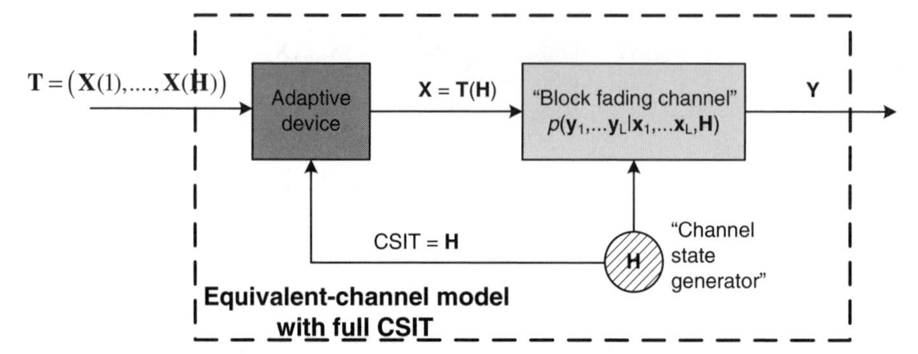

Figure 2.15. Block diagram of the equivalent-channel model with full CSIT only.

For easier illustration, assume that the set of input symbols $\mathcal{X}$ and the set of channel states $\mathcal{H}$ are discrete sets.[6] The adaptive encoding function in Equation (2.44) is equivalent to the following *two-stage encoding* arrangement as illustrated in Figure 2.6:

Stage 1 Encoding. Given a message index ω, the first stage encoder produces a code symbol $\mathbf{T}_n = \{\mathbf{X}_n(1), \ldots, \mathbf{X}_n(|\mathcal{H}|)\}$ during the nth symbol duration where $|\mathcal{H}|$ is the cardinality of the channel state. For example, for a 1×2 vector channel where each element has only two possible states $\{G, B\}$, the CSI set $\mathcal{H} = \{(G, G), (G, B), (B, G), (B, B)\}$ has a cardinality of 4. Hence, the first stage output symbol $\mathbf{T}$ is itself a *minicodebook* containing four elements.

Stage 2 Encoding. Given the current CSIT $\mathbf{H}_n$ and the output symbol from stage 1 encoder $\mathbf{T}_n$, the output coded symbol from the stage 2 encoding is given by the $\mathbf{H}_n$th component of the minicodebook $\mathbf{T}_n$:

$$\mathbf{X} = \mathbf{T}_n(\mathbf{H}_n)$$

For example, if the current CSIT is $\mathbf{H}_n = (G, B)$, the output coded symbol from stage 2 encoder is $\mathbf{T}_n(2)$.

In other words, the entire codebook of the adaptive transmitter is a *multidescription codebook*,[7] where a message index is mapped to multiple codewords by the *stage 1 encoding*, and the specific codeword selected is determined by the *stage 2 encoding* based on the current CSIT $\mathbf{H}_n$.

Figure 2.15 illustrates the *equivalent-channel model* with perfect CSIT based on the two-stage encoding model. The equivalent channel has $\mathbf{T}_n$ as the channel input and $\mathbf{Y}_n$ as the channel output. The adaptive device, which is the

[6]The results can be generalized to continuous $\mathcal{X}$ and $\mathcal{H}$ following standard arguments [44].
[7]Multidescription codebook is analogous to a multidescription constellation where an information bit is mapped to multiple constellation points.

stage 2 encoder in Figure 2.6, becomes part of the equivalent channel. Since the first stage encoding is independent of the CSIT $\mathbf{H}_n$, the equivalent channel can be regarded as a memoryless channel without CSIT. The *channel transition probability* of the *equivalent channel* is given by

$$p(\mathbf{y}|\mathbf{t}) = \sum_{\mathbf{h}} p(\mathbf{h})p(\mathbf{y}|\mathbf{X} = \mathbf{t}(\mathbf{h}), \mathbf{h}) \tag{2.45}$$

Furthermore

$$p(\mathbf{y}_1^N|\mathbf{t}_1^N) = \sum_{\mathbf{h}_1^N} p(\mathbf{h}_1^N)p(\mathbf{y}_1^N|\mathbf{x}_1^N = \mathbf{t}_1^N(\mathbf{h}_1^N), \mathbf{h}_1^N) \tag{2.46}$$

If the original channel is memoryless $p(\mathbf{y}_1^N|\mathbf{x}_1^N, \mathbf{h}_1^N) = \Pi_n p(\mathbf{y}_n|\mathbf{x}_n, \mathbf{h}_n)$, we have

$$p(\mathbf{y}_1^N|\mathbf{t}_1^N) = \prod_{n=1}^N \sum_{\mathbf{h}_n} p(\mathbf{y}_n|t_n, \mathbf{h}_n)p(\mathbf{h}_n) = \prod_{n=1}^N p(\mathbf{y}_n|t_n) \tag{2.47}$$

Hence, the equivalent channel is a memoryless channel with no CSIT and no CSIR with input alphabet $\mathbf{T}$ and output alphabet $\mathbf{Y}$. The channel capacity of the derived channel is given by

$$C_{\text{csit}} = \max_{p(\mathbf{T})} I(\mathbf{T}; \mathbf{Y}) \tag{2.48}$$

Following similar arguments for block fading channels with i.i.d. channel states across different fading blocks, the channel capacity (bits per symbol) of block fading channels with perfect CSIT is given by

$$C_{\text{csit}}(L) = \frac{1}{L} \max_{p(\mathbf{T}_1^L)} I(\mathbf{T}_1^L; \mathbf{Y}_1^L) \tag{2.49}$$

where $\mathbf{T}_1^L = [\mathbf{T}_1, \ldots, \mathbf{T}_L]$ and $\mathbf{Y}_1^L = [\mathbf{Y}_1, \ldots, \mathbf{Y}_L]$ are the L-tuples of the input symbols and output symbols in one fading block.

Although the general capacity formula is given by Equation (2.49), it is quite inconvenient to evaluate because the capacity is not expressed directly as the input alphabet of the original channel $\mathbf{X}$. Yet, we do not have a general formula for the capacity in terms of $\mathbf{X}$. The evaluation usually depends on the specific cases considered. A systematic approach for the capacity evaluation of symmetric channels is described elsewhere [38]. We shall illustrate applications of the formula with the following examples.

2.6.1 Discrete Block Fading Channels

Consider a single-antenna discrete block fading channel (with memoryless transition probability) with two output letters (Y), any number a of input

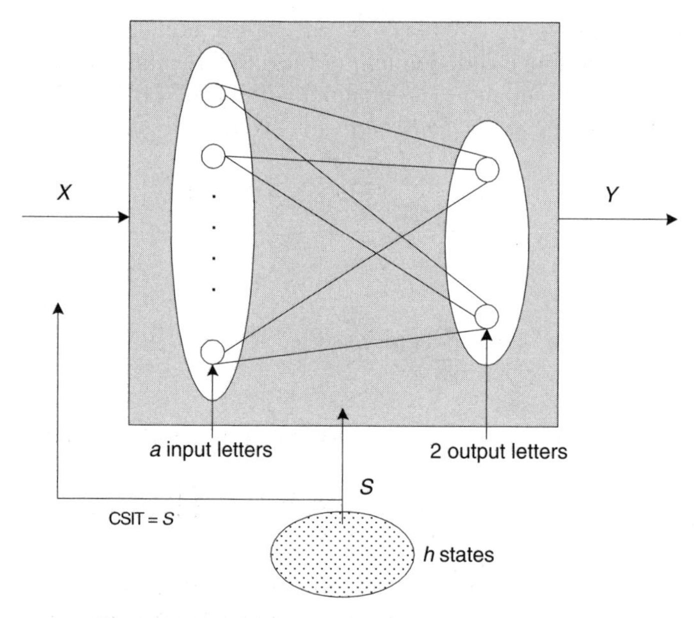

Figure 2.16. A discrete memoryless channel with CSIT.

letters (X), and any number h of states (S) as shown in Figure 2.16. We would like to find the channel capacity with perfect CSIT at the transmitter only.

The *equivalent channel* is a discrete memoryless channel with two output letters (Y) and a^h input letters (T). Let g_s be the probability that the channel is at state $S = s$. The channel capacity is given by

$$C_{\text{csit}} = \max_{p(T)} I(T;Y) \tag{2.50}$$

For a discrete memoryless channel with two outputs and $a^h > 2$ inputs, the maximizing input distribution, $p(T)$, should use two input letters $\{t_0, t_1\}$ with maximum and minimum transition probabilities to one of the two letters. Hence, the resulting transition probabilities of the derived channel are given by

$$\gamma_0 = p(Y = 0|T = t_0) = \sum_{s=1}^{h} g_s \max_{x \in [1,a]} p(Y = 0|S = s, X = x) \tag{2.51}$$

and

$$\gamma_1 = p(Y = 0|T = t_1) = \sum_{s=1}^{h} g_s \min_{x \in [1,a]} p(Y = 0|S = s, X = x). \tag{2.52}$$

The *equivalent channel* is identical to a binary input binary output channel with transition probabilities γ_0, $1 - \gamma_0$, γ_1, and $1 - \gamma_1$ as illustrated in Figure 2.17.

The channel capacity is therefore obtained by standard methods.

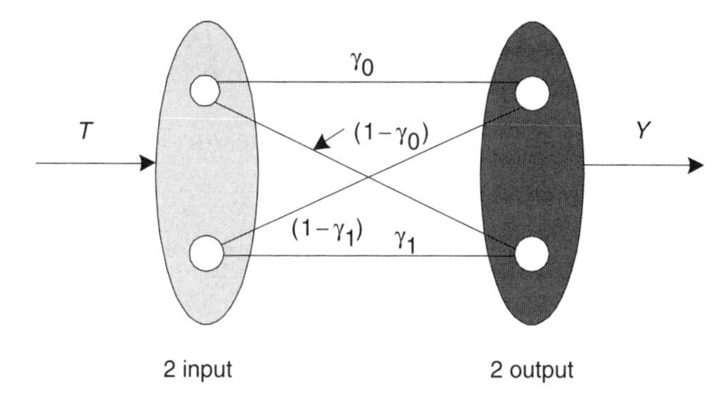

Figure 2.17. Capacity achieving configuration for the *equivalent channel*.

2.6.2 Discrete Channel with Three States

Consider a single-antenna discrete channel with three output letters, two input letters, and three states. The three states are equiprobable, each having a probability of $\frac{1}{3}$. The channel transition matrix is given by

State $S = 1$			State $S = 2$			State $S = 3$		
1	0	0	0	1	0	0	0	1
0	$\frac{1}{2}$	$\frac{1}{2}$	$\frac{1}{2}$	0	$\frac{1}{2}$	$\frac{1}{2}$	$\frac{1}{2}$	0

In this case, there are $2^3 = 8$ input letters (T) in the derived channel. The transition probability of the derived channel is therefore given by

$\frac{1}{2}$	$\frac{1}{2}$	0
0	$\frac{1}{2}$	$\frac{1}{2}$
$\frac{1}{2}$	0	$\frac{1}{2}$
$\frac{2}{3}$	$\frac{1}{6}$	$\frac{1}{6}$
$\frac{1}{6}$	$\frac{2}{3}$	$\frac{1}{6}$
$\frac{1}{6}$	$\frac{1}{6}$	$\frac{2}{3}$
$\frac{1}{3}$	$\frac{1}{3}$	$\frac{1}{3}$
$\frac{1}{3}$	$\frac{1}{3}$	$\frac{1}{3}$

where the rows represent the eight channel inputs T and the columns represent the three channel outputs (Y) in the derived channel.

Similarly, if there are only three output letters, the capacity achieving input distribution $p(T)$ will use only three input letters. From exhaustive search, we see that the first three input letters could be used. By symmetry, the capacity achieving input distribution is equiprobable with value $\frac{1}{3}$. The resulting channel capacity with perfect CSIT is $\log_2(\frac{3}{2})$.

On the other hand, if CSIT is not available, the *unconditional* channel transition probability is given by

$\frac{1}{3}$	$\frac{1}{3}$	$\frac{1}{3}$
$\frac{1}{3}$	$\frac{1}{3}$	$\frac{1}{3}$

This channel has zero capacity. Hence, the availability of full CSIT increases the channel capacity.

2.6.3 Fast Flat Fading MIMO Channels

Consider a fast flat fading MIMO channel where the channel output $\mathbf{Y}_n$ is given by

$$\mathbf{Y}_n = \mathbf{H}_n \mathbf{X}_n + \mathbf{Z}_n$$

and $\mathbf{H}_n$ is an i.i.d. complex Gaussian fading sequence with unit variance and $\mathbf{Z}_n$ is i.i.d. complex Gaussian channel noise with covariance $\sigma_z^2 \mathbf{I}_{n_R}$. The original channel is memoryless because $p(\mathbf{y}_1^N | \mathbf{x}_1^N, \mathbf{h}_1^N) = \Pi_n p(\mathbf{y}_n | \mathbf{x}_n, \mathbf{h}_n)$ and the transition probability of the equivalent channel is given by

$$p(\mathbf{y}|t) = \int_{\mathbf{h}} f(\mathbf{h}) p(\mathbf{y}|\mathbf{X} = \mathbf{t}(\mathbf{h}), \mathbf{h}) d\mathbf{h} = \int_{\mathbf{h}} f(\mathbf{h}) \frac{1}{(\pi \sigma_z^2)^{nR}} \exp\left(-\frac{1}{\sigma_z^2} (\mathbf{y} - \mathbf{t}(\mathbf{h})) * (\mathbf{y} - \mathbf{t}(\mathbf{h})) \right)$$

The channel capacity in *ergodic sense* is given by

$$C_{\text{csit}} = \max_{p(\mathbf{T})} I(\mathbf{T}; \mathbf{Y}) \tag{2.53}$$

In general, the capacity achieving distribution $p(\mathbf{T})$ is no longer Gaussian, and no rate adaptation is needed at the transmitter.

2.6.4 Slow Flat Fading MIMO Channels

Consider a slow flat fading MIMO channel with channel output $\mathbf{Y}_n$ given by

$$\mathbf{Y}_n = \mathbf{H}\mathbf{X}_n + \mathbf{Z}_n$$

where $\mathbf{H}$ is a complex Gaussian fading coefficient and $\mathbf{Z}_n$ is an i.i.d. complex Gaussian noise sequence with covariance $\sigma_z^2 \mathbf{I}_{n_R}$. In this case, the entire encoding frame spans across a single realization of channel fading $\mathbf{H}$ and therefore the channel is a nonergodic memory channel. Specifically, such channels belong to the class of *compound channels* [80]. The instantaneous capacity conditioned on the channel fading realization $\mathbf{h}$ is given by

$$C_{\text{csit}}(\mathbf{h}) = \max_{p(\mathbf{X}|\mathbf{h})} I(\mathbf{X}; \mathbf{Y})$$

Due to the absence of CSIR, the capacity-achieving input distribution is no longer Gaussian. However, unlike the case of slow fading without CSIT, nonzero ergodic capacity can be achieved with both transmit symbol adaptation $p(\mathbf{X}|\mathbf{h})$ as well as rate adaptation. Given the CSIT $\mathbf{h}$, the distribution of the transmitted codewords $p(\mathbf{X}|\mathbf{h})$ is optimized with respect to the CSIT. Moreover, rate adaptation is needed at the transmitter so that the transmission rate $R(\mathbf{h})$ is always less than the instantaneous channel capacity $C_{\text{csit}}(\mathbf{h})$ by an arbitrarily small value ε. In other words, the packet transmission is always guaranteed to be error-free. The average capacity (averaged over many frame transmissions) is given by

$$\overline{C}_{\text{csit}} = \varepsilon[C_{\text{csit}}(\mathbf{h})] = \int_{\mathbf{h}} f(\mathbf{h}) C_{\text{csit}}(\mathbf{h}) dh \tag{2.54}$$

2.7 CHANNEL CAPACITY WITH PERFECT CSIR AND PERFECT CSIT

In this section, we shall consider the fourth special case with perfect CSIT and perfect CSIR. Specifically, we have $U = V = \mathbf{H}$ and $V = \mathbf{H}$. With perfect CSIT, adaptation in the channel encoding is possible. With perfect CSIR, channel matched decoding is possible and hence, we expect the capacity of channels with perfect CSIR and perfect CSIT to be the highest among the four cases.

Consider a channel with memoryless transition probability $p(\mathbf{y}_1^N|\mathbf{x}_1^N, \mathbf{h}_1^N) = \Pi_n p(\mathbf{y}_n|\mathbf{x}_n, \mathbf{h}_n)$ and the i.i.d. state sequence distribution $p(\mathbf{h}_1^N) = \Pi_n p(\mathbf{h}_n)$. When we have perfect CSIR, we treat $(\mathbf{Y}_n, \mathbf{H}_n)$ as an *output supersymbol*. The equivalent channel becomes a memoryless probabilistic channel [with input $\mathbf{X}_n$ and output $(\mathbf{Y}_n, \mathbf{H}_n)$] with CSIT only. Following an argument similar to that in Section 2.6, the channel capacity is given by

$$C_{\text{csit,csir}} = \max_{p(\mathbf{T})} I(\mathbf{T}; \mathbf{Y}, \mathbf{H}) = \max_{p(\mathbf{T})} I(\mathbf{T}; \mathbf{Y}|\mathbf{H}) \tag{2.55}$$

where $\mathbf{T}$ represents the $|\mathcal{H}|$-tuple *input mapping rules* as described in Section 2.6 and the last equality is due the fact that $I(\mathbf{T}; \mathbf{H}) = 0$.

Extension of the arguments above to block fading channels with channel inputs $\mathbf{X}_1^N = [\mathbf{X}_{1,1}, \ldots, \mathbf{X}_{l,n}, \ldots, \mathbf{X}_{L,N}]$, channel outputs $\mathbf{Y}_1^N = [\mathbf{Y}_{1,1}, \ldots, \mathbf{Y}_{l,n}, \ldots, \mathbf{Y}_{L,N}]$, and channel states $\mathbf{H}_1^N = [\mathbf{H}_1, \ldots, \mathbf{H}_N]$ is straightforward. If the block fading channel has the transition probability given by $p(\mathbf{y}_1^N, \mathbf{h}_1^N|\mathbf{x}_1^N) = \Pi_n p(\mathbf{y}_n, \mathbf{h}_n|\mathbf{x}_n)$, where $\mathbf{y}_n = [\mathbf{y}_{1,n}, \ldots, \mathbf{y}_{L,n}]$ and $\mathbf{x}_n = [\mathbf{x}_{1,n}, \ldots, \mathbf{x}_{L,n}]$, the channel capacity (bits per symbol) is given by

$$C_{\text{csit,csir}}(L) = \frac{1}{L} \max_{p(\mathbf{T}_1^L)} I(\mathbf{T}_1^L; Y_1^L, \mathbf{H}) = \frac{1}{L} \max_{p(\mathbf{T}_1^L)} I(\mathbf{T}_1^L; \mathbf{Y}_1^L|\mathbf{H}) \tag{2.56}$$

where $\mathbf{T}_1^L = (\mathbf{T}_1, \ldots, \mathbf{T}_L)$ is the L-tuple input to the equivalent channel in a fading block and $\mathbf{T}_l$ represents the *minicodebook* with cardinality $|\mathcal{H}|$ as described in Section 2.6. Similar to Lemma 2.26, we have $\lim_{L\to\infty} C_{\text{csit}}(L) =$

$\lim_{L\to\infty} C_{\text{csit,csir}}(L)$. Hence, as the fading block memory (L) increases, the capacity with perfect CSIT only approaches the capacity with perfect CSIT and perfect CSIR for the same reason as in Lemma 2.26.

Note that in the special case of block fading channel with *memoryless transition* within a fading block ($p(\mathbf{y}_n|\mathbf{x}_n, \mathbf{h}_n) = \Pi_{l=1}^{L} p(\mathbf{y}_{l,n}|\mathbf{x}_{l,n}, \mathbf{h}_n)$), the channel capacity is independent of the block length L and is given by

$$C_{\text{csit,csir}}(L) = \overline{C}_{\text{csit,csir}} = \max_{p(\mathbf{T})} I(\mathbf{T}; \mathbf{Y}|\mathbf{H}) \tag{2.57}$$

Unlike the case with perfect CSIT only, we can further express the capacity formula in Equations (2.55) and (2.56) in terms of the input alphabets $\mathbf{X}$ and the channel states $\mathbf{H}$ of the original channel. The result is summarized in the following lemma.

Lemma 2.4 (Capacity of MIMO Channels with CSIT and CSIR) The channel capacity $C_{\text{csit,csir}}$ of MIMO channels with full CSIT and perfect CSIR in Equation (2.55) is given by

$$C_{\text{csit,csir}} = \varepsilon[\max_{p(\mathbf{X}|\mathbf{H}=\mathbf{h})} I(\mathbf{X}; \mathbf{Y}|\mathbf{H} = \mathbf{h})] \tag{2.58}$$

The channel capacity $C_{\text{csit,csir}}(L)$ of block fading channels with perfect CSIT and perfect CSIR in Equation (2.56) is given by

$$C_{\text{csit,csir}}(L) = \frac{1}{L}\varepsilon[\max_{p(\mathbf{X}|\mathbf{H}=\mathbf{h})} I(\mathbf{X}; \mathbf{Y}|\mathbf{H} = \mathbf{h})] \tag{2.59}$$

Proof From Equation (2.55), the channel capacity is given by

$$\begin{aligned} C_{\text{csit,csir}} &= \max_{p(\mathbf{T})} (\mathbf{T}; \mathbf{Y}|\mathbf{H}) = \max_{p(\mathbf{T})} \sum_{h} p(\mathbf{h}) I(\mathbf{T}; \mathbf{Y}|\mathbf{H} = \mathbf{h}) \\ &= \sum_{\mathbf{h}} p(\mathbf{h}) \max_{p(\mathbf{X}|\mathbf{h})} I(\mathbf{T}; \mathbf{Y}|\mathbf{H} = \mathbf{h}) \end{aligned}$$

where the last equality is due to the fact that given $\mathbf{H} = \mathbf{h}, \mathbf{X} = \mathbf{T}(\mathbf{h})$, and hence, any distribution on T induces a distribution on $\mathbf{X}$.

We consider a few examples below to illustrate these concepts.

2.7.1 Fast Flat Fading MIMO Channels

Consider a flat fading MIMO channel with ($n_R \times 1$)-dimensional channel output $\mathbf{Y}_n = \mathbf{H}_n\mathbf{X}_n + \mathbf{Z}_n$ as in Equation (2.2). The ergodic capacity is given by

$$\begin{aligned} C_{\text{csit,csir}} &= \varepsilon\left(\max_{p(\mathbf{X}|\mathbf{h})} I(\mathbf{X}; \mathbf{Y}|\mathbf{H} = \mathbf{h})\right) \\ &= \varepsilon\left(\max_{p(\mathbf{X}|\mathbf{h})} [H(\mathbf{Y}|\mathbf{h}) - H(\mathbf{Y}|\mathbf{X}, \mathbf{h})]\right) \\ &= \varepsilon\left(\max_{p(\mathbf{X}|\mathbf{h})} H(\mathbf{Y}|\mathbf{h})\right) - H(\mathbf{Z}). \end{aligned}$$

Since the channel noise is Gaussian, the capacity achieving distribution $p(\mathbf{X}|\mathbf{H} = \mathbf{h})$ is complex Gaussian with covariance $\varepsilon[\mathbf{XX}^*|\mathbf{H} = \mathbf{h}] = \mathbf{Q}(\mathbf{h})$. The ergodic capacity of fast flat fading MIMO channels with full CSIT and perfect CSIR is as given by Equation (2.58):

$$C_{\text{csit,csir}} = \varepsilon\left[\max_{\mathbf{Q}(\mathbf{h})} \log_2 \left| \mathbf{I}_{nR} + \frac{\mathbf{hQh}^*}{\sigma_z^2} \right|\right]$$

The channel input symbols must satisfy a transmit power constraint. We consider a simple case with average power constraint (averaged over an encoding frame) given by

$$\varepsilon[\text{tr}(\mathbf{Q}(\mathbf{h}))] \leq P_0 \tag{2.60}$$

Hence, the optimization problem is to choose the input covariance matrix $\mathbf{Q}$ with respect to every channel fading realization $\mathbf{h}$ (see Problem 2.1).

Problem 2.1 (Temporal and Spatial Power Water Filling) Select $\mathbf{Q}$ to optimize

$$L(\mathbf{Q}, \lambda) = \log_2 \left| \mathbf{I}_{nR} + \frac{\mathbf{hQh}^*}{\sigma_z^2} \right| - \mu\, \text{tr}(\mathbf{Q})$$

where μ is the Lagrange multiplier chosen to satisfy the average transmit power constraint in Equation (2.60).

The solution of the preceding optimization problem is summarized in the following theorem.

Theorem 2.2 (Temporal and Spatial Power Water Filling) Given any fading matrix realization $\mathbf{h}$, the optimizing input covariance $\mathbf{Q}$ is given by

$$\mathbf{Q} = \mathbf{B}\tilde{\mathbf{Q}}\mathbf{B}^* \tag{2.61}$$

where $\mathbf{B}$ is the $n_T \times n_T$ eigenvector matrix of $\mathbf{h}^*\mathbf{h}$ and $\tilde{\mathbf{Q}}$ is a $n_T \times n_T$ diagonal matrix with $m^* = \min[n_T, n_R]$ nonzero elements given by

$$\tilde{q}_{n,n} = \left(\frac{1}{\mu/n_2} - \frac{\sigma_z^2}{\lambda_n}\right)^+ \quad \forall n \in [1, m^*] \tag{2.62}$$

where μ is a constant chosen to satisfy $\Sigma_n \varepsilon[\tilde{q}_{n,n}] \leq P_0$. The corresponding ergodic capacity is given by

$$C_{\text{csit,csir}} = \sum_{n=1}^{m^*} \varepsilon\left[\log_2\left(\frac{\lambda n}{\sigma_z^2/n2}\right)^+\right] \tag{2.63}$$

Note that no rate adaptation is needed at the transmitter to achieve the ergodic capacity.

Compared with the solution for scalar channels, the optimal input covariance $\mathbf{Q}$ in Equation (2.61) suggests that the solution has both *spatial water-filling* and *temporal water-filling* components. For any i.i.d. $n_R \times n_T$ channel matrix $\mathbf{h}$, there is a probability 1 that the rank of $\mathbf{h}$ is $m^* = \min[n_T, n_R]$. Hence, the $n_R \times n_T$ channel matrix contains m^* spatial channels (or eigenchannels). *Spatial water filling* refers to the optimal distribution of power across the m^* *spatial channels*. Specifically, more power is allocated to the spatial channel(s) having a better channel condition (eigenvalues λ_n). Less power (or even zero power) is allocated to those spatial channels that have poor channel condition (small λ_n). Figure 2.18a illustrates the concept of spatial water filling. The resource is determined by the *water level*, which is given by $\frac{1}{\mu}$. Hence, the larger the Lagrange multiplier μ is, the lower the water level (or resource level) will be. The instantaneous power allocation over the spatial channels is given by the vertical level between the *water level* $\frac{1}{\mu}$ and the *noise level* σ_z^2/λ_n of the nth spatial channel. When there are some spatial channels with very high noise level (exceeding the water level), zero power will be allocated to save power for better channels.

On the other hand, *temporal water filling* refers to the allocation of *total power* across various fading symbols temporally. Smaller (or zero) total power is allocated to the n_T transmit antennas for spatial water filling during the time instance when $\Sigma_t \dfrac{\sigma_z^2}{\lambda_t}$ is large so as to save power for some other *better time instance*. The adaptation of the total power allocation at each time is automatically accommodated the constant *water level* $\frac{1}{\mu}$ in Figure 2.18a. The larger the average power constraint, the smaller the μ would be and therefore, the higher the water level will be.

If the average transmit power constraint is replaced by the instantaneous transmit power constraint $\mathrm{tr}(\mathbf{Q}(\mathbf{h})) \leq P_0$ for all $\mathbf{h}$, we do not need to have temporal power water filling. In that case, the water level $\frac{1}{\mu}$ will change accordingly during different fading realizations in order to ensure that the total transmit power is always given by P_0. Hence, in that case, we still have spatial power water-filling component but we do not have temporal water-filling component in the adaptation policy.

To derive the optimal transmitter structure, we observe that the optimal input distribution $p(\mathbf{X}|\mathbf{h})$ is circular symmetric complex Gaussian with covariance matrix $\mathbf{B}\tilde{\mathbf{Q}}\mathbf{B}^*$. This can be achieved by

$$\mathbf{X} = \underbrace{\mathbf{B}}_{n_T \times m^*} \underbrace{\sqrt{\tilde{\mathbf{Q}}}}_{m^* \times m^*} \underbrace{\mathbf{\Psi}}_{m^* \times 1} \tag{2.64}$$

where $\mathbf{\Psi}$ is i.i.d. complex Gaussian distributed with covariance matrix $\varepsilon[\mathbf{\Psi}\mathbf{\Psi}^*] = \mathbf{I}_{m^*}$. Figure 2.18b illustrates the optimal transmitter and receiver structure. The transmitter consists of a bank of m^* channel encoders. The outputs of

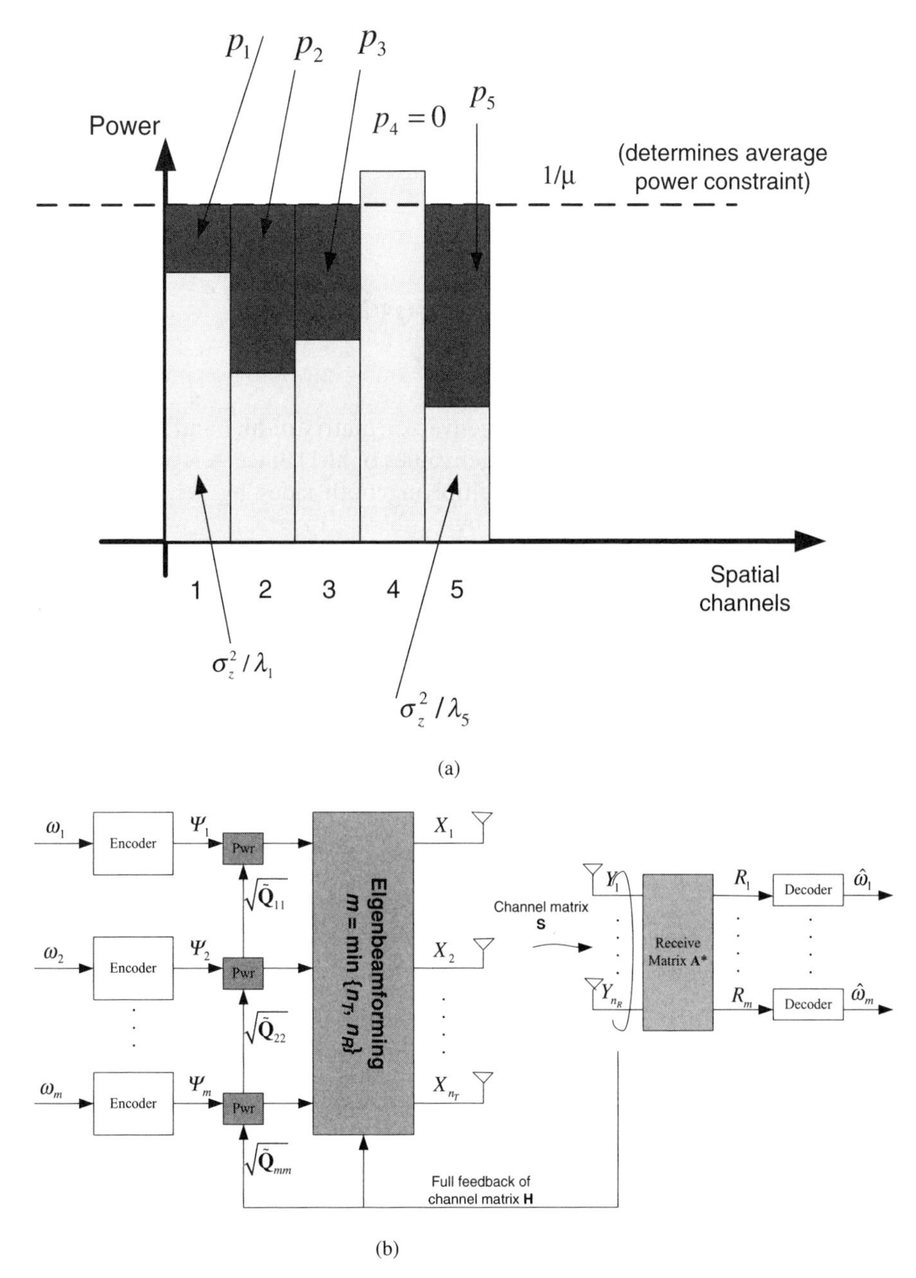

Figure 2.18. Ergodic capacity of MIMO channels with ideal CSIT and CSIR: (a) illustration of spatial water-filling of transmit power (the allocated power on each spatial channel is proportional to the shaded area); (b) optimal transmitter–receiver structure.

the channel encoders $\mathbf{\Psi} = [\Psi_1, \ldots, \Psi_{m^*}]$ are passed to the power adaptation matrix $\sqrt{\tilde{\mathbf{Q}}}$ to adjust the power allocated to individual spatial channels and then to the *eigenbeamforming* matrix $\mathbf{B}$ to decouple the MIMO channel into m^* independent vector channels.

At the receiver, the received symbols (dropping the symbol index n) can be expressed as

$$\begin{aligned}\mathbf{Y} &= \mathbf{HX} + \mathbf{Z} \\ &= \mathbf{A}\sqrt{\mathbf{D}}\mathbf{B}^*\left(\mathbf{B}\sqrt{\tilde{\mathbf{Q}}}\mathbf{\Psi}\right) + \mathbf{Z} \\ &= \mathbf{A}\sqrt{\mathbf{D}\tilde{\mathbf{Q}}}\mathbf{\Psi} + \mathbf{Z}\end{aligned}$$

where $\mathbf{B}$, $\mathbf{A}$ are unitary matrices (eigenvector matrix of $\mathbf{hh}^*$ and $\mathbf{h}^*\mathbf{h}$, respectively) and $\mathbf{D}$ is a diagonal matrix (eigenvalues of $\mathbf{hh}^*$) Since $\mathbf{A}$ is unitary, there is no loss of information by premultiplying both sides by $\mathbf{A}^*$. Hence, the processed received signal $\mathbf{R}$ is given by

$$\mathbf{R} = \mathbf{A}^*\mathbf{Y} = \sqrt{\mathbf{D}\tilde{\mathbf{Q}}}\mathbf{\Psi} + \mathbf{Z}' \tag{2.65}$$

where $\mathbf{Z}' = \mathbf{A}^*\mathbf{Z}$ is the equivalent-channel noise, which has the same distribution and covariance matrix as $\mathbf{Z}$. From Equation (2.65), the equivalent channel with input $\mathbf{\Psi}$ and output $\mathbf{R}$ can be decoupled into $m^* = \text{rank}(\mathbf{h})$ spatial channels (or eigenchannels). Hence, the optimal receiver structure (as illustrated in Figure 2.18b) consists of a *beamforming matrix* $\mathbf{A}^*$ followed by m^* isolated channel decoders.

2.7.2 Slow Flat Fading MIMO Channels

Consider a slow flat fading MIMO channel with channel input $\mathbf{X}_n$. The channel output $\mathbf{Y}_n$ is given by

$$\mathbf{Y}_n = \mathbf{HX}_n + \mathbf{Z}_n$$

where the channel fading $\mathbf{H}$ remains quasistatic for the entire encoding frame. For a given channel fading realization $\mathbf{H} = \mathbf{h}$, the instantaneous channel capacity is given by

$$C_{\text{csit,csir}}(\mathbf{h}) = \max_{p(\mathbf{T}(\mathbf{h}))} I(\mathbf{T}; \mathbf{Y}, \mathbf{H} = \mathbf{h}) = \max_{p(\mathbf{X}|\mathbf{h})} I(\mathbf{X}; \mathbf{Y}|\mathbf{H} = \mathbf{h}) \tag{2.66}$$

where the second equality is due to the fact that given $\mathbf{h}$, the distribution on $\mathbf{T}$ induces a distribution on $\mathbf{X}$.

In the Rayleigh fading example we considered, the capacity achieving input distribution is circular symmetric complex Gaussian with input covariance matrix $\varepsilon[\mathbf{XX}^*|\mathbf{h}] = \mathbf{Q}(\mathbf{h})$. Note that since we have slow fading, the entire frame spans across a quasistatic fading realization. Together with the average power constraint (averaged over an encoding frame), there is no temporal power adaptation across an encoding frame. Hence, the instantaneous channel capacity is given by

$$C_{\text{csit,csir}}(\mathbf{h}) = \max_{\mathbf{Q}:\text{tr}(\mathbf{Q})\le P_0} \log_2\left|\mathbf{I}_{n_R} + \frac{\mathbf{hQh}^*}{\sigma_z^2}\right| \tag{2.67}$$

Using an approach similar to that in the fast fading example, the capacity-achieving input covariance matrix $\mathbf{Q}(\mathbf{h})$ is given by the *spatial power water filling*

$$\mathbf{Q} = \mathbf{B}\overline{\mathbf{Q}}\mathbf{B}^* \tag{2.68}$$

where $\mathbf{B}$ is the $n_T \times m^*$ eigenvector matrix of $\mathbf{h}^*\mathbf{h}$ and $\tilde{\mathbf{Q}}$ is a $m^* \times m^*$ diagonal matrix with $m^* = \min[n_T, n_R]$ diagonal elements given by

$$\tilde{q}_{n,n} = \left(\frac{1}{\mu/n_2} - \frac{\sigma_z^2}{\lambda_n}\right)^+ \quad \forall n \in [1, m^*] \tag{2.69}$$

where μ is a constant chosen to satisfy $\Sigma_n \tilde{q}_{n,n} \le P_0$.

Since the slow fading channel can be regarded as a *compound channel* [80], we have to make use of CSIT to perform both rate adaptation as well power adaptation.[8] The power adaptation is done according to Equation (2.69), which consists of spatial and temporal power water filling. The rate adaptation is done in such a way that the transmission rate $R(\mathbf{h})$ is always smaller than the instantaneous channel capacity $C_{\text{csit,csir}}(\mathbf{h})$ by ε for arbitrarily small ε. In this way, all the frame transmission is guaranteed to have zero outage probability. In other words, the presence of perfect CSIT in slow fading channels can guarantee nonzero ergodic capacity. The average capacity over multiple frame transmissions is given by

$$C_{\text{csit,csir}} = \varepsilon\left[\max_{\mathbf{Q}:tr(\mathbf{Q})\le P_0} \log_2\left|\mathbf{I}_{nR} + \frac{\mathbf{hQh}^*}{\sigma_z^2}\right|\right]$$

where the expectation is taken over $\mathbf{h}$.

2.8 SUMMARY

In this chapter, we have considered the optimal channel adaptation design for a point-to-point MIMO link based on an information-theoretic framework. We introduce the concept of *ergodic capacity* and *outage capacity* for general

[8] This is in contrast to the fast fading situation (which is an ergodic channel), where only power adaptation is sufficient to achieve the optimal capacity.

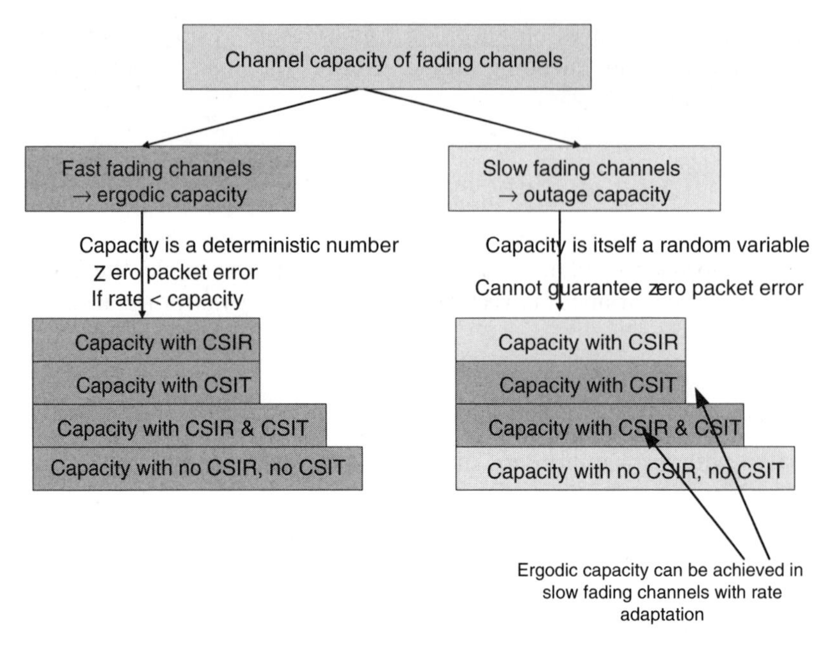

Figure 2.19. A summary of various channel capacities in fading channels.

probabilistic channels with states. A general transmission–feedback model with CSIT and CSIR is introduced.

We have discussed ideal CSI scenarios where various combinations of CSIT and CSIR configurations are considered. For example, we illustrate four special cases of *no CSIT and no CSIR*, *perfect CSIT and no CSIT*, *perfect CSIR and no CSIT*, and *perfect CSIT and perfect CSIR*. Figure 2.19 illustrates a summary of various channel capacities in fading channels. For fast fading channels, we see that full knowledge of CSI at the transmitter and the receiver could both enhance the channel capacity. When CSIR is present, *channel-matched decoding* can be done at the receiver, and this contributes to the capacity gain over that without CSIR. However, asymptotically in the limit of large block length (L), the capacity of channel with no CSIR approaches the channel capacity with perfect CSIR because perfect CSIR can be obtained with negligible overhead of embedding pilot symbols. When there is perfect CSIT, adaptation can be performed at the transmitter, and this contributes to the capacity gain. For a single-antenna link, the transmitter adaptation involves power water filling across the temporal domain. For multiple antenna link, the optimal adaptation strategy is temporal and involves spatial power water filling where the transmitter distributes the transmitted power across the spatial channels and temporal symbols. Specifically, the transmitter increases the transmit power when the channel condition is good and reduces the transmit power (or even shuts down the transmission) when the channel condition is poor. No variable rate adaptation is needed to realize the ergodic capacity of fast fading channels with CSIT.

On the other hand, for slow fading channels, the channel capacity (without CSIT) is itself a random variable and has to be specified by the cdf (instead of a single capacity number). For the same reason, the capacity of slow fading channels with CSIR is larger than that without CSIR at the same *outage probability*. However, for MIMO channels with a large number of transmit antennas, there is a *channel hardening effect*, which the outage capacity with CSIR converges to a deterministic value (ergodic capacity). In other words, there will be no packet outage as long as the information rate is less than the ergodic capacity even for slow fading channels with CSIR only. When there is CSIT, adaptation at the transmitter can improve the link performance. Unlike the fast fading case in which only power adaptation is done, the transmitter has to perform both power and rate adaptation for slow fading channels to eliminate packet outage.

EXERCISES

1. *Spatial Channels in MIMO Systems* Consider a wireless communication system with n_T transmit antennas and n_R receive antennas. In other words, the transmitted signal is of n_T dimensional space and the receive signal is of n_R space. Explain *geometrically* using the concept of signal space why there are only $\min(n_T, n_R)$ spatial channels. Consider the following cases separately.

(a) $n_T \geq n_R$

(b) $n_R \geq n_T$

2. *Jointly Gaussian Distribution* Suppose there is a fast fading channel with input $\mathbf{x} = x_1, x_2, \ldots, x_N$ with N symbols and the corresponding output is $\mathbf{y} = y_1, \ldots, y_N$. Define the channel states to be $\mathbf{h} = h_1, \ldots, h_N$ and the noise $\mathbf{z}$ is i.i.d Gaussian process. The input-output relationship of the channel is characterized by $Y = HX + Z$. Prove the following equalities.

$$p\left(\mathbf{y}|\mathbf{x}, \mathbf{h} = \frac{1}{(\pi|\Sigma_y|)^N} \exp\left(-\sum_n (y_n - h_n x_n) \Sigma_y^{-1} (y_n - h_n x_n)\right)\right. \quad (1)$$

$$p(\mathbf{y}|\mathbf{x}, \mathbf{h}) = \prod_n p(y_n|x_n, h_n) \quad (2)$$

3. *Fading Channels* Suppose a channel encoder with frame duration T_f and coding block length of N is used at the transmitter. Assuming powerful code is used at the transmitter and maximum likelihood decoding is done at the receiver. Comment on the following statements

(a) In a flat fading channel with coherence time $T_c/T_f \to 0$, the frame error rate will approach zero as $N \to \infty$.

(b) In a flat fading channel with $T_c \gg T_f$, the frame error rate will hit an error floor as $N \to \infty$.

(c) In a flat fading channel with $T_c = 0.2T_f$, the frame error rate will approach zero as $N \to \infty$.

(d) In a flat fading channel with $T_c \ll T_f$, rate and power adaptation should be done if the transmitter has knowledge of CSIT to maximize the capacity.

4. *Circular Symmetric Gaussian Distributions* Recall that a complex Gaussian $n \times 1$ random vector x is said to be circularly symmetric if

$$E\{(\hat{x} - E[\hat{x}])(\hat{x} - E[\hat{x}])^*\} \frac{1}{2}\begin{pmatrix} Re(Q) & -Im(Q) \\ Im(Q) & Re(Q) \end{pmatrix}$$

where $\hat{x} = [Re(x), Im(x)]^T]$, $Q = E\{(\hat{x} - \mu)(\hat{x} - \mu)^*\}$ and $\mu = E\{x\}$. Prove the followings

(a) $H(x) \leq \frac{1}{2}\log_2(2\pi e^n |Q|)$ where x is a $n \times 1$ real vector.

(b) $H(x) \leq \frac{1}{2}\log_2(\pi e^{2n} |Q|)$ where x is a $n \times 1$ complex vector.

(c) verify that equality holds in part (b) when x is circular symmetric Gaussian distributed.

5. *The Role of CSIR*

(a) Suppose the transmitter is using QPSK modulation (no differential encoding) to transmit information over flat fading channel. The received signal is given by $Y = HX + Z$ where X is the transmitted signal, H is the flat fading channel coefficient (zero mean complex Gaussian) and Z is the complex Gaussian channel noise. Suppose the receiver does not have knowledge of H. Can the receiver decode the information bits carried in X? Explain your choice.

(b) Suppose the transmitter is using the constellation (no differential encoding) in figure 1. Can the receiver decode the information based on observation on Y without knowledge of H? (Assume that receiver knows about the statistics of H only). Explain how bits are mapped to the 16 points if your answer is yes.

6. *CSIT v.s. CSIR* Let the received signal be given by $Y = HX + Z$ where X is the lowpass equivalent transmitted signal, H is the complex Gaussian fading coefficient and Z is the zero-mean complex Gaussian noise. Suppose the receiver has perfect knowledge of H and the transmitter is not capable of changing the transmit power dynamically.

(a) Suppose we have fast fading where the channel fading H_n is i.i.d. at different transmit symbols. Find out the channel capacities with and without CSIT.

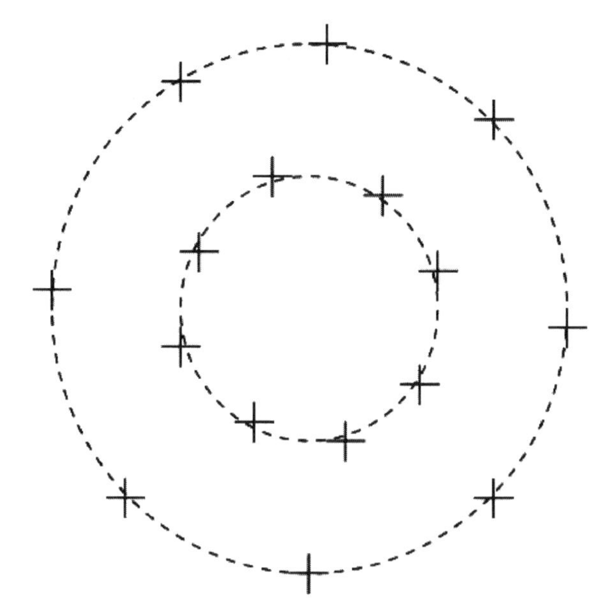

Figure 1. 16-point constellation.

(b) Suppose now, we have slow fading where the channel fading is quasi-static within a frame. Find out the capacity with and without CSIT.

(c) Comparing the answers in (a) and (b), what can be said about the roles of CSIT in fast fading and slow fading?

(*d) In realizing the channel capacity with CSIT in (b), does the transmitter need to inform the receiver about the coding rate per packet?

7. *Temporal and Spatial Water Filling* Recall that adaptation on data rates or power can be done at the transmitter side if full CSIT is assumed. Water filling is one of the well known optimization methods. Given any fading matrix realization h, define B to be the $n_T \times n_T$ eigenvector matrix of h^*h, $\tilde{Q}$ is a $n_T \times n_T$ diagonal matrix with m^* $\min(n_T, n_R)$ non-zero elements given by

$$\hat{q}_{n,n} = \left(\frac{1}{\mu} - \frac{\sigma_z^2}{\lambda_n}\right)^+ \quad \forall n \in [1, m^*]$$

where μ is a constant chosen to satisfy the power constraint $\Sigma_n E\{\hat{q}_{n,n}\} \leq P_0$. Prove that the optimal input covariance Q is of the form $B\tilde{Q}B^*$.

8. *CSIT v.s. CSIR in point-to-point MIMO channel* Consider a MIMO point-to-point system with n_T transmit antennas and n_R receive antennas. Assume the channel is fast fading with respect to an encoding frame and define $m^* = \min(n_T, n_R)$.
   (a) If the system has full CSIR, prove that there are only m^* spatial channels.
   (b) If the system has both perfect CSIT and CSIR, prove that there are m^* spatial channels.
   (c) Prove that the capacity obtained in (b) is larger than that in (a) and based on that, elaborate on the role of CSIT in MIMO channels.

3

MIMO LINK WITH IMPERFECT CHANNEL STATE INFORMATION

3.1 OVERVIEW

In Chapter 2 we discussed the MIMO link capacity with perfect CSI. We discussed the role of CSI at the transmitter and the receiver on MIMO link capacity. In this chapter, we discuss the role of CSI on MIMO link capacity when the knowledge of CSI is not perfect. As we discussed in the last chapter, there are in general two functions of CSI. The availability of CSI at the receiver (CSIR) serves the purpose of detection. On the other hand, the availability of CSI at the transmitter (CSIT) serves the purpose of adaptation. Hence, it is not surprising to see that the requirement on the accuracy of CSIR and CSIT can be quite different.

In practice, to estimate CSIR, a *pilot channel* containing known symbols will be transmitted along with the payload from the transmitter to facilitate the measurement of CSI at the receiver. For MIMO channels, there are $n_R \times n_T$ complex fading coefficients to estimate, and this induces special requirements on the design of the pilot symbols. For finite-power and finite-duration pilot symbols, the CSIR can never be perfectly estimated. On the other hand, a feedback channel is usually needed to convey the instantaneous CSI to the transmitter to realize the CSIT scenario[1] in FDD systems. For perfect CSIT, the requirement on the feedback link capacity to carry the CSI can be very large, especially for MIMO channels because there are $n_R \times n_T$ complex values in

[1] An exception is time-division duplex (TDD) systems where the uplink and downlink are transmitted using the same frequency band. For sufficiently slow mobility, the uplink channel state and the downlink channel state will be symmetric and no feedback will be needed for the transmitter to obtain CSIT.

Channel-Adaptive Technologies and Cross-Layer Designs for Wireless Systems with Multiple Antennas: Theory and Applications. By V. K. N. Lau and Y.-K. R. Kwok
ISBN 0-471-64865-5

the MIMO CSI. Hence, in this chapter, we focus on studying the effects of imperfect CSIR and imperfect CSIT on the MIMO channel capacity. For imperfect CSIR, we refer to the case when there is CSIR estimation error at the receiver due to the limited pilot power in the presence of channel noise. For imperfect CSIT, we refer to the case of *limited feedback*, where the number of bits allocated for CSI feedback is limited. The limited feedback is an important concern because a finite number of bits are allocated for CSI feedback in the physical frames of most practical systems. For example, there are 4 bits allocated for CSI feedback in the DPCCH of UMTS systems. Hence, it is important to know what is the best feedback strategy and the best transmission adaptation strategy given the 4-bit feedback constraint.

In Section 3.2, we focus on the effect of imperfect CSIR on the detection aspects. While there is no closed-form expression for the MIMO channel capacity with imperfect CSIR, we shall introduce a lower bound and an upper bound for the channel capacity. The bounds are shown to be asymptotically tight at high SNR. In Section 3.3, we discuss the effect of imperfect CSIT (limited feedback) for optimizing the SNR. In Section 3.4, we extend the discussion of limited feedback design for optimizing the ergodic capacity. In both cases, we consider the general problem of the optimal partial CSI feedback design and the optimal transmission strategy design when there is limited feedback. Finally, in Section 3.5, we conclude with a brief summary of the main points.

3.2 EFFECT OF IMPERFECT CSI ESTIMATION

In this section, we shall focus on the effect of imperfect CSIR at the receiver on the MIMO link capacity. As we have mentioned, the role of CSI at the receiver is to enhance the signal detection. For this purpose, the requirement on the accuracy of the CSIR is usually quite high. We first discuss the CSI estimation for MIMO channels. Next, we elaborate on the MIMO capacity with imperfect CSIR.

3.2.1 CSI Estimation for MIMO Channels

In general, channel estimation at the receiver is based on pilot channel where the transmitter inserts known pilot symbols in the frame. The $(n_R \times T_p)$-dimensional received symbol $\mathbf{Y}_p$ is given by

$$\mathbf{Y}_p = \mathbf{H}\mathbf{X}_p + \mathbf{Z} \tag{3.1}$$

where T_p is the number of pilot symbols transmitted in the frame, $\mathbf{X}_p$ is the $n_T \times T_p$ transmitted pilot symbols, and $\mathbf{Z}$ is the $n_R \times T_p$ i.i.d. complex Gaussian channel noise with variance σ_z^2. We assume that $\mathbf{H}$ remains quasistatic within the T_p pilot symbols. The loglikelihood function of the observations $\mathbf{Y}_p$ is given by

$$\log p(\mathbf{Y}_p|\mathbf{H}) = A\sum_{n=1}^{T_p}\|\mathbf{Y}_p(n) - \mathbf{H}\mathbf{X}_p(n)\|^2 + B \tag{3.2}$$

where A and B are some constants independent of $\mathbf{H}$ and $\mathbf{Y}_p(n)$ denotes the nth column of the matrix $\mathbf{Y}_p$. Hence, the maximum-likelihood (ML) estimate of the channel matrix $\mathbf{V}$ is given by

$$\mathbf{V} = \mathbf{Y}_p\mathbf{X}_p^*(\mathbf{X}_p\mathbf{X}_p^*)^{-1} = \mathbf{H} + \varepsilon \tag{3.3}$$

where $\varepsilon = \mathbf{Z}\mathbf{X}_p^*(\mathbf{X}_p\mathbf{X}_p^*)^{-1}$ is the channel estimation noise. Note that for $\mathbf{X}_p\mathbf{X}_p^*$ to be invertible, we must have $T_p \geq n_T$ and $\mathbf{X}_p$ is a full rank n_T matrix.

3.2.2 Capacity Bounds of MIMO Link

Consider a block fading MIMO link where the encoding frame spans over ergodic realizations of channel fading. The channel capacity (ergodic capacity) of the MIMO channels with imperfect CSIR is given by

$$C_{\text{csir}} = \max_{p(\mathbf{X})} I(\mathbf{X}; \mathbf{Y}, \mathbf{V}) = \max_{p(\mathbf{X})} I(\mathbf{X}; \mathbf{Y}|\mathbf{V}) \tag{3.4}$$

because $I(\mathbf{X}; \mathbf{V}) = 0$.

Consider an example of block flat fading MIMO channel where $\mathbf{H}_n$ is i.i.d. complex Gaussian with unit variance and $\mathbf{Y}_n = \mathbf{H}_n\mathbf{X}_n + \mathbf{Z}_n$ and $\mathbf{Z}_n$ is i.i.d. complex Gaussian channel noise with covariance $\sigma_z^2\mathbf{I}_{n_R}$. Assume that the CSIR, $\mathbf{V}_n$, is given by (3.3). The equivalent channel model is given by

$$\mathbf{Y}_n = (\mathbf{V}_n - \varepsilon_n)\mathbf{X}_n + \mathbf{Z}_n = \mathbf{V}_n\mathbf{X}_n + \mathbf{Z}'_n \tag{3.5}$$

where $\mathbf{Z}'_n = \mathbf{Z}_n - \varepsilon_n\mathbf{X}_n$ is the equivalent-channel noise. Note that the equivalent-channel noise $\mathbf{Z}'_n$ is non-Gaussian and finding the optimizing input distribution $p(\mathbf{X})$ is nontrivial. Hence, we shall derive an upper bound and lower bound [53,147] of the mutual information $I(\mathbf{X}; \mathbf{Y}|\mathbf{V})$.

Lower Bound. The mutual information $I(\mathbf{X}; \mathbf{Y}|\mathbf{V})$ is given by

$$I(\mathbf{X}; \mathbf{Y}|\mathbf{V}) = H(\mathbf{X}|\mathbf{V}) - H(\mathbf{X}|\mathbf{Y}, \mathbf{V})$$

Choosing $p(\mathbf{X})$ to be complex Gaussian distribution with covariance $\mathbf{Q}$ (which may not be the capacity-achieving distribution because of the non-Gaussian equivalent noise), we have $H(\mathbf{X}|\mathbf{V}) = \log_2|\pi e\mathbf{Q}|$. The second term is upper-bounded by the entropy of a Gaussian random variable $\tilde{\mathbf{X}}$ whose covariance is equal to that of $\mathbf{X}|\mathbf{Y}, \mathbf{V}$. In other words, we have $H(\mathbf{X}|\mathbf{Y}, \mathbf{V} = \mathbf{v}) \leq \log_2|\pi e\tilde{\mathbf{Q}}|$, where $|\tilde{\mathbf{Q}}|$ is given by

$$|\tilde{\mathbf{Q}}| = |\varepsilon[\mathbf{XX}^*|\mathbf{Y}, \mathbf{V} = \mathbf{v}]| = \frac{|\mathbf{Q}|(\sigma_z^2 + \sigma_\varepsilon^2 P_0)}{(\sigma_z^2 + \sigma_\varepsilon^2 P_0)\mathbf{I}_{nR} + \mathbf{vQv}^*} \tag{3.6}$$

where $\sigma_\varepsilon^2 = \varepsilon[|\varepsilon(i,j)|^2]$ is the error variance of the CSIR estimation. Hence, the mutual information is lower-bounded by

$$I(\mathbf{X}; \mathbf{Y}|\mathbf{V}) \geq \varepsilon\left[\log_2\left|\mathbf{I}_{nR} + \frac{\mathbf{vQv}^*}{\sigma_z^2 + \sigma_\varepsilon^2 P_0}\right|\right] \tag{3.7}$$

where the expectation is with respect to $\mathbf{v}$. Comparing with the case for perfect CSIR in Equation (2.32) of Chapter 2, there is a SNR loss factor of $\delta_\gamma = (\sigma_z^2 - \sigma_\varepsilon^2)/(\sigma_z^2 + \sigma_\varepsilon^2 P_0)$.

Upper Bound. To obtain the upper bound, the mutual information is expressed as

$$I(\mathbf{X}; \mathbf{Y}|\mathbf{V}) = H(\mathbf{X}|\mathbf{V}) - H(\mathbf{Y}|\mathbf{X}, \mathbf{V})$$

From (3.5), $\mathbf{Y}$ is a complex Gaussian random variable given $\mathbf{X}$ and $\mathbf{V}$ and the conditional covariance is given by

$$\begin{aligned}|Q(\mathbf{X} = \mathbf{x}, \mathbf{V} = \mathbf{v})| &= |\varepsilon[\mathbf{YY}^*|\mathbf{X} = \mathbf{x}, \mathbf{V} = \mathbf{v}]| \\ &= |\sigma_z^2\mathbf{I}_{nR} + \varepsilon[\varepsilon\mathbf{XX}^*\varepsilon^*]| \\ &= |\sigma_z^2\mathbf{I}_{nT} + \sigma_\varepsilon^2\mathbf{X}^*\mathbf{X}\mathbf{I}_{nT}|.\end{aligned} \tag{3.8}$$

Hence, the second term is given by

$$H(\mathbf{Y}|\mathbf{X}, \mathbf{V} = \mathbf{v}) = \varepsilon_X\left[\log_2\left|\pi e\left(\sigma_z^2 + \sigma_\varepsilon^2\|\mathbf{X}\|^2\right)^{nT}\right|\right] \tag{3.9}$$

On the other hand, the first term is bounded by the entropy of a circular symmetric complex Gaussian random variable with the same covariance. Hence, we have

$$H(\mathbf{Y}|\mathbf{V} = \mathbf{v}) \leq \log_2|\pi e(\mathbf{vQv}^* + (\sigma_z^2 + \sigma_\varepsilon^2 P_0)\mathbf{I}_{nR})| \tag{3.10}$$

Hence, the mutual information is upper-bounded by

$$I(\mathbf{X}; \mathbf{Y}|\mathbf{V}) \leq \varepsilon_V\varepsilon_X\left[\log_2\left|\frac{\mathbf{VQV}^* + (\sigma_z^2 + \sigma_\varepsilon^2 P_0)\mathbf{I}_{nR}}{\left(\sigma_z^2 + \sigma_\varepsilon^2\|\mathbf{X}\|^2\right)^{nT}}\right|\right] \tag{3.11}$$

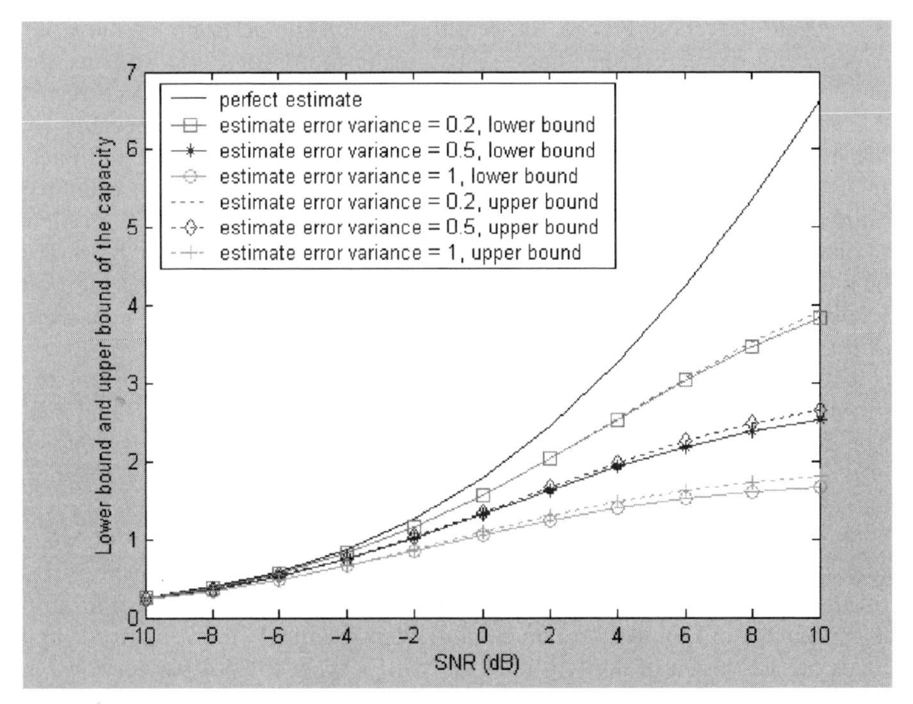

Figure 3.1. Lower bound and upper bound of 4 × 4 MIMO ergodic capacity versus SNR at various CSIR estimation errors σ_ε^2.

Comparing the upper bound and the lower bound, we have

$$C_{upper} = C_{lower} + n_R \varepsilon_X \left[\log_2 \frac{\sigma_z^2 + P_0 \sigma_\varepsilon^2}{\sigma_z^2 + \sigma_\varepsilon^2 \|\mathbf{X}\|^2} \right] \tag{3.12}$$

Observe that the gap between the upper bound and the lower bound is small for Gaussian input unless $n_R \gg n_T$. In the limit of high SNR, the second term in (3.12) approaches $(n_R/n_T)\log_2(\sqrt{e}) \approx 0.72(n_R/n_T)$. Figure 3.1 illustrates the MIMO ergodic capacity versus SNR at various CSIR estimation error σ_ε^2. As observed, both the C_{upper} and C_{lower} saturate as the SNR P_0/σ_z^2 increases. This is because the effective SNR in (3.7) (taking into account of the CSI estimation error) $\gamma_{\text{eff}} = \dfrac{\text{tr}(\mathbf{vQv}^*)}{\sigma_z^2 + \sigma_\varepsilon^2 P_0}$ converges to a constant as P_0/σ_z^2 increases.

3.3 EFFECT OF LIMITED FEEDBACK—OPTIMIZING FOR SNR

In this section, we shall consider the effect of imperfect CSIT on the MIMO link performance. Recall that the role of CSIT is for signal adaptation rather than signal detection. Hence, the requirement on the *accuracy* of the CSIT is generally more relaxing than CSIR. Furthermore, for FDD systems, the method of delivering CSI to the transmitter is through feedback. As a result, the cost associated with CSI feedback is usually quite high, especially for MIMO channels because there are $n_R \times n_T$ complex values to be fed back. In practical systems such as HSDPA and 3G1X, there are usually a limited number of bits in the frame assigned for feedback. Hence, the general problem of optimal imperfect CSI feedback design is very important. As will be shown later in this section, we shall propose a generic framework to find the best feedback strategy as well as the best adaptation strategy under the constraint of limited CSI feedback.

Generally speaking, there are two possible objective functions on the problem of the optimal CSI feedback design and the optimal MIMO adaptation design when the state feedback link has a capacity constraint: the *overall effective SNR* and the *MIMO link capacity*. The former objective function allows easier problem formulation, but the result does not directly imply a higher MIMO link capacity because of the mutual interference between spatial channels. On the other hand, the latter objective function produces results that will directly optimize the MIMO link capacity, but the optimization problem and solution are much more complicated. In this section, we first focus on the design of limited feedback with respect to SNR. In Section 3.4, we elaborate on the design of limited feedback with respect to capacity.

3.3.1 Introduction to Optimizing Effective SNR

It is well known that SNR is an important parameter in a communication system because of its close relationship to channel capacity and system symbol error rate (SER). In this section, we discuss the design of the optimal codebook for vector channel beamforming that achieves highest overall effective SNR and introduce a quantized maximum signal-to-noise ratio (SNR) beamforming technique where the receiver only sends the label of the best beamforming vector in a predetermined codebook to the transmitter. We then show that the codebook design problem is equivalent to the problem of packing onedimensional subspaces known as *Grassmannian* line packing, which takes advantage of the distribution of the optimal unquantized beamforming vector. The connection between Grassmannian line packing and quantized beamforming is important because it allows us to obtain results from the subspace packing literature to find constructive methods for deriving codebooks, which are found to be a function of the number of transmit antennas and the size of the codebook but are independent of the number of receive antennas. We first briefly outline the Grassmannian line packing problem in the next subsection.

3.3.2 Grassmannian Line Packing

Grassmannian line packing is the problem of spacing N lines that pass through the origin in order to maximize the sine of the minimum angular separation between any two lines. Grassmannian line packing forms the mathematical basis for our quantized beamforming codebook design. In the following text, we present a summary of key results of Grassmannian line packing.

Consider the space of unit-norm transmit beamforming vectors $\Omega_m \subset \mathbf{C}^m$ where $\mathbf{C}^m$ is the m-dimensional vector space over the complex field. Define an equivalence relation between two unit vectors $\mathbf{w_1} \in \Omega_m$ and $\mathbf{w_2} \in \Omega_m$ by $\mathbf{w_1} \equiv \mathbf{w_2}$ if for some $\theta \in [0, 2\pi)$ $\mathbf{w_1} = e^{j\theta}\,\mathbf{w_2}$. This equivalence relation says that two vectors are equivalent if they are on the same line in $\mathbf{C}^m$. The complex Grassmannian manifold $\mathcal{G}(m, 1)$ is the set of all one-dimensional subspaces of the vector space C^m. We define a distance function on $\mathcal{G}(m, 1)$ by letting the distance between the two lines generated from unit vectors $\mathbf{w}_1$ and $\mathbf{w}_2$ be the sine of the angle $\theta_{1,2}$ between the two lines. This distance is expressed as

$$d(\mathbf{w_1}, \mathbf{w_2}) = \sin(\theta_{1,2}) = \sqrt{1 - |\mathbf{w_1}^H \mathbf{w_2}|^2} \tag{3.13}$$

The Grassmannian line packing problem consists in finding the set (or packings) of N lines in C^m that has maximum minimum distance between any pairs of lines. Because of the relation to Ω_m, the problem simplifies down to arranging N unit vectors so that the magnitude correlation between any two unit vectors is as small as possible. We represent a packing of N lines in $\mathcal{G}(m, 1)$ by an $m \times N$ matrix $\mathbf{W} = [\mathbf{w_1}\ \mathbf{w_2} \ldots \mathbf{w_N}]$, where $\mathbf{w}_i$ is the vector in Ω_m whose column space is the ith line in the packing. The packing problem is of interest only in nontrivial cases where $N > m$ because when $N \leq m$, the packing lines can be simply picked as N orthogonal basis of Ω_m. The minimum distance of a packing is the sine of the smallest angle between any pair of lines. This is written as

$$\delta(\mathbf{W}) = \min_{1 \leq k \leq l \leq N} \sqrt{1 - |\mathbf{w}_k^H \mathbf{w}_l|^2} = \sin(\theta_{\min}) \tag{3.14}$$

where $\theta_{\min}$ is the smallest angle between any pair of lines in the packing. The Rankin bound gives an upper bound on the minimum distance of line packing as a function of m and $m \leq N$ and is given by

$$\delta(\mathbf{W}) \leq \sqrt{\frac{(m-1)N}{m(N-1)}} \tag{3.15}$$

The problem of finding algorithms to design packings for arbitrary m and N has been studied by many researchers in applied mathematics and information theory [8,28,124]. However, finding the global optimal solution of the minimum distance for arbitrary m and N is not easy either analytically or numerically. For this reason it is often more practical to rely on random computer searches. For example, Figure 3.2 illustrates a tabulation of the optimal search from Reference 123 for the real case. In some cases closed-form solutions are available, such as when $N = 2m = p^{\alpha} + 1$, where p is prime and α is a positive integer, conference matrices allow explicit constructions of packings [124].

$N \backslash m$	3	4	5	6	7	8	9
3	90.0000	90.0000	90.0000	90.0000	90.0000	90.0000	90.0000
4	70.5288[2]	90.0000	90.0000	90.0000	90.0000	90.0000	90.0000
5	63.4349[6]	75.5225[4]	90.0000	90.0000	90.0000	90.0000	90.0000
6	63.4349[3]	70.5288[2]	78.4630[4]	90.0000	90.0000	90.0000	90.0000
7	54.7356[6]	67.0213	73.3689	80.4059[4]	90.0000	90.0000	90.0000
8	49.6399[6]	65.5302	70.8039	76.0578	81.7868[4]	90.0000	90.0000
9	47.9821[6]	64.2619	70.5288	73.8437	78.4630	82.8192[4]	90.0000
10	46.6746[6]	64.2619	70.5288[2]	73.6935	76.3454	79.4704	83.6206[4]
11	44.4031	60.0000	67.2543	71.5651	75.0179	77.8695	80.6204
12	41.8820[6]	60.0000[1]	67.0213	71.5651	74.1734	76.6050	79.4704
13	39.8131	55.4646	65.7319	70.5288	73.8979	76.1645	77.9422
14	38.6824	53.8376	65.7241	70.5288	73.8979[3]	75.0349	77.2382
15	38.1349[6]	52.5016	65.5302	70.5288	71.5678	74.3318	76.5006
16	37.3774[6]	51.8273	63.4349[5]	70.5288[2]	70.9861	74.1005	75.9638
17	35.2353	50.8870	61.2551	68.1088	70.5926	73.1371	75.9638
18	34.4088	50.4577	61.0531	67.3744	70.5527	72.7464	75.9638[3]
19	33.2115	49.7106	60.0000	67.3700	70.5288	72.0756	74.4577
20	32.7071	49.2329	60.0000[6]	67.0996	70.5288	71.6706	74.2278
21	32.2161	48.5479	57.2025	67.0213	70.5288	71.3521	73.7518
22	31.8963	47.7596	56.3558	65.9052[5]	70.5288	71.0983	73.1894
23	30.5062	46.5104	55.5881	63.6744	70.5288	70.7720	72.7488
24	30.1628	46.0478	55.2279	63.6122	70.5288	70.6027	72.6547
25	29.2486	44.9471	54.8891	62.4240	70.5288	70.5490	72.3124
26	28.7126	44.3536	54.2116	61.7377	70.5288	70.5432	72.1763
27	28.2495	43.5530	53.5402	61.4053	70.5288	70.5392	71.6650
28	27.8473	43.1566	53.2602	60.5276	70.5288[2]	70.5322	71.5794
29	27.5244	42.6675	53.0180	60.1344	66.7780	70.5288	71.5175
30	26.9983	42.2651	52.7812	60.0213	65.7563	70.5288	71.5175
31	26.4987	42.0188	52.4120	60.0000	65.1991	70.5288	70.8508
32	25.9497	41.9554	52.3389	60.0000	64.7219	70.5288	70.7437
33	25.5748	41.4577	52.2465	60.0000	64.6231	69.3203	70.6940
34	25.2567	40.9427	51.8537	60.0000	64.6231	69.1688	70.6512
35	24.8702	40.7337	51.8273	60.0000	64.6231	69.0752	70.6337
36	24.5758	40.6325	51.8273	60.0000[1]	64.6231	69.0752[5]	70.5864
37	24.2859	40.4486	51.8273	57.6885	62.3797	67.7827	70.5695
38	24.0886	40.4419	50.3677	57.1057	62.1435	67.3835	70.5571
39	23.8433	39.6797	50.0611	56.8357	61.7057	67.1387	70.5443
40	23.3293	39.0236[5]	49.5978	56.0495	61.3792	66.3815	70.5288
41	22.9915	38.5346	49.2600	55.8202	61.1630	65.9282	70.5288
42	22.7075	38.3094	48.6946	55.6160	60.8232	65.8166	70.5288
43	22.5383	37.7833	48.4030	55.3981	60.5193	65.2885	70.5288
44	22.2012	37.3474	48.0955	55.1259	60.3623	65.2422	70.5288
45	22.0481	37.1198	47.7723	54.9980	60.2282	64.7476	70.5288
46	21.8426	36.9997	47.3753	54.9858	60.1101	64.4007	70.5288
47	21.6609	36.5952	47.0323	54.7356	60.0433	64.1542	70.5288
48	21.4663	36.3585	46.7105	54.5940	60.0116	63.8846	70.5288[2]
49	21.1610	36.1369	46.4345	54.5031	60.0000	63.4849	68.0498
50	20.8922	36.0754	46.1609	54.3191	60.0000[1]	63.1527	67.7426

***Key*:** 1,2: from sphere packings; 3: from conference matrices; 4: diplo-simplex; 5: described below; 6: described in Section 3.3.

Figure 3.2. A table of the best packing [123] found for $N \leq 50$ lines in $\mathcal{G}(m, 1)$ where $m \leq 9$. The values indicate the maximum angle separations.

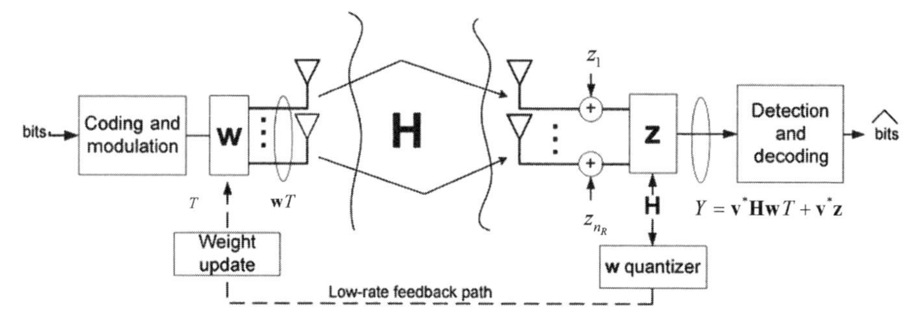

Figure 3.3. Block diagram of a MIMO system for spatial diversity.

3.3.3 Grassmannian Precoding for MIMO Systems—Spatial Diversity

We consider the problem of quantized maximum SNR beamforming for an independent and identically distributed (i.i.d.) MIMO system where beamforming and maximum ratio combining (MRC) are employed at the transmitter using n_T transmit antennas and at the receiver using n_R receive antennas respectively as illustrated in Figure 3.3. Here, the transmitter has access to a low-bandwidth feedback channel from the receiver. In addition, spatial diversity is employed to increase the resilience to fading in the system that we are going to study. Because of the limited capacity of the feedback channel, we assume that the use of a codebook of possible beamforming vectors known to both the transmitter and receiver and the codebook is restricted to having a limited cardinality of Q and is designed offline. To simplify the problem, the feedback channel is assumed to be error-free and zero-delay.

Suppose that the channel is flat fading; thus the discrete-time equivalent channel can be modeled by a $n_R \times n_T$ channel matrix $\mathbf{H}$. Given a transmitted symbol T, the received signal for this system after diversity combining Y is given by

$$Y = \mathbf{v}^*\mathbf{H}\mathbf{w}T + \mathbf{v}^*\mathbf{z} \tag{3.16}$$

The vectors $\mathbf{w}$ and $\mathbf{v}$ are the $n_T \times 1$ beamforming and $n_R \times 1$ combining vectors, respectively. The $n_R \times 1$ noise vector $\mathbf{z}$ has i.i.d. entries distributed according to $N(0, \sigma_z^2)$. We model the channel $\mathbf{H}$ as having i.i.d. entries distributed according to $\mathcal{N}(0, 1)$. The channel is assumed to be known perfectly at the receiver. The symbol energy is given by $E_T[|T|^2] = E_t$.

In a beamforming–combining system, the key question is how to design $\mathbf{w}$ and $\mathbf{v}$ to maximize the SNR, which in turn minimizes the average probability of error and maximizes the capacity. For the proposed system, the SNR γ_r after combining at the receiver is given by

$$\gamma_r = \frac{E_t|\mathbf{v}^H\mathbf{h}\mathbf{w}|^2}{\|\mathbf{v}\|_2^2\sigma_z^2} = \frac{\left(E_t\|\mathbf{w}\|_2^2\right)\left|\frac{\mathbf{v}}{\|\mathbf{v}\|_2}^H \mathbf{h}\frac{\mathbf{w}}{\|\mathbf{w}\|_2}\right|^2}{\sigma_z^2} \tag{3.17}$$

where **h** represents a realization of the channel. Note that in (3.17), the effect of $\|\mathbf{v}\|_2$ cancels each other between the numerator and the denominator. Hence, without loss of generality, we fix $\|\mathbf{v}\|_2 = 1$. Similarly, since the transmitter transmits with total energy $E_t\|\mathbf{w}\|_2^2$, we assume that $\|\mathbf{w}\|_2 = 1$ and the total transmit energy is given by E_t. Using these assumptions, we obtain

$$\gamma_r = \frac{E_r}{\sigma_z^2} = \frac{E_t|\mathbf{v}^H\mathbf{hw}|^2}{\sigma_z^2} = \frac{E_t\Gamma_r}{\sigma_z^2} \tag{3.18}$$

where $\Gamma = |\mathbf{v}^H\mathbf{hw}|^2$ is the effective channel gain.

In a MIMO system both a transmit beamforming vector and a receive combining vector need to be chosen. Areceiver with **v** maximizes $|\mathbf{v}^H\mathbf{hw}|$ given **w** is called a *maximumratio combining* (MRC) receiver. The MRC vector **v** follows from the vector norm inequality

$$|\mathbf{v}^H\mathbf{hw}|^2 \le \|\mathbf{v}\|_2^2\|\mathbf{hw}\|_2^2 \tag{3.19}$$

Since $\|\mathbf{v}\|_2^2 = 1$, thus the MRC vector must satisfy

$$|\mathbf{v}^H\mathbf{hw}|^2 = \|\mathbf{hw}\|_2^2 \tag{3.20}$$

Hence, the MRC vector **v** given **w** is easily seen to be the unit vector $\mathbf{v} = \mathbf{hw}/\|\mathbf{hw}\|_2$. The remaining optimization parameter is **w**. The optimal **w** without any design constraints is given by

$$\mathbf{w} = \arg\max_{\mathbf{x}\in\Omega_{n_T}} \|\mathbf{hx}\|_2 \tag{3.21}$$

where Ω_{n_T} refer to the set of unit vectors in C^{n_T}. However, recall that the channel state information is not available to the transmitter. Instead, there is a low-rate, error-free, zerodelay feedback link for the purpose of conveying the optimal **w** to the transmitter. Since **w** can be any unit vector in the space Ω_{n_T}, it is essential to introduce some method of quantization due to the limited feedback channel. A reasonable solution is to let both the receiver and transmitter use a codebook of Q beamforming vectors. The receiver then quantizes the beamforming vector by selecting the best [according to Equation (3.21)] beamforming vector from the codebook and feeding the index of this vector back to the transmitter via the limited feedback link of capacity $C_{fb} = \log_2 Q$. Unfortunately, it is not obvious which Q vectors should be included in the codebook. So, the remaining design problem is to select an optimal codebook for the beamforming vector **w** given by the following problem.

Problem 3.1 (Maximum SNR) Find the codebook **W** in which $\{\mathbf{w}_1, \mathbf{w}_2, \ldots, \mathbf{w}_Q\}$ denote beamforming vectors and Q represents the size of codebook to maximize the average receiving SNR: $\varepsilon_{\mathbf{H}}[\arg\max_{\mathbf{x}\in\mathbf{W}}\|\mathbf{hx}\|_2]$.

It has been shown [31,130] that an optimal beamforming vector for the maximum ratio transmission (MRT) systems is the dominant right singular vector of $\mathbf{H}$ with $\mathbf{H}$ defined as in the beginning of this section. The optimal vector is given by

$$\mathbf{w} = \arg \max_{\mathbf{x} \in \Omega_{nT}} \|\mathbf{x}^H \mathbf{h}^H \mathbf{h}\mathbf{x}\|^2 \tag{3.22}$$

Note that the optimal solution in this equation is not unique. For instance, if $\mathbf{w}$ is an optimal solution, then $e^{j\phi}\mathbf{w}$ is also an optimal solution because $|\mathbf{w}^H\mathbf{h}^H\mathbf{h}\mathbf{w}|^2 = |e^{-j\phi}\mathbf{w}^H\mathbf{h}^H\mathbf{h}e^{j\phi}\mathbf{w}|^2$. Hence, we can deduce that if $\mathbf{w} \equiv \tilde{\mathbf{w}}$ (using the equivalence relation defined in Section 3.3.2), then $\mathbf{w}$ and $\tilde{\mathbf{w}}$ are both optimal solutions and will provide the same SNR performance. In other words, all the optimal beamforming vectors $\mathbf{w}$ form a line over C^{n_T}. This result is summarized in the following lemma [93].

Lemma 3.1 (Isotropic Beamforming Vectors) The line generated by the optimal beamforming vectors for a MIMO Rayleigh fading channel is an isotropically oriented line in C^{n_T} passing through the origin.

Therefore, the problem of the quantized transmit beamforming in a MIMO communication system reduces to quantizing an isotropically oriented line in C^{n_T} and the codebook designing problem is equivalent to the Grassmannian line packing problem with the following criterion.

Grassmannian Beamforming Criterion. Design the set of codebook vectors $\{\mathbf{w}_i\}_{i=1}^{Q}$ such that the corresponding codebook matrix $\mathbf{W}$ maximizes

$$\delta(\mathbf{W}) = \max_{1 \le k \le l \le Q} \sqrt{1 - |\mathbf{w}_k^H \mathbf{w}_l|^2} \tag{3.23}$$

Figure 3.4 summarizes the codebook search results [124] at various n_T values.

Conclusion. From the preceding discussion above, we can summarize that the MIMO beamforming codebook can be designed offline using the Grassmannian line packing method mentioned in Section 3.3.2. Within the online process, the optimal beamforming vector is chosen from the codebook $\mathbf{W}$, based on $\mathbf{w}_{\text{opt}} = \arg \max_{\mathbf{x} \in \mathbf{W}} \|\mathbf{h}\mathbf{x}\|_2$. Index of the optimal vector is fed back to the transmitter. Figure 3.5 provides further intuition into the beamforming codebook quality by showing a plot of the 4×2 MIMO capacity using unquantized beamforming and three different types of Grassmannian beamforming systems (codebook size $Q = 64$, $Q = 16$, and selection diversity). The system with $Q = 64$ provides approximately a 1.5 dB SNR gain compared to selection diversity (SDT) and a 0.5 dB SNR gain compared to the system with $Q = 16$.

Packings \ Case	$n_T = 2, Q = 2$ (1 bit)	$n_T = 2, Q = 4$ (2 bits)	$n_T = 2, Q = 8$ (3 bits)	$n_T = 3, Q = 4$ (2 bits)
w_1	(1,0)	(-0.1612 - 0.7348j, -0.5135 - 0.4128j)	(0.8393 - 0.2939j, -0.1677 + 0.4256j)	(0.5774, 0.5774, 0.5774)
w_2	(0,1)	(-0.0787 - 0.3192j, -0.2506 + 0.9106j)	(-0.3427 + 0.9161j, 0.0498 + 0.2019j)	(0.5774j, -0.5774, -0.5774j)
w_3	N/A	(-0.2399 + 0.5985j, -0.7641 - 0.0212j)	(-0.2065 + 0.3371j, 0.9166 + 0.0600j)	(-0.5774, 0.5774, -0.5774)
w_4	N/A	(-0.9541, 0.2996)	(0.3478 - 0.3351j, 0.2584 + 0.8366j)	(-0.5774j, -0.5774, 0.5774j)
w_5	N/A	N/A	(0.1049 + 0.6820j, 0.6537 + 0.3106j)	N/A
w_6	N/A	N/A	(0.0347 - 0.2716j, 0.0935 - 0.9572j)	N/A
w_7	N/A	N/A	(-0.7457 + 0.1181j, -0.4553 - 0.4719j)	N/A
w_8	N/A	N/A	(-0.7983 + 0.3232j, 0.5000 + 0.0906j)	N/A

Figure 3.4. Table of generated codebook based on Grassmannian search for $n_T = 2,3$ and $Q = 2,4,8$.

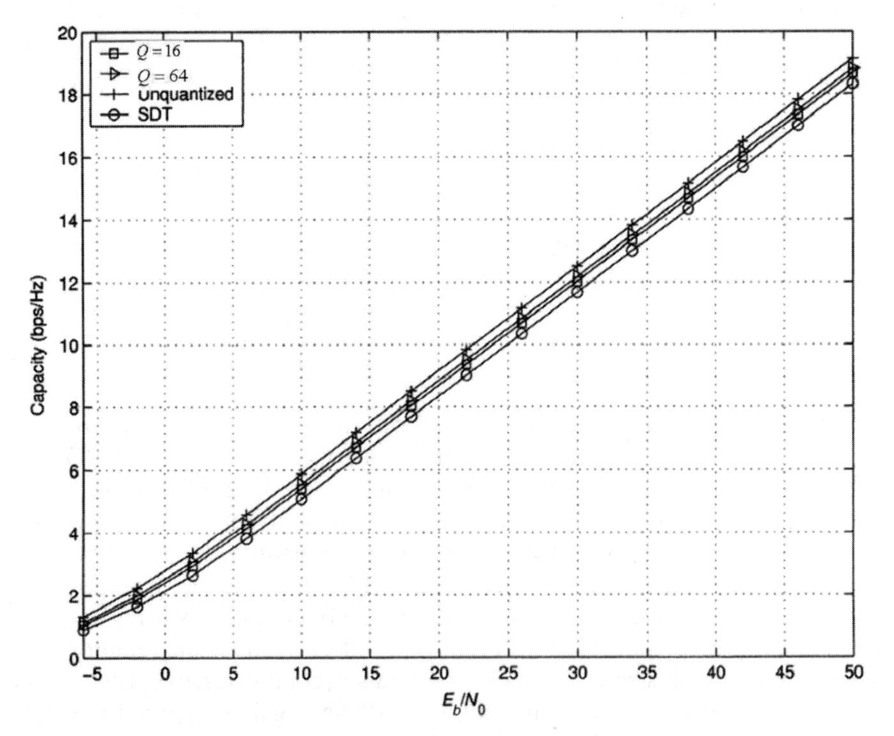

Figure 3.5. Capacity comparison of unquantized beamforming with three Grassmannian beamforming schemes [93] for a 4 × 2 MIMO system.

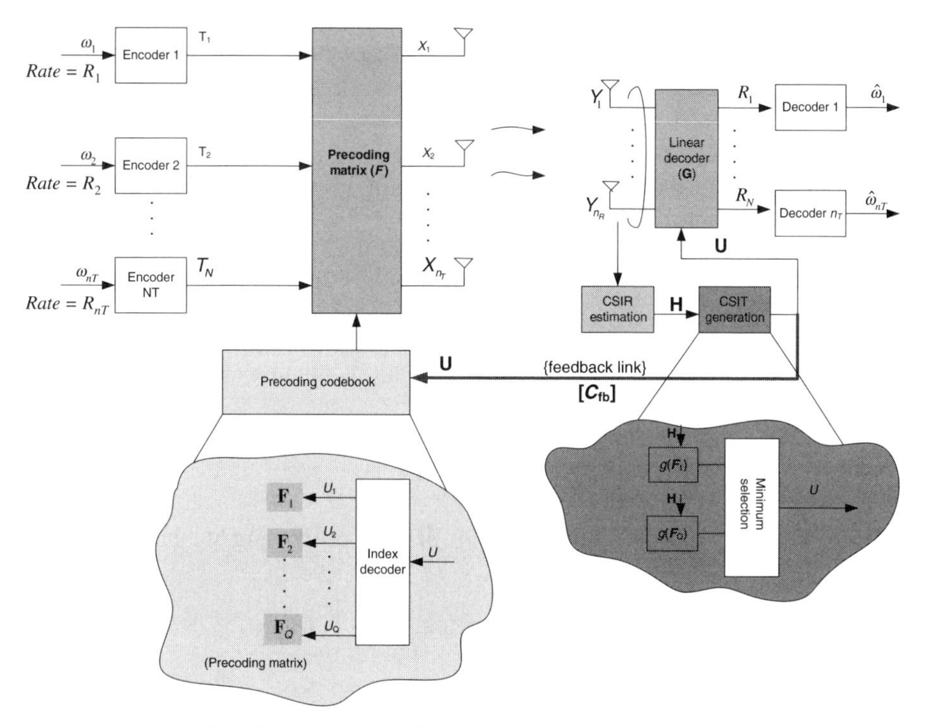

Figure 3.6. Block diagram of MIMO systems with spatial multiplexing.

3.3.4 Grassmannian Precoding for MIMO Systems—Spatial Multiplexing

In previous sections, we illustrated the optimization of effective SNR for MIMO system (spatial diversity) and that the optimization problem is equivalent to the Grassmannian line packing problem. In this section, we extend the Grassmannian precoding method to the spatial multiplexing of MIMO systems.

Figure 3.6 illustrates the block diagram of a MIMO link with spatial multiplexing. A bitstream is fed into a vector encoder and modulator block. The vector encoder produces N different symbol streams. Each of the N bitstreams is modulated independently using the same constellation. This yields a symbol vector $\mathbf{T} = [T_1 T_2 \ldots T_N]^T$. For convenience, we assume $\varepsilon[\mathbf{TT}^*] = \mathbf{I}_N$, where $\mathbf{I}_N$ is the $N \times N$ identity matrix. The symbol vector $\mathbf{T}$ is then precoded by an $n_T \times N$ matrix $\mathbf{F}$ yielding a $n_T \times 1$ transmit vector $\mathbf{X} = \sqrt{\frac{E_s}{N}}\mathbf{FT}$, where E_s is the total transmit energy over all transmit antennas and $n_T \geq N$ is the number of transmit antennas. We restrict $\mathbf{F} \in \mathcal{U}(n_T, N)$, the set of all $n_T \times N$ matrices with orthonormal columns. The $n_R \times 1$ received signal vector $\mathbf{Y}$ is given by

$$\mathbf{Y} = \sqrt{\frac{E_s}{N}}\mathbf{HFT} + \mathbf{Z} \tag{3.24}$$

where $\mathbf{Z}$ is the $n_R \times 1$ AWGN with covariance $\sigma_z^2 \mathbf{I}_{n_R}$.

At the receiver, we assume linear processing in which the receiver obtains an N received vector $\mathbf{R}$ by

$$\mathbf{R} = \mathbf{GY} \tag{3.25}$$

where $\mathbf{G}$ is the $N \times n_R$ linear matrix. The observations of the N substreams are contained in the N components of $\mathbf{R}$ and are decoded separately. For *zero-forcing* processing, we have

$$\mathbf{G}_{ZF} = \mathbf{HF}^{\dagger}$$

where † denotes the pseudoinverse. For *MMSE* processing, we have

$$\mathbf{G}_{\text{MMSE}} = \left[\mathbf{F}^*\mathbf{H}^*\mathbf{HF} + \frac{\sigma_z^2 N}{E_s}\mathbf{I}_N\right]^{-1}\mathbf{F}^*\mathbf{H}^*$$

In the systems considered, the matrix $\mathbf{F}$ is designed at the receiver and sent back to the transmitter over a limited capacity feedback link. In this case, the limited feedback capacity requires the use of a codebook with Q entries $\mathcal{F} = \{\mathbf{F}_1, \mathbf{F}_2, \ldots, \mathbf{F}_Q\}$ of possible matrices that is known to both the transmitter and receiver. The receiver chooses an optimal precoding matrix from $\mathcal{F}$ using some performance criteria such as *minimum singular value* criterion (MSV), *minimum mean-squared error* criterion (MMSE), and *capacity selection* criterion (CS). These criteria are elaborated below [94].

Minimum Singular Value. It is shown [94] that the SNR of the kth substream for zeroforcing is given by

$$\text{SNR}_k^{(\text{ZF})} = \frac{E_s}{N\sigma_z^2[\mathbf{F}^*\mathbf{H}^*\mathbf{HF}]_{k,k}^{-1}} \tag{3.26}$$

The SNR of the kth substream for MMSE is given by

$$\text{SNR}_k^{(\text{MMSE})} = \frac{E_s}{N\sigma_z^2\left[\mathbf{F}^*\mathbf{H}^*\mathbf{HF} + \frac{\sigma_z^2 N}{E_s}\mathbf{I}_N\right]_{k,k}^{-1}} \tag{3.27}$$

where $\mathbf{A}_{k,k}$ denotes the (k, k)th entry of the matrix $\mathbf{A}$. Since the symbol error probability is upper-bounded by the minimum SNR over the N substreams, we would like to express the minimum SNR in terms of a single expression. For ZF receiver, we have

$$\begin{aligned}\max_{k\in[1,N]}[\mathbf{F}^*\mathbf{H}^*\mathbf{H}\mathbf{F}]^{-1}_{k,k} &= \max_{k\in[1,N]} \mathbf{e}_k^*[\mathbf{F}^*\mathbf{H}^*\mathbf{H}\mathbf{F}]^{-1}\mathbf{e}_k \\ &\le \max_{\mathbf{z}:\|\mathbf{z}\|^2=1} \mathbf{z}^*[\mathbf{F}^*\mathbf{H}^*\mathbf{H}\mathbf{F}]^{-1}\mathbf{z} \\ &= \lambda_{\min}^{-2}\{\mathbf{H}\mathbf{F}\}\end{aligned}$$

where $\lambda_{\min}^{-2}\{\mathbf{A}\}$ is the minimum eigenvalue of $\mathbf{A}$ and $\mathbf{e}_k$ is the kth column of $\mathbf{I}_N$. Hence, the minimum SNR is given by

$$\mathrm{SNR}_{\min}^{(\mathrm{ZF})} = \min_{k\in[1,N]} \mathrm{SNR}_k^{(\mathrm{ZF})} \ge \lambda_{\min}^{2}\{\mathbf{H}\mathbf{F}\}\frac{E_s}{N\sigma_z^2} \tag{3.28}$$

Similarly, it can be shown that $\mathrm{SNR}_{\min}^{(\mathrm{MMSE})}$ is related to $\lambda_{\min}^{-2}\{\mathbf{H}\mathbf{F}\}$ in a similar way. Therefore, we have the *minimum singular value criterion* given by the following problem.

Problem 3.2 (Minimum Singular Value Criterion) Pick $\mathbf{F}$ such that

$$\mathbf{F} = \arg\max_{\mathbf{F}_i\in\mathcal{F}} \lambda_{\min}\{\mathbf{H}\mathbf{F}_i\} \tag{3.29}$$

Then

Minimum Mean-Squared Error Criterion. Pick $\mathbf{F}$ such that

$$\mathbf{F} = \arg\max_{\mathbf{F}_i\in\mathcal{F}} \mathcal{M}\left(\overline{\mathrm{MSE}}(\mathbf{F}_i)\right) \tag{3.30}$$

where $\mathcal{M}(\cdot)$ is either the trace or determinant of a matrix and $\overline{\mathrm{MSE}}(\mathbf{F})$ is the mean squared error given by

$$\overline{\mathrm{MSE}}(\mathbf{F}) = \frac{E_s}{N}\left(\mathbf{I}_N + \frac{E_s}{N\sigma_z^2}\mathbf{F}^*\mathbf{H}^*\mathbf{H}\mathbf{F}\right)^{-1} \tag{3.31}$$

Capacity Selection Criterion. Pick $\mathbf{F}$ such that

$$\mathbf{F} = \arg\max_{\mathbf{F}_i\in\mathcal{F}} C(\mathbf{F}_i) \tag{3.32}$$

where

$$C(\mathbf{F}_i) = \log_2\left(\left|\mathbf{I}_N + \frac{E_s}{N\sigma_z^2}\mathbf{F}^*\mathbf{H}^*\mathbf{H}\mathbf{F}\right|\right) \tag{3.33}$$

is the channel capacity. In Section 3.4, we shall elaborate further on the capacity aspects with limited feedback.

To derive the optimal precoder matrix **F**, we will use the singular value decomposition of **H**:

$$\mathbf{H} = \mathbf{V}_L \Sigma \mathbf{V}_R^* \tag{3.34}$$

It can be proved that the optimal precoder matrix **F** for all three criteria is given by $F = \overline{\mathbf{V}}_R$, where $\overline{\mathbf{V}}_R$ is a matrix constructed from the first N columns of $\mathbf{V}_R$ [94]. However, since we have a limited feedback constraint, we have to optimize the *codebook* $\mathcal{F}$ instead of **F** itself. We will attempt to minimize the degradation in channel power introduced by quantizing the precoder $\|\mathbf{HF}_{\text{opt}}\|_F^2 - \|\mathbf{HF}\|_F^2$, where $\|\cdot\|_F$ is the Frobenius matrix norm. This is equivalent to design our codebook $\mathcal{F}$ by attempting to minimize the distortion metric

$$E_{\overline{\mathbf{V}}_R}\left[\min_{i\in\{1,2\ldots Q\}} \frac{1}{2}\| \overline{\mathbf{V}}_R \overline{\mathbf{V}}_R^* - \mathbf{F}_i \mathbf{F}_i^* \|_F^2\right] \tag{3.35}$$

To understand how to design $\mathcal{F}$, we must understand some common properties of finite subsets of $\mathcal{U}(n_T, N)$. Matrices in $\mathcal{U}(n_T, N)$ can be characterized as representing N-dimensional subspaces of the complex n_T-dimensional vector space. Thus we will adopt the common Grassmannian packing notation and define the set of all column spaces of the matrices in $\mathcal{U}(n_T, N)$ to be the complex Grassmannian space $\mathcal{G}(n_T, N, \mathbf{C})$. Thus, if $\mathbf{F}_1, \mathbf{F}_2 \in \mathcal{U}(n_T, N)$, then the column spaces of $\mathbf{F}_1$ and $\mathbf{F}_2$, $\mathcal{P}_1$ and $\mathcal{P}_2$ respectively, are contained in $\mathcal{G}(n_T, N, \mathbf{C})$. The *chordal distance* between two subspaces $\mathcal{P}_1$ and $\mathcal{P}_2$ is defined as

$$d(\mathbf{F}_1, \mathbf{F}_2) = \frac{1}{\sqrt{2}}\|\mathbf{F}_1\mathbf{F}_1^* - \mathbf{F}_2\mathbf{F}_2^*\|_F \tag{3.36}$$

A precoding codebook $\mathcal{F}$ represents a packing in $\mathcal{G}(n_T, N, \mathbf{C})$, $S = \{\mathcal{P}_1, \mathcal{P}_2, \ldots, \mathcal{P}_Q\}$, where $\mathcal{P}_i$ is the column space of $\mathbf{F}_i$. A packing can be described by the minimum distance:

$$\delta = \min_{i,j\in[1,N]} d(\mathbf{F}_i, \mathbf{F}_j) \tag{3.37}$$

It can be shown that the distortion metric in Equation (3.35) is a decreasing function of the minimum distance in Equation (3.37). Hence, the distortion metric is minimized by maximizing δ.

We have now established that designing low-distortion codebooks is equivalent to packing subspaces in the Grassmannian space using the chordal distance metric. Finding good Grassmannian packings for arbitrary n_T, N, and Q is actually quite difficult. In general, iterative optimization algorithms

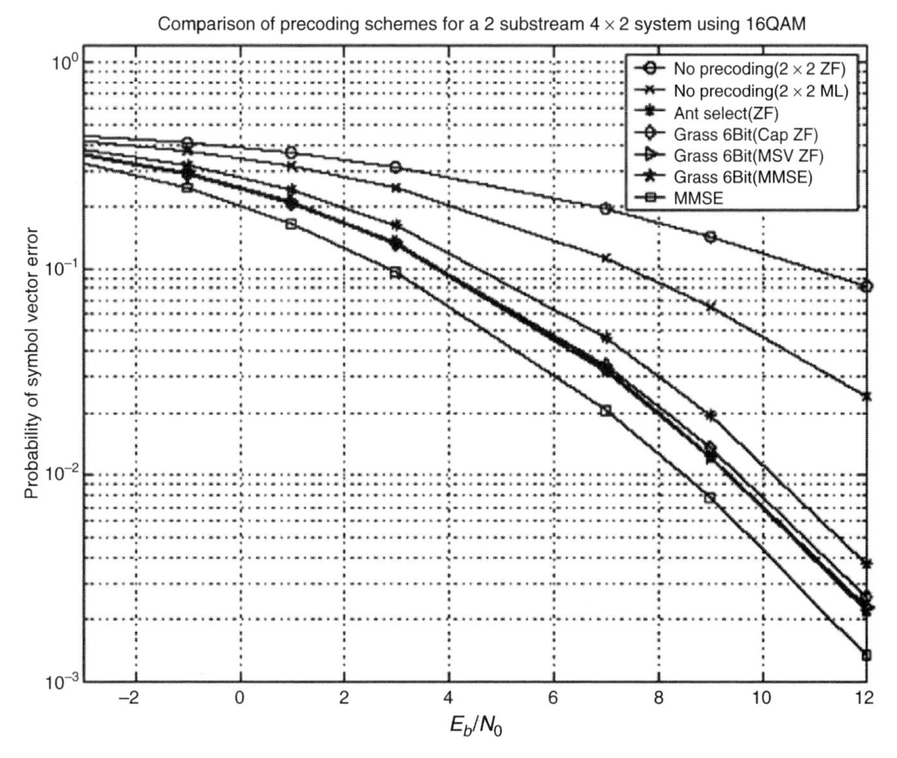

Figure 3.7. Symbol error probability of 4 × 2 MIMO link based on *chordal distance* precoding with 6 bits of feedback [94].

or random searches must be employed to find good packings [28]. Once the codebook has been designed, the receiver chooses the codebook $\mathbf{F} \in \mathcal{F}$ such that

$$\mathbf{F} = \arg\max_{\mathbf{F}_i \in \mathcal{F}} g(\mathbf{F}_i) \tag{3.38}$$

where $g(\cdot)$ is the cost function to be maximized for the chosen selection criterion (MSV, MSE, capacity). The choice of $\mathbf{F}$ can then be sent to the transmitter via a $C_{\text{fb}} = \log_2 Q$ bit representation as illustrated in Figure 3.6.

Figure 3.7 illustrates the 16-QAM symbol error probability of the 4 × 2 MIMO link based on *chordal distance* precoding with 6 bits of feedback using MSV, MMSE, and capacity criteria. All three selection criteria give approximately the same results. Each provides approximately a 1-dB improvement over antenna selection and lags behind the unquantized MMSE performance by 1 dB.

3.4 EFFECT OF LIMITED FEEDBACK—OPTIMIZING FOR ERGODIC CAPACITY

In order to derive the optimal partial CSI feedback and the optimal MIMO adaptation schemes, we have to first express the MIMO capacity in terms of general CSI feedback and adaptation strategies and then optimize the capacity with respect to the CSI feedback and adaptation algorithms. To derive the MIMO link capacity, we make use of the general CSIT and CSIR model described in Section 2.2.2 in Chapter 2. Specifically, the receiver has knowledge of CSIR $\mathbf{V}$, which is statistically related to the actual channel state $\mathbf{H}$ through $p(\mathbf{V}|\mathbf{H})$. The transmitter has knowledge of CSIT U, which is statistically related to the CSIR $\mathbf{V}$ through $p(U|\mathbf{V})$. The CSI feedback link to convey the CSIT U is assumed to be noiseless but has a capacity constraint C_{fb}; that is, the number of bits that can be fed back to the transmitter is limited to C_{fb} bits per fading symbol.

3.4.1 Channel Capacity with Partial CSIT

Consider a block fading channel with N fading blocks and L symbols within each fading block. The channel transition probability is given by $p(\mathbf{y}_1, \ldots, \mathbf{y}_N|\mathbf{x}_1, \ldots, \mathbf{x}_N, \mathbf{h}_1, \ldots, \mathbf{h}_N) = \Pi_n p(\mathbf{y}_n|\mathbf{x}_n, h_n)$, where $\mathbf{X}_n = [\mathbf{X}_{1,n}, \ldots, \mathbf{X}_{L,n}]$ and $\mathbf{Y}_n = [\mathbf{Y}_{1,n}, \ldots, \mathbf{Y}_{L,n}]$ are the L-tuple transmit and receive symbols, respectively, during the nth fading block. The channel state sequence is i.i.d. given by $p(\mathbf{h}) = \Pi_{n=1}^{N} p(\mathbf{h}_n)$. Using a similar argument of *two-stage encoding* in Chapter 2, the channel encoder can be split into a first-stage encoder (which is independent of the CSIT U) together with a second-stage *adaptive device* that selects the transmit symbol on the basis of the CSIT U. Given any specific form of feedback strategy $p(U|\mathbf{V})$ and partial CSIR $p(\mathbf{V}|\mathbf{H})$, the equivalent channel model is as illustrated in Figure 3.8 with channel inputs $\{\mathbf{T}_n\}$ and channel outputs $\{\mathbf{Y}_n, \mathbf{V}_n\}$.

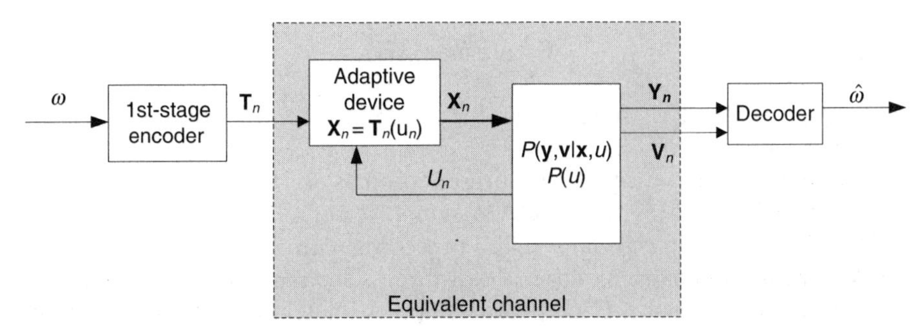

Figure 3.8. Equivalent-channel model of the communication system with partial CSIT U and partial CSIR $\mathbf{V}$.

Note that the equivalent channel is memoryless with the channel transition probability given by

$$\begin{aligned} p(\mathbf{y}_n, \mathbf{v}_n | \mathbf{t}_n) &= \sum_{u_n} p(u_n) p(\mathbf{y}_n, \mathbf{v}_n | \mathbf{t}_n, u_n) \\ &= \sum_{u_n} p(u_n) \sum_{\mathbf{h}_n} p(\mathbf{h}_n, \mathbf{v}_n | u_n) p(\mathbf{y}_n | \mathbf{t}_n, u_n, \mathbf{h}_n) \\ &= \sum_{u_n} \sum_{\mathbf{h}_n} p_{\mathbf{H},U,\mathbf{V}}(\mathbf{h}_n, u_n, \mathbf{v}_n) \prod_{l=1}^{L} p_{\mathbf{Y}|\mathbf{X},\mathbf{H}}(\mathbf{y}_{n,l} | t_{n,l}(u_n), \mathbf{h}_n) \end{aligned}$$

Therefore, using techniques similar to those in Chapter 2, the forward channel capacity is given by

$$C_{\text{fwd}}(L) = \frac{1}{L} \max_{p(\mathbf{T})} I(\mathbf{T}; \mathbf{Y} | \mathbf{V}) \tag{3.39}$$

where $\mathbf{T}$ and $\mathbf{Y}$ are the L-tuple input symbols and output symbols of the *derived channel* in a fading block similar to that described Section 2.6.

In case the channel is memoryless within fading blocks and we have perfect CSIR, that is $p(\mathbf{y}_n|\mathbf{x}_n, \mathbf{h}_n) = \Pi_{l=1}^{L} p(\mathbf{y}_{l,n}|\mathbf{x}_{l,n}, \mathbf{h}_n)$ and $\mathbf{V} = \mathbf{H}$, the channel capacity can be further simplified as

$$C_{\text{fwd}} = \max_{p(T)} I(T; \mathbf{Y} | \mathbf{H}) \tag{3.40}$$

In fact, the channel capacity of $C_{\text{fwd}}(L)$ in (3.39) approaches C_{fwd} in (3.40) as $L \to \infty$ because at very large block length L, perfect CSIR can be obtained with negligible overhead of pilot symbols.

3.4.2 Coding Theorem with Partial CSIT

We have derived the general forward channel capacity formula for channels with partial CSIT and partial CSIR. The next step is to optimize the channel capacity with respect to both the *block transmission strategy* $\mathbf{T}$ and the feedback strategy $p(U|\mathbf{V})$. The feedback design means that we design the channel state feedback $\mathbf{U}$ on the basis of the channel state estimate $\mathbf{V}$. This implies that $\mathbf{H} \to \mathbf{V} \to \mathbf{U}$ forms a Markov chain. The relation among $\mathbf{H}$, $\mathbf{V}$, and $\mathbf{U}$ is $p(\mathbf{h}, \mathbf{v}, \mathbf{u}) = \Pi_n p(\mathbf{h}_n) p(\mathbf{v}_n|\mathbf{h}_n) p(\mathbf{u}|\mathbf{v})$. Yet, before we proceed with the optimization, two important issues need to be addressed.

Optimizing CSIT Feedback Strategy. Note that it is not clear whether an i.i.d. relation between U_n and $\mathbf{V}_n$ is optimal when we have the freedom to design the feedback strategy. Furthermore, it is not clear whether the relationship between the CSIT U_n and the CSIR $\mathbf{V}_n$ has to be deterministic or random. However, in the coding theorem (Theorem 3.1), we shall illustrate that an i.i.d. and deterministic relation $p(\mathbf{u}|\mathbf{v}) = \Pi_n \delta(u_n - h(\mathbf{v}_n))$ achieves the capacity.

Optimizing Transmission Strategy. The transmission strategy design means that we have to design an adaptive channel encoder f such that

$$\mathbf{X}_n = f(\omega, U_1, \ldots, U_n)$$

where ω is the message index and $(U_1, \ldots, U_n)$ are the block of causal CSIT feedback. Note that when we design the adaptive channel encoder, it is not clear whether the transmitted signal $\mathbf{X}_n$ is related to the past state feedback $[U_1, \ldots, U_{n-1}]$. The coding theorem below shows that a simple two-stage memoryless encoding structure similar to that discussed in Section 2.6 of Chapter 2 is sufficient to achieve the capacity in channels with general partial CSIT and partial CSIR.

The following mutual information identity is useful in the proof of the lemma and the coding theorem. The unconditional mutual information $I(A; B)$ and the conditional mutual information $I(A; B|C)$ is related by the identity

$$I(A; B) = I(A; B|C) + I(A; C) - I(A; C|B) \tag{3.41}$$

This identity is obtained by expanding $I(B, C; A)$ using the chain rule of mutual information:

$$\begin{aligned} I(B, C; A) &= I(B; A) + I(C; A|B) \\ &= I(C; A) + I(B; A|C) \end{aligned}$$

To prove the coding theorem, we first prove a simple lemma. The further implication of the lemma is discussed in the remarks following the theorem.

Lemma 3.2 (Deterministic Feedback) For every statistical relation $p(u|\mathbf{v})$, we can always find a deterministic relation $u = h(\mathbf{v})$ such that

$$I(\mathbf{T}; \mathbf{Y}|\mathbf{V}) = I(\mathbf{T}; \mathbf{Y}|U, \mathbf{V}) - I(U; \mathbf{T}|\mathbf{Y}, \mathbf{V}) \tag{3.42}$$

$$\leq \sum_{\mathbf{v}} p(\mathbf{v}) I(\mathbf{T}; \mathbf{Y}|u = h(\mathbf{v}), \mathbf{v}) \tag{3.43}$$

Proof Employing the identity (3.41), we have

$$\begin{aligned} I(\mathbf{T}; \mathbf{Y}|\mathbf{V}) &= I(\mathbf{T}; \mathbf{Y}|U, \mathbf{V}) + I(U; \mathbf{T}|\mathbf{V}) - I(U; \mathbf{T}|\mathbf{Y}, \mathbf{V}) \\ &= I(\mathbf{T}; \mathbf{Y}|U, \mathbf{V}) - I(U; \mathbf{T}|\mathbf{Y}, \mathbf{V}) \end{aligned} \tag{3.44}$$

where we have used the fact that $I(U; \mathbf{T}|\mathbf{V}) = 0$ because U and $\mathbf{T}$ are independent given $\mathbf{V}$.

The first term of (3.44) can be upper-bounded as

$$I(\mathbf{T};\mathbf{Y}|U,\mathbf{V}) = \sum_{v} p(\mathbf{v}) \sum_{u} p(u|\mathbf{v}) I(\mathbf{T};\mathbf{Y}|u,\mathbf{v})$$
$$\leq \sum_{v} p(\mathbf{v}) \max_{u} I(\mathbf{T};\mathbf{Y}|u,\mathbf{v})$$

If $u = h(\mathbf{v})$, $I(U;\mathbf{T}|\mathbf{Y},\mathbf{V}) = 0$. Therefore, $u = h(\mathbf{v}) \triangleq \text{argmax}_u\, I(\mathbf{T};\mathbf{Y}|u,\mathbf{v})$ simultaneously maximizes the first term of (3.44) and minimizes the second term of (3.44).[2]

The following coding theorem gives the channel capacity when we have the freedom to design feedback and transmission strategies. The achievability part shows how to achieve the capacity with a deterministic memoryless feedback function and a deterministic memoryless device at the transmitter to select the transmitted symbol from a codeword with an expanded alphabet. Let $\mathcal{E}_{\mathbf{v}\in\mathcal{H}}[g(\mathbf{v})]$ denote $\Sigma_{\mathbf{v}\in\mathcal{H}} p(\mathbf{v})g(\mathbf{v})$.

Theorem 3.1 (Coding Theorem with Partial CSIT) Consider a discrete memoryless block fading MIMO channel with block memory L having input $\mathbf{X}_1^L \in \mathcal{X}^L$ and output $\mathbf{Y}_1^L \in \mathcal{Y}^L$, $\mathbf{V} \in \mathcal{V}$. The channel transition probability is $p(\mathbf{y}_1^L, \mathbf{v}|\mathbf{x}_1^L) = p(\mathbf{v})p(\mathbf{y}_1^L|\mathbf{x}_1^L, \mathbf{v})$, that is, the input $\mathbf{X}_1^L$ and the output $\mathbf{V}$ are independent. For the nth channel use, the transmitter receives a delayless channel state feedback $U_n \in \mathcal{U}$ from the receiver. The cardinality constraint is $|\mathcal{U}| = Q = 2^{C_{fb}}$. The feedback $U_1^n = [U_1, \ldots, U_n]$ is related to the channel output $\mathbf{V}_1^n = [\mathbf{V}_1, \ldots, \mathbf{V}_n]$ through a causal relation $p(u_1^n|\mathbf{v}_1^n)$. The channel output $\mathbf{Y}_{1,n}^{L,n}$ is conditionally independent of the sequence $\mathbf{U} = \{U_1, U_2, \ldots\}$ given $\mathbf{V}_n$ and $\mathbf{X}_{1,n}^{L,n}$. When the relation $p(u_1^n|\mathbf{v}_1^n)$ can be designed and we count a vector channel use as L channel uses, the capacity of this channel (bits per channel use) is

$$C = \frac{1}{L} \max_{\mathcal{H}_1^Q} \sum_{q=1}^{Q} \max_{p(\mathbf{X}_1^L|u_q)} \mathcal{E}_{\mathbf{v}\in\mathcal{R}_q}[I(\mathbf{X}_1^L;\mathbf{Y}_1^L|u_q,\mathbf{v})] \tag{3.45}$$

where $\mathcal{H}_1^Q$ is a partition of the CSIR space $\mathcal{V}$. The partition is related to function $u = h(\mathbf{v})$ through $\mathcal{R}_q = \{\mathbf{v}: h(\mathbf{v}) = u_q\}$, where $u_q \in \mathcal{U}$. The maximizations are over all the possible partitions, or equivalently all the possible deterministic memoryless functions $u = h(\mathbf{v})$, and distributions $\{p(\mathbf{x}_1^L|u_q), q = 1, \ldots, Q\}$.

Proof (Achievability) Let $h^*(\mathbf{v})$ be the function corresponding to the optimal partition in (3.45). Let the feedback be $U_n = h^*(\mathbf{V}_n)$. This reduces the channel to a memoryless channel with the transmitter-side information. This case is studied in References 120 and 23 as well. A memoryless device

[2] The upper bound can also be proved using the convexity of the mutual information as a function of the channel transition probability $p(\mathbf{y}|\mathbf{v},\mathbf{t}) = \Sigma_u p(u|\mathbf{v})p(\mathbf{y}|u,\mathbf{v},\mathbf{t})$ [30].

$\{\mathbf{X}_n = \mathbf{T}_n(U_n), \mathbf{T}_n \in \mathcal{X}^{L \times |\mathcal{U}|}\}$ produces the transmitted symbol $\mathbf{X}_n$ according to current feedback U_n. The equivalent channel $p(\mathbf{y}, \mathbf{v}|\mathbf{t})$ is a discrete memoryless channel (DMC) without side information. The capacity can be achieved according to the DMC coding theorem [30]. The codebook is constructed according to $p(\mathbf{t}(u_q)) = p^*_{\mathbf{X}|U}(\mathbf{t}(u_q)|u_q)$, where $p^*_{\mathbf{X}|U}(\cdot|\cdot)$ is the optimal distribution in (3.45).

Converse The converse is proved using Fano's lemma [30]. Let the message be $W \in \mathcal{W}$ with a uniform distribution and $|\mathcal{W}| = 2^{LNR}$. The message is related to the channel input $\mathbf{X}_1^N$ and the corresponding channel output $\mathbf{Y}_1^N$. Let $\hat{W}$ be the detected message at the receiver and define the error probability as $P_e^{(LN)} = \Pr\{\hat{W} \neq W\}$. The rate R satisfies

$$\begin{aligned} LNR &= H(W) \\ &\overset{a}{=} H(W|\mathbf{Y}_1^N, \mathbf{V}_1^N) + I(W; \mathbf{Y}_1^N, \mathbf{V}_1^N) \\ &\overset{b}{=} H(W|\mathbf{Y}_1^N, \mathbf{V}_1^N) + I(W; \mathbf{Y}_1^N, \mathbf{V}_1^N) \end{aligned} \tag{3.46}$$

where (a) follows from the definition of the mutual information and (b) follows from the independence between W and $\mathbf{V}_1^N$. By Fano's lemma [30], the first term of (3.46) is upper-bounded as

$$H(W|\mathbf{Y}_1^N, \mathbf{V}_1^N) \le 1 + P_e^{(LN)} LNR \tag{3.47}$$

In general, $\mathbf{X}_n$ and W are related through $p(\mathbf{x}_n|w, u_1^n)$ because of the causality of the feedback. A key step in the proof of the converse is to introduce an intermediate random variable $\mathbf{T}_n$ with conditional distribution $p(t_n|w, u_1^{n-1})$ and assume that $\mathbf{X}_n$ is independent of W and U_1^{n-1} given $\mathbf{T}_n$ and U_n, that is, $p(\mathbf{x}_n|w, u_1^n, \mathbf{t}_n) = p(\mathbf{x}_n|u_n, \mathbf{t}_n)$. The assumption has no loss of generality because we can always let $\mathbf{T}_n = \{W, U_1^{n-1}\}$ to restore the original relation.

The second term of (3.46) can be upper-bounded as

$$\begin{aligned} I(W; \mathbf{Y}_1^N, \mathbf{V}_1^N) &\overset{a}{=} I(W; \mathbf{Y}_1^N|\mathbf{V}_1^N, U_1^N) + I(U_1^N; W|\mathbf{V}_1^N) - I(U_1^N; W|\mathbf{Y}_1^N, \mathbf{V}_1^N) \\ &\overset{b}{\le} I(W; \mathbf{Y}_1^N|\mathbf{V}_1^N, U_1^N) \\ &\overset{c}{=} \sum_{i=1}^N I(W; \mathbf{Y}_i|\mathbf{V}_1^N, U_1^N, \mathbf{Y}_1^{i-1}) \\ &\overset{d}{\le} \sum_{i=1}^N I(\mathbf{T}_i; \mathbf{Y}_i|\mathbf{V}_1^N, U_1^N, \mathbf{Y}_1^{i-1}) \\ &= \sum_{i=1}^N H(\mathbf{Y}_i|\mathbf{V}_1^N, U_1^N, \mathbf{Y}_1^{i-1}) - H(\mathbf{Y}_i|\mathbf{V}_1^N, U_1^N, \mathbf{Y}_1^{i-1}, \mathbf{T}_i) \\ &\overset{e}{\le} \sum_{i=1}^N H(\mathbf{Y}_i|\mathbf{V}_i, U_i) - H(\mathbf{Y}_i|\mathbf{V}_1^N, U_1^N, \mathbf{Y}_1^{i-1}, \mathbf{T}_i) \end{aligned} \tag{3.48}$$

$$\begin{aligned}
&\stackrel{f}{=} \sum_{i=1}^{N} H(\mathbf{Y}_i|\mathbf{V}_i, U_i) - H(\mathbf{Y}_i|\mathbf{V}_i, U_i, \mathbf{T}_i) \\
&= \sum_{i=1}^{N} I(\mathbf{T}_i, \mathbf{Y}_i|\mathbf{V}_i, U_i) \\
&\stackrel{g}{\leq} \sum_{i=1}^{N} \max_{h(\cdot)} \sum_{\mathbf{v}_i} p(\mathbf{v}_i) I(\mathbf{T}_i, \mathbf{Y}_i|\mathbf{v}_i, u_i = h(\mathbf{v}_i)) \\
&\stackrel{h}{\leq} \sum_{i=1}^{N} \max_{h(\cdot)} \sum_{\mathbf{v}_i} p(\mathbf{v}_i) I(\mathbf{X}_i, \mathbf{Y}_i|\mathbf{v}_i, u_i = h(\mathbf{v}_i)) \\
&\stackrel{i}{\leq} LNC,
\end{aligned} \tag{3.49}$$

where a follows from the mutual information identity (3.41), b follows from the independence between U_1^N and W given $\mathbf{V}_1^N$ and the nonnegativity of mutual information $I(U_1^N;W|\mathbf{Y}_1^N,\mathbf{V}_1^N)$, c follows from the chain rule of the mutual information, d follows from the data processing inequality [30] and the fact that $W \rightarrow \mathbf{T}_i \rightarrow \mathbf{Y}_i$ is a Markov chain given $\{\mathbf{V}_1^N, U_1^N, \mathbf{Y}_1^{i-1}\}$, e follows from the fact that the conditioning reduces the entropy, f follows from the fact that $\mathbf{Y}_i$ is independent from the other variables given $\{\mathbf{V}_i, U_i, \mathbf{T}_i\}$, g follows from Lemma 3.2, h follows from the fact that $\mathbf{T}_i \rightarrow \mathbf{X}_i \rightarrow \mathbf{Y}_i$ is a Markov chain given u_i, $\mathbf{v}_i$, and i follows from Equation (3.45).

Combining Equations (3.46), (3.47), and (3.49), we obtain

$$R \leq \frac{1}{LN} + P_e^{(LN)} R + C$$

If $P_e^{(LN)} \rightarrow 0$ as $N \rightarrow \infty$, then the rate R must be less than or equal to C.

Remark 1. If there is a memoryless feedback error, the relation between U_n and $\mathbf{V}_n$ is an i.i.d. statistical relation $p_{U|\mathbf{V}}(u_n|\mathbf{v}_n)$. Comparing Equation (3.42) in Lemma 3.2 and the capacity C in (3.45), we observe that the feedback error causes two penalties. The first penalty is due to the averaging [i.e., $\Sigma_U p(u|\mathbf{v}) I(\mathbf{T};\mathbf{Y}|u,\mathbf{v}) \leq \max_u I(\mathbf{T};\mathbf{Y}|u,\mathbf{v})$], and the second penalty is due to the fact that the receiver is uncertain about U. Therefore, Lemma 3.2 will be useful for the design of the index assignment of the feedback when there is feedback error.

Remark 2. Theorem 3.1 is proved with the assumption that the receiver does not know the feedback U. Of course, since the optimal feedback is a deterministic function of $\mathbf{V}$, the receiver knows the feedback implicitly. The capacity is the same even if we start by assuming that the receiver knows the feedback. In this case, the proof of the converse part will start from Equation (3.48). The rest of the proof is the same.

For the case of perfect CSIR $\mathbf{V} = \mathbf{H}$, the capacity C in (3.45) can be simplified to a scalar form as

$$C_{\mathbf{V}=\mathbf{H}} = \max_{\mathcal{H}_1^Q} \sum_{q=1}^{Q} \max_{p_{\mathbf{X}|U}(\mathbf{x}|u_q)} \mathcal{E}_{\mathbf{h}\in\mathcal{H}_q}[I(\mathbf{X};\mathbf{Y}|u_q,\mathbf{h})] \tag{3.50}$$

For the case of $L = 1$ or the case of the low-complexity receiver, the capacity C in (3.45) can be simplified to a scalar form as

$$C_{L=1} = \max_{\mathcal{H}_1^Q} \sum_{q=1}^{Q} \max_{p_{\mathbf{X}|U}(\mathbf{x}|u_q)} \mathcal{E}_{\mathbf{v}\in\mathcal{H}_q}[I(\mathbf{X};\mathbf{Y}|u_q,\mathbf{v})] \tag{3.51}$$

For the case of perfect CSIR $\mathbf{V} = \mathbf{H}$, the capacity C in (3.45) can be simplified to a scalar form as

$$C = \max_{\mathcal{H}_1^Q} \sum_{q=1}^{Q} \max_{p(\mathbf{x}|u_q)} \mathcal{E}_{\mathbf{h}\in\mathcal{R}_q}[I(\mathbf{X};\mathbf{Y}|u_q,\mathbf{h})] \tag{3.52}$$

3.4.3 Equivalence with Vector Quantization Problem

In applications, we might have additional constraints on the input distribution of X. For example, a common constraint is the average transmit power of codewords. In this case, we need to choose $p(\mathbf{x}|u)$ to maximize the mutual information under the constraint $\mathcal{E}[|\mathbf{X}|^2] \le P_0$. To characterize the benefit of the state feedback, the following optimization problem needs to be solved.

Problem 3.3 (Optimal MIMO Transmission and Feedback Design) Find the feedback strategy with partition $\{\mathcal{R}_1, \ldots, \mathcal{R}_Q\}$, and the transmission strategy $\{p(\mathbf{x}|u_1), \ldots, p(\mathbf{x}|u_Q)\}$, to achieve the capacities in (3.45), (3.50), and (3.51).

3.4.4 Fast Flat Fading MIMO Channels

Although the framework described above is general enough to cover imperfect CSIR, we are especially interested in the simpler case of perfect CSIR ($\mathbf{V} = \mathbf{H}$). We shall illustrate the optimal partial feedback and MIMO transmission design with an important example of fast ($L = 1$) flat fading MIMO channels. The received symbols is given by

$$\mathbf{Y}_n = \mathbf{H}_n\mathbf{X}_n + \mathbf{Z}_n \tag{3.53}$$

where $\mathbf{Z}_n$ is the i.i.d. complex Gaussian channel noise with covariance $\sigma_z^2\mathbf{I}_{n_R}$. For simplicity, we assume perfect CSIR $\mathbf{V} = \mathbf{H}$. From (3.45) and the AWGN channel noise, the capacity-achieving input distribution $p(\mathbf{X}|u_q)$ with CSIT $U = u_q$ is circular symmetric complex Gaussian with covariance

$$P_q = \mathcal{E}[\mathbf{X}\mathbf{X}^*|u_q] \tag{3.54}$$

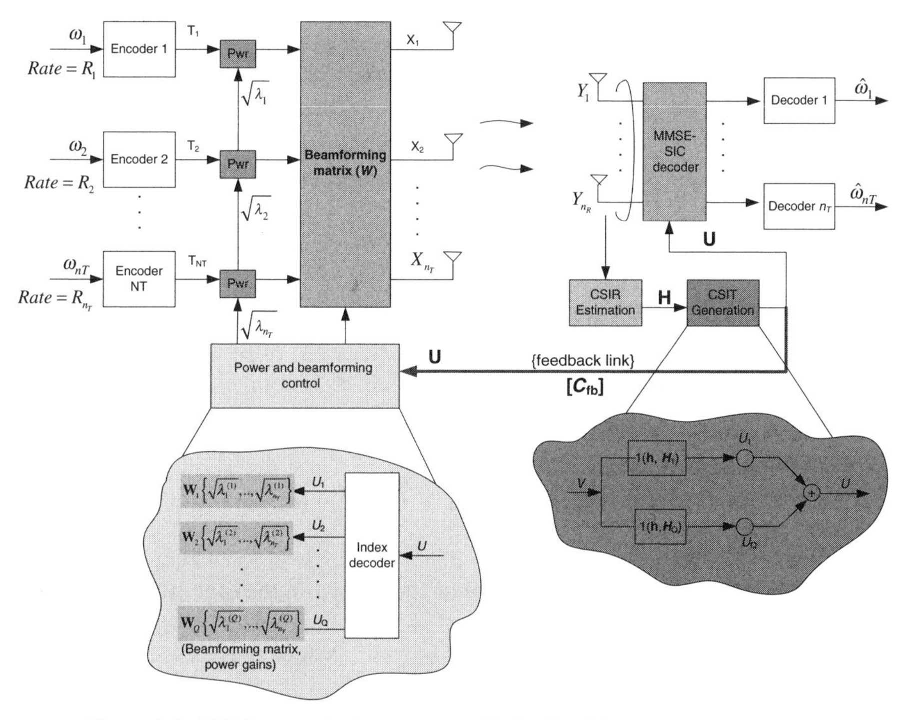

Figure 3.9. MIMO transmission strategy with feedback link capacity constraint.

Since the covariance matrix of $\mathbf{X}$ is Hermitian, by singular value decomposition, we have

$$\rho_q = \varepsilon[\mathbf{XX}^* | U = u_q] = \mathbf{W}_q \mathbf{\Lambda}_q \mathbf{W}_q \tag{3.55}$$

where $\mathbf{\Lambda}_q = \text{diag}\{\lambda_1^{(q)}, \ldots, \lambda_{nT}^{(q)}\}$ is the diagonal matrix containing eigenvalues of ρ_q and $\mathbf{W}_q$ is the corresponding eigenmatrix. If we define a $n_T \times 1$ dimensional matrix $\mathbf{T}$ to be complex Gaussian with covariance $\mathbf{I}_{nT}$, we obtain the capacity achieving transmit symbol $\mathbf{X}_q$ given CSIT $U = u_q$ as

$$\mathbf{X}_q = \mathbf{W}_q \mathbf{\Lambda}_q \mathbf{T} \tag{3.56}$$

The optimal transmitter structure and the receiver structure are illustrated in Figure 3.9. At the receiver, the CSIR space is partitioned into Q regions $\{\mathcal{H}_1, \ldots, \mathcal{H}_Q\}$, where $Q = 2^{C_{fb}}$. When the current CSIR, $\mathbf{H}$, is in the qth region, $\mathcal{H}_q$, the associated partition index, u_q, is generated and fed back to the transmitter using the feedback channel. (This consumes the feedback link capacity of C_{fb}.) At the transmitter, the bank of N isolated channel encoders produces independent outputs $\mathbf{T} = [T_1, \ldots, T_N]$. The channel encoder output $\mathbf{T}$ is passed to the power control block $\{\sqrt{\lambda_1}, \ldots, \sqrt{\lambda_{n_T}}\}$, where spatial power water filling is done across the N spatial channels. After the power control block, the transmit signal is passed to the *eigenbeamforming* block. There is a table of Q entries [power control matrix ($\sqrt{\Lambda_q}$), beamforming matrix ($\mathbf{W}_q$)] associated with feedback CSIT indices $u_q \in [1, Q]$. The power control matrix is a diagonal matrix with nonnegative elements. The beamforming matrix is a unitary matrix. When the current CSIT is u_q, the corresponding power matrix Λ_q and beamforming matrix $\mathbf{W}_q$ are selected for the current fading symbol. Hence, the feedback strategy is characterized by the CSIR partition $\tilde{\mathcal{H}} = \{\mathcal{H}_1, \ldots, \mathcal{H}_Q\}$. The MIMO transmission strategy is characterized by $\{\rho_1, \ldots, \rho_Q\}$. The optimization problem on the design of the optimal feedback strategy and the optimal MIMO transmission strategy is thus equivalent to the design of a vector quantizer with a modified distortion measure $d_{\text{cap}}(\mathbf{h}, \rho_q)$ given by

$$d_{\text{cap}}(\mathbf{h}, \rho_q) = -\left[\log_2\left|\mathbf{I}_{nT} + \frac{\mathbf{h}^* \mathbf{h}\rho_{\mathbf{q}}}{\sigma_z^2}\right| - \alpha \operatorname{tr} \rho_{\mathbf{q}}\right] \tag{3.57}$$

where α is a Lagrange multiplier chosen to satisfy the average transmit power constraint

$$\sum_{q=1}^{Q} \operatorname{tr}[\rho_q] P_q \leq P_0 \tag{3.58}$$

where $P_q = \Pr[U = u_q]$. Observe that the distortion measure is a function of $\mathbf{h}$ and ρ_q only. The partition index u_q becomes a dummy variable and is omitted in the formulation.

Hence, the optimization problem (Problem 3.3) is equivalent to selecting $\{\rho_1, \ldots, \rho_Q\}$ and $\tilde{\mathcal{H}} = \{\mathcal{H}_1, \ldots, \mathcal{H}_Q\}$ so as to minimize the *average distortion* D_{cap} given by

$$D_{\text{cap}} = \sum_{q=1}^{Q} \varepsilon_{\mathbf{h} \in \mathcal{H}_q}[d_{\text{cap}}(\mathbf{h}, \rho_q)] P_q. \tag{3.59}$$

3.4.5 Lloyd's Algorithm

Hence, the optimization problems can be solved by Lloyd's algorithm, which is outlined below. Steps 1 and 2 are repeatedly applied until convergence.[3] Note that the algorithm below is an offline optimization algorithm and hence,

[3]In general, there is no guarantee that Lloyd's algorithm will converge to the global optimal [95]. In the simulation, we repeat each partition iteration 10 times, each time starting with a random set of initial *partition centroids*. At the end of the algorithm, we pick the one that gives us the largest capacity.

complexity has a relatively small impact on implementation. The online strategy is given by Section 3.4.7 and in general has a very low *runtime complexity*.

Step 1: Determine the optimal transmission strategy $\{\rho_1, \ldots, \rho_Q\}$ *given a certain partition* $\{\mathcal{H}_1, \ldots, \mathcal{H}_Q\}$. The optimal transmission strategy, ρ_q, is given by the generalized *partition centroid*:

$$\rho_q = \arg\min_{\rho_q} \varepsilon_{\mathbf{h}\in\mathcal{H}_q}[d(\mathbf{h}, \rho_q)]P_q \tag{3.60}$$

Step 2: Determine the optimal partition $\{\mathcal{H}_q\}$ *given a transmission strategy* $\{\rho_q\}$. The optimal partition is given by the *nearest-neighbor rule*:

$$\mathcal{H}_q = \{\mathbf{h} : d(\mathbf{h}, \rho_q) \leq d(\mathbf{h}, \rho_j); \forall j, q \in [1, \ldots, Q], j \neq q\} \tag{3.61}$$

3.4.6 Approximate Closed-Form Solution for Step 1

To solve for (3.60), we let $f = \varepsilon_{\mathbf{h}\in\mathcal{H}_q}[d_{\text{cap}}(\mathbf{h}, \rho_q)]P_q$. As a result of the expectation operator, the optimization of Equation (3.61) is quite tedious. To obtain closed-form solutions for step 1, we shall consider the following *heuristic approximation*

$$f = \varepsilon_{\mathbf{h}\in\mathcal{H}_q}\left[\log_2\left|\mathbf{I}_{n_T} + \frac{\mathbf{h}^*\mathbf{h}\rho_q}{\sigma_z^2}\right| - \lambda\,\text{tr}(\rho_q)\right]P_q \approx P_q\left[\log_2\left|\mathbf{I}_{n_T} + \frac{\bar{\mathbf{s}}\rho_q}{\sigma_z^2}\right| - \lambda\,\text{tr}(\rho_q)\right] \tag{3.62}$$

where $\bar{\mathbf{s}} = \varepsilon_{\mathbf{h}\in\mathcal{H}_q}[\mathbf{h}^*\mathbf{h}]$ is a conditional average matrix and $P_q = \Pr[\mathbf{h} \in \mathcal{H}_q]$ for the qth partition.

The approximate closed-form solution derived below as a result of the approximation in (3.62) serves as a *realizable, nontrivial capacity lower bound*.

Since $\bar{\mathbf{s}}$ is Hermitian, it can be uniquely expressed as $\bar{\mathbf{s}} = \overline{\mathbf{w}}\overline{\boldsymbol{\Psi}}_s\overline{\mathbf{w}}^*$, where $\overline{\boldsymbol{\Psi}}_v$ is the diagonal matrix of eigenvalues of $\bar{\mathbf{s}}$ and $\overline{\mathbf{w}}$ is the unitary matrix of the corresponding eigenvectors. Multiplying $\overline{\mathbf{w}}^*$ (to the left) and $\overline{\mathbf{w}}$ (to the right) of the determinant on the right-hand side of equation (3.62), we have

$$\left|\mathbf{I}_{n_T} + \frac{\bar{\mathbf{s}}\rho_q}{\sigma_z^2}\right| = \left|\overline{\mathbf{w}}^*\overline{\mathbf{w}} + \frac{\overline{\mathbf{w}}^*\bar{\mathbf{s}}\rho_q\overline{\mathbf{w}}}{\sigma_z^2}\right| = \left|\mathbf{I}_{n_T} + \frac{\overline{\boldsymbol{\Psi}}_s\overline{\mathbf{w}}^*\bar{\mathbf{s}}\rho_q\overline{\mathbf{w}}}{\sigma_z^2}\right|$$

Using Equation (3.55), the optimal ρ_q has the form

$$\rho_q = \overline{\mathbf{w}}\boldsymbol{\Lambda}_q\overline{\mathbf{w}}^* \tag{3.63}$$

because this would diagonalize the argument of the determinant and therefore maximize the determinant. Hence, we have $\text{tr}(\rho_q) = \text{tr}(\boldsymbol{\Lambda}_q)$. Let $\mathcal{A}$ be the set of indices with nonzero entries in $\overline{\boldsymbol{\Psi}}_s$:

$$\mathcal{A} = \{t : \overline{\boldsymbol{\Psi}}_s(t,t) > 0\} \tag{3.64}$$

Equation (3.62) can be simplified as follows:

$$f = \sum_{t \in \mathcal{A}} \log_2\left(1 + \frac{\overline{\boldsymbol{\Psi}}_s(t,t)\boldsymbol{\Lambda}_q(t,t)}{\sigma_z^2}\right) - \alpha \operatorname{tr}(\boldsymbol{\Lambda}_q) \tag{3.65}$$

By observation, $\boldsymbol{\Lambda}_q(t, t) = 0$ if $t \notin \mathcal{A}$. For $t \in \mathcal{A}$, the necessary condition for optimal point is given by $df/d\boldsymbol{\Lambda}_q(t, t) = 0$. This is equivalent to

$$\left(1 + \frac{\overline{\boldsymbol{\Psi}}_s(t,t)\boldsymbol{\Lambda}_q(t,t)}{\sigma_z^2}\right)^{-1} \overline{\boldsymbol{\Psi}}_s(t,t) - \alpha = 0 \tag{3.66}$$

for all $t \in \mathcal{A}$. The optimal solution is given by

$$\boldsymbol{\Lambda}_q = \left[\frac{\mathbf{I}}{\alpha} - \sigma_z^2 \overline{\boldsymbol{\Psi}}_s^{-1}\right]^+ \tag{3.67}$$

where $[x]^+ = \max(x, 0)$. The corresponding optimal input covariance matrix ρ_q is given by $\rho_q = \overline{\mathbf{w}}\boldsymbol{\Lambda}_q\overline{\mathbf{w}}^*$. The average transmitted power over the coding block is given by

$$\overline{\rho} = \sum_{q=1}^{Q} P_q \operatorname{tr}[\rho_q]. \tag{3.68}$$

The corresponding average SNR is given by $\frac{\overline{\rho}}{\sigma_z^2}$. Note that from (3.67), the optimal feedback strategy with partial feedback constraint also has the form of temporal water filling.

3.4.7 Complexity of the Online Adaptation Strategy

While the optimization of the modified Lloyd algorithm is in general quite time-consuming, the algorithm is an offline process. In this section, we shall elaborate the complexity of the realtime (online) adaptation operation at the MIMO transmitter and the receiver.

Figure 3.9 illustrates the block diagram of the online adaptation processing. At the receiver, the CSIR space is partitioned into Q regions $\{\mathcal{H}_1, \ldots, \mathcal{H}_Q\}$, where $Q = 2^{C_{\text{fb}}}$. The partition regions are obtained from the offline Lloyd algorithm. When the current CSIR $\mathbf{H}$ is in the qth region $\mathcal{H}_q$, the associated partition index u_q is generated and fed back to the transmitter using the feedback channel. (This consumes the feedback link capacity of C_{fb}.) Hence, instead of feeding back a quantized value of the CSIR **H** on the feedback channel, we just need to feedback a partition index. The complexity of the online processing is much lower compared with the offline Lloyd's optimization. At the receiver, the complexity lies on the generation of the partition index q given a CSIR **H**. In general, it requires Q *distance metric* computations and sorting per CSIR **H**. Hence, the advantage of the processing structure in Figure 3.9 is the low-complexity realtime processing. Furthermore, this structure is the optimal structure with respect to ergodic capacity.

At the transmitter, there is a table of Q entries [power control matrix ($\mathbf{\Lambda}_q$), beamforming matrix ($\mathbf{W}_q$)] associated with feedback indices $u_q \in [1, Q]$. Similarly, the table is obtained from the results of the offline Lloyd algorithm. The power control matrix is a diagonal matrix with nonnegativeelements. The beamforming matrix is a unitary matrix. When the current CSIT is u_q, the corresponding power matrix and beamforming matrix are selected for the current fading block. Let **T** be the $n_T \times 1$ vector from the outputs of the bank of n_T encoders. The transmitted symbol **X** at the current fading block, after the application of the power control matrix and the beamforming matrix, is given by

$$\mathbf{X} = \mathbf{W}_q \sqrt{\mathbf{\Lambda}_q}\,\mathbf{T} \tag{3.69}$$

Hence, the complexity involved at the transmitter is just a lookup table of Q entries that is usually negligible.

While the computational loading at the receiver is small for moderate Q, the complexity may also grow to a large value for large Q. For example, in OFDM systems with a large number of subcarriers, Q may be large because for the same percentage overhead of the feedback capacity, the absolute number of bits allocated for CSI feedback can increase as the number of subcarriers. Hence, Q for OFDM can be quite large, making the partition search quite complicated. To further simplify the partition search, we can introduce some artificial structures into the codebook so that the search for the *best-fit partition region* with respect to a CSIR **H** can be more efficient. This is similar to the vector quantization problems.

3.4.8 MMSE-SIC Receiver Structure

At the receiver, minimum mean-square error spatial processing with successive interference cancellation (MMSE-SIC) can be used to achieve the optimal channel capacity. From (3.53), the received symbol **Y** (dropping the symbol index n) is given by

$$\begin{aligned}\mathbf{Y} &= \mathbf{H}\mathbf{X} + \mathbf{Z} = \mathbf{H}\mathbf{W}\sqrt{\mathbf{\Lambda}}\,\mathbf{T} + \mathbf{Z} \\ &= \sum_{m=1}^{n_T} \mathbf{F}_m \sqrt{\lambda_m}\, T_m + \mathbf{Z}\end{aligned}$$

where $\mathbf{F} = \mathbf{HW}$ represents the aggregate effect of matrix channel and beamforming matrix, $\mathbf{T}$ is the $n_T \times$ one-dimensional i.i.d. complex Gaussian input vector from the channel encoder bank, T_m is the mth component of $\mathbf{T}$, and $\lambda_m = \mathbf{\Lambda}(m, m)$ is the mth diagonal element representing the power control action.

In fact, the MMSE spatial processing [136] stems from the chain rule of mutual information, which is illustrated below. From the structure in Figure 3.9, the channel capacity is given by

$$\begin{aligned} C_{\text{fwd}} &= \sum_{q=1}^{Q} \varepsilon_{\mathbf{h} \in \mathcal{H}_q}[I(\mathbf{T}; \mathbf{Y}|\mathbf{H})] \\ &= \sum_{m=1}^{n_T} \sum_{q=1}^{Q} \varepsilon_{\mathbf{h} \in \mathcal{H}_q}[I(T_m; \mathbf{Y}|T_1, \ldots, T_{m-1}, \mathbf{H})] \\ &= \sum_{q=1}^{Q} \sum_{m=1}^{n_T} R_m \end{aligned} \tag{3.70}$$

where R_m is the encoding rate of the mth encoding branch at the transmitter given by

$$\begin{aligned} R_m &= \sum_{q=1}^{Q} \varepsilon_{\mathbf{h} \in \mathcal{H}_q}[I(T_m; \mathbf{Y}|T_1, \ldots, T_{m-1}, \mathbf{H})] \\ &= \sum_{q=1}^{Q} \varepsilon_{\mathbf{h} \in \mathcal{H}_q}\left[\log_2\left(1 + \frac{\lambda_m}{\sigma_z^2} \mathbf{F}_m^* \left(\mathbf{I}_{n_R} + \sum_{i=m+1}^{n_T} \frac{\lambda_i}{\sigma_z^2} \mathbf{F}_i \mathbf{F}_i^*\right)^{-1} \mathbf{F}_m\right)\right] \end{aligned} \tag{3.71}$$

Hence, the MMSE-SIC processing at the receiver is based on successive interference cancellation. The information from encoder 1 (with rate R_1) is first decoded, and then the signal is subtracted from the received symbol $\mathbf{Y}$. The information from encoder 2 (with rate R_2) is then decoded without the interference from encoder 1. The process continues until all the information from the n_T channel encoders is decoded. Note that the decoding order is irrelevant in the formulation but has to be agreed on between the transmitter and the receiver. Furthermore, because the coding frame spans over ergodic realization of fading symbols, the rate information $\{R_1, \ldots, R_{n_T}\}$ is deterministic and can be precomputed at the transmitter without extra feedback.

3.4.9 Numerical Results and Discussion

In this section we shall evaluate the performance of the partial feedback design in Section 3.4.2 with numerical results. We shall consider the SNR gain of the forward channel capacity with various n_T and n_R configurations and various SNRs. When feedback capacity $C_{\text{fb}} = 0$, this corresponds to the case

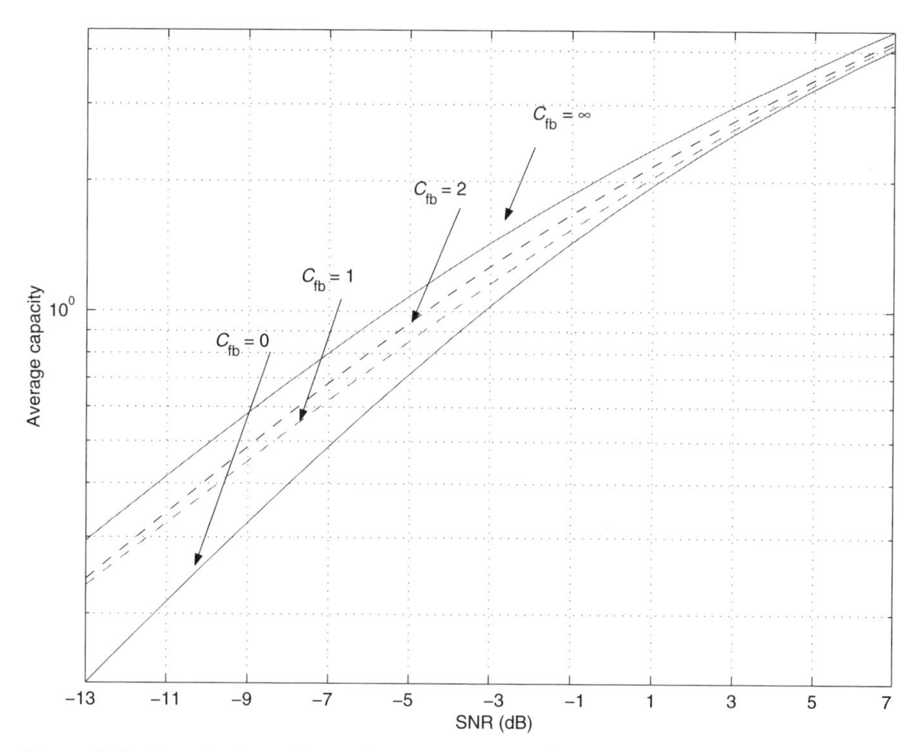

Figure 3.10. Ergodic channel capacity versus average SNR of 2 × 2 system at various feedback channel capacities $C_{fb} = 0,1,2,\infty$.

with no CSIT. On the other hand, when the feedback capacity $C_{fb} = \infty$, it corresponds to the case with perfect CSIT. Note that C_{fb} has a unit of bits per fading block.

3.4.9.1 Performance of MIMO Link. We shall illustrate the contribution of temporal power water filling[4] to the overall SNR gain in this section. A 2 × 2 MIMO link performance is studied. Figure 3.10 illustrates the 2 × 2 forward MIMO channel capacity versus the average forward SNR with feedback capacity constraint, which varies from $C_{fb} = 0$ to $C_{fb} = \infty$. For low SNRs, there is a significant SNR gain of around 3.5 dB in forward channel capacity between perfect CSIT and no CSIT. Furthermore, there is a gain of 2.5 and 3 dB for partial feedback with feedback capacity $C_{fb} = 1$and $C_{fb} = 2$, respectively. In other words, $C_{fb} = 2$ realized about 85% of the feedback gain. On the other hand, the effectiveness of feedback is reduced (but still significant) for high SNRs. This is reasonable because when the average SNR is large, the penalty

[4]Although the transmit power and beam-forming matrix is constant within a fading block, they are adaptive to the partial feedback once every fading block and therefore, we have temporal adaptation over a time horizon of a coding frame.

of transmitting power less efficiently is small compared to the case when the average SNR is small.

3.4.9.2 Performance of MISO Link. In this section, we consider $n_T = 2,4$ and $n_R = 1$. Figure 3.11a, b illustrates the forward channel capacity versus average SNR with $C_{fb} = 0$ to $C_{fb} = \infty$ for $(n_T, n_R) = (2,1)$ and $(n_T, n_R) = (4,1)$, respectively. At low SNRs, there are significant SNR gains of around 5 and 8 dB in forward channel capacity between perfect CSIT and no CSIT for $n_T = 2$ and $n_T = 4$, respectively. For $n_T = 2$, there is a gain of 2.2, 4, and 4.4 dB for partial feedback with feedback capacity $C_{fb} = 1,2,3$, respectively. For $n_T = 4$, the corresponding SNR gains are 2.4, 4, 5, and 5.5 dB for feedback capacity of $C_{fb} = 1,2,3,4$, respectively.

At high SNRs, there are also significant SNR gains of 3 and 6 dB between the perfect feedback and no-feedback case for $n_T = 2$ and $n_T = 4$, respectively. These results demonstrate that the *spatial power water filling* is very effective for high and low SNRs for 2×1 and 4×1 systems. There is a large incremental SNR gain from $C_{fb} = 1$ to $C_{fb} = 2$.

3.5 SUMMARY

In this chapter, we have considered the effect of imperfect CSI on MIMO channel capacity. For instance, there are two roles of CSI in the MIMO wireless link. The CSI at the receiver is used to enhance the signal detection and has a more stringent requirement on the accuracy. The CSI at the transmitter, on the other hand, is used for transmission adaptation and therefore has a more relaxed requirement on the accuracy.

In the first part of the chapter, we consider the effect of imperfect CSIR on the MIMO channel capacity. An asymptotically tight upper bound and lower bound on the capacity are derived. In the second part of the chapter, we consider the effect of limited feedback link capacity on the partial CSI feedback strategy and the transmission adaptation strategy. We found that the optimal feedback and transmission design with the constraint of limited feedback is equivalent to the classical vector quantization problem. When we optimize for the *effective SNR*, the optimal solution requires Grassmannian partition on the CSIR space. On the other hand, when we optimize for the forward MIMO link capacity, the optimal solution requires a partition defined over the *generalized distortion measure*. In either case, the optimal transmission strategy has temporal and spatial water-filling components. At the receiver, there are Q partitions on the CSIR space $\mathcal{H}$. On every received CSIR $\mathcal{H}$, the receiver has to search for the best-fit partition region and feedback the partition index. Hence, instead of carrying the quantized CSIR on the feedback link, we found that the optimal structure just needs to carry the partition index on the feedback link. At the transmitter, there is a table of Q entries. Each entry from the table corresponds to a specific power allocation (across

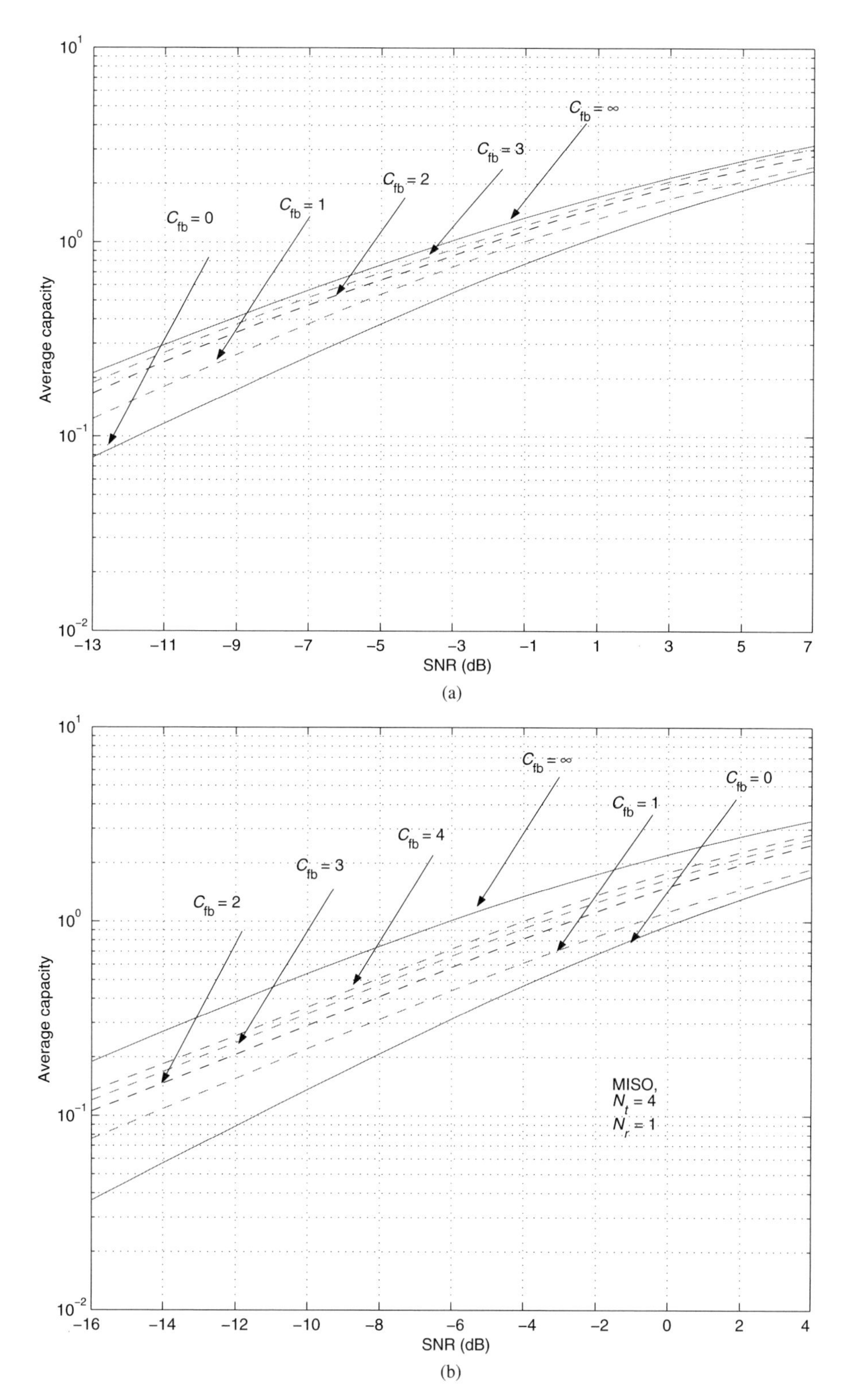

Figure 3.11. Forward channel capacity versus average SNR with ideal CSIR for 2×1 (a) and 4×1 (b) systems and $C_{fb} = 0,1,2,3,4,\infty$.

n_T antennas) as well as a beamforming matrix. Instead of adapting to the instantaneous channel state, the transmitter adapts to the instantaneous *partition index*. The receiver partition and the transmitter table are obtained from the offline Lloyd algorithm in much the same way as the classical vector quantization problems.

EXERCISES

1. *Capacity Lower Bound of MIMO Links* We assume imperfect CSIR V which is an erroneous measurement of the channel, H.

$$V = H + \varepsilon$$

We consider a block flat fading MIMO channels characterized by

$$Y_n = H_n X_n + Z_n$$

As illustrated in the text, the ergodic capacity can be written as,

$$C_{csir} = \max_{p(X)} I(X;Y|V) = \max_{p(X)} H(X|V) - H(X|Y,V)$$

(a) prove that $H(X|V) = \log_2(|\pi e \mathbf{Q}|)$ where $\mathbf{Q} = E\{xx^*|H\}$ where $H(.)$ denotes the entropy and $E\{.\}$ denotes the expectation.

(b) prove that the variance of $X - AY$ is minimized for any A if AY is the linear minimum mean square estimation of X.

(c) prove that $H(X|Y, V) \leq H(X - AY)$ for any A. *hint:* adding a constant does not change the differential entropy and conditioning reduces entropy.

(d) prove that $H(S - AY) \leq ln(\pi e var(S - AY))$ where $var(.)$ denotes the variance of a random variable.

(e) use (c) to (d) to obtain an upper bound of $H(X|Y, V)$

(f) Hence, prove that

$$I_{lower}(X;Y|V) = E\left\{\log_2\left|I + \frac{1}{1+\sigma_E^2 P} V^* VQ\right|\right\}$$

where σ_E^2 is the variance of the CSIR error ε.

(g*) Suppose we have quasi-static fading, is the lower bound on the instantaneous channel capacity in (f) still valid?

2. *Maximum Likelihood Estimator* Maximum Likelihood Estimator (MLE) is one popular approach in channel estimation. Consider a block fading channel characterized by

$$Y = HX + Z$$

where X is a $n_T \times T_p$ pilot symbol and Z is the $n_R \times T_p$ i.i.d zero mean complex guassian noise with variance σ_z^2. Y is of dimension $n_R \times T_p$ where T_p denotes the number of pilot symbols.

The maximum likelihood estimator of the channel matrix is an estimate which maximizes the probability of Y given a measurement of the channel.

$$p(Y|H) = \frac{1}{\pi^R |C_z|} \exp\left[-(Y - HX)^H C_z^{-1} (Y - HX)\right]$$

and to simplify this, the log likelihood function can be obtained.

$$\log p(Y|H) = \log \frac{1}{\pi^R |C_z|} - (Y - HX)^H C_z^{-1} (Y - HX)$$

The following identities may be useful.

$$\frac{\partial}{\partial \mathbf{X}} a^* \mathbf{X} b = ab^*$$

$$\frac{\partial}{\partial \mathbf{X}} a^* \mathbf{X}^* b = ba^*$$

$$\frac{\partial}{\partial \mathbf{X}} a^* \mathbf{X}^* C \mathbf{X} b = (C + C^*) \mathbf{X} aa^*$$

where a, b are column vectors; C is a matrix; * denotes complex conjugate.

(a) prove that $p(Y|H) = \dfrac{1}{\pi^R |C_z|} \, exp[-(Y - HX)^H C_z^{-1} (Y - HX)]$

(b) differentiate the log-likelihood function with respect to H.

(c) Henc, prove that $V = YX^*(XX^*)^{-1}$

(d) verify that for XX^* to be invertible, we must have $T_p \geq n_T$ and X is of full rank.

4

SPACETIME CODING AND LAYERED SPACETIME CODING FOR MIMO WITH PERFECT CHANNEL STATE INFORMATION

4.1 OVERVIEW

In Chapter 2, we discussed the channel capacity of MIMO channels with perfect CSIT and/or perfect CSIR. We showed that with the perfect CSIR, the channel capacity is increased significantly by multiple antennas at both the transmitter and receiver. The capacity gain is due to the additional degree of freedom in the spatial dimension of the resource space. When perfect CSIT is available, the $n_T \times n_R$ MIMO channels can be decomposed into $m^* = \min[n_T, n_R]$ parallel independent spatial channels. With perfect CSIT, transmit power adaptation on both the spatial and temporal domains is possible, which introduces additional capacity gain on top of the MIMO capacity with perfect CSIR. This demonstrates the importance of adaptation in MIMO channels, especially when $n_T > n_R$.

In this chapter, our focus is on the practical transmitter and receiver architecture and design issues to realize the capacity gains predicted by information theory. For instance, we shall first focus on the case with perfect CSIR in which the design of the optimal spacetime coding design at the transmitter to realize the MIMO capacity is considered. Specifically, with the $n_T \times n_R$ MIMO channels, there are additional degrees of freedom in the spatial domain. On one hand, to enhance the reliability of the packet transmission, we can exploit the spatial diversity of the m^* spatial channels. Basically, independent copies of the same information are carried on the m^* spatial channels. Reliability is

Channel-Adaptive Technologies and Cross-Layer Designs for Wireless Systems with Multiple Antennas: Theory and Applications. By V. K. N. Lau and Y.-K. R. Kwok
ISBN 0-471-64865-5

enhanced by making a joint detection based on the m^* observations of the information bits at the receiver. On the other hand, to increase the transmission bit rate, we can exploit the spatial multiplexing of the m^* spatial channels. Basically, m^* independent streams of information are multiplexed onto the m^* spatial channels to increase the bit rate. We shall address the transmitter and receiver designs with respect to the spatial diversity and the spatial multiplexing dimensions utilizing perfect knowledge of CSIR. For example, we shall discuss the spacetime code design criteria, the maximal-likelihood decoding, as well as the layered spacetime coding designs. Finally, we shall also discuss the fundamental tradeoff between the spatial diversity and the spatial multiplexing.

In the second part of the chapter, we explore additional design issues associated with the availability of CSIT. In practice, with perfect CSIT, the transmitter has to perform adaptation with respect to the encoding rate and the modulation constellation. For instance, one has to deal with the following practical design considerations with respect to an adaptive transmitter.

Optimal Choice of Transmission Modes. In order for the receiver to decode the packet using the appropriate transmission mode, a field of overhead control bits is needed in every transmit frame to carry the information about the current transmission mode. Because of the finite number of overhead bits, the transmitter can support only a finite number of distinct modes. Hence, a first design issue is that given a fixed number of distinct modes M at the transmitter, what are the best choice of the encoding rate and modulation constellation with respect to each of the M modes?

Optimal Transmitter Architecture to Support M *Distinct Modes.* Given that the transmitter has M distinct modes, what is the best architecture to support these M distinct modes?

Optimal Design of Adaptation Thresholds. What should be the optimal choice of the adaptation thresholds? The challenge lies in the fact that the CSIT is a $n_R \times n_T$ matrix (instead of a scaler as in SISO case). Hence, given a CSIT, the adaptation threshold is no longer a scalar but in general a partition on the CSIT space. Hence, the design of the *adaptation thresholds* is not trivial.

We shall address these design issues and formulate a systematic design framework regarding the adaptation design of the MIMO transmitter in the second part of the chapter.

4.2 DESIGN OF MIMO LINKS WITH PERFECT CSIR

In this section, we shall focus on the design of transmitter and receiver to realize the performance gains of the MIMO channels with perfect CSIR only as predicted by information theory. From Chapter 2, we illustrate that the optimal transmitter architecture with respect to maximizing the ergodic capacity with perfect CSIR is given by isolated encoding on each transmit antenna. In other words, there is no need for joint encoding across the n_T transmit antennas, and the capacity gain is realized by *spatial multiplexing* on each spatial channel. In addition, because of the lack of CSIT, no adaptation can be done at the transmitter.

However, as we have introduced, information theory is an asymptotic theory, and the conclusion applies only when the codelength approaches infinity. Hence, the *optimal transmitter architecture* in Figure 2.7 (of Chapter 2) can approach the optimal performance only if the component codes along each transmit antenna are powerful codes or *capacity achieving codes* (such as LDPC or turbo codes). However, in practice, for less powerful component codes or codes with finite length, such isolated encoding structure may not achieve an optimal performance because there may be packet transmission error due to finite codelength.

For instance, when the component codes are capacity-achieving, the packet transmission has a very low error probability and therefore, the isolated encoding structure can approach the ergodic capacity through spatial multiplexing. On the other hand, when the codes are not capacity-achieving, packet transmission may be in error and therefore, spatial diversity is an important attribute to improve the reliability of packet transmission. This is particularly important for slow fading channels where diversity cannot be exploited on the temporal dimension.

In general, an effective and practical way to approach the capacity of MIMO channels is to employ *spacetime coding* as illustrated in Figure 4.1. Similar to the coding framework for single-antenna systems, the spacetime encoder is characterized by a vector codebook Ω that is a table of size 2^{NR}. The mth entry in the codebook table consists of a vector codeword

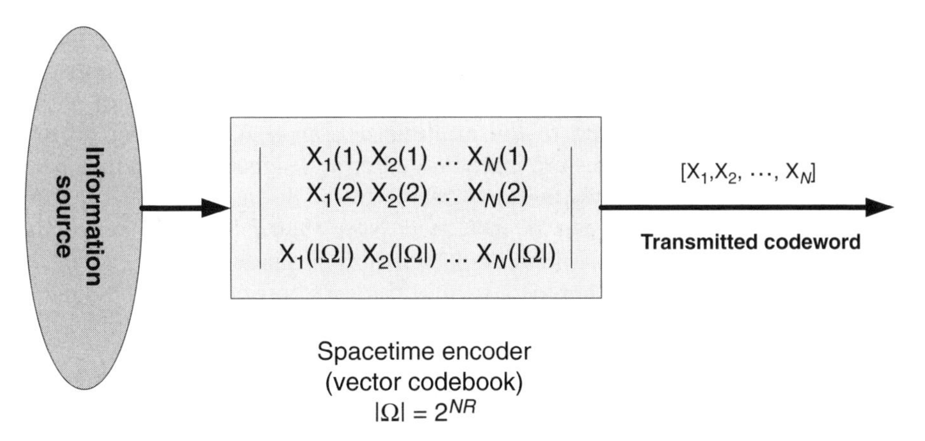

Figure 4.1. Structure of a generic spacetime code at the transmitter.

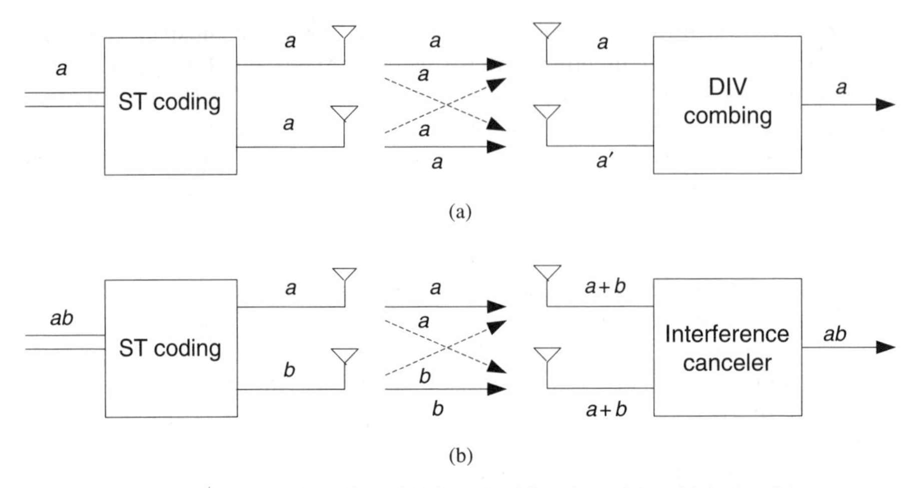

Figure 4.2. Illustration of spatial diversity (a) and spatial multiplexing (b).

$[\mathbf{x}_1(m), \ldots, \mathbf{x}_N(m)]$, where $\mathbf{x}_n(m)$ is the nth transmitted vector symbol (with dimension $n_T \times 1$) and N is the length of the codeword. A message index $m \in [1, 2^{NR}]$ is fed into the spacetime encoder to select a codeword $[\mathbf{x}_1(m), \ldots, \mathbf{x}_N(m)]$ for transmission on the MIMO channels. Hence, redundancy is added in both the space and time dimensions. Transmission of a message index m involves N vector channel uses.

There are two things we can exploit on the MIMO channels through spacetime coding: *spatial diversity* and *spatial multiplexing*. The concepts of spatial diversity and spatial multiplexing are illustrated in Figure 4.2, parts (a) and (b), respectively.

To exploit spatial diversity, the transmitted signals from multiple antennas are jointly encoded and are related to the same copy of information. When the signals reach the receiver through the MIMO channels, the receiver obtains multiple independent copies of the same information and performs diversity combining so as to increase the reliability of the packet transmission. On the other hand, to exploit spatial multiplexing, different streams of information bits are transmitted to the multiple antennas. At the receiver, the various information streams are mixed together and signal processing is performed to separate the information streams from the mixture. With spatial multiplexing, the bit rate can be increased. Note that if the level of coding between transmit antennas is increased, the amount of independence between signals decreases. Hence, there is a natural tradeoff between spatial diversity and spatial multiplexing. The set of encoding schemes allowing adjustment and optimization of joint encoding of multiple transmit antennas are called *spacetime codes* (STCs). There are various approaches in the spacetime coding structures, including spacetime block codes (STBCs), spacetime trellis codes (STTCs), spacetime turbo codes, as well as layered spacetime codes (LST).

The first three structures are targeted to exploit the full-diversity order of the spatial channels so as to maximize the reliability of packet transmission. The fourth structure (LST) is a different design target used to exploit spatial multiplexing gain of the spatial channels so as to maximize the bit rate or capacity. In Section 4.2.1, we shall focus on the designs and structures of spacetime coding for spatial diversity with emphasis on STBC. In Section 4.2.2, we shall discuss the designs and structures of the layered spacetime coding structure for spatial multiplexing. Finally, we discuss the fundamental tradeoff between spatial diversity and spatial multiplexing in MIMO channels.

4.2.1 Spacetime Coding—Spatial Diversity

In wireless mobile communications, diversity techniques are widely used to reduce the effect of fading so as to reduce the error probability of packet transmission without increasing the transmit power. Diversity techniques require multiple *independent* observations (with respect to the same information) to be available at the receiver. In most wireless systems, a number of methods are used to exploit the diversity, namely, *time diversity*, *frequency diversity*, and *spatial diversity*. Time diversity can be achieved by transmitting identical messages in different timeslots (longer than the coherence time of the channels). In practice, replicas of the information signals are usually provided to the receiver in the form of redundancy in the time domain introduced by the error correction coding. In frequency diversity, a number of different frequencies (separated by more than the coherence bandwidth of the channels) are used to transmit the same message. For example, direct-sequence spread-spectrum, multicarrier modulation as well as frequency hopping are practical systems utilizing frequency diversity to improve the reliability of packet transmission. In space diversity, independent observations of the same message are provided to the receiver through different antennas (separated by more than the coherence distance of the scattering channels). There are two types of antenna diversity: *receive diversity* and *transmit diversity*. In this subsection, we shall discuss a general design framework to fully exploit both transmit diversity and receive diversity for slow fading MIMO channels.

4.2.1.1 Transmit Diversity with Alamouti Codes. In transmit diversity, messages to be transmitted are usually processed at the transmitter and then sent from multiple transmit antennas. In contrast to receive diversity, transmit diversity has some technical difficulties. Since the transmit signals from multiple antennas are mixed spatially before they arrive at the receiver, some additional signal processing is required at both the transmitter and the receiver in order to separate the received signals and exploit diversity. Therefore, the design of transmit diversity is not straightforward.

Alamouti [9] proposed a first simple transmit diversity scheme for $n_T = 2$ and $n_R = 1$ or $n_R = 2$ based on STBC structure. Figure 4.3 is a block diagram of the transmit diversity scheme based on Alamouti codes.

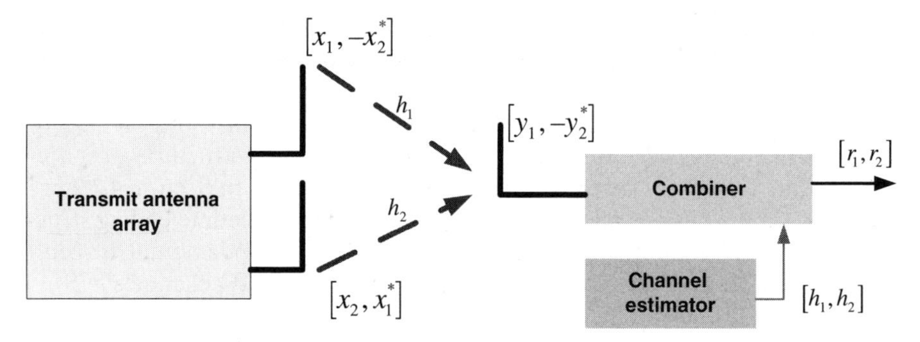

Figure 4.3. Block diagram of transmit diversity based on the Alamouti code.

The modulation outputs $\{x_1, x_2\}$ are transmitted on the basis of two consecutive channel uses. During the first transmission period, the two signals x_1, x_2 are transmitted simultaneously from antennas 1 and 2, respectively. During the second channel use, the signal $-x_2^*$ is transmitted from transmit antenna 1 and signal x_1^* is transmitted from transmit antenna 2. Hence, the codeword of the STBC is given by

$$X = \begin{pmatrix} x_1 & -x_2^* \\ x_2 & x_1^* \end{pmatrix}$$

The code rate (bits per channel use) is given by

$$R_b = \frac{2b}{2} = b \tag{4.1}$$

where x_i is drawn from a constellation of 2^b points.

For a single-antenna receiver, the received signals over the two channel uses are given by

$$\begin{bmatrix} y_1 \\ y_2 \end{bmatrix} = \begin{bmatrix} h_1 x_1 + h_2 x_2 \\ -h_1 x_2^* + h_2 x_1^* \end{bmatrix} + \begin{bmatrix} z_1 \\ z_2 \end{bmatrix}$$

Expressing the received signals $\mathbf{y}$ in vector form in terms of the transmit signals $\mathbf{x}$, we have

$$\mathbf{y} = \begin{bmatrix} y_1 \\ y_2^* \end{bmatrix} = \underbrace{\begin{bmatrix} h_1 & h_2 \\ h_2^* & -h_1^* \end{bmatrix}}_{\tilde{\mathbf{H}}} \underbrace{\begin{bmatrix} x_1 \\ x_2 \end{bmatrix}}_{\mathbf{x}} + \underbrace{\begin{bmatrix} z_1 \\ z_2 \end{bmatrix}}_{\mathbf{z}} = \tilde{\mathbf{H}}\mathbf{x} + \mathbf{z} \tag{4.2}$$

where y_i denotes the received signal from the ith channel use.

Observe that the channel matrix $\tilde{\mathbf{H}}$ is nonsingular and $\tilde{\mathbf{H}}^{-1} = \frac{\tilde{\mathbf{H}}^*}{\|\mathbf{h}\|^2}$, where $\|\mathbf{h}\|^2 = (|h_1|^2 + |h_2|^2)$. Hence, without loss of information, we can premultiply the equation with $\tilde{\mathbf{H}}^*$ on both sides:

$$\mathbf{r} = \tilde{\mathbf{H}}^* \mathbf{y} = \left(\|\mathbf{h}\|^2\right)\mathbf{x} + \tilde{\mathbf{H}}^* \mathbf{z} \tag{4.3}$$

The maximum-likelihood (ML) detection with perfect CSIR $\mathbf{h}$ is given by

$$\hat{\mathbf{x}} = \arg\max_{\mathbf{x}} p(\mathbf{r}|\mathbf{x}, \tilde{\mathbf{H}})$$

The covariance of the noise vector is given by

$$\mathbf{Q}_z = \varepsilon[\tilde{\mathbf{H}}^*\mathbf{z}\mathbf{z}\tilde{\mathbf{H}}] = \tilde{\mathbf{H}}^*\varepsilon[\mathbf{z}\mathbf{z}^*]\tilde{\mathbf{H}} = \|\mathbf{h}\|^2 \sigma_z^2 \mathbf{I}$$

Hence, the conditional density of $\mathbf{r}$ given $\mathbf{x}$ is jointly Gaussian with mean $(\|\mathbf{h}\|^2)\mathbf{x}$ and covariance $\|\mathbf{h}\|^2\sigma_z^2\mathbf{I}$. The ML detection rule can be simplified as

$$\hat{\mathbf{x}} = \arg\min_{\mathbf{x}} \left|\mathbf{r} - \|\mathbf{h}\|^2 \mathbf{x}\right|^2 \tag{4.4}$$

Because of the diagonal covariance matrix of the noise term, the ML detection can be decoupled into

$$\hat{x}_i = \arg\min_{x_i} \left|r_i - \|\mathbf{h}\|^2 x_i\right|^2 \tag{4.5}$$

for $i \in \{1, 2\}$. Suppose that there are 2^b constellation points for x_1 and x_2. There ML detection complexity reduces from $2^b \times 2^b$ to $2^b + 2^b$ because of the decoupling. Furthermore, the sufficient statistic $\mathbf{r}$ can be expressed as

$$\mathbf{r} = \tilde{\mathbf{H}}^*\mathbf{y} = \|\mathbf{h}\|^2 \mathbf{x} + \underbrace{\tilde{\mathbf{H}}^*\mathbf{z}}_{\mathbf{v}}$$

where the noise term $\mathbf{v}$ has covariance given by $\|\mathbf{h}\|^2\sigma_z^2\mathbf{I}$. The signal-to-noise ratio (SNR) of the ML decision metric with respect to x_i is given by

$$\gamma_i = \frac{\|\mathbf{h}\|^2 \sigma_x^2(i)}{\sigma_z^2}$$

for i = 1,2. Hence, there is two-order diversity in the decision metric in Equation (4.5) due to the factor $\|\mathbf{h}\|^2 = |h_1|^2 + |h_2|^2$.

Similarly, for a receiver with n_R receive antennas, the received signals (over the two channel uses) $\mathbf{y}$ in vector form is given by

$$\underbrace{\begin{bmatrix} y_{1,1} \\ y_{1,2}^* \\ \vdots \\ y_{nR,1} \\ y_{nR,2}^* \end{bmatrix}}_{\mathbf{y}} = \underbrace{\begin{bmatrix} h_{11} & h_{12} \\ h_{12}^* & -h_{11}^* \\ \vdots & \vdots \\ h_{nR1} & h_{nR2} \\ h_{nR2}^* & -h_{nR1}^* \end{bmatrix}}_{\tilde{\mathbf{H}}} \underbrace{\begin{bmatrix} x_1 \\ x_2 \end{bmatrix}}_{\mathbf{x}} + \underbrace{\begin{bmatrix} z_{1,1} \\ z_{1,2}^* \\ \vdots \\ z_{nR,1} \\ z_{nR,2}^* \end{bmatrix}}_{\mathbf{z}}$$

where $y_{i,j}$ denotes the received signal from the ith received antenna during the jth channel use and $h_{i,k}$ denotes the channel fading between the ith receive antenna and the kth transmit antenna. Unlike the single-receive-antenna case, we cannot follow the same approach directly by claiming that $\tilde{\mathbf{H}}^*\mathbf{y}$ is a sufficient statistic with respect to $\mathbf{x}$ because $\tilde{\mathbf{H}}$ is singular. Since the covariance matrix of the Gaussian vector noise $\mathbf{z}$ is $\sigma_z^2\mathbf{I}$, the ML detection is given by

$$\begin{aligned}\hat{\mathbf{x}} &= \arg\min_{\mathbf{x}}(\mathbf{y}-\tilde{\mathbf{H}}\mathbf{x})^*(\mathbf{y}-\tilde{\mathbf{H}}\mathbf{x}) \\ &= \arg\min_{\mathbf{x}}(\mathbf{y}^*\mathbf{y}-\mathbf{x}^*\tilde{\mathbf{H}}^*\mathbf{y}-\mathbf{y}^*\tilde{\mathbf{H}}\mathbf{x}+\mathbf{x}\tilde{\mathbf{H}}^*\tilde{\mathbf{H}}\mathbf{x}) \\ &= \arg\min_{\mathbf{x}}(-\mathbf{x}^*\tilde{\mathbf{H}}^*\mathbf{y}-\mathbf{y}^*\tilde{\mathbf{H}}\mathbf{x}+\mathbf{x}^*\tilde{\mathbf{H}}^*\tilde{\mathbf{H}}\mathbf{x})\end{aligned}$$

Since $\tilde{\mathbf{H}}^*\mathbf{H} = \|\mathbf{h}\|^2\mathbf{I}_2$, where $\|\mathbf{h}\|^2 = (|h_{11}|^2 + |h_{12}|^2 + \cdots + |h_{n_R1}|^2 + |h_{n_R2}|^2)$, the ML detection rule can be further simplified as

$$\begin{aligned}\hat{\mathbf{x}} &= \arg\min_{\mathbf{x}}\left(-\mathbf{x}^*\tilde{\mathbf{H}}^*\mathbf{y}-\mathbf{y}^*\tilde{\mathbf{H}}\mathbf{x}+\|\mathbf{h}\|^2\|\mathbf{x}\|^2\right) \\ &= \arg\min_{\mathbf{x}}\left(-\mathbf{x}^*\mathbf{r}-\mathbf{r}^*\mathbf{x}+\|\mathbf{h}\|^2\|\mathbf{x}\|^2\right)\end{aligned}$$

where $\mathbf{r} = \tilde{\mathbf{H}}^*\mathbf{y}$. In other words, it turns out that $\mathbf{r}$ is a sufficient statistic with respect to the detection of $\mathbf{x}$. In addition, ML detection can be decoupled into two ML detection searches as

$$\hat{x}_i = \arg\min_{x_i}\left[|x_i|^2\|\mathbf{h}\|^2 - r_i x_i^* - r_i^* x_i\right]$$

for $i = \{1, 2\}$. Finally, similar to the single antenna situation, we have $2 \times n_R$ order of diversity due to the term $\|\mathbf{h}\|^2$.

4.2.1.2 Error Analysis of Spacetime Codes. Consider a general spacetime codebook Ω that is a table of 2^{NR} entries of codewords. The mth spacetime codeword is denoted by $\tilde{\mathbf{x}}(m) = [\mathbf{x}_1(m), \ldots, \mathbf{x}_N(m)]$, where $\mathbf{x}_n(m)$ is a $n_T \times 1$ vector with covariance $\mathbf{I}$. The $n_R \times N$ received signal $\tilde{\mathbf{y}}$ as a result of transmitting the codeword through slow flat fading channels is given by

$$\tilde{\mathbf{y}} = \sqrt{\frac{E_s}{n_T}}\mathbf{h}\tilde{\mathbf{x}}(m) + \tilde{\mathbf{z}} \tag{4.6}$$

where E_s is the total transmit symbol energy (over all transmit antennas), $\tilde{\mathbf{z}} = [\mathbf{z}_1, \ldots, \mathbf{z}_N]$ is the channel noise matrix over the N channel uses, and $\mathbf{z}_n$ is the $n_R \times 1$ complex channel noise with covariance $\eta_0\mathbf{I}$. The likelihood density function is given by

$$p(\tilde{\mathbf{y}}|\tilde{\mathbf{x}}(m),\mathbf{h}) = K \exp\left\{-tr\left[\left(\tilde{\mathbf{y}} - \sqrt{\frac{E_s}{n_T}}\mathbf{h}\tilde{\mathbf{x}}(m)\right)\left(\tilde{\mathbf{y}} - \sqrt{\frac{E_s}{n_T}}\mathbf{h}\tilde{\mathbf{x}}(m)\right)^*\right]\right\}$$

where K is a constant. Hence, the ML decoding rule is given by

$$\hat{\tilde{x}} = \arg\min_{\tilde{\mathbf{x}}} \operatorname{tr}\left[\left(\tilde{\mathbf{y}} - \sqrt{\frac{E_s}{n_T}}\mathbf{h}\tilde{\mathbf{x}}\right)\left(\tilde{\mathbf{y}} - \sqrt{\frac{E_s}{n_T}}\mathbf{h}\tilde{\mathbf{x}}\right)^*\right] \tag{4.7}$$

Without loss of generality, assume that $\tilde{\mathbf{x}}(i)$ is transmitted. $\tilde{\mathbf{x}}(j)$ will be selected if

$$\begin{aligned}
&\operatorname{tr}(\tilde{\mathbf{y}} - \mathbf{h}\mathbf{x}(i))(\tilde{\mathbf{y}} - \mathbf{h}\tilde{\mathbf{x}}(i))^* > \operatorname{tr}(\tilde{\mathbf{y}} - \mathbf{h}\tilde{\mathbf{x}}(j))(\tilde{\mathbf{y}} - \mathbf{h}\tilde{\mathbf{x}}(j))^* \\
&\Rightarrow \operatorname{tr}(\tilde{\mathbf{z}}\tilde{\mathbf{z}}^*) > \operatorname{tr}(\tilde{\mathbf{z}} - \mathbf{h}\tilde{\mathbf{x}}_{ji})(\tilde{\mathbf{z}} - \mathbf{h}\tilde{\mathbf{x}}_{ji})^* \\
&\Rightarrow \operatorname{tr}(\tilde{\mathbf{z}}\tilde{\mathbf{z}}^*) > \operatorname{tr}(\tilde{\mathbf{z}}\tilde{\mathbf{z}}^*) + \frac{E_s}{n_T}\operatorname{tr}(\mathbf{h}\tilde{\mathbf{x}}_{ji}\tilde{\mathbf{x}}_{ji}^*\mathbf{h}^*) - 2\operatorname{tr}\operatorname{Re}\left[\sqrt{\frac{E_s}{n_T}}\mathbf{h}\tilde{\mathbf{x}}_{ji}\tilde{\mathbf{z}}^*\right] \\
&\Rightarrow \frac{E_s}{n_T}\operatorname{tr}(\mathbf{h}\tilde{\mathbf{x}}_{ji}\tilde{\mathbf{x}}_{ji}^*\mathbf{h}^*) < 2\sqrt{\frac{E_s}{n_T}}\operatorname{tr}\operatorname{Re}[\mathbf{h}\tilde{\mathbf{x}}_{ji}\tilde{\mathbf{z}}^*]
\end{aligned} \tag{4.8}$$

where $\tilde{\mathbf{x}}_{ji} = \tilde{\mathbf{x}}(j) - \tilde{\mathbf{x}}(i)$. Define $d^2(\tilde{\mathbf{x}}(i), \tilde{\mathbf{x}}(j)) = \frac{E_s}{n_T}\operatorname{tr}\left(\mathbf{h}\tilde{\mathbf{x}}_{ji}\tilde{\mathbf{x}}_{ji}^*\mathbf{h}^*\right)$. The left side in Equation (4.8) is a constant given $\mathbf{h}$, and the right side is a zero-mean Gaussian random variable with variance $2\eta_0 d^2(\tilde{\mathbf{x}}(i), \tilde{\mathbf{x}}(j))$. Hence, the conditional probability of decoding error $P_e(i \to j|\mathbf{h})$ is given by

$$\begin{aligned}
P_e(i \to j|\mathbf{h}) &= Q\left(\sqrt{\frac{d^4(\tilde{\mathbf{x}}(i), \tilde{\mathbf{x}}(j))}{2\eta_0 d^2(\tilde{\mathbf{x}}(i), \tilde{\mathbf{x}}(j))}}\right) \\
&= Q\left(\sqrt{\frac{d^2(\tilde{\mathbf{x}}(i), \tilde{\mathbf{x}}(j))}{2\eta_0}}\right) \le \frac{1}{2}\exp-\left[\frac{E_s}{4\eta_0 n_T}\operatorname{tr}\left(\mathbf{h}\tilde{\mathbf{x}}_{ji}\tilde{\mathbf{x}}_{ji}^*\mathbf{h}^*\right)\right]
\end{aligned} \tag{4.9}$$

Define a signal matrix $\mathbf{A}_{ij} = \tilde{\mathbf{x}}_{ji}\tilde{\mathbf{x}}_{ji}^* = \mathbf{V}\boldsymbol{\Lambda}\mathbf{V}^*$, where $\mathbf{V}$ is the eigenvector matrix and $\boldsymbol{\Lambda}$ is the eigenvalue diagonal matrix of the $n_T \times n_T$ Hermitian matrix $\mathbf{A}_{ij}$. We have

$$\begin{aligned} d^2(\tilde{\mathbf{x}}(i),\tilde{\mathbf{x}}(j)) &= \frac{E_s}{n_T}\sum_{j=1}^{n_T}\mathbf{h}_j\mathbf{V}\mathbf{\Lambda}\mathbf{V}^*\mathbf{h}_j^* \\ &= \frac{E_s}{n_T}\sum_{j=1}^{n_T}\mathbf{w}_j\mathbf{\Lambda}\mathbf{w}_j^* \\ &= \frac{E_s}{n_T}\sum_{j=1}^{n_T}\sum_{i=1}^{n_T}\lambda_i|w_{ij}|^2 \end{aligned}$$

where

$$\mathbf{h} = \begin{bmatrix} \mathbf{h}_1 \\ \vdots \\ \mathbf{h}_{nR} \end{bmatrix}$$

and $\mathbf{w}_j = \mathbf{h}_j\mathbf{V}$ and it has the same distribution as $\mathbf{h}_j$ because $\mathbf{V}$ is unitary. Therefore, the conditional error probability can be written as follows:

$$P_e(i \to j|\mathbf{h}) \le \frac{1}{2}\prod_{j=1}^{n_R}\exp-\left(\frac{E_s}{4\eta_0 n_T}\sum_{i=1}^{n_T}\lambda_i|w_{ij}|^2\right) \tag{4.10}$$

Unconditioning with respect to $|w_{ij}|^2$, the unconditional error probability is given by

$$\begin{aligned} P_e(i\to j) &\le \frac{1}{2}\prod_{j=1}^{n_R}\prod_{i=1}^{n_T}\int_0^\infty \exp-\left(\frac{E_s\lambda_i|w_{ij}|^2}{4\eta_0 n_T}\right)f\left(|w_{ij}|^2\right)d|w_{ij}|^2 \\ &= \frac{1}{2}\prod_{j=1}^{n_R}\prod_{i=1}^{n_T}\left(\frac{1}{\frac{E_s\lambda_i}{4\eta_0 n_T}+1}\right) \\ &\le \left(\prod_{i=1}^{m}\lambda_i\right)^{-n_R}\left(\frac{E_s}{4\eta_0 n_T}\right)^{-mn_R} \end{aligned} \tag{4.11}$$

where the upper bound is valid for high SNR and $m \le n_T$ is the rank of the signal matrix $\mathbf{A}_{ij}$, which is the number of nonzero eigenvalues $\{\lambda_i\}$. The exponent of the SNR term in Equation (4.11), mn_R, represents the diversity gain of the spacetime code. Hence, when $m = n_T$, the spacetime code is said to achieve *full diversity*. Diversity order is important for the BER performance at high SNR because it determines the slope of the BER curves. On the other hand, the term $\Pi_{i=1}^{m}\lambda_i$ in Equation (4.11) represents the coding gain of the spacetime code and is important to BER performance at small to moderate SNR. Hence, the code design criteria of the spacetime code can be summarized as follows:

The Rank Criteria. To achieve the maximum diversity order $n_T n_R$, the spacetime code should be designed such that the signal matrix $\mathbf{A}_{ij}$ from any two distinct codewords $\tilde{\mathbf{x}}(i)$ and $\tilde{\mathbf{x}}(j)$ always have full rank.

The Determinant Criteria. The product of the eigenvalues of the signal matrix $\mathbf{A}_{ij}$ determines the coding gain. When the spacetime code is full rank, the coding gain is given by $|\mathbf{A}_{ij}|$.

Similar to the coding design for scalar channels, a useful parameter to characterize the amount of information transmitted per codeword transmission is the *code rate*. Let $B(n_T, d_H)$ be the set of spacetime codewords with n_T transmit antennas and minimum Hamming distance d_H. The Hamming distance $d_H(\tilde{\mathbf{x}}(i), \tilde{\mathbf{x}}(j))$ between two distinct codewords is given by

$$d_H(\tilde{\mathbf{x}}(i), \tilde{\mathbf{x}}(j)) = \sum_{n=1}^{N} \mathbf{1}(\mathbf{x}_n(i) \neq \mathbf{x}_n(j))$$

where $\mathbf{1}(E)$ is an indicator function about event E. The code rate R_c of the spacetime code (bits per channel use) is upper-bounded by

$$R_c \le \frac{\log_2 |\mathcal{B}(n_T, d_H)|}{N} \tag{4.12}$$

where $|\mathcal{B}|$ denotes the cardinality of the set $\mathcal{B}$. In particular, if we require full-rank spacetime code, we need to have $d_H = n_T$ and $R_c(\max) \le \frac{\log_2 |\mathcal{B}(n_T, n_T)|}{N} = b$ (bits per channel use) where there are 2^b constellation points per transmit symbol. A spacetime code is called *full rate* if the bit rate of the spacetime code achieves the uppder bound b bits per channel use when the diversity order is $d_H = n_T$. We shall elaborate on the tradeoff between rate and diversity order in Section 4.2.6. We consider several examples of full-diversity spacetime codes below.

Example 4.1 (Real Orthogonal Spacetime Codes) Consider a 4×4 real orthogonal spacetime block code design given by

$$\tilde{\mathbf{x}} = \begin{bmatrix} x_1 & x_2 & x_3 & x_4 \\ -x_2 & x_1 & -x_4 & x_3 \\ -x_3 & x_4 & x_1 & -x_2 \\ -x_4 & -x_3 & x_2 & x_1 \end{bmatrix}$$

where the rows of the codebook are mutually orthogonal and four channel uses are needed to transmit independent information $\{x_1, x_2, x_3, x_4\}$ through the four transmit antennas. Note that x_i are assumed to be real transmit symbols. Observe that the determinant of the signal matrix is given by

$$|\mathbf{A}(\tilde{\mathbf{x}},\tilde{\mathbf{x}})| = |(\tilde{\mathbf{x}}-\tilde{\mathbf{x}})(\tilde{\mathbf{x}}-\tilde{\mathbf{x}}')^*| = \left(\sum_i |x_i - x_i'|^2\right)^2 \neq 0$$

Hence, the 4 × 4 orthogonal spacetime block code can achieve full diversity order of $4n_R$. The code rate is given by $R_c = 4b/4 = b$ bits per channel use and therefore, it's also full-rate. Similar to the Alamouti codes, the ML detection rule for the orthogonal spacetime code can be decoupled into four individual ML rules and therefore, the complexity is given by 4×2^b rather than $2^b \times 2^b \times 2^b \times 2^b$. However, it has been shown [140] that real orthogonal codes exist only for $n_T = 2,4,8$.

Example 4.2 (Generalized Complex Orthogonal Spacetime Codes) Let $\tilde{\mathbf{x}}$ be a codeword of the generalized complex orthogonal spacetime codes where the transmit symbols are drawn from a complex constellation. We have $\tilde{\mathbf{x}}\tilde{\mathbf{x}}^* = \alpha\mathbf{I}$. It can be shown that for rate $R_c \leq b/2$ (bits per use), there exists such generalized complex orthogonal design [140]. For example, a rate $b/2$ generalized complex orthogonal design with $n_T = 3$ is given by

$$\begin{bmatrix} x_1 & -x_2 & -x_3 & -x_4 & x_1^* & -x_2^* & -x_3^* & -x_4^* \\ x_2 & x_1 & x_4 & -x_3 & x_2^* & x_1^* & x_4^* & -x_3^* \\ x_3 & -x_4 & x_1 & x_2 & x_3^* & -x_4^* & x_1^* & x_2^* \end{bmatrix}$$

This code achieves the full diversity order of $3n_R$ but cannot achieve full rate.

Although we use spacetime block codes for illustration purposes in this section, the same design criteria (rank and determinant criteria) can be extended to the spacetime trellis code in a straightforward manner [140].

4.2.2 Layered Spacetime Coding—Spatial Multiplexing

In Section 4.2.1, we have discussed the design of spacetime coding structures to exploit the spatial diversity so as to enhance the reliability of packet transmission. To achieve full order diversity of n_T, full-rank signal matrix $\mathbf{A}_{ij}$ is required. In addition, an ML receiver is also required to exploit the full-diversity order. Sometimes, when full-diversity order is not required, we can increase the bit rate of the encoder by exploiting spatial multiplexing. However, regular spacetime coding and decoding structures impose a potential drawback in that the ML decoding complexity in general grows exponentially with the number of bits per channel use, thus limiting the achievable data rate. In this section, we focus on the design of a special spacetime coding struc-

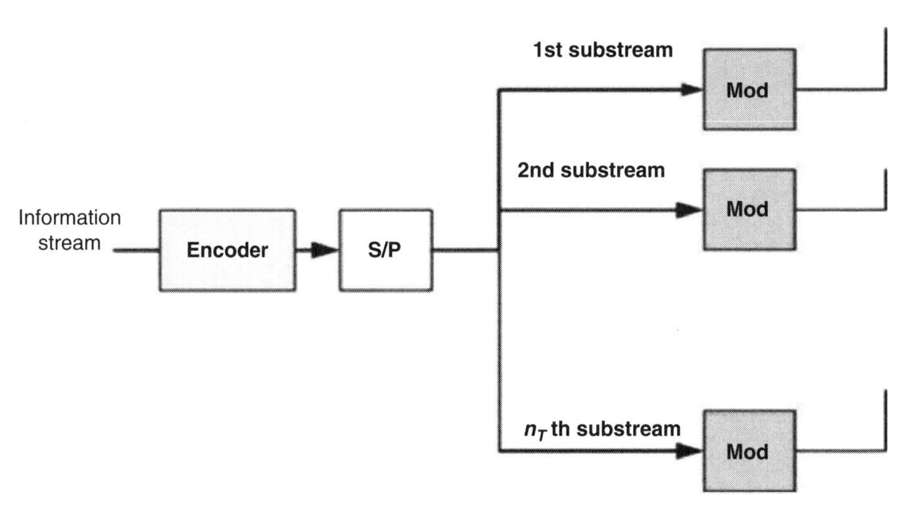

Figure 4.4. Block diagram of the V-BLAST transmitter.

ture, namely, *layered spacetime coding*, as well as the corresponding receiver structures to exploit spatial multiplexing with the objective of increasing bit rate rather than diversity order.

Foschini and Gans [41] proposed a layered spacetime architecture, namely, the *BLAST* architecture, which can achieve significant spatial multiplexing gains with simple receiver processing. Here, n_T independent streams of information are transmitted at the transmitter. The receiver uses $n_R \geq n_T$ antennas to separate the signals from the n_T streams of information by spatial signal processing. One distinctive feature of the BLAST design is that it allows processing of multidimensional signals in the space domain by successive one-dimensional processing steps in space. There are various possible designs in the BLAST architecture, reviewed below.

4.2.2.1 Vertical BLAST. Figure 4.4 illustrates the design of V-BLAST transmitter. Information bits are first encoded and split into n_T substreams. Each substream of bits is modulated by an M-ary modulation, and the modulated symbols are transmitted by the respective antenna. The transmission matrix of the V-BLAST transmitter after N channel uses is given by

$$\tilde{\mathbf{x}} = \begin{bmatrix} x_1(1) & x_2(1) & \dots & x_N(1) \\ x_1(2) & x_2(2) & \dots & \vdots \\ \vdots & \vdots & \vdots & \vdots \\ x_1(n_T) & x_2(n_T) & \dots & x_N(n_T) \end{bmatrix}$$

where $x_t(n)$ is the coded and modulated symbol transmitted at the nth antenna and the tth symbol duration. Hence, the encoded information bits are transmitted across multiple antennas and experience different fading even in slow fading channels. The data rate or spectral efficiency of the V-BLAST system is given by

$$r_b = r_c \log_2(M) n_T \quad \text{bits per channel use} \tag{4.13}$$

where r_c is the encoding rate of the channel encoder and M is the modulation level.

4.2.2.2 Horizontal BLAST. In H-BLAST design, information bits are split into n_T substreams. Each information substream is encoded and modulated separately. Each modulated substream is carried by one of the n_T transmit antennas. The channel encoders of various layers can be identical or different as illustrated in Figure 4.5a. In the H-BLAST design, the coded bits are carried through one transmit antenna only and therefore, all the coded bits will see the same fading in slow fading channels.

The signal matrix of the H-BLAST transmitter is shown in Figure 4.5b. The transmission matrix $\tilde{\mathbf{x}}$ is partitioned into *layers* where the first layer corresponds to the coded symbols from encoder 1 (delivered by antenna 1) and the second layer corresponds to the coded symbols from encoder 2 (delivered by antenna 2) and so on. Finally, the spectral efficiency of H-BLAST is given by $r_c \log_2(M) n_T$ bits per channel use.

4.2.2.3 Diagonal BLAST. As we have illustrated, there is a disadvantage in H-BLAST in that the encoded symbols in any encoder are transmitted through one transmit antenna only. In D-BLAST, the encoded symbols are distributed evenly across the n_T transmit antennas as illustrated in Figure 4.6a.

Similar to the case for the H-BLAST design, information bits are split into n_T substreams. Each substream is separately encoded and modulated. However, the modulated symbols from the n_T branches are interleaved by a *spatial interleaver* so that the coded symbols from any coded stream are evenly distributed across the n_T antennas. The signal matrix of D-BLAST design is illustrated in Figure 4.6b, and the spatial interleaving operation is summarized as follows. The transmission matrix is formed by delaying the ith row by $(i-1)$ symbols. Entries below the diagonal are padded by zero. The first diagonal is transmitted from the first antenna, followed by the second diagonal (transmitted from the second antenna) and so on. Hence, the codeword of the mth encoder, $[x_1(m), x_2(m), \ldots, x_{n_T}(m)]$, is transmitted on different antennas as illustrated in the figure. Hence, the diagonal layering introduces space diversity in the codeword of each layer and therefore achieves a better performance compared with H-BLAST. Compared with H-BLAST, there is a spectral efficiency loss due to the zero padding in the signal transmission matrix.

In the original version of D-BLAST, the span of each codeword is restricted to the width of a layer, which is n_T, in order to facilitate layer-by-layer receiver

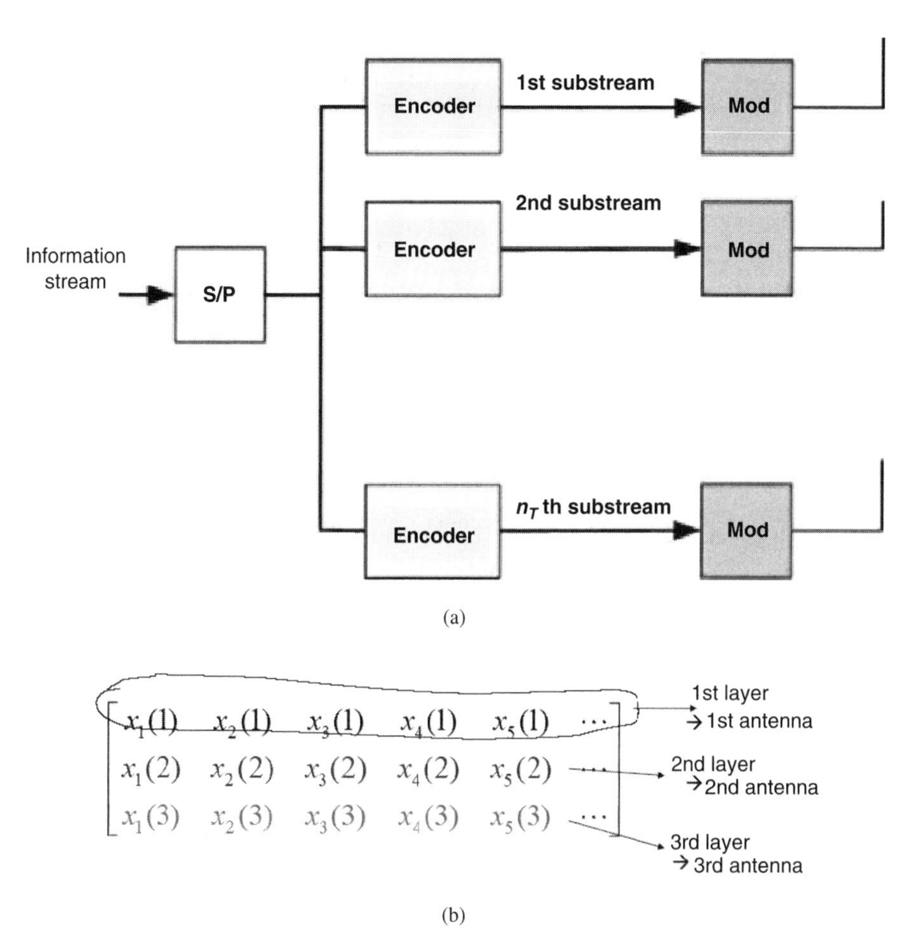

Figure 4.5. Block diagram of the H-BLAST transmitter: (a) H-BLAST transmitter architecture; (b) H-BLAST signal matrix.

processing. In other words, a codeword from one encoder is not allowed to span over more than one cycle of the layering in Figure 4.6b. This limitation can be relaxed by considering a hybrid design between H-BLAST and D-BLAST in the spatial interleaver. This is illustrated in Figure 4.7. For example, layer 1 is given by the *block diagonal* strip consisting of $[x_1(1), x_2(1), \ldots, x_5(1), x_6(1)]$. With the hybrid design, we can support codewords of lengths that are multiples of n_T. Finally, the spectral efficiency of the D-BLAST design is given by

$$r_b = r_c \log_2(M) n_T \frac{N - n_T + 1}{N} \tag{4.14}$$

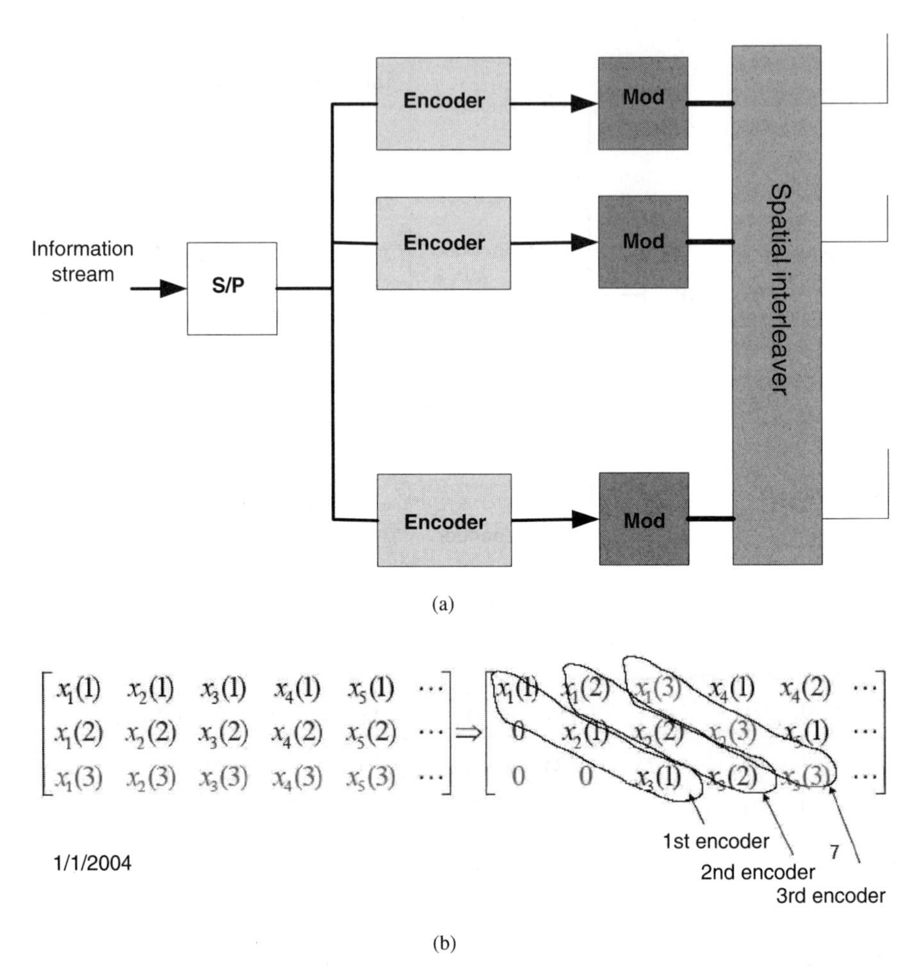

Figure 4.6. Block diagram of the D-BLAST transmitter: (a) D-BLAST transmitter architecture; (b) D-BLAST signal matrix.

$$\begin{bmatrix} x_1(1) & x_2(1) & x_1(2) & x_2(2) & x_1(3) & x_2(3) \\ 0 & 0 & x_3(1) & x_4(1) & x_3(2) & x_4(2) \\ 0 & 0 & 0 & 0 & x_5(1) & x_6(1) \end{bmatrix}$$

Figure 4.7. Spatial interleaving of hybrid H-BLAST/D-BLAST.

$$\begin{bmatrix} x_1(1) & x_2(1) & x_3(1) & x_4(1) & x_5(1) \\ x_1(2) & x_2(2) & x_3(2) & x_4(2) & x_5(2) \\ x_1(3) & x_2(3) & x_3(3) & x_4(3) & x_5(3) \end{bmatrix} \Rightarrow \begin{bmatrix} x_1(1) & x_2(3) & x_3(2) & x_4(1) & x_5(3) \\ x_1(2) & x_2(1) & x_3(3) & x_4(2) & x_5(1) \\ x_1(3) & x_2(2) & x_3(1) & x_4(3) & x_5(2) \end{bmatrix}$$

Figure 4.8. Illustration of threaded LST design.

where N is the total number of channel uses for the D-BLAST codeword. Note that the spectral efficiency is slightly smaller than that of the V-BLAST or H-BLAST because of the overhead of zero padding but as $N \to \infty$, the spectral efficiency approaches that of the V-BLAST or H-BLAST.

4.2.2.4 Threaded Layered Spacetime Code. The transmitter design of threaded layered spacetime coding (TLST) [45] is similar to the format in Figure 4.6a. The only difference is the way spatial interleaver operates. In TLST, we have a generalized notion of layer called *thread*. A generalized layer is defined by an index set

$$\mathcal{L} = \{(a,t) : a \in [1, n_T], t \in [1, N]\}$$

such that if $(a_1, t_1) \in \mathcal{L}$ and $(a_2, t_2) \in \mathcal{L}$, then either $a_1 = a_2$ or $t_1 \neq t_2$; that is, at any given time, no symbols from the same layer occupy more than one antenna. A generalized layer is called a *thread* if

- Each layer is active during the entire transmission interval.
- Over time, each layer uses the n_T transmit antennas equally often.

For example, the generalized layers $\{\mathcal{L}_1, \ldots, \mathcal{L}_{n_T}\}$ defined below are *threads*:

$$\mathcal{L}_i = \{(a,t) : a = [(i + t - 2) \bmod n_T] + 1\}.$$

This is illustrated in Figure 4.8 for $n_T = 3$.

Unlike D-BLAST, the encoded symbols from the ith thread is distributed to the transmit antenna a and time t according to the thread index $\mathcal{L}_i$ without the need for zero padding. In addition, the encoded block length can be longer than n_T. Since the encoded symbols from each thread are delivered to the receiver through different transmit antennas, the TLST architecture allows full diversity within each codeword and is therefore outage-optimal in slow MIMO fading channels as will be shown in Section 4.2.5. Yet, TLST design requires more sophisticated receive processing. The spectral efficiency of the TLST design is given by $r_b = r_c \log_2(M) n_T$ bits per channel use.

4.2.3 Receiver Designs for Layered Spacetime Codes

In this section, we shall consider the receiver structures for layered spacetime architectures. The received signal is given by

$$\mathbf{y} = \mathbf{hx} + \mathbf{z} = \sum_{k=1}^{n_T} \mathbf{h}_k x_k + \mathbf{z}$$

where $\mathbf{h}_k$ is the kth column of the $n_R \times n_T$ channel matrix $\mathbf{h}$ and x_k is the transmitted symbol from the kth transmit antenna. Hence, the LST receiver is in general very similar to a multiuser detector in which the number of transmit antennas is equal to the number of users. Hence, the optimal receiver for uncoded LST systems is a ML multiuser detector [136] operating on a trellis and it has a complexity that is exponential in the number of transmit antennas n_T and modulation throughput $\log_2(M)$. For coded LST systems, the optimal receiver performs joint ML detection and decoding over the entire spatial domain and the entire temporal domain. It has an overall complexity given by the number of transmit antennas n_T, modulation throughput $\log_2(M)$, as well as the channel code memory order. For a moderate number of transmit antennas, modulation level, and code memory, the optimal receiver becomes impractical. Hence, in this section, we shall discuss various reduced-complexity receiver structures.

4.2.3.1 Zero-Forcing Receiver. Since a special *layering* structure is introduced at the transmitter in various LST designs, the receiver can exploit this layering structure to simplify the processing complexity. Figure 4.9a illustrates a linear receiver structure based on the zero-forcing (ZF) technique.

At the kth receive branch, the received vector $\mathbf{y}$ is projected onto the orthogonal subspace spanned by vectors $\{\mathbf{h}_1, \ldots, \mathbf{h}_{k-1}, \mathbf{h}_{k+1}, \ldots, \mathbf{h}_{n_T}\}$ (denoted by $\mathcal{V}_k$). Mathematically, the projected received vector $\mathbf{v}_k$ is given by

$$\mathbf{v}_k = \mathbf{P}_k \mathbf{y} = \mathbf{P}_k \mathbf{h}_k x_k + \mathbf{P}_k \mathbf{z} \tag{4.15}$$

Suppose that the dimension of the orthogonal subspace $\mathcal{V}_k$ is d_k, which is given by $d_k = n_R - \min[n_R, (n_T - 1)]$. The projection matrix is thus a $d_k \times n_R$ linear matrix with rows forming an orthonormal basis of $\mathcal{V}_k$. For $d_k \geq 1$, we must have $n_R \geq n_T$. After the zero-forcing processing, the equivalent channel between the information x_k and the observation $\mathbf{v}_k$ has the SNR given by $\frac{P_0 \|\mathbf{P}_k \mathbf{h}_k\|^2}{n_T \sigma_z^2}$. The ergodic capacity (fast fading channels) is given by

$$C_{\text{ZF}} = \sum_{k=1}^{n_T} \varepsilon \left[\log_2 \left(1 + \frac{P_0 \|\mathbf{P}_k \mathbf{h}_k\|^2}{n_T \sigma_z^2} \right) \right] \tag{4.16}$$

Since $\mathbf{v}_k$ is a projection, $\|\mathbf{P}_k \mathbf{h}_k\|^2$ is a chi-sqaure distribution $\chi^2_{2d_k}$ with degree $2d_k$. Also, by symmetry, the projection $\|\mathbf{P}_k \mathbf{h}_k\|^2$ is i.i.d for i.i.d. $\mathbf{h}$. Hence, the zero-forcing capacity can be written as

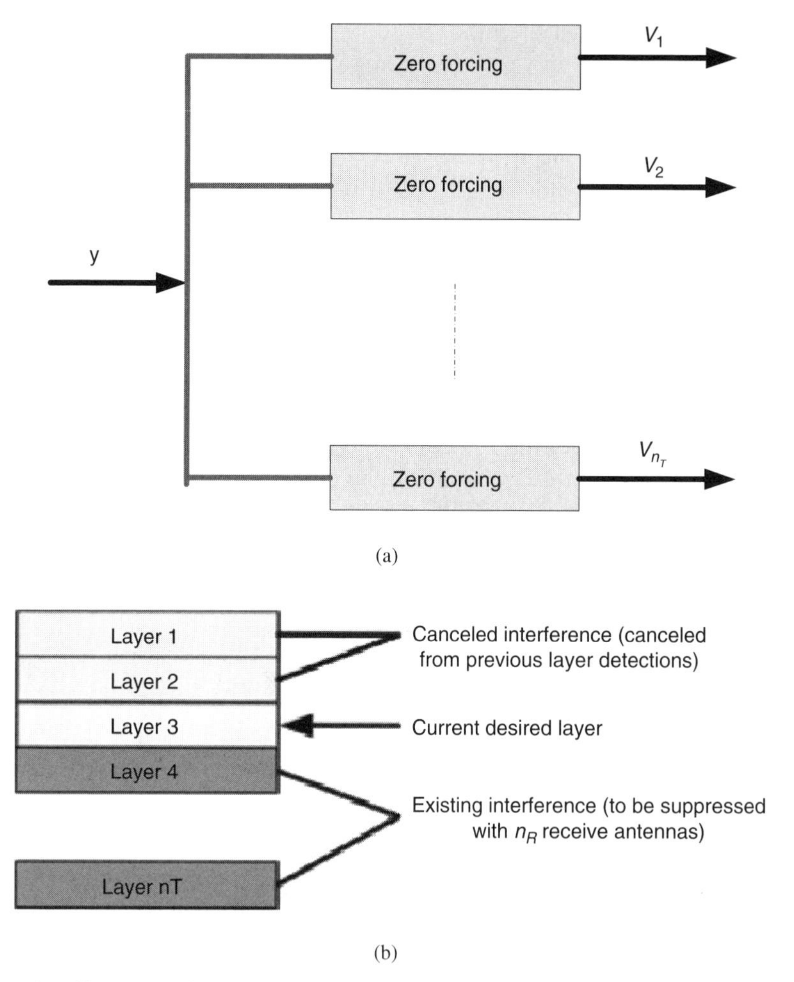

Figure 4.9. Illustration of zero-forcing receiver: (a) ZF receiver; (b) ZF with successive interference cancellation receiver.

$$\begin{aligned} C_{\mathrm{ZF}} &= \sum_{k=1}^{n_T} \varepsilon\left[\log_2\left(1+\frac{\mathrm{SNR}}{n_T}\chi_{2d}^2\right)\right] \\ &\approx n_T \log_2\left(\frac{\mathrm{SNR}}{n_T}\right) + n_T \varepsilon[\log_2(\chi_{2d}^2)] \end{aligned} \tag{4.17}$$

where $\mathrm{SNR} = \dfrac{P_0}{n_T}$ and $d = n_R - n_T + 1$. Comparing the capacity equation with the optimal ergodic capacity, we found that the first-order term is the same and therefore, the zero-forcing receiver is able to fully exploit the spatial degrees of freedom. On the other hand, the second term illustrates the performance degradation of the zero-forcing receiver compared to the optimal receiver. Hence, we can deduce that the zero-forcing receiver achieves the optimal ergodic capacity to the first order in high SNR. This is expected as zero forcing is well known to approach optimal performance at high SNR. The penalty of the zero-forcing structure is severe only in the low SNR region.

In fact, we can do more than the regular zero-forcing processing as illustrated in Figure 4.9b, where we have successive interference cancellation (SIC) together with a zero-forcing front end. For instance, interference from the other *unprocessed* substream is first suppressed (or nulled out in the zero-forcing case) by zero-forcing processing. Signals from one layer are extracted and decoded. If the decoding is succcessful, the contribution can be subtracted from the received signal $\mathbf{y}$ so that when we process the next layer, the interference that the receiver has to deal with is reduced by one. When processing the last layer, all the interference from the previous layers has been canceled and the current layer sees only the channel noise. Mathematically, with SIC structure, the dimension of the projection operation $\mathbf{P}_k$ at the kth stage of processing can be increased to $n_R - n_T + k$ because we need to project $\mathbf{y}$ onto the orthogonal subspace of $\{\mathbf{h}_{k+1}, \ldots, \mathbf{h}_{n_T}\}$ only. Therefore, the ergodic capacity of ZF-SIC receiver is given by

$$C_{\text{ZF-SIC}} \approx n_T \log_2\left(\frac{\text{SNR}}{n_T}\right) + \sum_{k=1}^{n_T} \left[\log_2\left(\chi^2_{2(n_R - n_T + k)}\right)\right] \tag{4.18}$$

In comparison with ZF capacity, the first-order term is the same and this indicates that SIC does not increase the degrees of freedom. Yet, the second term provides a constant gain [with respect to (w.r.t.) SNR] over the ZF capacity.

The ZF-SIC receiver is suitable for uncoded V-BLAST, coded H-BLAST, or coded D-BLAST. For uncoded V-BLAST, demodulation of the received signal from the current layer is based on "hard decision." The contribution of the layer is regenerated by modulation based on the hard decision and subtracted from the received signal. On the other hand, for H-BLAST, the entire length of received signal $[\mathbf{y}_1, \ldots, \mathbf{y}_N]$ is buffered. The soft information belonging to the kth layer, $[\mathbf{v}_k(1), \ldots, \mathbf{v}_k(N)]$, is passed to the channel decoder corresponding to that layer. The decoded bits are reencoded and remodulated to generate the contributions for further processing on the $(n + 1)$th layer. Hence, the layer corresponding to kth-stage processing enjoys a diversity order of $n_R - n_T + k$ uniformly across the codeword. For D-BLAST, the processing is similar to the H-BLAST. For example, consider the D-BLAST in Figure 4.6b. In order to extract the observation containing $x_3(1)$ of the first layer, the receiver has to suppress interference contributed by $x_1(3)$ and $x_2(2)$. Hence, there is only $n_R - n_T + 1$ order of diversity. Yet, to extract the observation containing $x_2(1)$ of the first layer, the receiver has to suppress interference contributed by $x_1(2)$ due only to zero padding at the transmitter. Hence, there is a diversity order of $n_R - n_T + 2$ for this symbol. Similarly, there is no interference when extracting $x_1(1)$ and there is a diversity order of $n_R - n_T + 3$. Once all the coded symbols of layer 1 is collected, they are passed to the channel decoder. The decoded bits are reencoded and remodulated. When we want to extract $x_1(2)$ for layer 2, the contribution from $x_2(1)$ is known and can be subtracted. Hence, in a similar manner, each layer sees the same composition of diversity order from $n_R - n_T + 1$ to n_R.

Since ZF-SIC relies on the successful decoding of the previous layers, there is error propagation problem; therefore, it is critical to make sure that the decoding order is in descending order of $\|\mathbf{P}_k\mathbf{h}_k\|$. In other words, the optimal decoding order is given by $\pi = (\pi_1, \ldots, \pi_{n_T})$ such that

$$\|\mathbf{P}_{\pi_k}\mathbf{h}_{\pi_k}\| \le \|\mathbf{P}_{\pi_{k-1}}\mathbf{h}_{\pi_{k-1}}\| \quad \forall k \in [2, n_T]$$

Note that the ZF-SIC structure is not applicable to the coded V-BLAST architecture. This is because to successfully decode a particular layer, we need to collect the soft information from the entire frame. Yet, the entire coded frame is delivered through various transmit antennas. Hence, it is not possible to cancel the contribution from any particular transmit antenna before the entire coded symbols are collected for soft decision.

4.2.3.2 MMSE Receiver. In addition to zero forcing, another possible linear processing receiver structure is the minimum mean-square error (MMSE) receiver. While ZF is asymptotically optimal at high SNR, there is a performance penalty at small SNR. As we shall illustrate in this section, MMSE can achieve a better performance even in the small SNR region. The MMSE receiver for LST designs is similar to the format in Figure 4.9a except that the ZF blocks are replaced by MMSE blocks. The basic MMSE processing is illustrated below.

The soft information corresponding to the transmit symbol x_k, namely, v_k, is given by premultiplying the received signal $\mathbf{y}$ with a $n_R \times 1$ weight vector $\mathbf{w}$:

$$v_k = \mathbf{w}_k^* \mathbf{y} \tag{4.19}$$

The optimal weight is the one that minimizes the mean-square error $e_k = \varepsilon[|\mathbf{w}_k^*\mathbf{y} - x_k|^2]$. The MMSE weight is given by the following optimization problem:

$$\mathbf{w}_k = \arg\min_{\mathbf{w}} \varepsilon\left[\left|\mathbf{w}_k^*\mathbf{y} - x_k\right|^2\right] \quad \forall k \in [1, n_T] \tag{4.20}$$

The MMSE receiver makes an optimal tradeoff between suppressing the spatial interference and minimizing the channel noise. The optimal solution of the weight vector $\mathbf{w}_k$ is given by

$$\mathbf{w}_k = \left(\sigma_z^2 \mathbf{I}_{n_R} + \frac{P_0}{n_T}\sum_{i\neq k}^{n_T} \mathbf{h}_i \mathbf{h}_i^*\right)^{-1} \mathbf{h}_k \tag{4.21}$$

Hence, the equivalent channel between the transmitted symbol x_k and the MMSE receiver processing v_k is given by

$$v_k = \mathbf{w}_k^* \mathbf{y} = \mathbf{h}_k^* \mathbf{A}_k^{-1} \mathbf{h}_k x_k + \mathbf{A}_k^{-1} \mathbf{z} \tag{4.22}$$

where $\mathbf{A}_k = \mathrm{h}_k^* \left(\sigma_z^2 \mathbf{I}_{n_R} + \frac{P_0}{n_T} \sum_{i\neq k}^{n_T} \mathbf{h}_i \mathbf{h}_i^*\right)^{-1}$. Hence, the overall ergodic capacity is given by

$$C_{\mathrm{MMSE}} = \sum_{k=1}^{n_T} \varepsilon\left[\log_2\left(1 + \frac{P_0}{n_T} \mathbf{h}_k^* \mathbf{A}_k^{-1} \mathbf{h}_k\right)\right] \tag{4.23}$$

Since MMSE optimizes the SINR $\frac{P_0}{n_T} \mathbf{h}_k^* \mathbf{A}_k^{-1} \mathbf{h}_k$, the capacity C_{MMSE} is strictly larger than the zero-forcing capacity C_{ZF}.

Similarly, the MMSE receiver can be modified into MMSE-SIC structure as illustrated in Figure 4.9b. With interference cancellation, the equivalent channel at the kth-stage processing is given by

$$v_k = \mathbf{w}_k^* \mathbf{y} = \mathbf{h}_k^* \tilde{\mathbf{A}}_k^{-1} \mathbf{h}_k x_k + \tilde{\mathbf{A}}_k^{-1} \mathbf{z} \tag{4.24}$$

where $\tilde{\mathbf{A}}_k = \mathbf{h}_k^* \left(\sigma_z^2 \mathbf{I}_{n_R} + \frac{P_0}{n_T} \sum_{i=k+1}^{n_T} \mathbf{h}_i \mathbf{h}_i^*\right)^{-1}$. The overall ergodic capacity is given by

$$C_{\mathrm{MMSE\text{-}SIC}} = \sum_{k=1}^{n_T} \varepsilon\left[\log_2\left(1 + \frac{P_0}{n_T} \mathbf{h}_k^* \tilde{\mathbf{A}}_k^{-1} \mathbf{h}_k\right)\right] \tag{4.25}$$

Yet, from the chain rule of mutual information, we have

$$I(\mathbf{X};\mathbf{Y}|\mathbf{H}) = \sum_{k=1}^{n_T} I(X_k;\mathbf{Y}|\mathbf{H}, X_1, \ldots, X_{k-1})$$

For AWGN noise and complex Gaussian transmitted signal, we have

$$I(X_k;\mathbf{Y}|\mathbf{H}, X_1, \ldots, X_{k-1}) = \log_2\left(1 + \frac{P_0}{n_T} \mathbf{h}_k^* \tilde{\mathbf{A}}_k^{-1} \mathbf{h}_k\right)$$

Hence, the MMSE-SIC ergodic capacity in Equation (4.25) actually achieves the optimal MIMO ergodic capacity $\log_2\left|\mathbf{I}_{n_R} + \frac{P_0}{n_T\sigma_z^2}\mathbf{hh}^*\right|$. In other words, the MMSE-SIC structure is information-theoretically optimal.

4.2.3.3 Iterative Receiver. As mentioned in the beginning of Section 4.2.3, the received signals in both the spatial and temporal dimensions contain information about the transmitted message in coded LST systems. The true optimal receiver should be jointly ML with extremely high decoding complexity. In the following, we introduce a suboptimal but powerful receiver structure based on iterative detection. The basic idea is to have iteration on observations in both the spatial and temporal domains to enhance the overall soft decision. This is illustrated in Figure 4.10.

The received matrix $[\mathbf{y}_1, \ldots, \mathbf{y}_N]$ is first buffered. Soft-information metrics are fed into the front-end block, which performs *symbol-by-symbol* spatial interference suppression. For example, the ZF, ZF-SIC, MMSE, and MMSE-SIC as well as the MAP (maximal aposteria processing) structures can be used as the spatial interference suppression front-end block. This block offers *enhanced soft information* to be fed to the soft-input soft-output (SISO) channel decoder by suppressing the spatial interference between layers. Soft outputs from the bank of n_T independent encoders are fed back to the front end for spatial interference suppression again. This completes a single iteration. At sufficiently high SNR, the iterative algorithm will converge and the soft information will be enhanced continuously as the number of iterations increases.

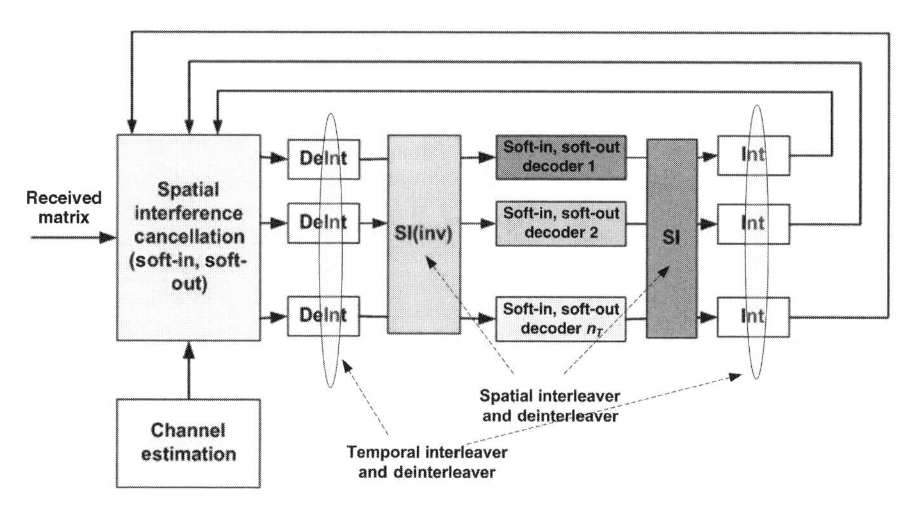

Figure 4.10. Illustration of iterative LST receiver.

4.2.4 Optimal Architecture for Fast Flat Fading Channels

For fast fading channels with CSIR only, there is no outage problem if capacity-achieving component codes are used. This is because there is an ergodic realization of random fading in the temporal domain across the encoding frame. In this case, we have shown that the ergodic capacity is

$$C_{\text{csir}} = \varepsilon\left[\log_2\left|\mathbf{I}_{n_R} + \frac{P_0}{n_T \sigma_z^2}\mathbf{hh}^*\right|\right]$$

where the capacity-achieving input covariance matrix $\mathbf{Q}_x = \varepsilon[\mathbf{xx}^*]$ is given by $\frac{P_0}{n_T}\mathbf{I}_{n_T}$. Given the optimizing input covariance matrix $\mathbf{Q}_x$, we shall try to deduce the optimal transmitter architecture. In Chapter 2, we mentioned that the optimal transmitter architecture of MIMO fast fading channels with perfect CSIR is given by the H-BLAST architecture.

In fact, it is not straightforward to deduce directly that H-BLAST is the optimal transmitter architecture simply because the optimizing $\mathbf{Q}_x$ is diagonal. The problem arises from the fact that in deriving the ergodic capacity described above, we have used the standard random codebook argument [but in this case, each code symbol in the codewords is generated i.i.d. according to $p(\mathbf{x})$] and shown that the mutual information is maximized when $p(\mathbf{x})$ is circularly symmetric complex Gaussian with input covariance $\mathbf{Q}_x = \alpha\mathbf{I}_{n_T}$. However, more than one possible architectures can have such $\mathbf{Q}_x$. For example, the H-BLAST, D-BLAST, and TLST architectures all have $\varepsilon[\mathbf{xx}^*] = \alpha\mathbf{I}_{n_T}$. Since the random codebook argument indicates that there exists at least a vector codebook realization that can achieve the capacity, we cannot conclude directly that H-BLAST is the optimal transmitter architecture from the fact that the optimal $\mathbf{Q}_x = \alpha\mathbf{I}$. Moreover, from the typical set concepts, the vector codebook as a result of H-BLAST design has a *staggered structure* as illustrated in Figure 4.11b. In comparison with an arbitrarily realization of the random vector codebook satisfying $\mathbf{Q}_x = \alpha\mathbf{I}$ as illustrated in Figure 4.11a, although both vector codebooks satisfy $\mathbf{Q}_x = \alpha\mathbf{I}$, the former one (H-BLAST) is actually in the *atypical set* of the random vector codebook realizations. In other words, the chance of *hitting* such a H-BLAST vector codebook realization from the random vector codebook generation process is arbitrarily small as the codeword length increases.

On the other hand, the H-BLAST is indeed a capacity-achieving architecture for fast fading MIMO channels, and we shall illustrate this from another point of view (the multiaccess capacity). Considering the H-BLAST architecture, the received signal can be written as

$$\mathbf{y} = \mathbf{hx} + \mathbf{z} = \sum_k \mathbf{h}_k x_k + \mathbf{z} \tag{4.26}$$

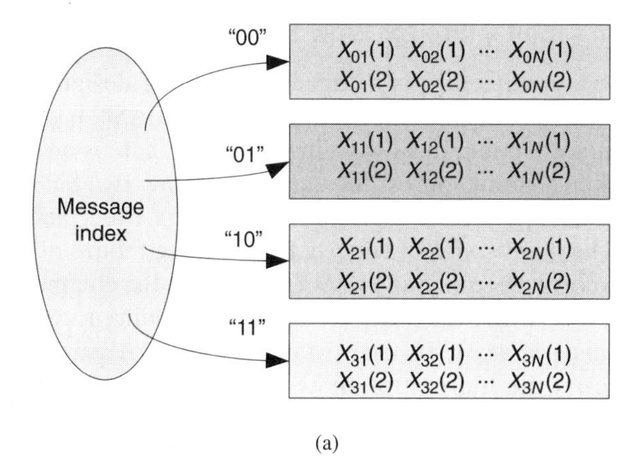

(a)

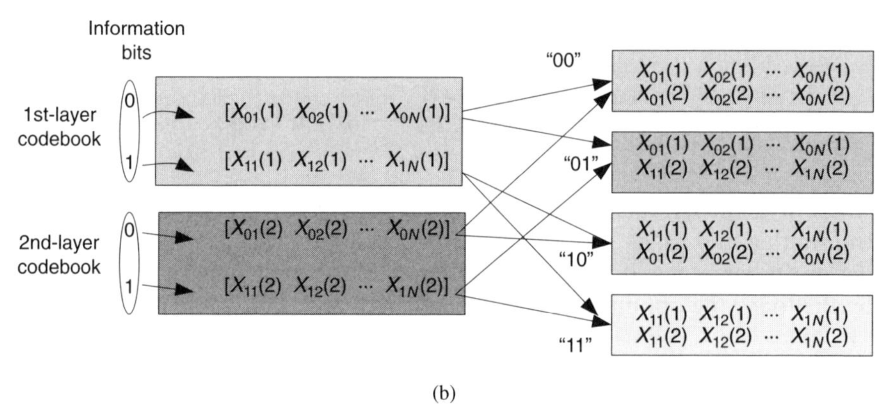

(b)

Figure 4.11. Comparison of the vector codebooks between random generation (a) and H-BLAST (b) architectures.

where $\mathbf{h}_k$ is the kth column of $\mathbf{h}$ and x_k is the transmitted signal from the kth antenna. This is exactly the same channel model as in multiaccess communications, where n_T transmitters encode information independently and the n_T users have the same transmit power constraint P_0/n_T. The sum capacity from multiaccess communication theory is given by

$$C_{\text{csir}} = \sum_{k=1}^{n_T} C_{k,\text{csir}} = \varepsilon[I(X_1, X_2, \ldots, X_{n_T}; \mathbf{Y}|\mathbf{h})] = \varepsilon\left[\log_2\left|\mathbf{I}_{n_R} + \frac{P_0}{n_T\sigma_z^2}\mathbf{h}\mathbf{h}^*\right|\right]$$

which is the same as the MIMO ergodic capacity. Hence, we can deduce that the H-BLAST architecture is optimal with respect to fast fading channels. At the receiver, we can adopt either the ML receiver or MMSE-SIC receiver because the MMSE is information-lossless.

4.2.5 Optimal Architecture for Slow Flat Fading Channels

While H-BLAST architecture is a capacity-achieving design for fast fading channels, we shall illustrate in this section that H-BLAST is strictly suboptimal in slow fading channels. In slow fading channels, there is no temporal variation of the fading channels across the encoding frame. Two factors contribute to packet transmission errors: the finite codelength effect and the channel outage effect. The *finite-codelength effect* refers to the failure of the error correction code to correct errors due to AWGN noise in the channel. The *channel outage effect* refers to the fact that the current encoding rate is higher than the actual channel capacity. In fact, even if we employ powerful (capacity-achieving) codes at the transmitter, there is still a finite probability of packet error due to the second factor (channel outage effect).

In general, following the standard random coding argument, the outage probability of general MIMO channels with CSIR only in slow fading channels is given by

$$P_{\text{out}}(R, \mathbf{Q}_x) = \Pr\left[\log_2\left|\mathbf{I}_{n_R} + \frac{1}{\sigma_z^2}\mathbf{h}\mathbf{Q}_x\mathbf{h}^*\right| < R\right] \tag{4.27}$$

where R is the overall information throughput (bits per channel use) at the transmitter and $\mathbf{Q}_x$ is the input covariance matrix. Hence, the transmitter design (parameterized by $\mathbf{Q}_x$) that minimizes the outage probability is given by the following optimization problem.

Problem 4.1 (Minimum Outage Probability)

$$P_{\text{out}}(R) = \min_{\mathbf{Q}_x:\text{tr}(\mathbf{Q}_x)\le P_0} P_{\text{out}}(R, \mathbf{Q}_x)$$

It has been shown that in MISO channels ($n_R = 1$), the optimizing covariance matrix [which achieves the minimum outage probability $P^*_{\text{out}}(R)$] is given by $\mathbf{Q}_x = \frac{P_0}{n_T}\mathbf{I}_{n_T}$. In general, Telatar [126] conjectured that it is also the optimizing input covariance matrix in the outage sense at high SNR. While the optimizing $\mathbf{Q}_x$ is a scaled identity matrix, we shall illustrate that this does not imply that all different transmitter designs that lead to the same optimizing $\mathbf{Q}_x$ can also optimize the outage probability. In Theorem 4.1 we shall illustrate that the minimum outage probability $P_{\text{out}}(R)^*$ decays like $\frac{1}{\text{SNR}^{n_T n_R}}$ at high SNR. In other words, the maximum diversity order that can be achieved by slow fading MIMO channels with CSIR only is $n_T n_R$.

We first consider the H-BLAST design performance in slow fading channels. Because of the slow fading channels, there is only a single realization of $\mathbf{h}$ across the temporal domain of the encoding frame. Even if powerful component codes are used in the H-BLAST transmitter, there is still a finite probability of packet outage due to *channel outage*. In H-BLAST, the encoded symbols from one stream are always transmitted out from a single-transmit antenna. The encoded stream will have a diversity order of at most n_R. Hence, if the channel gains from the n_R receive antennas happen to be in deep fade, the encoded stream cannot be successfully decoded at the receiver, and this results in packet outage. On the other hand, we know that the optimal diversity order of the MIMO channels is at most $n_T n_R$. Hence, the H-BLAST design is strictly suboptimal in the outage sense. This argument is quantitatively elaborated below.

In terms of the decomposition of the instantaneous capacity given **h** according to MMSE processing, we have

$$\log_2\left|\mathbf{I}_{nR}+\frac{1}{\sigma_z^2}\mathbf{h}\mathbf{Q}_x\mathbf{h}^*\right|=\sum_{k=1}^{nT}\log_2(1+\gamma_k) \tag{4.28}$$

where γ_k is the SINR of the kth stage of the MMSE-SIC receiver as discussed in Section 4.2.3.2. In other words, the MIMO channels can be considered as having n_T parallel channels parameterized by n_T SINR $\{\gamma_1, \ldots, \gamma_{n_T}\}$. Hence, outage (with jointly ML decoding) occurs if

$$\sum_{k=1}^{nT}\log_2(1+\gamma_k)<R \tag{4.29}$$

However, consider the outage situation in the H-BLAST architecture. If we split the rate R into n_T substreams, each with rate R_k, we have $\Sigma_{k=1}^{n_T} R_k = R$. Since each substream is independently encoded and always transmitted over the same transmit antenna, we have an outage event whenever the capacity of any of the n_T substreams is less than the allocated rate: $\log_2(1 + \gamma_k) < R_k$ for some k. Since $\Pr[C_1 + C_2 < R_1 + R_2] \leq \Pr[(C_1 < R_1) \cup (C_2 < R_2)]$, we can conclude that the outage probability of H-BLAST is greater than or equal to the minimum outage probability at the same rate R. Therefore, H-BLAST is strictly suboptimal in the outage sense.

From the analysis above, we see that the fundamental reason for the suboptimality of H-BLAST in outage consideration is the fact that encoded streams always transmit through the same transmit antenna and therefore, the SINR of the *MMSE channel* that a coded substream sees is always γ_k across the coded frame. To break the suboptimal performance, one should try to send the encoded substreams through all the n_T transmit antennas so that the encoded symbols of any substreams can see all the SINR of the n_T MMSE subchannels. On one hand, one can jointly encode the information bits in a similar way as regular spacetime codes and rely on jointly ML decoding at the receiver. However, this architecture no longer has the notion of *layers* or *substreams*. Hence, D-BLAST is a good choice in the sense that it still keeps the *layering* structure but at the same time, the encoded symbols from any sub-streams are carried through various n_T transmit antennas. Ignoring the over-head due to zero padding in D-BLAST (which can be justified at large n_T), the first-layer codeword sees the entire n_T parallel channels as a whole and thus are the subsequent layer codewords. Therefore, the outage event is given by

$$\sum_{k=1}^{n_T} \log_2(1+\gamma_k) < \sum_k R_k = R$$

This matches exactly with the optimal outage event in Equation (4.29). Hence, the D-BLAST is an outage optimal structure. Similarly, TLST is also outage-optimal.

4.2.6 Fundamental Tradeoff between Spatial Diversity and Spatial Multiplexing

In fast fading MIMO channels, the benefit of multiple antennas is obviously the spatial multiplexing gain that increases the ergodic capacity. This is because the encoded frame spans across ergodic realization of channel fading, and therefore, when powerful capacity-achieving codes are used at the transmitter, the packet transmission error (due to the decoding error w.r.t. AWGN) vanishes and there is no risk of packet outage. However, in slow fading MIMO channels, packet outage is possible even with the application of powerful codes at the transmitter. With respect to the packet outage, the MIMO channels offer another dimension of gain, namely, the *spatial diversity gain* to improve the reliability of packet transmission. Obviously, we cannot simultaneously enjoy the maximum diversity gain as well as the maximum spatial multiplexing gain. It is therefore the focus of this section to study the fundamental tradeoff between spatial diversity and spatial multiplexing in MIMO channels [151].

Consider a generic MIMO transmitter with a random vector codebook and a jointly ML receiver. The outage event is given by

$$\log_2 \left| \mathbf{I}_{nR} + \frac{1}{\sigma_z^2} \mathbf{h}\mathbf{Q}_x\mathbf{h}^* \right| < R \tag{4.30}$$

where R is the information rate (bits per channel use) and $\mathbf{Q}_x$ is the input covariance matrix. As conjectured by Telatar [126], the input covariance matrix that minimizes the outage probability at a given rate R and at high SNR is given by $\mathbf{Q}_x = \dfrac{P_0}{n_T}\mathbf{I}_{nT}$. In fact, for the purpose of asymptotic analysis used to study the tradeoff between spatial diversity and spatial multiplexing, it is sufficient to use this covariance matrix as a good approximation to the actual outage minimizing input covariance matrix because the decay rate of the minimum outage probability is the same as that by using $\mathbf{Q}_x = \frac{P_0}{n_T}\mathbf{I}_{n_T}$. We first define the spatial multiplexing and spatial diversity gains at high SNR as follows.

Definition 4.1 (Spatial Multiplexing Gain) A MIMO link is said to exhibit an *asymptotic spatial multiplexing gain* of r if $R = r\log(\mathrm{SNR})$, where $\mathrm{SNR} = \frac{P_0}{\sigma_z^2}$.

As shown in Chapter 2, we expect the spatial multiplexing gain r to always lie between 0 (no spatial multiplexing gain) and $\min\{n_T, n_R\}$ (full spatial multiplexing gain). Note that the *asymptotic spatial multiplexing gain* here refers to how the bit rate R scales with SNR. In other words, if we have a MIMO link that delivers a constant bit rate R irrespective of SNR, then the asymptotic spatial multiplexing gain according to Definition 4.1 is zero. Unless otherwise specified, the spatial multiplexing gain below is in the asymptotic sense.

Definition 4.2 (Spatial Diversity Gain) A MIMO link is said to exhibit a spatial diversity of order $d(r)$ if the outage probability $p_{\text{out}} \approx \mathrm{SNR}^{-d(r)}$, or more precisely

$$\lim_{\mathrm{SNR}\to\infty} \frac{\log(p_{\text{out}}(R))}{\log(\mathrm{SNR})} = -d(r)$$

where $\mathrm{SNR} = \frac{P_0}{\sigma_z^2}$.

To study the fundamental tradeoff between r and $d(r)$, we consider the following optimization problem on the outage performance at a given spatial multiplexing gain r.

Problem 4.2 (Tradeoff between Spatial Multiplexing and Spatial Diversity) Given a certain spatial multiplexing gain r, what is the largest $d(r)$ that can be achieved?

A plot of $d(r)$ versus r will give the optimal tradeoff between spatial diversity and spatial multiplexing gains.

For the slow MIMO fading channels, the minimum outage probability at a rate $R = r\log(\mathrm{SNR})$ is given by

$$p_{\text{out}}(r) = \Pr\left(\log_2 \left| \mathbf{I}_{n_R} + \frac{\mathrm{SNR}}{n_T}\mathbf{hh}^* \right| < r\log(\mathrm{SNR}) \right)$$

The outage probability can be further simplified by asymptotic analysis at high SNR [151] as

$$p_{\text{out}}(r) = \Pr\left(\log_2\left|\mathbf{I}_{n_R} + \frac{\text{SNR}}{n_T}\mathbf{hh}^*\right| < r\log(\text{SNR})\right) \approx \frac{1}{\text{SNR}^{(n_t-r)(n_R-r)}} \tag{4.31}$$

As a result, the optimal tradeoff result is summarized in the following theorem [151].

Theorem 4.1 (Tradeoff between Spatial Multplexing and Spatial Diversity) For sufficiently large encoding block length N, the optimal tradeoff curve $d^*(r)$ is given by the piecewise-linear function connecting the points $(r, d^*(r))$ where $r = 0, 1, \ldots, \min\{n_T, n_R\}$ and $d^*(r) = (n_T - r)(n_R - r)$. In particular, we have $d^*_{\max} = n_T n_R$ and $r^*_{\max} = \min\{n_T, n_R\}$.

Figure 4.12a illustrates the optimal tradeoff curve as described by Theorem 4.1. Note that we have constellation of varying sizes to achieve the intermediate points between the line segments. To understand the operational meaning of the tradeoff curve, we consider the following examples, namely, the repetition spacetime code and the Alamouti code.

Example 4.3 (Repetition Spacetime Code) Consider a 2×2 MIMO link utilizing a simple repetition scheme to exploit the maximum diversity gain ($d^*(0) = 2 \times 2 = 4$). The transmission matrix is given by

$$\tilde{\mathbf{x}} = \begin{bmatrix} x_1 & 0 \\ 0 & x_1 \end{bmatrix}$$

For a fixed constellation, the diversity gain of such coding scheme is 4, but this can be achieved only with a multiplexing gain $r = 0$ (since with a fixed constellation, the data rate is upper-bounded by a constant and therefore $r = R/\log(\text{SNR}) \to 0$ as $\text{SNR} \to \infty$). However, if we increase the size of the constellation for the symbol x_1 as SNR increases to support a data rate $R = r\log(\text{SNR})$ bits per channel use, the distance between constellation points will shrink with SNR and the achievable diversity gain will be decreased. For instance, the error probability of the repetition code (with QAM constellation of size 2^R) is given by

$$p_e \approx \left(\frac{1}{D^2_{\min}}\right)^{n_T n_R}$$

where $D_{\min}$ is the minimum distance between any two constellation points and is given by $D_{\min} = \sqrt{\dfrac{\text{SNR}}{2^R}}$. Setting $R = r\log(\text{SNR})$, we have

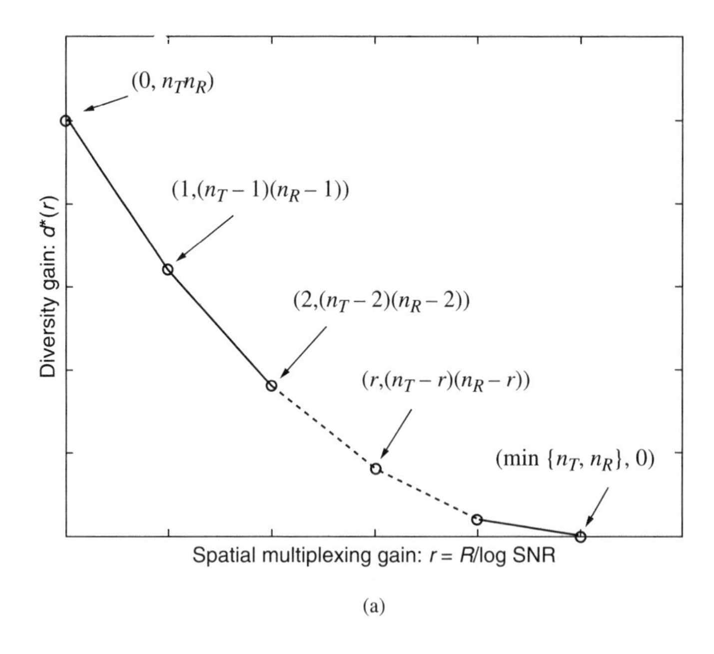

(a)

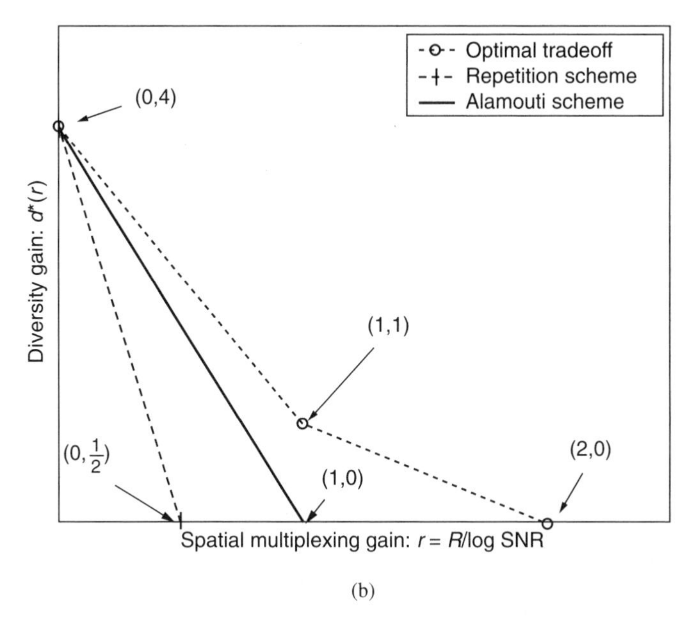

(b)

Figure 4.12. Tradeoff between spatial diversity and spatial multiplexing in slow MIMO fading channels: (a) optimal tradeoff; (b) examples.

$$p_e \approx \frac{1}{\text{SNR}^{(1-2r)n_T n_R}}$$

The maximum spatial multiplexing gain achieved by this scheme is only $\frac{1}{2}$ since only one symbol is transmitted per two channel uses. To achieve $r = \frac{1}{2}$, we have to increase the constellation size accordingly with increasing SNR and therefore, the diversity gain is 0, achieving the $(\frac{1}{2}, 0)$ point in the tradeoff curve. Figure 4.12b illustrates the tradeoff performance of the repetition code.

Example 4.4 (Alamouti Code) Consider the 2×2 Alamouti code. The transmission matrix is given by

$$\tilde{\mathbf{x}} = \begin{bmatrix} x_1 & -x_2^* \\ x_2 & x_1^* \end{bmatrix}$$

The tradeoff performance of the Alamouti code is also shown in Figure 4.12b. The fractional spatial multiplexing gain r along the tradeoff line is achieved by varying the modulation constellation levels of x_1 and x_2. Similarly, with fixed constellations on x_1 and x_2, the diversity order is shown to be $n_T \times n_R = 4$. This corresponds to the point (0,4) in the tradeoff curve. On the other hand, the maximal spatial multiplexing gain is 1 because two symbols are transmitted over two channel uses. Yet, to achieve the spatial multiplexing gain of $r = 1$, the constellation size has to increase according to increasing SNR. For instance, the error probability of the Alamouti code (with QAM constellation of size $2^{R/2}$) is given by

$$p_e \approx \left(\frac{1}{D_{\min}^2} \right)^{n_T n_R}$$

where $D_{\min}$ is the minimum distance between any two constellation points and is given by $D_{\min} = \frac{\sqrt{\text{SNR}}}{2^{R/2}}$. Setting $R = r\log(\text{SNR})$, we have

$$p_e \approx \frac{1}{\text{SNR}^{(1-r)n_T n_R}}$$

Hence, when $r = 1$, the overall diversity gain becomes 0, achieving the point (1,0) in the tradeoff curve. Note that the Alamouti code is strictly better than the repetition code because it yields a higher diversity gain for any positive spatial multiplexing gain r. However, the tradeoff curve of the Alamouti code is still below the optimal tradeoff curve.

Example 4.5 (D-BLAST) In Section 4.2.5, we show that the D-BLAST architecture together with ML receiver (or MMSE-SIC) receiver can achieve the optimal diversity order at any finite rate R when n_T is large (so that the zero-padding overhead can be neglected). In this example, we shall illustrate that the D-BLAST with MMSE-SIC receiver is also *tradeoff-optimal* when n_T is large. From Section 4.2.3.2, we show that D-BLAST architecture with MMSE-SIC receiver transforms the MIMO channels into n_T subchannels, and that each has SINR given by γ_k:

$$\log_2\left|\mathbf{I}_{n_R} + \frac{1}{\sigma_z^2}\mathbf{h}\mathbf{Q}_x\mathbf{h}^*\right| = \sum_{k=1}^{n_T}\log_2(1+\gamma_k)$$

Since the codewords from each layer are transmitted over the n_T subchannels, the outage probability is given by

$$p_{\text{out}} = \Pr\left(\sum_{k=1}^{n_T}\log_2(1+\gamma_k) < r\log(\text{SNR})\right)$$

Hence, the corresponding diversity gain of the D-BLAST architecture with MMSE-SIC receiver is given by the decay rate of the outage probability with respect to the SNR. On the other hand, the optimal outage probability with random vector codebook transmitter and jointly ML receiver is given by

$$p_{\text{out}}^* = \Pr\left(\log_2\left|\mathbf{I}_{n_R} + \frac{1}{\sigma_z^2}\mathbf{h}\mathbf{Q}_x\mathbf{h}^*\right| < r\log(\text{SNR})\right)$$

Hence, the optimal outage probability p_{out}^* is the same as that of the D-BLAST with MMSE-SIC receiver. Hence, we can conclude that D-BLAST achieves the optimal tradeoff performance as well (in the limit of large n_T).

4.3 SWITCHING THRESHOLD DESIGN FOR MIMO ADAPTATION WITH PERFECT CSIT AND PERFECT CSIR

In previous sections, we have focused on the design of MIMO links with perfect CSIR only. Lacking CSIT, no adaptation can be done at the transmitter, and we have to endure packet outage in slow fading channels even if powerful codes are employed. On the other hand, when perfect CSIT is available at the transmitter, rate and power adaptation can be employed and nonzero ergodic capacity can be achieved even for slow fading channels. Hence, when we have perfect CSIT, the design criteria at the transmitter are very different.

However, in practice, we can support a fixed combination of the encoding modes and modulation constellation levels only at the transmitter. Hence, even if we obtain perfect CSIT,[1] we cannot support continuously varying rate

[1] In FDD systems, the CSIT has to be fed back from the receiver. Usually, feedback channels can support only a finite number of bits for CSI feedback. On the other hand, for TDD systems, full CSIT may be obtained from the channel reciprocal properties. Yet, there will be CSIT estimation error due to channel noise. For the design issues associated with imperfect CSIT, please refer to Chapter 5.

and modulation level adaptation. Therefore, it is important to consider adaptation design issues.

Specifically, three issues need to be addressed in designing a channel-adaptive transmitter.

1. How many different adaptation modes Q are needed at the transmitter? We may not want to have too many modes because of the finite overhead in the control field needed to communicate to the receiver on the current transmission mode.
2. Given that the transmitter can support Q different encoding rate and modulation level combinations, what is the best choice of the M transmission modes in terms of encoding rate and modulation level?
3. Given a current CSIT, howdoes the transmitter figure out the right transmission mode (out of the Φ possible modes) to use?

In regular SISO channels, the adaptation strategy (point 3) is relatively straightforward as the CSIT is a complex value. However, in the MIMO channels, the CSIT is a matrix and therefore, the adaptation strategy is more tricky. Furthermore, there is the issue of power allocation and precoding matrix computation across the spatial channels. In this section, we discuss a systematic design framework to address the abovementioned design issues. In addition, we focus on slow fading (quasistatic fading) channels, which is more challenging due to channel outage.

4.3.1 MIMO Transmitter and Adaptation Designs

As illustrated in Chapter 2, the MIMO channels with perfect CSIT can be decomposed into $m^* = \min[n_T, n_R]$ decoupled parallel channels. Hence, the adaptation design problem of the MIMO transmitter can be transformed into the adaptation design problem for m^* parallel channels as follows. The received signal $\mathbf{y}$ of the original MIMO channels can be expressed as $\mathbf{y} = \mathbf{hx} + \mathbf{z}$. If the channel matrix $\mathbf{h}$ is i.i.d., it will have full rank with probability 1. Applying SVD on $\mathbf{h}$, we have $\mathbf{h} = \mathbf{U\Lambda V}^*$, where $\mathbf{U}$ and $\mathbf{V}$ are $n_R \times n_R$ and $n_T \times n_T$ eigenvector matrices of $\mathbf{hh}^*$ and $\mathbf{h}^*\mathbf{h}$, respectively, and $\mathbf{\Lambda}$ is the $n_R \times n_T$ diagonal matrix with the diagonal entries given by the eigenvalues of $\mathbf{h}$. Since $\mathbf{U}$ and $\mathbf{V}$ are nonsingular, there is no loss of information if we pre-multiply $\mathbf{y}$ with $\mathbf{U}^*$, as follows

$$\tilde{\mathbf{y}} = \mathbf{U}^*\mathbf{y} = \mathbf{\Lambda}\tilde{\mathbf{x}} + \tilde{\mathbf{z}} \tag{4.32}$$

where $\tilde{\mathbf{x}} = \mathbf{V}^*\mathbf{x}$ is the $n_T \times 1$ equivalent channel input vector and $\tilde{\mathbf{z}} = \mathbf{U}^*\mathbf{z}$ is the transformed noise having the same distribution as the original noise. Figure 4.13 illustrates a generic MIMO transmitter with CSIT based on Equation (4.32). The adaptation parameters include the encoding rates $\mathbf{r}_c = (r_c(1), \ldots, r_c(n_T))$ (coded bits per channel use), the modulation levels

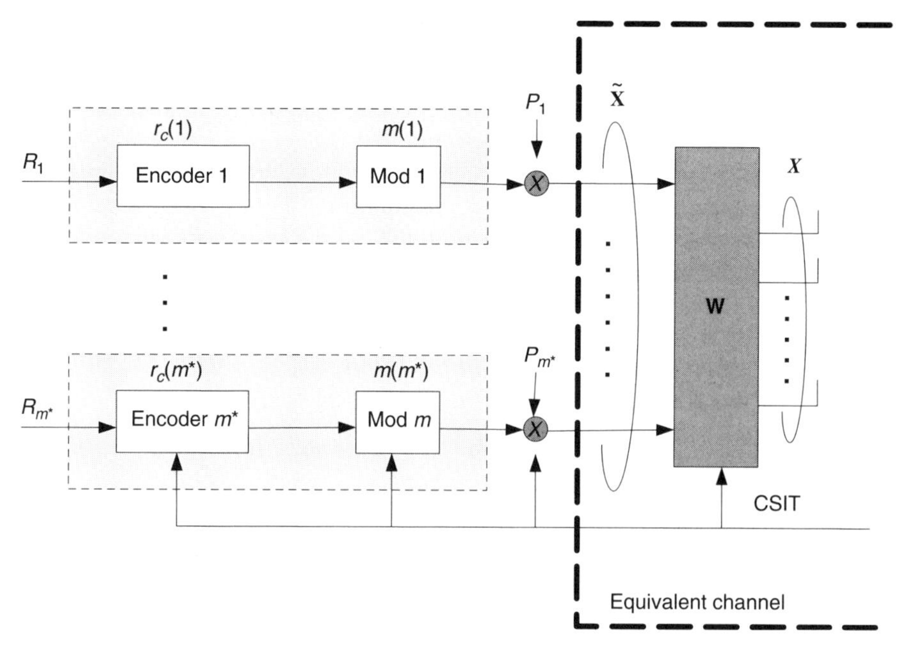

Figure 4.13. Illustration of MIMO transmitter adaptation.

$\mathbf{M} = (M_1, \ldots, M_{n_T})$, and the power allocation $\mathbf{p} = (p_1, \ldots, p_{n_T})$. For purposes of practical implementation, we can afford to have a total of only Q combinations of the encoding rate and modulation level per spatial channel. With Q transmission modes per spatial channel, we require a total of $n_T \log_2(Q)$ overhead control bits in the frame to indicate the transmission modes in the packet. The qth transmission mode $\mathcal{M}(q)$ is a collective description of the encoding rate vector $\mathbf{r}_c(q)$ and the modulation constellation $\mathbf{M}(q)$. Note that if $n_R < n_T$ such that $m^* = n_R$, the observations $\tilde{\mathbf{y}}$ do not contain information about the $(n_R + 1), \ldots, (n_T)$th components of the equivalent-channel input vector $\tilde{\mathbf{x}}$. Hence, the power and rate allocated to these components must be zero. For notation convenience, we assume $n_T \leq n_R$ such that $m^* = n_T$.

Since the original channel matrix $\mathbf{h}$ is transformed to the space of n_T eigenvalues $\mathbf{\Lambda} = \{\lambda_1, \ldots, \lambda_{n_T}\}$, a general adaptation architecture of the MIMO transmitter can be specified by the *CSIT partition* $\{\mathcal{H}(1), \mathcal{R}(2), \ldots, \mathcal{H}(Q)\}$, where the union of the partition regions gives the entire CSIT space (containing $\{\lambda_1, \ldots, \lambda_{n_T}\}$) and the partition regions are mutually disjoint. Given a CSIT $\mathbf{\Lambda}$, transmission mode $\mathcal{M}(q)$ is selected if $\mathbf{\Lambda} \in \mathcal{H}(q)$. Our focus is to select the transmission modes $\{\mathcal{M}(1), \ldots, \mathcal{M}(Q)\}$ and CSIT partition $\{\mathcal{H}(1), \ldots, \mathcal{H}(Q)\}$ so as to maximize the overall throughput of the MIMO link in slow fading channels. This can be cast into an optimization problem to be elaborated in the next section.

4.3.2 Optimization Problem—Quasistatic Fading Channels

Given the parallel channels in Figure 4.13, the joint ML detection can be decoupled into n_T-independent ML detection on each of the parallel channels. Therefore, the outage events[2] of the n_T parallel channels are decoupled from each other. Assuming that capacity-achieving codes and Gaussian constellations are used in each subchannel, the information rate of the nth subchannel at a transmission mode $\mathcal{M}n(q)$ is given by $R_n(q) = r_c(q, n)\log_2(M(q, n))$. The overall throughput of the MIMO link is given by

$$\begin{aligned}\overline{C} &= \sum_{q=1}^{Q} \varepsilon_{\Lambda}\left[\max_{(p_1,\ldots,p_{n_T})} \sum_{n=1}^{n_T} R_n(q) 1(C(p_n, \lambda_n) > R_n(q)) | \mathbf{\Lambda} \in \mathcal{H}(q)\right] \Pr[\mathbf{\Lambda} \in \mathcal{H}(q)] \\ &= \sum_{q=1}^{Q}\sum_{n=1}^{n_T} R_n(q) \varepsilon_{\lambda_n}\left[\max_{(p_1,\ldots,p_{n_T})} 1(C(p_n, \lambda_n) > R_n(q)) | \lambda_n \in \mathcal{H}_n(q)\right] \alpha(q, n) \end{aligned} \quad (4.33)$$

where $C(p_n, \lambda_n) = \log_2\left(1 + \frac{p_n|\lambda_n|^2}{\sigma_z^2}\right)$, $\alpha(q, n) = \Pr[\lambda_n \in \mathcal{H}_n(q)]$ and the n_T-dimensional partition region $\mathcal{H}(q)$ can be decoupled into n_T one-dimensional partition region $\mathcal{H}_n(q)$ for $n = 1, \ldots, n_T$.

Assume an average transmit power constraint $\varepsilon[\Sigma_n p_n] \leq P_0$. The inner optimization solution with respect to the power allocation $(p_1, \ldots, p_{n_T})$ is given by the standard power water-filling solution

$$p_n^* = \left(\frac{1}{\mu} - \frac{\sigma_z^2}{|\lambda_n|^2}\right)^+ \quad (4.34)$$

where $x^+ = \max(0, x)$ and μ is a Lagrange multiplier (independent of $\{\lambda_n\}$) chosen to satisfy the average transmit power constraint P_0. Hence, the adaptation design can be cast into the following n_T-independent optimization problems.

Problem 4.3 (MIMO Adaptation Design with Perfect CSIT) Choose the rate allocations $\{R_n(1), \ldots, R_n(Q)\}$ and the adaptation thresholds $\{\mathcal{H}_n(1), \ldots, \mathcal{H}_n(Q)\}$ such that the overall throughput of the nth subchannel is optimized:

$$\overline{C}_n^* = \max_{\{R_n(1),\ldots,R_n(Q)\}\{\mathcal{H}_n(1),\ldots,\mathcal{H}_n(Q)\}} \sum_{q=1}^{Q} R_n(q) \varepsilon_{\lambda_n}[1(C^*(\lambda_n) > R_n(q)) | \lambda_n \in \mathcal{R}_n(q)] \alpha(q, n)$$

[2]The outage in the presence of perfect CSIT is due to fact that transmitter fails to adapt the encoding rate continuously with the CSIT.

where

$$C^*(\lambda) = \begin{cases} \log_2\left(\frac{|\lambda|^2}{\sigma_z^2\mu}\right), & \text{if} |\lambda|^2 \geq \sigma_z^2\mu \\ 0, & \text{otherwise} \end{cases}$$

4.3.3 Equivalence to the Classical Vector Quantization Problem

Note that the optimization problem in Problem 4.3 is equivalent to the classical vector quantization problem if we define a *modified distortion measure* $d(\lambda_n, q)$ as

$$d(\lambda_n, q) = R_n(q)1(C^*(\lambda_n) > R_n(q)) \tag{4.35}$$

Hence, the optimization problem becomes

$$\overline{C}_n^* = \max_{\{R_n(1),\dots,R_n(Q)\}\{\mathcal{H}_n(1),\dots,\mathcal{H}_n(Q)\}} \sum_{q=1}^{Q} \varepsilon_{\lambda_n}[d(\lambda_n, q)|\lambda_n \in \mathcal{H}_n(q)]\alpha(q,n)$$

which is identical to the classical vector quantization problem. It can be solved by the classical Lloyd iterative algorithm.[3]

Step 1: Given the set of adaptation thresholds $\{\mathcal{H}_n(1), \dots, \mathcal{H}_n(Q)\}$, *find the optimal rate allocation* $\{R_n(1), \dots, R_n(Q)\}$. The optimal transmission mode allocation (or equivalently, the optimal rate allocation $R_n^*(q)$) is given by

$$\frac{\partial \varepsilon_{\lambda_n}[d(\lambda_n, q)|\lambda_n \in \mathcal{H}_n(q)]}{\partial R_n(q)} = 0$$

Observe that

$$\begin{aligned} \varepsilon_{\lambda_n}[d(\lambda_{n,q})|\lambda_n \in \mathcal{H}_n(q)] &= R_n(q)\Pr[C^*(\lambda_n) > R_n(q)|\lambda_n \in \mathcal{H}_n(q)] \\ &= R_n(q)\Pr\left[|\lambda_n|^2 > 2^{R_n(q)}\mu\sigma_z^2|\lambda_n \in \mathcal{H}_n(q)\right] \end{aligned} \tag{4.36}$$

Hence, the optimal transmission mode allocation [or equivalently, the optimal rate allocation $R_n^*(q)$] is given by the root of the nonlinear equation with respect to the variable R

$$\int_{\gamma > 2^R\mu\sigma_z^2 \cap \gamma \in \mathcal{H}_n(q)} f(\gamma)d\gamma - \sigma_z^2\mu\log(2)R2^R f(\sigma_z^2\mu 2^R) = 0$$

[3] Careful choice of the initial point is needed for the Lloyd's algorithm to converge to the global optimal solution.

where $\gamma = |\lambda_n|^2$ and $f(\gamma)$ is the pdf of $|\lambda_n|^2$ given by [126]

$$f(\gamma) = \frac{1}{m^*} \sum_{i=1}^{m^*} \varphi_i(\gamma)^2 \gamma^{n^*-m^*} \exp(-\gamma)$$

where $n^* = \max[n_T, n_R]$, $m^* = \min[n_T, n_R]$ and

$$\varphi_{i+1}(\gamma) = \sqrt{\left[\frac{i!}{(i+n^*-m^*)!}\right]} L_i^{n^*-m^*}(\gamma) \quad \forall i = \{0, 1, \ldots, m^*-1\}$$

and $L_i^{n^*-m^*}(x)$ is the associated Laguerre polynomial of order k given by

$$L_i^{n^*-m^*}(x) = \frac{1}{i!} \exp(x) x^{m^*-n^*} \frac{d^i}{dx_i} (\exp(-x) x^{n^*-m^*+i})$$

Step 2: Given the set of rate allocations $\{R_n(1), \ldots, R_n(Q)\}$, *find the optimal adaptation thresholds* $\{\mathcal{H}_n(1), \ldots, \mathcal{H}_n(Q)\}$. The optimal adaptation threshold is given by

$$\mathcal{H}_n(q) = \{\lambda : d(\lambda, q) \geq d(\lambda, q') \quad \forall q' \neq q\}$$

After the optimal partition $\mathcal{H}_n(q)$ and the optimal rate allocation $R_n^*(q)$ are obtained for all $q = 1, \ldots, Q$ and $n = 1, \ldots, m^*$, the optimal transmission modes $\mathcal{M}_n(q) = (r_c(n, q), M(n, q))$ is obtained by first forming a set of all possible combinations of encoding rate $r_c(n, q)$ and constellation level $M(n, q)$, $S_n(q)$, such that $r_c(n, q)\log_2(M(n, q)) = R_n^*(q)$:

$$S_n(q) = \{(r_c, M) : r_c \log_2(M) = R_n^*(q)\}$$

The choice depends on a many factors, such as implementation complexity as well as robustness with respect to CSIR errors. For instance, we can choose the transmission mode $(r_c, M) \in S_n(q)$ with the lowest constellation level M because this will be most robust to CSIR error. For example, if $R_n^*(q) = 1$ bits per channel use, the set of feasible transmission modes is given by $S_n(q) = \{(\frac{1}{4}, \text{16QAM}), (\frac{1}{3}, \text{8PSK}), (\frac{1}{2}, \text{QPSK})\}$. For robustness consideration, we might like to pick $(\frac{1}{2}, \text{QPSK})$ to realize this transmission mode.

4.3.4 Results and Discussion

We apply the design framework described in the previous sections to 2×2 slow fading MIMO channels as examples to illustrate the design performance. We consider the adaptation designs with $Q = 2,4,8,16$, consuming 1, 2, 3, and 4 bits of forward control overhead for each parallel channel to indicate the transmission mode to the receiver. At the transmitter, we can afford to have constellation levels including only BPSK, QPSK, 8PSK, and 16QAM. We employ powerful component codes, namely, the turbo codes, at the transmitter.

Figure 4.14a illustrates the information-theoretic goodput versus SNR of the adaptive MIMO systems at various Q, Gaussian constellation, and capacity-achieving component codes per parallel channel. We observe that the curve with $Q = \infty$ corresponds to the ergodic capacity with perfect CSIT as illustrated in Chapter 2. The performance degradation values of limiting the adaptation to $Q = 2$ modes and $Q = 16$ modes are 2 and 0.25 dB, respectively. Hence, we can conclude that $Q = 16$ different transmission modes per parallel channel in the 2×2 MIMO link is about sufficient.

As an example for illustration purposes, we consider the performance of the practical adaptation design for $Q = 4$. Figure 4.14b illustrates the overall throughput of the actual adaptive system based on turbo component codes and finite constellations (BPSK, QPSK, 8PSK, 16QAM). For instance, the optimal encoding rates and modulation levels of the Q transmission modes as well as the corresponding optimal adaptation thresholds (or partition) $\{\mathcal{H}_1(1), \ldots, \mathcal{H}_1(Q), \mathcal{H}_2(1), \ldots, \mathcal{H}_2(Q)\}$ are listed in Table 4.1. The realtime adaptation process is described below.

- Given the 2×2 CSIT, $\mathbf{H}$, perform SVD on $\mathbf{H}$ to obtain λ_1 and λ_2 (the singular values of the two spatial channels). For example, letting $\mathbf{H} = \begin{bmatrix} 0.3 & 1 \\ 1.2 & -0.7 \end{bmatrix}$, we have $\lambda_1^2 = 2.0504$ and $\lambda_2 = 0.9696$.
- From Table 4.1, we deduce the adaptation modes for spatial channels 1 and 2 independently on the basis of λ_1 and λ_2. For example, since $\lambda_1^2 = 2.0504$ belongs to region 3, transmission mode 3 (rate $-\frac{3}{4}$ encoding + 8PSK) is chosen for channel 1. Similarly, since $\lambda_2^2 = 0.9696$ belongs to region 2, transmission mode 2 (rate $\frac{2}{3}$ + QPSK) is chosen for channel 2.
- Two bits of control information (per spatial channel) are embedded into the forward packet to indicate the current transmission modes adopted by the transmitter.

It is shown from Figure 4.14 that the performance of the actual system matches closely with the theoretical optimal performance for the same Q at small to medium SNR. At high SNR, there is a gap in the performance of the actual system because discrete constellation fails to deliver a higher throughput as the Gaussian constellation does at high SNR.

4.4 SUMMARY

In this chapter, we have considered practical MIMO link design with perfect CSIR and perfect CSIT. In the first part, we discussed the case of perfect CSIR design. There are two important benefits from the spatial degrees of freedom in the MIMO channels, namely, spatial diversity and spatial multiplexing. While *spatial diversity* is useful for enhancing the reliability of packet

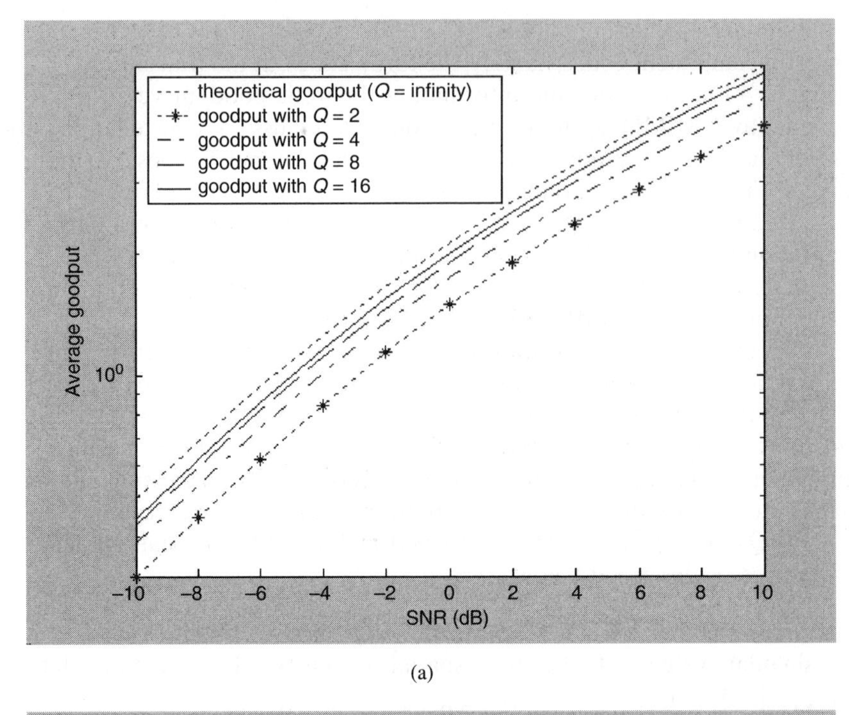

(a)

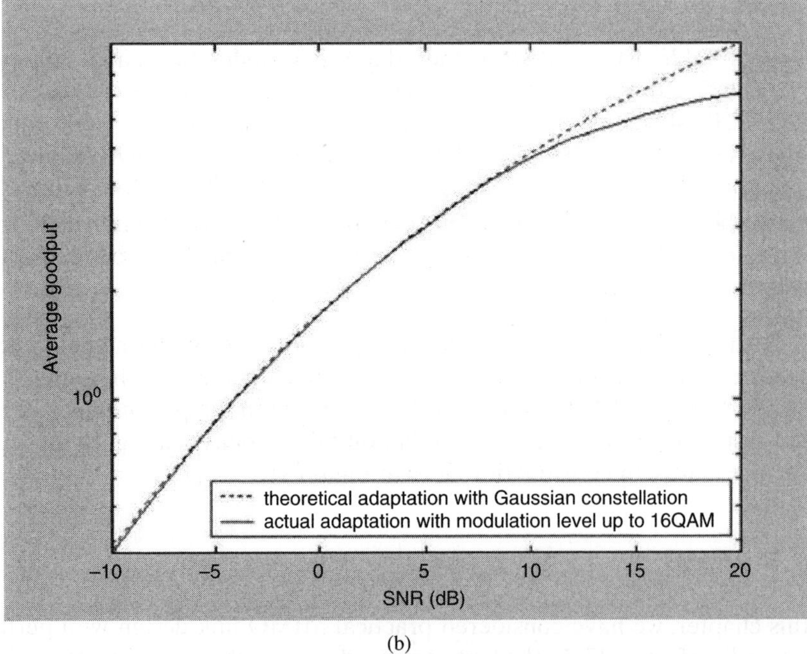

(b)

Figure 4.14. Performance of adaptive MIMO systems with $n_T = n_R = 2$ and perfect CSIT: (a) theoretical performance at $Q = 2,4,8,16$ and Gaussian constellation; (b) actual performance at $Q = 4$ with discrete constellations (BPSK/QPSK/8PSK/16QAM).

TABLE 4.1. A List of Selected Encoding Rate and Modulation Levels of the Four Transmission Modes in a 2 × 2 MIMO System (SNR = 5 dB)

	Adaptation Modes for Channel n		
	Modulation Level	Encoding Rate	Switching Threshold (λ_n^2)
Mode 1	BPSK	$\frac{3}{5}$	0.6222
Mode 2	QPSK	$\frac{2}{3}$	1.0341
Mode 3	8PSK	$\frac{3}{4}$	1.9526
Mode 4	16QAM	$\frac{3}{4}$	3.2839

transmission, *spatial multiplexing* is useful for increasing the transmission bit rate of the MIMO link.

We have reviewed the spacetime coding design framework as a systematic approach to exploit spatial diversity. For instance, the maximum spatial diversity order that one can achieve is $n_T n_R$. In order to achieve this full diversity order, the spacetime codebook has to satisfy the full-rank criterion in the signal matrix $\mathbf{A}_{i,j}$.

On the other hand, to exploit spatial multiplexing gain, a special form of spacetime codes, namely, layered spacetime coding (LST), is needed. We have discussed the design of various LST schemes, namely, V-BLAST, H-BLAST, and D-BLAST as well as T-LST. Because of the vast complexity involved, various suboptimal receiver architectures are discussed. For example, we have introduced the ZF receiver structure and pointed out that the ZF receiver is asymptotically optimal at high SNR. On the other hand, the MMSE receiver is strictly optimal when compared with the ZF receiver. In addition, if successive interference cancellation (SIC) is applied on top of the MMSE linear receiver, the receiver will be information-theoretically optimal. Finally, we have briefly introduced the iterative LST receiver when the soft front end (spatial interference suppression based on soft input and soft output) exchanges soft information with the bank of soft channel decoders (channel decoders based on soft input and soft output) iteratively in such a way that the overall decisions improve as the number of iterations increases.

We have also discussed the fundamental tradeoff between spatial diversity and spatial multiplexing gain in the high-SNR region. For instance, the maximum diversity gain that a MIMO system can support at a spatial multiplexing gain of r is given by $d(r) = (n_T - r)(n_R - r)$. We have compared the tradeoff performance of various existing schemes such as Alamouti codes and repetition codes and shown that they are strictly suboptimal in the tradeoff sense. In fact, when we compare two systems at the same fixed rate R (bits per channel use), both systems may deliver the same full-order diversity of $n_T n_R$. We tend to conclude that the two systems are equally effective, but this may not be the case as one can have a better tradeoff performance than the others.

Finally, we discuss the optimal transmitter architectures in both fast fading channels and slow fading channels. In fast fading channels, one can achieve nonzero ergodic capacity when a powerful code is used at the transmitter. Hence, the focus is to exploit the spatial multiplexing gain as diversity order becomes less of a concern. In this context, we show that H-BLAST can be capacity-achieving in fast fading channels. However, in slow fading channels, there is a finite probability of packet transmission outage even if powerful codes are used. This is due to channel outage and the lack of CSIT at the transmitter. Hence, outage occurs whenever the information rate of the packet is larger than the instantaneous channel capacity of the fading channels. Hence, diversity is important and cannot be ignored. We have also shown that H-BLAST is not the optimal architecture in slow fading channels whereas D-BLAST and TLST are the optimal architectures.

In the second part of the chapter, we considered the transmitter design of slow fading MIMO channels when there is perfect CSIT. With perfect CSIT, one can decompose the MIMO channels into m^* parallel channels. However, in practice, we can support only a finite number of transmission modes because of the overhead in the forward control channel required to indicate to the receiver on the transmission mode of the current packet. Hence, the adaptation design is not trivial. For instance, we would like to ascertain exactly how many different transmission modes are needed. Given Q transmission modes, how should the modes be selected? Finally, given a current CSIT $\mathbf{h}$, how should we pick one out of the Q modes to use? We propose a systematic framework to answer these questions. Finally, we illustrate the design framework with an example of 2×2 MIMO with a total of 4 bits control overhead in the packet.

EXERCISES

1. [Space Time Coding]

a) Consider a generalized orthogonal space time code for a 3×1 MIMO link. The Space Time code is given below:

$$\begin{bmatrix} x_1 & -x_2 & -x_3 & -x_4 & x_1^* & -x_2^* & -x_3^* & -x_4^* \\ x_2 & x_1 & x_4 & -x_3 & x_2^* & x_1^* & x_4^* & -x_3^* \\ x_3 & -x_4 & x_1 & x_2 & x_3^* & -x_4^* & x_1^* & x_2^* \end{bmatrix}$$

where x_1, x_2, x_3 are three independent modulation symbols on the three transmit antenna. If QPSK is used in the 3 transmit antennas, what is the information bit rate of the above scheme (bits per channel use)?

b) Assuming slow flat fading where the channel fading is quasi-static within the space time codeword. What is the maximum likelihood decoding metric at the receiver with single receive antenna? How large is the search space?

c) Repeat b) if there are two receive antennas at the receiver.

2. [MIMO Capacity for Quasi-static Fading Channels]

Consider a MIMO link with 2 transmit antennas and 2 receive antennas. The received 2×1 signal vector at the receiver side is given by: $\underbrace{\begin{bmatrix} y_1 \\ y_2 \end{bmatrix}}_{\mathbf{Y}} = \underbrace{\begin{bmatrix} h_{11} & h_{12} \\ h_{21} & h_{22} \end{bmatrix}}_{\mathbf{H}} \underbrace{\begin{bmatrix} x_1 \\ x_2 \end{bmatrix}}_{\mathbf{X}} + \underbrace{\begin{bmatrix} z_1 \\ z_2 \end{bmatrix}}_{\mathbf{Z}}$ where $\mathbf{H}$ is the 2×2 channel matrix (known to the receiver), $\mathbf{X}$ is the 2×1 transmit vector and $\mathbf{Z}$ is the 2×1 circularly symmetric complex Gaussian noise with mean $\mathbf{0}$ and covariance $\mathbf{I}$.

a) Find the channel capacity of the 2×2 link (in bits/channel use) if $\mathbf{H} = \begin{bmatrix} 0.5 & 1 \\ 1 & 2 \end{bmatrix}$ over the entire packet and the transmit power allocated to each antenna is limited to 1 W each.

b) One intelligent student notice that it is more effective to allocate all the power to transmit antenna 2 because the antenna 2 sees a better channel condition. Suppose the transmitter allocates 2 W to antenna 2 and 0 W to antenna 1. (i.e. the transmitter uses antenna 2 only and ignore antenna 1). What is the channel capacity (in bits/channel use) of this scheme?

c) One diligent student wonders about the following. In part A, the capacity formula we used should give the optimal result (maximum capacity) because the capacity achieving power allocation for MIMO channel with CSIR should be uniform power allocation across the 2 transmit antennas. However, comparing with the result in part B, the channel capacity in part A is in fact smaller than that in part B. Explain why this is the case?

d) Another diligent student observed that rank of H is 1 and therefore the MIMO capacity is not increased substantially compared to the single antenna capacity. He deduces that in order to increase the MIMO link capacity, he has to increase the rank of H. He is wondering if we can increase the rank of the effective channel by introducing an artificial random phase at each of the two transmit antennas. What do you think?

3. [Hybrid Spatial Diversity and Spatial Multiplexing]

Figure 4.3.1 shows a block diagram of orthogonal space time block code. The transmit sequence from antenna 1 to antenna 4 are given by:

$$\begin{pmatrix} \mathbf{x}^{(1)} \\ \mathbf{x}^{(2)} \\ \mathbf{x}^{(3)} \\ \mathbf{x}^{(4)} \end{pmatrix} = \begin{bmatrix} x_1 & -x_2 & -x_3 & -x_4 & x_1^* & x_2^* & x_3^* & x_4^* \\ x_2 & x_1 & x_4 & -x_3 & x_2^* & x_1^* & x_4^* & x_3^* \\ x_3 & -x_4 & x_1 & x_2 & x_3^* & -x_4^* & x_1^* & x_2^* \\ x_4 & x_3 & -x_2 & x_1 & x_4^* & x_3^* & -x_2^* & x_1^* \end{bmatrix} \quad \text{where} \quad \begin{bmatrix} x_1 \\ x_2 \\ x_3 \\ x_4 \end{bmatrix} \quad \text{is}$$

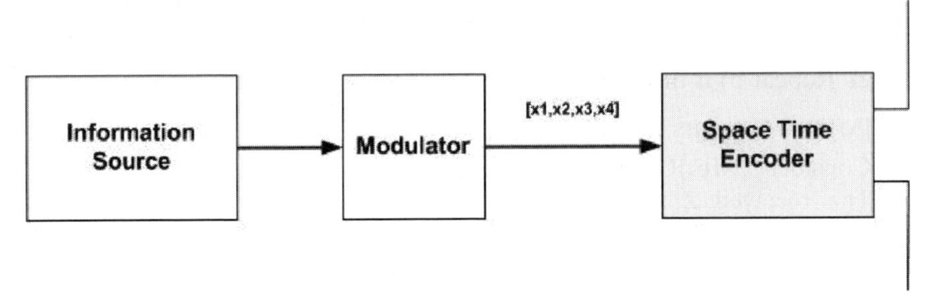

Figure 4.3.1. Block Diagram of Orthogonal Space Time Block Encoding with 4 transmit Antennas.

transmitted from the 4 antennas first followed by $\begin{bmatrix} -x_2 \\ x_1 \\ -x_4 \\ x_3 \end{bmatrix}$ and so on.

a) Prove that the above space time block code achieves full diversity. What is the bit rate (bits per channel use) of the space time block code.

b) Derive the maximal likelihood decoding rule at the receiver (with one antenna and perfect knowledge of channel matrix). If $x_i \in S$ where S is a 2D constellation set with cardinality |S|, express the complexity of the ML decoder in terms of |S|.

c) Derive an upper bound on the symbol error probability of the 4×1 system with respect to slow flat fading channel. Hence, comment on the performance advantage/disadvantage of the above space time block code with respect to the uncoded system with 1 transmit antenna and 4 receive antennas.

{Hint 1: possible performance dimensions are the diversity order, SNR, spectral efficiency}.

{Hint 2: The symbol error probability of a full diversity order space time code is given by: $P_e(A \to B) \leq \left(\frac{E_s}{4N_0}\right)^4 \frac{1}{|K|}$ where K is the signal matrix $|\mathbf{K}| = \|\mathbf{x}(A) - \mathbf{x}(B)\|^2$}.

d) A possible modification to increase the data rate of the system is to employ a double Alamouti encoding at the transmitter as illustrated in figure 4.3.2. For instance, the output symbols of antenna 1 and 2 are given by: $\begin{pmatrix} \mathbf{x}^{(1)} \\ \mathbf{x}^{(2)} \end{pmatrix} = \begin{pmatrix} x_1 & -x_2^* \\ x_2 & x_1^* \end{pmatrix}$. The output symbols of antenna 3 and antenna 4 are given by: $\begin{pmatrix} x^{(3)} \\ x^{(4)} \end{pmatrix} = \begin{pmatrix} x_3 & -x_4^* \\ x_4 & x_3^* \end{pmatrix}$. How many times could the

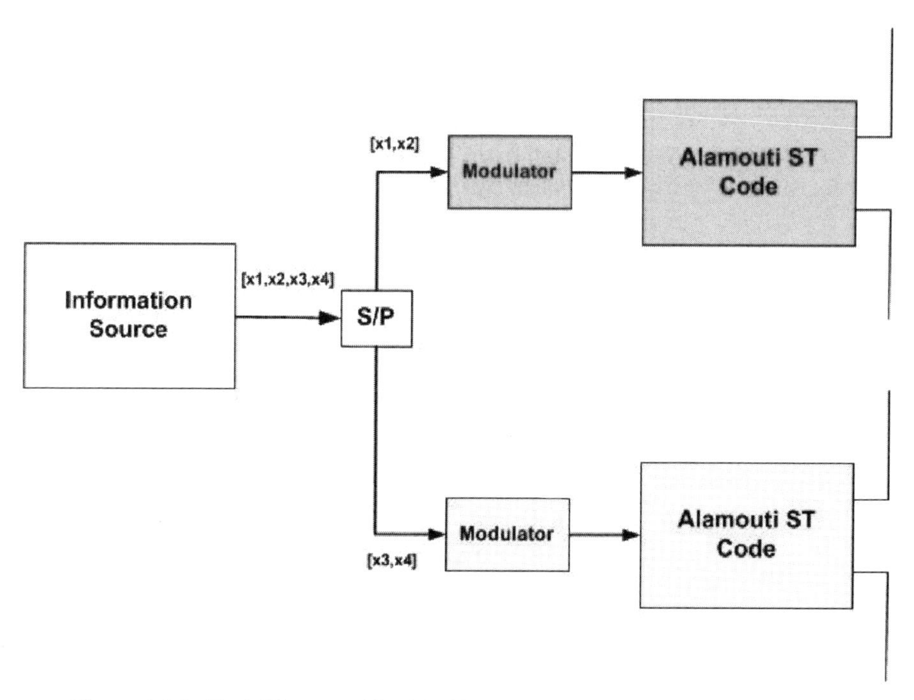

Figure 4.3.2. Block Diagram of Double STTD Encoding with 4 transmit Antennas.

transmitted data rate be increased compared to the scheme in Figure 4.1? Express the received signal in terms of the channel matrix and the transmitted symbols when the receiver has Nr receive antennas. Hence, suggest a receiver structure based on VBLAST successive interference cancellation approach.

5

MIMO CONSTELLATION DESIGN WITH IMPERFECT CHANNEL STATE INFORMATION

5.1 OVERVIEW

In Chapter 4, we discussed the MIMO design issues with perfect CSIR and perfect CSIT. Using multiple antennas at both the transmitter and receiver can dramatically increase the spectral efficiency. On one hand, spacetime coding is employed to exploit the spatial diversity, which is very important for slow fading channels because of channel outage. On the other hand, layered space-time coding structures such as V-BLAST, H-BLAST, D-BLAST, and TLST are proposed to exploit the spatial multiplexing gain. The fundamental tradeoff between spatial diversity and spatial multiplexing has been elaborated. However, all these results are derived under the assumption that the receiver has perfect side information about channel state (CSIR). In other words, coherent demodulation and channel-matched decoding are always assumed in these results. However, the perfect CSIR assumption may not be realistic in a fading environment or with a large number of antennas. For instance, CSI estimation error is related to the power and duration of the pilot symbols included in the packet. Given a limited pilot symbol energy, the estimated CSIR at the receiver can never be perfect. The problem is further complicated by the presence of multiple transmit antennas because longer training symbols are required for the same channel estimation performance. Motivated by these considerations, Marzetta and Hochwald studied the capacity of noncoherent multiple-antenna channels where neither the transmitter nor the receiver has knowledge on the CSI [99]. The channel capacity without knowledge of CSI still increases with the number of transmit and receive antennas, although the increase in capacity is somewhat smaller. Those same authors later proposed

Channel-Adaptive Technologies and Cross-Layer Designs for Wireless Systems with Multiple Antennas: Theory and Applications. By V. K. N. Lau and Y.-K. R. Kwok
ISBN 0-471-64865-5

a class of unitary spacetime signals that are well tailored for noncoherent channels [54,55]. Numerous methods have since been proposed to construct unitary spacetime signals that operate in the absence of channel state information. All these methods, however, focus mainly on unitary spacetime constellation design for uncoded systems.

To exploit the rich diversity and boost coding gains, a power/bandwidth-efficient trellis-coded unitary spacetime modulation (TC-USTM) was suggested [125,150]. It was shown that TC-USTM performs significantly better than does the uncoded unitary spacetime modulation (USTM) with the same spectral efficiency. Nonetheless, these schemes all employ heuristic design approaches. They directly use the unitary spacetime constellation and then employ a general set partition method similar to the conventional set partitions for PSK or QAM-type constellations. However, this is not optimal for systems with multiple antennas with imperfect CSIR.

In this chapter, we focus on a systematic framework [20] to investigate the spacetime coding and constellation design for MIMO channels with imperfect CSIR. In particular, we consider the asymptotically optimal spacetime coding and constellation design in the absence of perfect CSIR. With channel estimation errors, constellations whose design is based on the statistics of the CSI error are more desirable than those designed for perfect CSIR. Considering a slow fading MIMO link with imperfect CSIR, the $(n_R \times N)$-dimensional received signal $\mathbf{y}$ is given by

$$\mathbf{y} = \mathbf{h}\mathbf{x} + \mathbf{z} = \hat{\mathbf{h}}\mathbf{x} - \boldsymbol{\Delta}\mathbf{x} + \mathbf{z} \tag{5.1}$$

where $\mathbf{x}$ is the $n_T \times N$ transmitted spacetime codeword, $\boldsymbol{\Delta}$ is the $n_R \times n_T$ CSI estimation error that is modeled as i.i.d. complex Gaussian $CN(0, \sigma_e^2)$ and is independent of the CSIR $\mathbf{h}$, $\mathbf{z}$ is the $n_R \times N$ white channel noise with unit variance, and $\hat{\mathbf{h}}$ is the estimated CSIR given by

$$\hat{\mathbf{h}} = \mathbf{h} + \boldsymbol{\Delta} \tag{5.2}$$

The likelihood function $p(\mathbf{y}|\mathbf{x}, \hat{\mathbf{h}})$ is given by

$$\begin{aligned} p(\mathbf{y}|\mathbf{x}, \hat{\mathbf{h}}) &= \varepsilon_{\Delta}\left[p(\mathbf{y}|\mathbf{x}, \hat{\mathbf{h}}, \Delta)\right] \\ &= \frac{\exp\left(-tr\left[(\mathbf{I}_N + \sigma_z^2 \mathbf{x}^* \mathbf{x})^{-1} (\mathbf{y} - \hat{\mathbf{h}}\mathbf{x})^* (\mathbf{y} - \hat{\mathbf{h}}\mathbf{x})\right]\right)}{\pi^{n_R N} |\mathbf{I}_N + \sigma_z^2 \mathbf{x}^* \mathbf{x}|^{n_R}} \end{aligned} \tag{5.3}$$

The ML detection will choose the matrix $\mathbf{x}$ that optimizes the ML function in (5.3). Our goal is to design the best choice of the signal set $\mathbf{x}$ so as to minimize the average symbol error probability of ML detector.

The organization of this chapter is as follows. In Section 5.2, we discuss the constellation design framework for MIMO links with imperfect CSIR (partially coherent designs). In Section 5.3, we address the coded modulation for

MIMO links with imperfect CSIR (partially coherent systems). Finally, in Section 5.4, we summarize the main points discussed in this chapter.

5.2 CONSTELLATION DESIGN FOR MIMO CHANNELS WITH IMPERFECT CSIR

In practical systems, there are usually pilot symbols or pilot channels available to enable channel estimation at the receiver. Because of the limited energy of pilot symbols and the presence of multiple transmit antennas, the channel estimation can never be perfect. The conventional approach is to design the constellation on the basis of perfect CSIR assumption and then evaluate the performance degradation when there is CSIR estimation error. Yet, with this approach, we have no idea as to the best possible design in the presence of imperfect CSIR. In fact, constellations designed using the statistics of the CSIR error are more desirable than the ones designed for perfect CSIR. In this section, we derive a design criteria for spacetime constellations when only imperfect CSIR is available at the receiver. We design single- and multiple-antenna constellations based on this criterion and illustrate significant performance improvement over conventional constellations in the presence of imperfect CSIR.

5.2.1 System Model

We consider a MIMO system with n_T transmit and n_R receive antennas. The $n_R \times N$ received symbols $\mathbf{y}$ is given by (5.1), where N is the length of the spacetime code. With imperfect CSIR $\hat{\mathbf{h}}$ given by (5.2), the conditional probability density of the received signal, $p(\mathbf{y}|\hat{\mathbf{h}}, \mathbf{x})$, is given by

$$p(\mathbf{y}|\hat{\mathbf{h}}, \mathbf{x}) = \varepsilon_{\Delta}\left[p(\mathbf{y}|\hat{\mathbf{h}}, \Delta, \mathbf{x})\right]$$

which is evaluated in (5.3). Let M be the size of the signal set $\mathcal{M} = \{\mathbf{x}_1, \ldots, \mathbf{x}_M\}$. Define $p_m(\mathbf{y}) = p(\mathbf{y}|\hat{\mathbf{h}}, \mathbf{x}_m)$. Without loss of generality, assuming that $\mathbf{x}_i$ is the transmitted signal, the conditional pairwise error probability (with ML detection) of detecting $\mathbf{x}_j$ instead of $\mathbf{x}_i$ is given by

$$\Pr\left[\mathbf{x}_i \to \mathbf{x}_j|\hat{\mathbf{h}}\right] = \Pr[p_j(\mathbf{y}) > p_i(\mathbf{y})] \tag{5.4}$$

5.2.2 Design Criteria Based on Kullback–Leibler Distance

The conditional error probability of the ML detector for a signal set of cardinality M is upper-bounded by summing all conditional pairwise error probabilities

$$P_e(\hat{\mathbf{h}}) \leq \sum_{i=1}^{M} p(\mathbf{x}_i) \sum_{j \neq i} \Pr\left[\mathbf{x}_i \to \mathbf{x}_j|\hat{\mathbf{h}}\right] \tag{5.5}$$

where $p(\mathbf{x}_i)$ is the probability of transmitting the ith signal element. However, this error probability is usually dominated by the largest term inside the summation. Hence, the error probability is approximated by

$$P_e(\hat{\mathbf{h}}) \approx \max_{i,j,j\neq i} \Pr[\mathbf{x}_i \to \mathbf{x}_j | \hat{\mathbf{h}}] \tag{5.6}$$

The signal constellation design can be done by selecting the signal set $\mathcal{M}$ such that it minimizes the worst-case pairwise error probability as in (5.6). Unfortunately, even with the approximation above, the exact expression or the Chernoff bound for (5.6) is in general intractable. Hence, we shall seek another simple and effective performance criterion. We first review a very important lemma, namely, *Stein's lemma* [30], which relates the *Kullback–Leibler* (KL) distance to the pairwise error probabilities of hypothesis testing.

Lemma 5.1 (Kullback–Leibler Distance) Let $Y_1, Y_2, \ldots, Y_n \in \mathcal{Y}$ be n i.i.d. observations generated according to the probability density function $Q(y)$. Consider the hypothesis testing between two alternatives, $Q = P_1$ and $Q = P_2$, where $\mathcal{D}(P_1 \| P_2) < \infty$ and $\mathcal{D}(P_1 \| P_2)$ is the relative entropy or Kullback–Leibler distance given by

$$\mathcal{D}(P_1 \| P_2) = \sum_{y \in \mathcal{Y}} P_1(y) \log \frac{P_1(y)}{P_2(y)} = \varepsilon_{P_1}\left[\log \frac{P_1(y)}{P_2(y)}\right]$$

Let $\mathcal{A}_n \subseteq \mathcal{Y}^n$ be a decision region for hypothesis 1. Let the probability of error be $\alpha_n = \Pr(P_1 \to P_2) = P_1^n(\mathcal{A}_n^c)$ and $\beta_n = \Pr(P_2 \to P_1) = P_2^n(\mathcal{A}_n)$. Define $\beta_n^{\epsilon} = \min_{\mathcal{A}_n \subseteq \mathcal{Y}^n, \alpha_n < \varepsilon} \beta_n$. Then we have

$$\lim_{\varepsilon \to 0} \lim_{n \to \infty} \frac{1}{n} \log \beta_n^{\varepsilon} = -\mathcal{D}(P_1 \| P_2)$$

From this lemma, we can deduce that the best achievable error exponent for $\Pr(\mathbf{x}_j \to \mathbf{x}_i)$ with the constraint that $\Pr(\mathbf{x}_i \to \mathbf{x}_j) \leq \varepsilon$ is given by the KL distance $\mathcal{D}(p(\mathbf{y}|\mathbf{x}_1, \hat{\mathbf{h}}) \| p(\mathbf{y}|\mathbf{x}_2, \hat{\mathbf{h}}))$. Specifically, on the basis of Stein's lemma, we have the following lemma [30] showing that the error performance of the ML detector is related to the KL distance between the distributions.

Lemma 5.2 (Error Bound Based on KL Distance) Let $Y_1, Y_2, \ldots, Y_n \in \mathcal{Y}$ be n i.i.d. observations generated according to $p_0(y)$ on $\mathcal{Y}$. Given the true hypothesis p_0 and two probability density functions p_1 and p_2 on $\mathcal{Y}$, consider two different sets of hypothesis testings, namely, one between $Q = p_0$ and $Q = p_1$ and the other between $Q = p_0$ and $Q = p_2$, where $0 \leq \mathcal{D}(p_0 \| p_2) \leq \mathcal{D}(p_0 \| p_1) < \infty$. Denote $L_{1n} = \prod_{i=1}^{n} \frac{p_0(Y_i)}{p_1(Y_i)}$ and $L_{2n} = \prod_{i=1}^{n} \frac{p_0(Y_i)}{p_2(Y_i)}$ as the likelihood ratio for the two tests, so that the probabilities of mistaking p_0 for p_1 and p_2 using the ML detector are given by $P_{e1n} = \Pr\{L_{1n} < 1\}$ and $P_{e2n} = \Pr\{L_{2n} < 1\}$, respectively. Let $\Delta\mathcal{D}$ be the difference between the two KL distances:

$$\Delta\mathcal{D} = \mathcal{D}(p_0\|p_1) - \mathcal{D}(p_0\|p_2) > 0$$

With these assumptions, we have

$$\Pr_{p_0}\left\{L_{1n} > \exp\left(\frac{n\Delta\mathcal{D}}{2}\right)L_{2n}\right\} \to 1$$

as $n \to \infty$.

This lemma states that for sufficiently large n, there is a high probability that the likelihood ratio of the first test L_{1n} will be greater than the likelihood ratio of the second test L_{2n}. In addition, the ratio of these two likelihood ratios, $\frac{L_{1n}}{L_{2n}}$, grows exponentially with n. Recall that the error probability of each test is the probability that the corresponding likelihood ratio is smaller than one, which implies that for large n, the first test will have a lower probability of error than will the second test. Hence, the KL distance between a true hypothesis and a false hypothesis $\mathcal{D}(p_0\|p_1)$ is a good indication of the associated error probability. The larger the KL distance is, the smaller will be the error probability.

Assuming i.i.d. fading as seen by the n_R receive antennas, from Stein's lemma, the conditional pairwise error probability is approximated by

$$\Pr\left[\mathbf{x}_i \to \mathbf{x}_j \middle| \hat{\mathbf{h}}\right] \approx \exp(-n_R\mathcal{D}(p_i\|p_j)) \tag{5.7}$$

Hence, the unconditional pairwise error probability is approximated by

$$\begin{aligned}\Pr[\mathbf{x}_i \to \mathbf{x}_j] &= \mathcal{E}_{\hat{\mathbf{h}}}\left[\Pr\left[\mathbf{x}_i \to \mathbf{x}_j \middle| \hat{\mathbf{h}}\right]\right] \\ &\approx \mathcal{E}_{\hat{\mathbf{h}}}[\exp(-n_R\mathcal{D}(p_i\|p_j))] = \exp(-n_R\overline{\mathcal{D}}(\mathbf{x}_i;\mathbf{x}_j))\end{aligned} \tag{5.8}$$

where $\overline{\mathcal{D}}(\mathbf{x}_i; \mathbf{x}_j)$ is the *average KL distance* given by

$$\overline{\mathcal{D}}(\mathbf{x}_i;\mathbf{x}_j) = \frac{1}{n_R}\log_e[\exp(-n_R\mathcal{D}(p_i\|p_j))] \tag{5.9}$$

which is a function of the signal elements $\mathbf{x}_i$ and $\mathbf{x}_j$.

Hence, the signal set design problem can be cast into the following optimization problem.

Problem 5.1 (Signal Set Design) Choose the signal set $\mathcal{M} = \{\mathbf{x}_1, \ldots, \mathbf{x}_M\}$ to maximize

$$\min_{i,j} \overline{\mathcal{D}}(\mathbf{x}_i; \mathbf{x}_j)$$

where $\frac{1}{M}\sum_{m=1}^{M}\|\mathbf{x}_m\|^2$ and $\|\mathbf{x}_m\|^2$ is the matrix norm given by

$$\|\mathbf{x}_m\|^2 = \sum_{n=1}^{N}\sum_{t=1}^{n_T}|\mathbf{x}_m(t,n)|^2$$

It is shown [20] that the KL distance between the probability densities p_i and p_j for the MIMO channels described in (5.3) can be expressed as

$$\begin{aligned}\mathcal{D}(p_i\|p_j) = {} & n_R \mathrm{tr}\left\{\left(\mathbf{I}_N + \sigma_e^2\mathbf{x}_i^*\mathbf{x}_i\right)\left(\mathbf{I}_N + \sigma_e^2\mathbf{x}_j^*\mathbf{x}_j\right)^{-1}\right\} - n_R N \\ & - n_R \log_e\left|\left(\mathbf{I}_N + \sigma_e^2\mathbf{x}_i^*\mathbf{x}_i\right)\left(\mathbf{I}_N + \sigma_e^2\mathbf{x}_j^*\mathbf{x}_j\right)^{-1}\right| \\ & + \mathrm{tr}\left\{\left(\mathbf{I}_N + \sigma_e^2\mathbf{x}_j^*\mathbf{x}_j\right)^{-1}(\mathbf{x}_i - \mathbf{x}_j)^*\mathbf{h}^*\mathbf{h}(\mathbf{x}_i - \mathbf{x}_j)\right\} \end{aligned} \quad (5.10)$$

From (5.9), the *average KL distance* $\overline{\mathcal{D}}(\mathbf{x}_i; \mathbf{x}_j)$ is given by

$$\begin{aligned}\overline{\mathcal{D}}(\mathbf{x}_i; \mathbf{x}_j) = {} & n_R tr\left\{\left(\mathbf{I}_N + \sigma_e^2\mathbf{x}_i^*\mathbf{x}_i\right)\left(\mathbf{I}_N + \sigma_e^2\mathbf{x}_j^*\mathbf{x}_j\right)^{-1}\right\} - n_R N \\ & - n_R \log_e\left|\left(\mathbf{I}_N + \sigma_e^2\mathbf{x}_i^*\mathbf{x}_i\right)\left(\mathbf{I}_N + \sigma_e^2\mathbf{x}_j^*\mathbf{x}_j\right)^{-1}\right| \\ & + n_R \log_e\left|\mathbf{I}_{nT} + (1-\sigma_e^2)(\mathbf{x}_i - \mathbf{x}_j)\left(\mathbf{I}_N + \sigma_e^2\mathbf{x}_j^*\mathbf{x}_j\right)^{-1}(\mathbf{x}_i - \mathbf{x}_j)^*\right| \end{aligned} \quad (5.11)$$

Considering one extreme when $\sigma_e^2 = 0$, we have perfect CSIR, and the average KL distance reduces to

$$\overline{\mathcal{D}}(\mathbf{x}_i; \mathbf{x}_j) = n_R \log_e|\mathbf{I}_{nT} + (\mathbf{x}_i - \mathbf{x}_j)(\mathbf{x}_i - \mathbf{x}_j)^*|$$

Hence, Problem 5.1 is identical to the design criteria in coherent spacetime codes as described in Chapter 4, specifically, the rank and determinant criteria of $(\mathbf{x}_i - \mathbf{x}_j)(\mathbf{x}_i - \mathbf{x}_j)^*$. On the other hand, when $\sigma_e^2 = 1$, the average KL distance reduces to

$$\begin{aligned}\overline{\mathcal{D}}(\mathbf{x}_i; \mathbf{x}_j) = {} & n_R tr\left\{\left(\mathbf{I}_N + \mathbf{x}_i^*\mathbf{x}_i\right)\left(\mathbf{I}_N + \mathbf{x}_j^*\mathbf{x}_j\right)^{-1}\right\} - n_R N \\ & - n_R \log_e\left|\mathbf{I}_{nT} + \left(\mathbf{I}_N + \mathbf{x}_i^*\mathbf{x}_i\right)\left(\mathbf{I}_N + \mathbf{x}_j^*\mathbf{x}_j\right)^{-1}\right| \end{aligned}$$

which gives the design criteria in MIMO channels without CSIR (noncoherent). For intermediate values of σ_e^2, the performance criterion is a combination of the two extremes, which implies that for an optimal design, contributions from both perfect CSIR and no-CSIR designs have to be exploited to achieve better performance. We shall consider two examples in the following text. Since the term n_R does not affect the optimization in Problem 5.1, we assume $n_R = 1$ for simplicity.

5.2.3 Constellation Design Optimization

For a fixed spectral efficiency, the cardinality of the signal set $\mathcal{M}$ grows exponentially with respect to the codelength N. For example, to achieve a spectral efficiency of 4 bits per channel use, the size of the constellation set $\mathcal{M}$ reaches 2^{20} with $N = 5$. This makes the design problem almost impossible to solve. In this section, we consider the constellation design problem for the uncoded system where $N = 1$. In other words, the signal element $\mathbf{x}_i$ is a $n_T \times 1$ complex vector.

The average KL distance between $\mathbf{x}_i$ and $\mathbf{x}_j$ is given by [20]

$$\overline{\mathcal{D}}(\mathbf{x}_i;\mathbf{x}_j) = \frac{1+\sigma_e^2\|\mathbf{x}_i\|^2}{1+\sigma_e^2\|\mathbf{x}_j\|^2} - 1 - \log\left(\frac{1+\sigma_e^2\|\mathbf{x}_i\|^2}{1+\sigma_e^2\|\mathbf{x}_j\|^2}\right) + \log\left[1+(1-\sigma_e^2)\frac{\|\mathbf{x}_i-\mathbf{x}_j\|^2}{1+\sigma_e^2\|\mathbf{x}_j\|^2}\right] \tag{5.12}$$

The optimization of $\{\mathbf{x}_m\}$ in (5.12) is in general not straightforward. To simplify the optimization problem, we shall impose some extra constraints, namely, the *unitary constellation constraint*, on the signal set.

Definition 5.1 (Unitary Constellation) The constellation set $\mathcal{M}$ is unitary if there is a partition on $\mathcal{M}, \{\mathcal{M}_1, \ldots, \mathcal{M}_K\}$, such that there are M_k constellation points in the constellation partition $\mathcal{M}_k$ and $\Sigma_{k=1}^{k} M_k = M$. In addition, $\|\mathbf{x}_i\|^2 = \|\mathbf{x}_j\|^2$ for any $\mathbf{x}_i, \mathbf{x}_j \in \mathcal{M}_k$ where $\|\mathbf{x}\| = tr(\mathbf{x}^*\mathbf{x})$ is the norm of $\mathbf{x}$.

Figure 5.1a illustrates an example of a two-dimensional unitary constellation. Essentially, unitary constellation can be visualized as constellation points over concentric rings of sphere about the origin of the constellation space.

Using the definition of unitary constellation, we can see that if two constellation points $\mathbf{x}_i$, $\mathbf{x}_j$ lie on the same partition $\mathcal{M}_k$ for some k, then $\|\mathbf{x}_i\|^2 = \|\mathbf{x}_j\|^2$ and the average KL distance $\overline{\mathcal{D}}(\mathbf{x}_i; \mathbf{x}_j)$ can be reduced to

$$\overline{\mathcal{D}}(\mathbf{x}_i;\mathbf{x}_j) = \log\left[1+(1-\sigma_e^2)\frac{\|\mathbf{x}_i-\mathbf{x}_j\|^2}{1+\sigma_e^2\|\mathbf{x}_j\|^2}\right]$$

if $\sigma_e^2 < 1$.

On the other hand, if the two constellation points lie on different partitions $\mathcal{M}_k$ and $\mathcal{M}_{k'}$, the minimum average KL distance will be observed when they lie on a line that passes through the origin of the constellation space, which is determined by the radii of these two constellation spheres r_k and $r_{k'}$. Hence, the minimum average KL distance is given by

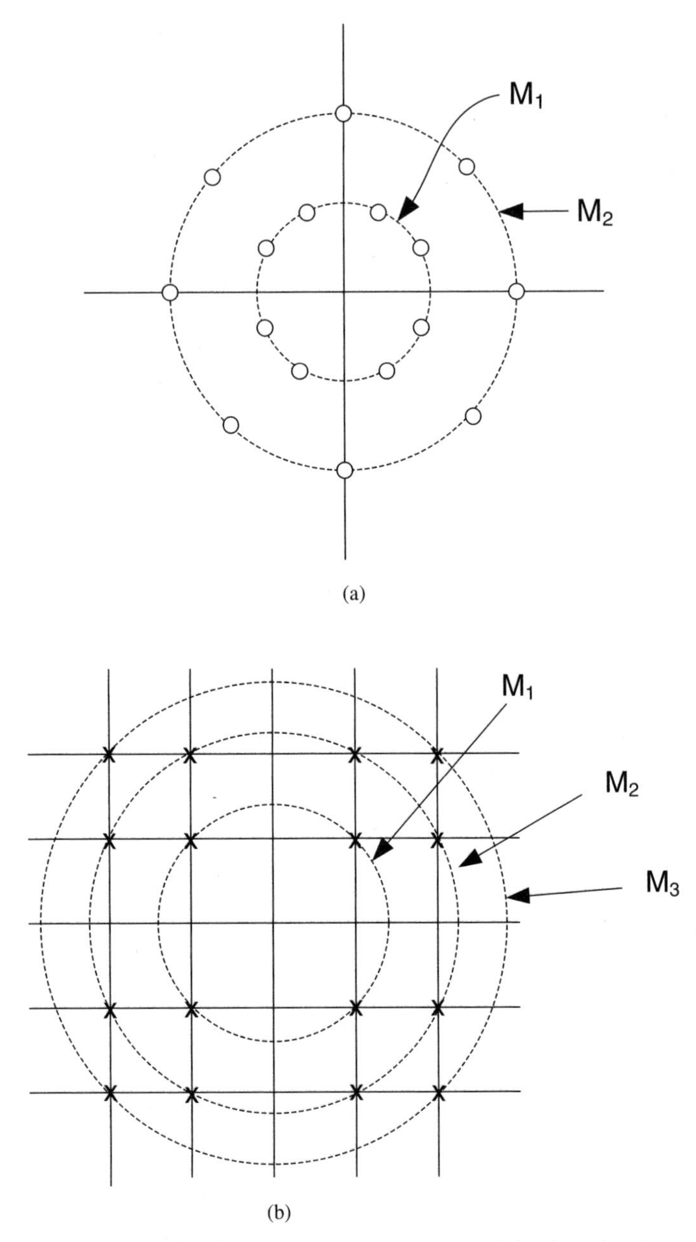

Figure 5.1. Illustration of 2D unitary constellation with $n_T = 1$: (a) unitary 8-point constellation: $\mathcal{M} = \{\mathcal{M}_1, \mathcal{M}_2\}$ and $M_1 = M_2 = 8$; (b) unitary 16-point constellation: $\{\mathcal{M}_1, \mathcal{M}_2, \mathcal{M}_3\}$ and $M_1 = 4$, $M_2 = 8$, $M_3 = 4$.

$$\min_{i,j} \overline{\mathcal{D}}(\mathbf{x}_i;\mathbf{x}_j) \geq \min\left\{\min_{k\in[1,K]} \mathcal{D}_{\text{intra}}(k), \min_{k=1,\ldots,K-1} \mathcal{D}_{\text{inter}}(k,k+1)\right\} \tag{5.13}$$

where $\mathcal{D}_{\text{intra}}(k)$ is the *intrasubset* distance defined as

$$\mathcal{D}_{\text{intra}}(k) = \min_{\mathbf{x}_i,\mathbf{x}_j\in\mathcal{M}_k} \log\left\{1+\frac{1-\sigma_e^2}{1+\sigma_e^2 r_k^2} 4r_k^2 \sin^2(\angle\tilde{\mathbf{x}}_i,\tilde{\mathbf{x}}_j)\right\} \tag{5.14}$$

and the $\mathcal{D}_{\text{inter}}(k, k')$ is the *intersubset* distance given by

$$\mathcal{D}_{\text{inter}}(k,k') = \frac{1+\sigma_e^2 r_k^2}{1+\sigma_e^2 r_{k'}^2} - 1 - \log\left(\frac{1+\sigma_e^2 r_k^2}{1+\sigma_e^2 r_{k'}^2}\right) + \log\left(1+\frac{1-\sigma_e^2}{1+\sigma_e^2 r_k^2}|r_k - r_{k'}|^2\right) \tag{5.15}$$

Here $\tilde{\mathbf{x}}$ is the vector obtained by staggering the real and imaginary parts of $\mathbf{x}$

$$\tilde{\mathbf{x}} = [\mathcal{R}(\mathbf{x}), \mathcal{Z}(\mathbf{x})]^* \tag{5.16}$$

and $\angle\tilde{\mathbf{x}}_i$, $\tilde{\mathbf{x}}_j$ is the angle between two signal vectors $\tilde{\mathbf{x}}_i$ and $\tilde{\mathbf{x}}_j$. Therefore, the optimization problem (Problem 5.1) can be simplified as follows.

Problem 5.2 (Signal Set Design) Select the unitary constellation parameters, $K \in [1, M]$, $\{M_1, \ldots, M_K\}$, $\{r_1, \ldots, r_K\}$ so as to maximize the minimum between the *intraset distance* $\mathcal{D}_{\text{intra}}(k)$ and the *interset distance* $\mathcal{D}_{\text{inter}}(k, k+1)$:

$$\max_{K\in[1,M],\frac{1}{M}M_k r_k^2,\sum_k M_k = M, 0\leq r_1\leq\cdots\leq r_K} \min\left\{\min_{k\in[1,k]} D_{\text{intra}}(k), \min_{k=1,\ldots,K-1} D_{\text{inter}}(k,k+1)\right\} \tag{5.17}$$

The optimization problem involves discrete variables (K and $M_1, \ldots, M_K$) as well as continuous variables $r_1, \ldots, r_K$. For any fixed value of K and M_1, $\ldots$, M_K satisfying the constraints in (5.17), the optimal ($r_1, \ldots, r_K$) can be obtained numerically.

5.2.4 Single-Transmit Antenna Example

Consider a simple case of *constellation design* ($N = 1$) for systems with single transmit antenna $n_T = 1$ in fading channels. Hence, the signal element $\mathbf{x}_i$ is a complex scalar. The optimization parameter is the constellation set $\mathcal{M} = \{x_1, \ldots, x_M\}$ where the mth constellation point, x_m, is a complex scalar.

Figure 5.2 illustrates the 8-point and 16-point optimal constellations with average energy P_0 of 10 dB for different values of σ_e^2.

The uncoded performance (average symbol error probability) of the 8-point and 16-point constellation designs is illustrated in Figure 5.3. The curves with the label *coherent* refer to the performance of unitary constellation designed for $\sigma_e^2 = 0$. Because of the larger minimum average KL distance of the new constellation, the exponential decay of the symbol error probability

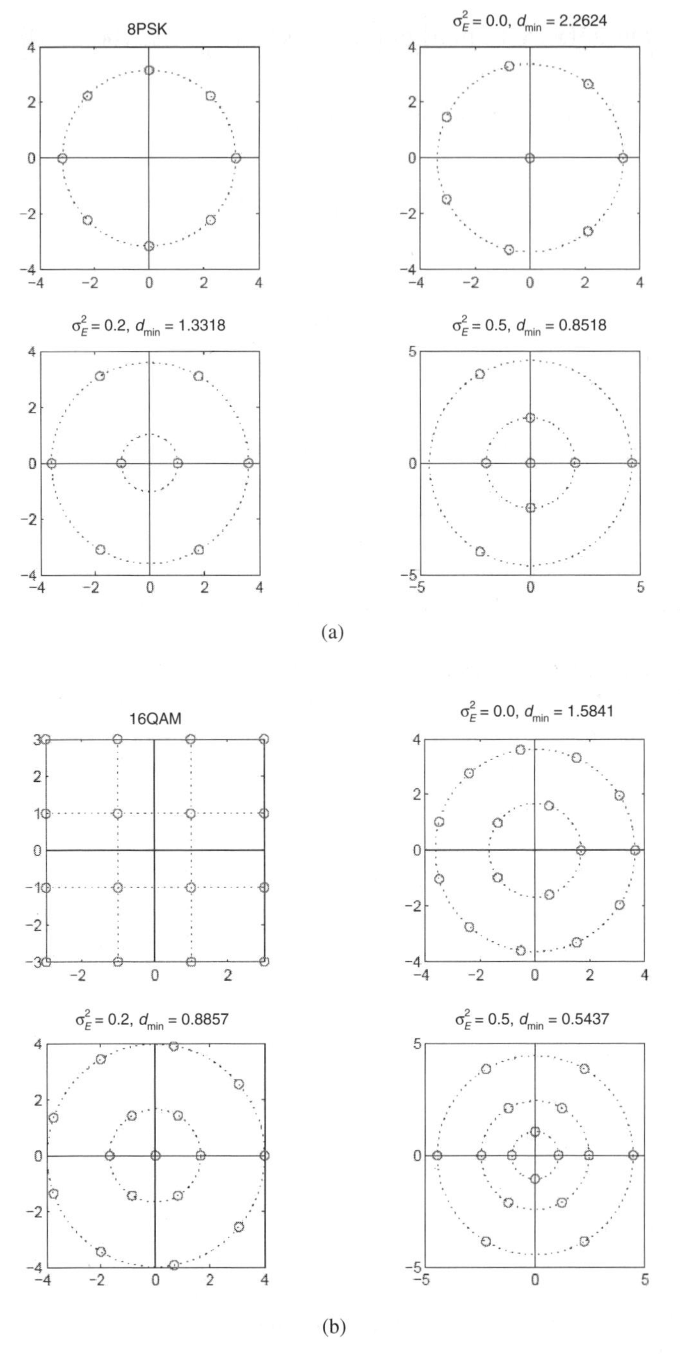

Figure 5.2. Illustration of the optimal constellation of point sizes 8 (a) and 16 (b) for $n_T = n_R = 1$ and $P_0 = 10\,\text{dB}$. [20]

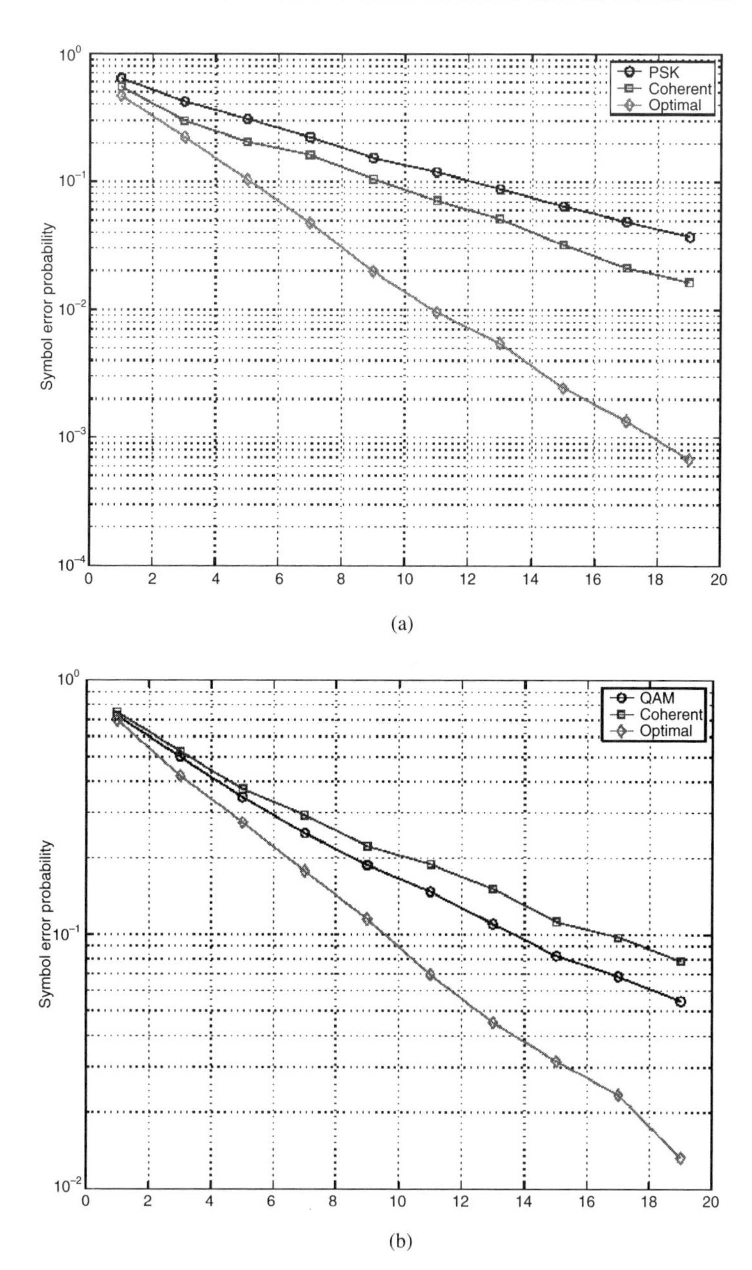

Figure 5.3. Average symbol error probability versus n_R of the optimal constellation of point sizes 8 (a) and 16 (b) for $n_T = 1$, $P_0 = 10$, and $\sigma_e^2 = 0.5$. [20]

versus n_R is much higher for the new constellation (comparing the optimal constellation with the regular 8PSK constellation).

5.2.5 Multitransmit Antenna Example

When there is more than one transmit antenna $n_T > 1$, the signal set consists of $n_T \times 1$ vectors. The constellation design is given by the optimization problem (Problem 5.2). For illustration purposes, we consider $n_T = n_R = 2$ uncoded MIMO with throughputs of 4 bits per channel use and 8 bits per channel use. Figure 5.4 illustrates the uncoded performance (average symbol error probability) of the 4-point constellation designed for $\sigma_e^2 = 0.05$ and $\sigma_e^2 = 0.1$. For comparison, we consider two baseline references. The first reference (labeled as "QPSK") is given by regular QPSK constellation per transmit antenna followed by a ML receiver for fair comparison. The second reference (labeled as "Alamouti QAM") is given by regular 16QAM constellation together with Alamouti spacetime coding [9]. These two reference systems have the same throughput of 4 bits per channel use. The performance improvement of the optimal constellations (designed for $\sigma_e^2 = 0.05$ and $\sigma_e^2 = 0.10$, respectively) is substantial compared with the two reference systems. Note that the Alamouti scheme suffers from severe performance degradation due to the CSIR error and its performance becomes worse than the conventional QPSK reference performance without any transmit diversity.

Figure 5.5 illustrates the uncoded performance (average symbol error probability) of the 16-point constellation designed for $\sigma_e^2 = 0.00$ (perfect CSIR) and $\sigma_e^2 = 0.01$, respectively. Similarly, we consider two baseline references. The first reference (labeled as "QAM") refers to the regular 16QAM constellation per transmit antenna followed by ML receiver for fair comparison. The second reference (labeled as "Alamouti QAM") refers to the regular 256QAM constellation together with Alamouti spacetime coding. These two reference systems have the same throughput of 8 bits per channel use for fair comparison. Observe that in the absence of CSIR error (perfect CSIR), the performance of the optimal design (designed for $\sigma_e^2 = 0$) and the regular 16QAM reference remains the same. Also, although Alamouti coding has a larger transmit diversity advantage, the performance is worse than the others because of smaller coding advantage. On the other hand, with $\sigma_e^2 = 0.01$, the optimal 16-point constellation (designed for $\sigma_e^2 = 0.01$) achieves significant performance relative to the other two references.

5.3 SPACETIME CODING FOR MIMO CHANNELS WITH IMPERFECT CSIR

In the previous sections, we focused on the constellation design problem for uncoded MIMO systems ($N = 1$) with imperfect CSIR. In practice, there is always coding on top of the modulation to improve the reliability of packet

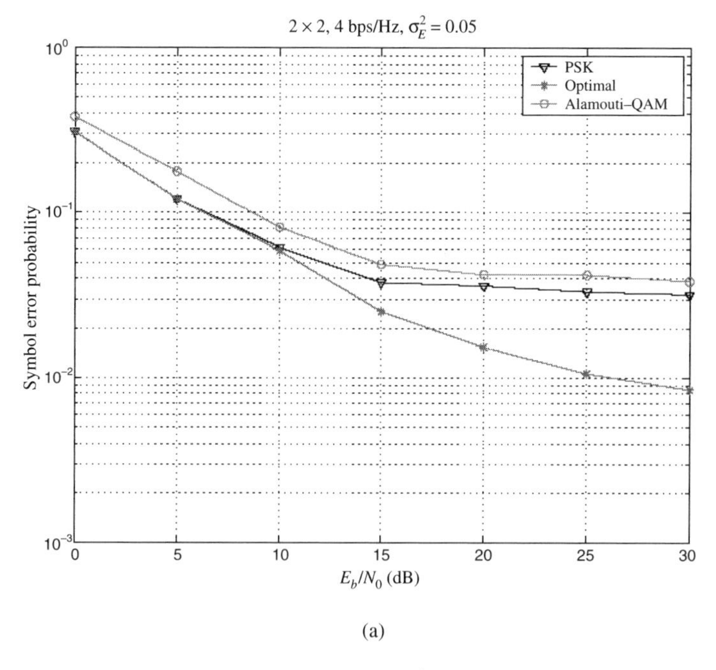

(a)

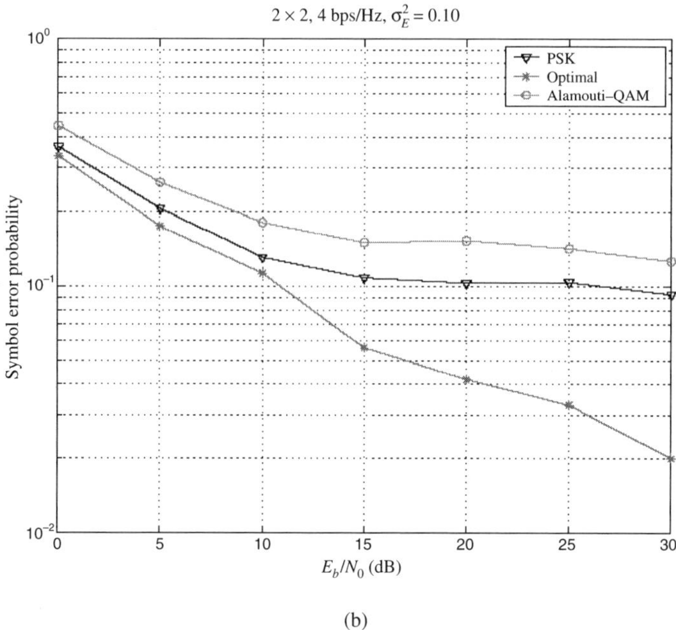

(b)

Figure 5.4. Average symbol error probability versus SNR of the optimal constellation design for $n_T = n_R = 2$, $\sigma_e^2 = 0.05$ (a), $\sigma_e^2 = 0.1$ (b) and uncoded throughput of 4 bits per channel use. [20]

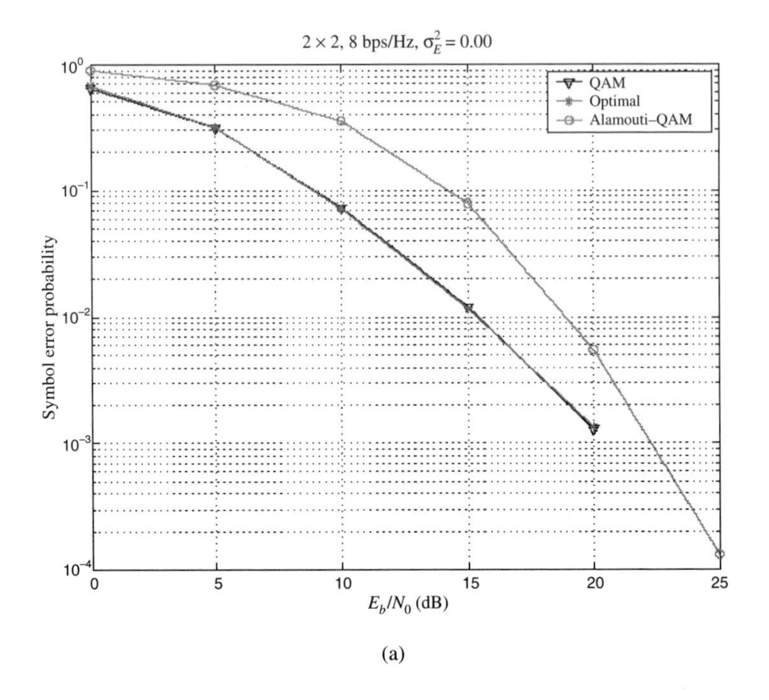

(a)

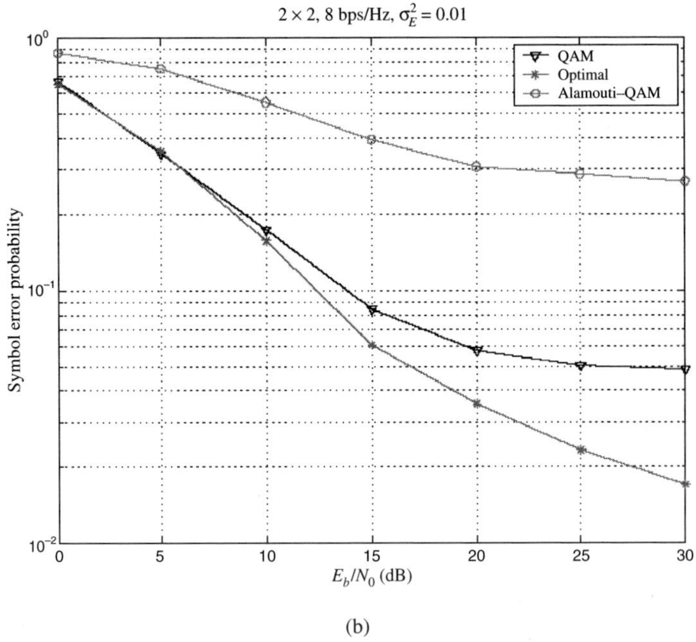

(b)

Figure 5.5. Average symbol error probability versus SNR of the optimal constellation design for $n_T = n_R = 2$, $\sigma_e^2 = 0.00$ (perfect CSIR) (a), $\sigma_e^2 = 0.01$ (imperfect CSIR) (b) and uncoded throughput of 8 bits per channel use. [20]

transmission. In the following text, we shall discuss the design criteria of coded modulation in MIMO systems with imperfect CSIR. We first review the conventional coded modulation [114] design criteria. In Section 5.3.2, we discuss the coded modulation design for MIMO fading channels with imperfect CSIR.

5.3.1 Overview of Coded Modulation in AWGN Channels

In this section, the general structure of trellis-coded modulation (TCM) schemes in AWGN channels and the principles of code construction are reviewed. TCM has evolved since the mid-1990s as a combined coding–modulation technique for digital transmission over bandlimited channels. Its main attraction is that it allows the achievement of significant coding gains over conventional uncoded multilevel modulation without any increase in bandwidth.

In general, when coding is applied to the bandlimited channel, a performance gain is desired without expanding the signal bandwidth. This goal can be achieved by increasing the number of signals over the corresponding uncoded system to compensate for the redundancy introduced by the code. If the modulation is treated as a separate operation independent of encoding, the use of very powerful codes is required to offset the loss and provide some significant coding gain. On the other hand, if the modulation is an integral part of the encoding process and is designed in conjunction with the code to increase the minimum Euclidean distance between pairs of coded signals, the loss from the expansion of the signal set is easily overcome and a significant gain is achieved with relatively simple codes. The key to this integrated modulation and coding approach is to devise an effective method for mapping the coded bits into signal points such that the minimum Euclidean distance is maximized. Such a method was developed by Ungerboeck [133], based on the principle of mapping by set partitioning.

In the method of mapping by set partitioning, the signal set is partitioned into several subsets of a relatively large minimum intrasubset squared Euclidean distance, while the minimum intersubset distance is the same as the minimum distance of the original signal set. For example, let us partition an eight-phase signal constellation shown in Figure 5.6 into subsets of increasing minimum Euclidean distance between signal points in a subset. In the eight-phase signal set, the signal points are located on a circle of radius $\sqrt{\varepsilon}$ and have a minimum distance separation of

$$d_0 = 2\sqrt{\varepsilon}\sin\frac{1}{8}\pi = \sqrt{\left(2-\sqrt{2}\right)\varepsilon} = 0.765\sqrt{\varepsilon} \tag{5.18}$$

In the first partition, the 8 points are subdivided into two subsets of 4 points, such that the minimum distance between neighboring points increases to $d_1 = \sqrt{2\varepsilon}$. In the second level of partitioning, each of the two subsets is subdivided into two subsets of 2 points such that the minimum distance increases to $d_2 = \sqrt{2\varepsilon}$. This results in four subsets of two points each. Finally, the last

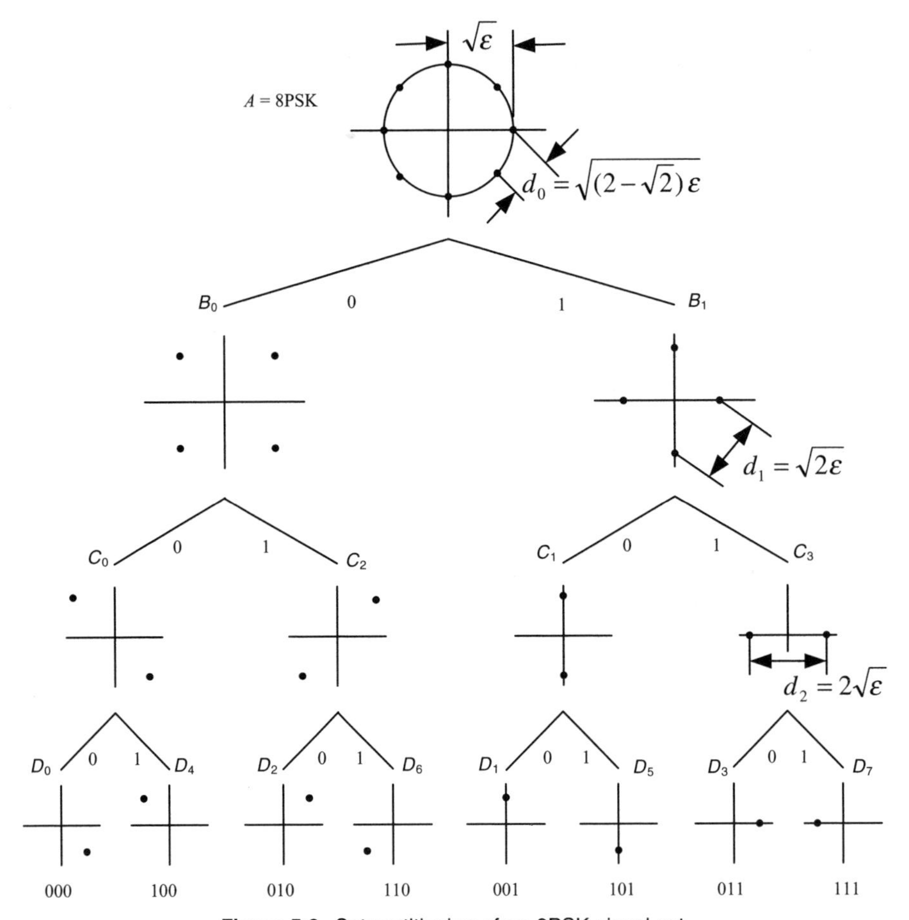

Figure 5.6. Set partitioning of an 8PSK signal set.

stage of partitioning leads to eight subsets, where each subset contains a single point. In general, it may not necessary to partition until each subset contains only 1 point. For example, the eight-phase PSK may be partitioned only twice, to yield four subsets of 2 points each. Note that each level of partitioning increases the minimum Euclidean distance between signal points.

With set partitioning in mind, the general structure of TCM encoder/modulator is depicted in Figure 5.7. Referring to the figure, TCM signals are generated as follows. Assume that the number of information bits to be transmitted per encoder/modulation operation is n. From the n-bit input vector $\mathbf{b} = [b_1\ b_2 \cdots b_n]^T$, $m \leq n$ bits are expanded by a rate $m/(m+1)$ convolutional encoder into $m + 1$ coded bits. These bits are used to select one of 2^{m+1} subsets of redundant 2^{m+1}-ary signal set. The remaining $n - m$ uncoded bits determine which of the 2^{n-m} signals in the subset is to be transmitted. Hence, one redundant bit is added every n input bits, which amounts to expanding the constel-

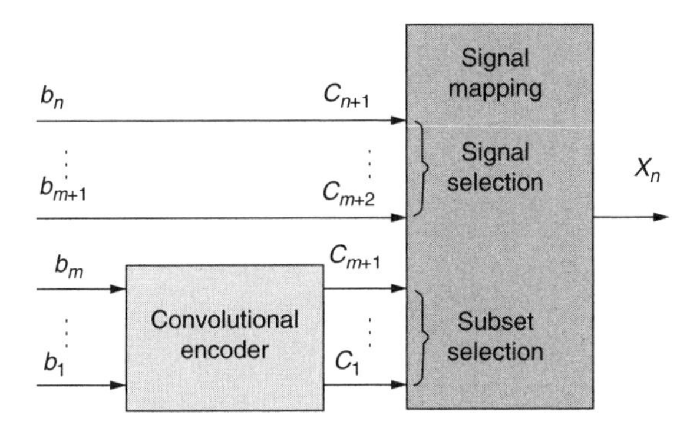

Figure 5.7. General structure of encoder/modulalor for trellis-coded modulation.

lation size to 2^{n+1}. If $m < n$, parallel transitions are allowed in the trellis; if $m = n$, no parallel transition exists.

The assignment of signal subsets to state transition in the trellis is based on the three heuristic rules devised by Ungerboeck [133], which are summarized as follows:

1. Parallel state transitions are assigned signal points separated by the greatest Euclidean distance. Note that parallel transitions in the trellis are the characteristics of TCM containing one or more uncoded information bits.
2. The transitions originating from the same state or merging into the same state in the trellis are assigned subsets of signals that are separated by the greatest Euclidean distance.
3. All signals are used in the trellis diagram with equal frequency, and with a certain degree of regularity and symmetry.

It should be pointed out that, for a trellis-coded modulation scheme, the performance will be dominated by the minimum free Euclidean distance between codewords that can be expressed as [134]

$$D_{\text{free}} = \min[\delta, d_{\text{free}}] \tag{5.19}$$

where δ is the minimum distance between parallel transitions and d_{free} denotes the minimum distance between nonparallel paths in the TCM trellis diagram. In the special case of $m = n$, the subsets contain only one signal, and hence there are no parallel transitions. The final TCM design criterion, thus, is to maximize the minimum free Euclidean distance.

Note that the set partitioning is performed with the goal of maximizing the first quantity δ, whereas the trellis of the constituent code is designed to maximize the second quantity d_{free}. With an appropriate set partitioning and trellis design, the overall minimum distance of the code will be large enough to overcome the loss from the constellation expansion (due to the redundancy in the code), and provide a significant coding gain. In the following example, we present the trellis-coded 8PSK modulation and determine the coding gain.

Example 5.1 (TCM Design Example; Example of 8PSK TCM Design in AWGN Channels) Let us consider the use of the 8PSK signal constellation in conjunction with rate-$\frac{2}{3}$ trellis codes. The coded 8PSK signal set employs the signal points in Figure 5.6. The uncoded 4PSK, which is used as a reference in measuring coding gain, employs the signal points in either subset B_0 or B_1. Figure 5.8 is a state transition (trellis) diagram for uncoded 4PSK modulation with (a) one and (b) four trellis states and (c) coded 8PSK modulation with four trellis states. The trivial one-state trellis diagram is shown only to illustrate uncoded 4PSK from the viewpoint of TCM. The subsets D_0, D_2, D_4, and D_6 are used as the signal points for illustration purposes. Every connected path through a trellis represents an allowable signal sequence. In both systems, starting from any state, four transitions can occur, as required to encode two information bits per modulation interval (2 bps/Hz).

The four parallel transitions in the one-state trellis diagram of Figure 5.8a for uncoded 4PSK do not restrict the sequences of 4PSK signals that can be transmitted; that is, there is no sequence coding. Hence, the optimum decoder can make independent nearest-signal decisions for each noisy received 4PSK signal. The smallest distance between the 4PSK signals is $\sqrt{2\varepsilon}$, which we denote it as d_0 and call the "free distance" of uncoded 4PSK modulation for the sake of using common terminology with sequence-coded systems. Each 4PSK signal has two nearest-neighbor signals at this distance.

In the four state trellis of Figure 5.8b, the 8PSK signals are assigned to the transitions in the four-state trellis in accordance with the principle of mapping by set partition as discussed above. Note that each branch in the trellis corresponds to one of the four subsets C_0, C_1, C_2, or C_3. For the 8-point constellation, each of the subsets C_0, C_1, C_2, and C_3 contains two signal points. As illustrated in Figure 5.6, the distances d_0, d_1, and d_2 are defined as the minimum distance of constellation points in the subsets A, B_0, and C_0, respectively. It can be easily seen that the squared Euclidean distance between parallel transitions is $d_2 = 2\sqrt{\varepsilon}$. On the other hand, any two signal paths in the trellis of Figure 5.8c that diverge in one state and remerge in another after more than one transition have at least squared Euclidean distance $d_0^2 + 2d_1^2 = d_0^2 + d_2^2$ between them. For example, the paths with signals 0–0–0 and 2–1–2 have this distance. The distance between such paths is greater than the distance between the signals assigned to parallel transitions. Hence, the minimum Euclidean distance separation between paths that diverge from any state and remerge at the same state in the four-state trellis is $d_2 = 2\sqrt{\varepsilon}$. This minimum distance in the trellis code is called the *free Euclidean distance* and is denoted by D_{free}.

When compared with the Euclidean distance $d_0 = \sqrt{2\varepsilon}$ for the uncoded 4PSK modulation, we observe that the four-state trellis code gives a coding gain of 3 dB.

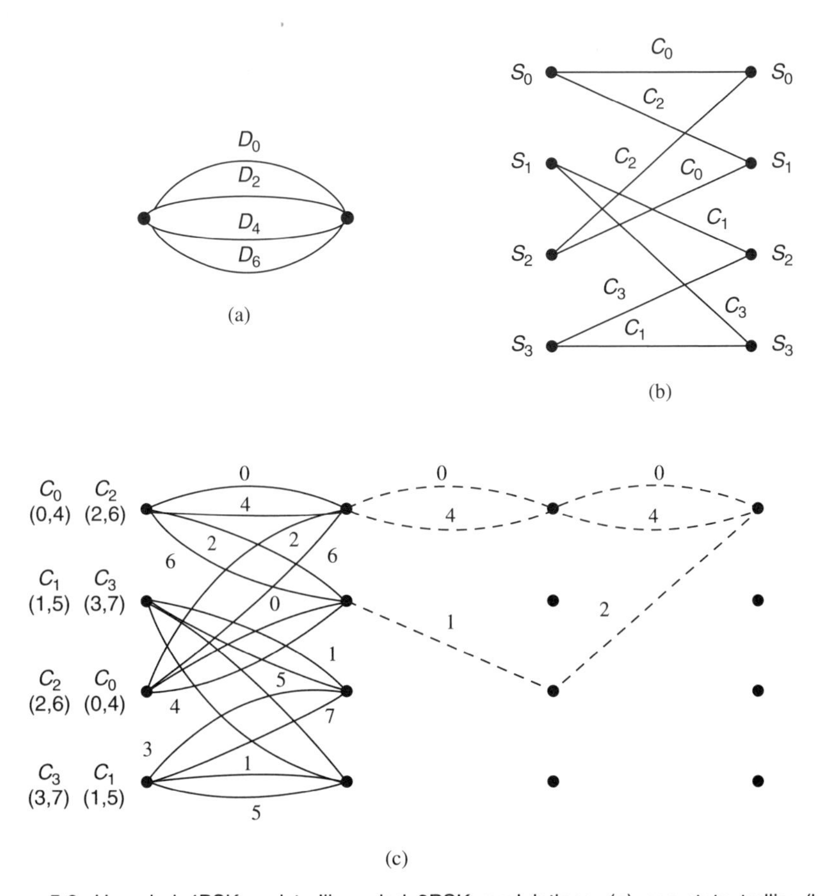

Figure 5.8. Uncoded 4PSK and trellis-coded 8PSK modulations: (a) one-state trellis; (b,c) four-state trellis.

Similar techniques have been developed for coded modulation in fast fading scenarios, when the receiver is assumed to have perfect channel state information. The design criteria in this case are maximization of both the symbol Hamming distance and the minimum product Euclidean distance between pairs of codewords. Therefore, the set partitioning and trellis design are performed to maximize the length of the shortest error event path and the product of the Euclidean distances along this path. In the next section, we apply an idea similar to trellis-coded modulation in the AWGN channels, to design a coded modulation scheme for MIMO channels with imperfect CSIR.

5.3.2 Coded Modulation Design for MIMO Channels with Imperfect CSIR

In this section, we derive the design criterion for coded modulation over MIMO fading channels with imperfect CSIR. To understand how coding can help combat fading over MIMO channels, we consider block fading channels where an encoding frame of length N is partitioned into B fading blocks (each of length T). The channel fading is assumed to be quasistatic within a fading block and i.i.d. between the B fading blocks.

During the bth fading block, the $n_R \times T$ received signal $\mathbf{y}_b$ is given by

$$\mathbf{y}_b = \mathbf{h}_b \mathbf{x}^b + \mathbf{z}_b \tag{5.20}$$

where $\mathbf{h}_b$ is the $n_R \times n_T$ fading matrix, $\mathbf{x}^b$ is the $n_T \times T$ transmitted signals, and $\mathbf{z}_b$ is the $n_R \times T$ channel noise (i.i.d. with unit noise variance) during the bth fading block. Similarly, we define the imperfect CSIR $\hat{\mathbf{h}}_b$ as

$$\hat{\mathbf{h}}_b = \mathbf{h}_b + \mathbf{\Delta}_b \tag{5.21}$$

The equivalent channel model is given by

$$\mathbf{y}_b = \hat{\mathbf{h}}_b \mathbf{x}^b - \mathbf{\Delta}_b \mathbf{x}^b + \mathbf{z}_b \tag{5.22}$$

The joint conditional likelihood function is given by

$$\begin{aligned} & p(\mathbf{y}_1, \ldots, \mathbf{y}_B | \mathbf{x}^1, \ldots, \mathbf{x}^B, \hat{\mathbf{h}}_1, \ldots, \hat{\mathbf{h}}_B) \\ & = \varepsilon_\Delta \left[p(\mathbf{y}_1, \ldots, \mathbf{y}_B | \mathbf{x}^1, \ldots, \mathbf{x}^B, \hat{\mathbf{h}}_1, \ldots, \hat{\mathbf{h}}_B, \Delta_1, \ldots, \Delta_B) \right] \\ & = \prod_{b=1}^{B} \frac{1}{\pi^{T n_R} | \mathbf{I}_T + \sigma_e^2 \mathbf{x}^{b*} \mathbf{x}^b |^{n_R}} \exp\left\{ -\mathrm{tr}\left[(\mathbf{I}_T + \sigma_e^2 \mathbf{x}^{b*} \mathbf{x}^b)^{-1} (\mathbf{y}_b - \hat{\mathbf{h}}_b \mathbf{x}^b)^* (\mathbf{y}_b - \hat{\mathbf{h}}_b \mathbf{x}^b) \right] \right\} \end{aligned} \tag{5.23}$$

Let p_i denote the conditional likelihood $p(\mathbf{y}_1, \ldots, \mathbf{y}_B | \mathbf{x}_i^1, \ldots, \mathbf{x}_i^B, \hat{\mathbf{h}}_1, \ldots, \hat{\mathbf{h}}_B)$. The conditional KL distance between p_i and p_j is given by

$$\mathcal{D}(p_i \| p_j) = \sum_{b=1}^{B} \mathcal{D}^b(\hat{\mathbf{h}}_b) \tag{5.24}$$

where

$$\begin{aligned} \mathcal{D}^b(\hat{\mathbf{h}}_b) = & \, n_R \mathrm{tr}\left\{ (\mathbf{I}_T + \sigma_e^2 \mathbf{x}_i^{b*} \mathbf{x}_i^b)(\mathbf{I}_T + \sigma_e^2 \mathbf{x}_j^{b*} \mathbf{x}_j^b)^{-1} \right\} - n_R T \\ & - n_R \log \left| (\mathbf{I}_T + \sigma_e^2 \mathbf{x}_i^{b*} \mathbf{x}_i^b)(\mathbf{I}_T + \sigma_e^2 \mathbf{x}_j^{b*} \mathbf{x}_j^b)^{-1} \right| \\ & + \mathrm{tr}\left\{ (\mathbf{I}_T + \sigma_e^2 \mathbf{x}_j^{b*} \mathbf{x}_j^b)^{-1} (\mathbf{x}_i^b - \mathbf{x}_j^b)^* \hat{\mathbf{h}}_b^* \hat{\mathbf{h}}_b (\mathbf{x}_i^b - \mathbf{x}_j^b) \right\} \end{aligned} \tag{5.25}$$

Similar to Section 5.2.2, we consider the pairwise error proabability. The conditional error probability is approximated by

$$P_e(\mathbf{x}_i \to \mathbf{x}_j|\hat{\mathbf{h}}) \approx \exp(-\mathcal{D}(p_i\|p_j)) = \exp\left(-\sum_{b=1}^{B}\mathcal{D}^b(\hat{\mathbf{h}}_b)\right) \tag{5.26}$$

where $\mathbf{x}_i = [\mathbf{x}_i^1, \ldots, \mathbf{x}_i^B]$ and $\hat{\mathbf{h}} = [\hat{\mathbf{h}}_1, \ldots, \hat{\mathbf{h}}_B]$ are the aggregate transmit signals and the aggregate imperfect CSIR over the B fading blocks. Since the CSIR is i.i.d. across the B fading blocks, we have

$$p(\hat{\mathbf{h}}) = \prod_{b=1}^{B} p(\hat{\mathbf{h}}_b)$$

The unconditioned error probability is then expressed as

$$\begin{aligned}
P_e(\mathbf{x}_i \to \mathbf{x}_j) &= \varepsilon_{\hat{\mathbf{h}}}\left[P_e(\mathbf{x}_i \to \mathbf{x}_j|\hat{\mathbf{h}})\right] \\
&\approx \varepsilon_{\hat{\mathbf{h}}}\left[-\sum_{b=1}^{B}\mathcal{D}^b(\hat{\mathbf{h}}_b)\right] \\
&= \prod_{b=1}^{B} \varepsilon_{\hat{\mathbf{h}}_b}\left[\exp(-\mathcal{D}^b(\hat{\mathbf{h}}_b))\right] \\
&= \prod_{b=1}^{B} \exp(-\overline{\mathcal{D}}^b(\mathbf{x}_i;\mathbf{x}_j)) \\
&= \exp\left(-\sum_{b=1}^{B}\overline{\mathcal{D}}^b(\mathbf{x}_i;\mathbf{x}_j)\right)
\end{aligned} \tag{5.27}$$

where $\overline{\mathcal{D}}^b(\mathbf{x}_i; \mathbf{x}_j)$ is the average KL distance of the bth fading block

$$\begin{aligned}
\overline{\mathcal{D}}^b(\mathbf{x}_i;\mathbf{x}_j) = {} & n_R tr\left\{\left(\mathbf{I}_T + \sigma_e^2\mathbf{x}_i^{b*}\mathbf{x}_i^b\right)\left(\mathbf{I}_T + \sigma_e^2\mathbf{x}_j^{b*}\mathbf{x}_j^b\right)^{-1}\right\} - n_R T \\
& - n_R \log\left|\left(\mathbf{I}_T + \sigma_e^2\mathbf{x}_i^{b*}\mathbf{x}_i^b\right)\left(\mathbf{I}_T + \sigma_e^2\mathbf{x}_j^{b*}\mathbf{x}_j^b\right)^{-1}\right| \\
& + n_R \log\left|\mathbf{I}_{nT} + (1-\sigma_e^2)(\mathbf{x}_i^b - \mathbf{x}_j^b)\left(\mathbf{I}_T + \sigma_e^2\mathbf{x}_j^{b*}\mathbf{x}_j^b\right)^{-1}(\mathbf{x}_i^b - \mathbf{x}_j^b)^*\right|
\end{aligned} \tag{5.28}$$

From Equation (5.27), the overall average KL distance $\overline{\mathcal{D}}(\mathbf{x}_i; \mathbf{x}_j)$ is given by

$$\overline{\mathcal{D}}(\mathbf{x}_i;\mathbf{x}_j) = \sum_{b=1}^{B}\overline{\mathcal{D}}^b(\mathbf{x}_i;\mathbf{x}_j) \tag{5.29}$$

Hence, the code design criterion is to maximize the minimum of the sum of the average KL distance between any two *vector codewords* $\mathbf{x}_i$ and $\mathbf{x}_j$. This is very similar to the conventional code design in AWGN channels where the design criterion is to maximize the minimum of Euclidean distance between any two codewords. For instance, the property of the overall average KL distance $\overline{\mathcal{D}}(\mathbf{x}_i; \mathbf{x}_j)$ is similar to the additive property of the Euclidean distance, which is used as the design criterion in conventional coded modulation for AWGN channels as discussed in Section 5.3.1. As a result, we can use similar

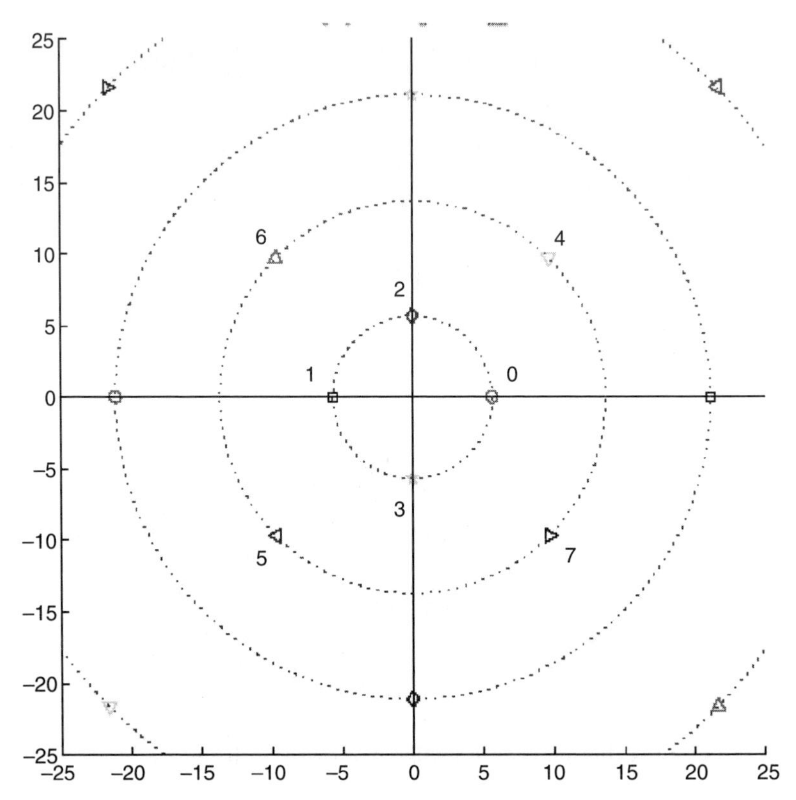

Figure 5.9. Illustration of 16-point constellation designed for MIMO channels with imperfect CSIR; $\sigma_e^2 = 0.01$ and SNR = 20 dB. This constellation is designed to optimize the uncoded error performance for $T = 1$. [20]

techniques as conventional coded modulation and the introduced set partitioning [134] to design good codes for MIMO block fading channels with imperfect CSIR. The only difference in the design procedure is to use the average KL distance $\overline{\mathcal{D}}^b(\mathbf{x}_i^b; \mathbf{x}_j^b)$ rather than the Euclidean distance. Therefore, we need to perform set partitioning and trellis design based on $\overline{\mathcal{D}}^b(\mathbf{x}_i^b; \mathbf{x}_j^b)$.

To reduce the search space, we consider one special case of fast fading channels with $T = 1$. In addition, we shall decouple the design of the coding from the constellation; otherwise, the search complexity will be enormous. For example, we consider the four-level 16-point constellation designed for a channel estimation error of $\sigma_e^2 = 0.01$ as illustrated in Figure 5.9. This constellation is designed to optimize the uncoded performance as discussed in Section 5.2.5. In the design of the constellation, we have allowed for appropriate rotations of the circular subsets to obtain a greater minimum KL distance. We partition this constellation into eight subsets, each one containing 2 points. The partitioning is performed with a goal of maximizing the minimum intrasubset KL distance.

For each block of three information bits, we encode 2 bits using a 16-state rate-$\frac{2}{3}$ convolutional code with octal transfer function [142,143] to produce three coded bits. These three coded bits are used to select one of the eight subsets of the constellation, and we use the remaining uncoded bit to select 1 point from the subset. In Figure 5.9, the numbers next to the eight innermost signal points represent the labels of their corresponding subsets in the partitioned constellation. The degree of set partitioning is determined by the average KL distance.

Figure 5.10a illustrates the performance of the uncoded and trellis-coded 16QAM at $\sigma_e^2 = 0.01$. The coded modulation scheme for the 16QAM constellation is designed with the assumption of perfect CSIR and is labeled as "coherent Rx". The label "Optimal Rx" refers to the trellis-coded modulation design based on KL distance. Observe from the figure that for $\sigma_e^2 = 0.01$, the optimal designs show similar performance gains over the coded and uncoded systems, respectively. Figure 5.10b illustrates the performance comparison at $\sigma_e^2 = 0.05$. The performance gain of the partial coherent design (based on KL distance) in both the coded and uncoded comparisons are more significant. In general, as the channel estimation error variance increases, larger performance gains can be obtained using the KL distance as the design criterion.

5.4 SUMMARY

In this chapter, we have introduced a systematic framework [20] to investigate the spacetime coding and constellation design for MIMO channels with imperfect CSIR. In particular, we considered the asymptotically optimal spacetime coding and constellation design in the absence of perfect CSIR. With channel estimation errors, constellations that are designed according to the statistics of the CSI error are more desirable than those designed for perfect CSIR. Using Stein's lemma, the conditional pairwise error probability can be approximated by $\exp(-\mathcal{D}(p_i \| p_j))$, where $\mathcal{D}(p_i \| p_j)$ is the *KL distance* between the likelihood functions $p_i = p(\mathbf{y}|\mathbf{x}_i, \hat{\mathbf{h}})$ and $p_j = p(\mathbf{y}|\mathbf{x}_j, \hat{\mathbf{h}})$. Hence, the constellation and code design for MIMO fading channels with imperfect CSIR can be expressed as the optimization problem to select the codewords $\{\mathbf{x}_i\}$ so as to maximize the minimum average KL distance $\overline{\mathcal{D}}(\mathbf{x}_i; \mathbf{x}_j)$ (which is the expectation of the exponential error bound over the CSIR estimation error $\mathbf{\Delta}$).

We applied the design criterion ($\overline{\mathcal{D}}(\mathbf{x}_i; \mathbf{x}_j)$) to the constellation design so as to optimize the uncoded MIMO performance with imperfect CSIR. To restrict the search space, we consider optimization over all unitary constellations. We found that the uncoded performance gains of the *optimal design* based on the average KL distance deliver substantial performance gains over the regular constellations designed for perfect CSIR.

Next, we extended the design framework to consider coded modulation design for MIMO with imperfect CSIR. For simplicity, we considered block fading MIMO channels. Similarly, the pairwise error probability can be

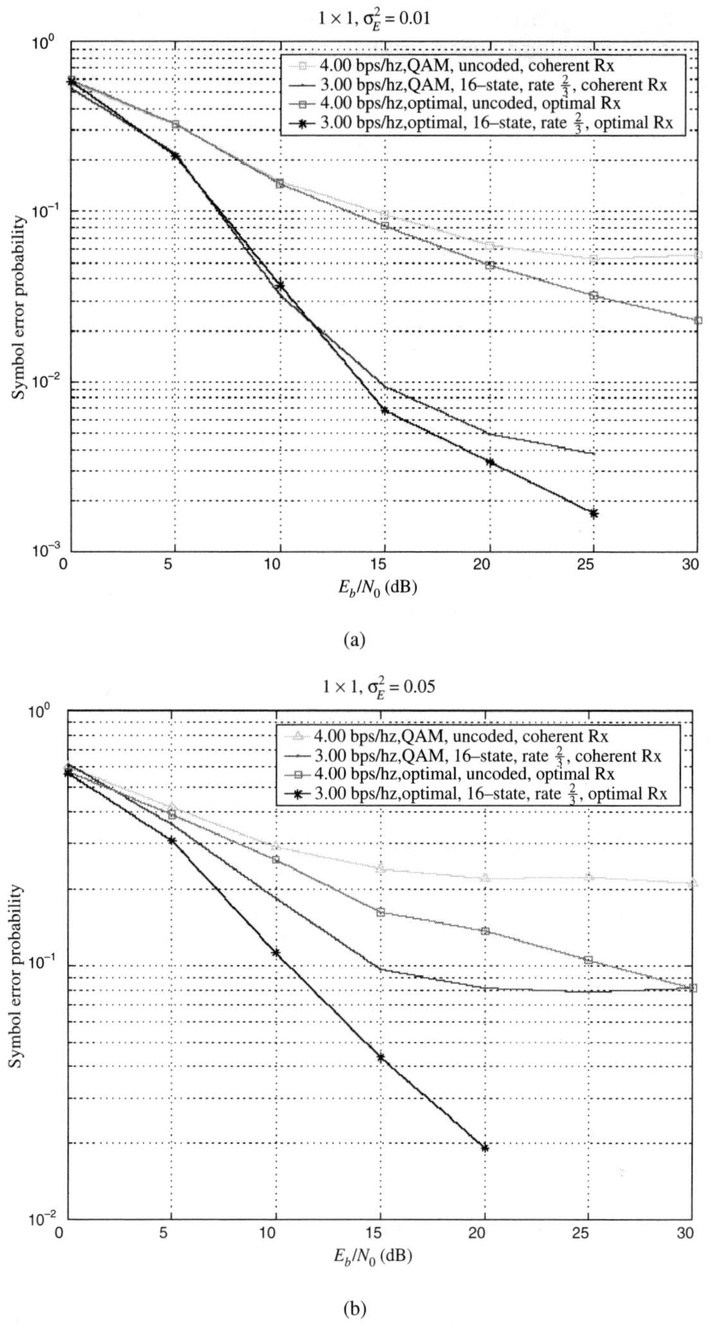

Figure 5.10. Performance comparison between the partially coherent design and the coherent design in the presence of imperfect CSIR: (a) $\sigma_e^2 = 0.01$; (b) $\sigma_e^2 = 0.05$. [20]

bounded with the exponent of the bound given by the average KL distance between the vector codewords $\mathbf{x}_i$ and $\mathbf{x}_j$. More interestingly, the overall average KL distance has similar additive property of the conventional Euclidean distance, which is the design criterion for coded modulation with perfect CSIR. Hence, similar techniques of set partitioning can be applied in our case except using the average KL distance rather than the Euclidean distance. The performance of the coded modulation with imperfect CSIR designed based on the KL distance is compared with that for the coded modulation schemes designed for perfect CSIR. It is shown that substantial SNR gain can be achieved with the coded modulation designed with reference to KL distance and the gain increases as the CSIR error σ_e^2 increases.

6

CROSS-LAYER SCHEDULING FOR MULTIUSER SYSTEMS WITH MULTIPLE ANTENNAS

6.1 OVERVIEW

Traditionally, the achievable bit rate of a communication link is limited by the available bandwidth and power. Given a fixed power budget, the only way to increase the bit rate is to increase the bandwidth of transmission. In the previous chapters, we focused on the performance of MIMO point-to-point link with CSI knowledge at the transmitter and the receiver. With multiple transmit and receive antennas, the bit rate can be significantly increased without increasing the bandwidth or power budget. This is due to spatial multiplexing over $m^* = \min[n_T, n_R]$ spatial channels, where n_T is the number of transmit antennas and n_R is the number of receive antennas.

For the point-to-point link, the performance measure is relatively simple. Specifically, we would like to optimize the link capacity or reduce the BER of the link. It has been shown in previous chapters that knowledge of CSI at the transmitter can significant increase the MIMO link capacity, especially when $n_T > n_R$. However, for multiuser systems, optimizing the individual link performance is not always the best approach from a system performance perspective. It is also very important to consider the upper-layer resource allocation together with the adaptive physical layer design so as to completely exploit the temporal dimension (scheduling) and the spatial dimension (multiple antennas) in the resource space to achieve good system-level performance. Two factors are unique to multiuser systems only: *multiuser diversity* and the *distributed spatial multiplexing*. When the base station has knowledge of CSI, an extra dimension of adaptation is possible for systems with multiple

Channel-Adaptive Technologies and Cross-Layer Designs for Wireless Systems with Multiple Antennas: Theory and Applications. By V. K. N. Lau and Y.-K. R. Kwok
ISBN 0-471-64865-5

users. For instance, it may be beneficial to give different priorities to users according to their instantaneous CSI. Hence, not only is the physical layer adaptive (with varying throughput) but the multiaccess layer can also be adaptive to exploit the *multiuser diversity*. This refers to cross-layer optimization.

On the other hand, we have shown in previous chapters that the number of independent spatial channels in a MIMO link is given by $m^* = \min[n_T, n_R]$. Hence, no spatial multiplexing gain can be expected when the mobile receiver has a single antenna only. However, the situation is different in multiuser systems. For instance, when there are multiple antennas at the base stations, the capacity of the multiuser systems can be increased substantially (even if the mobile users have single antenna only) by exploiting the distributed $n_T \times K$ configuration formed by the base station and K single-antenna mobile stations as illustrated in Figure 6.1. One can imagine that the virtual receiver has K receive antennas, but the processing is distributed over the K users.

In this chapter, we focus on the cross-layer design of multiuser systems with a multiantenna base station, exploiting both the *multiuser diversity* and *distributed spatial multiplexing*. Cross-optimization between the link layer and the MAC layer has been generally ignored in the traditional MAC layer design because of the enormous complexity involved. Some investigators have tried to take advantage of the optimization gap by cooperative scheduling, which factors the link-level metrics into scheduling decisions [16,60]. For example, the scheduling efficiency is shown to be greatly enhanced in a single-input single-output (SISO) system with a variable-throughput adaptive physical layer [84] by a jointly adaptive approach in which a scheduling decision is

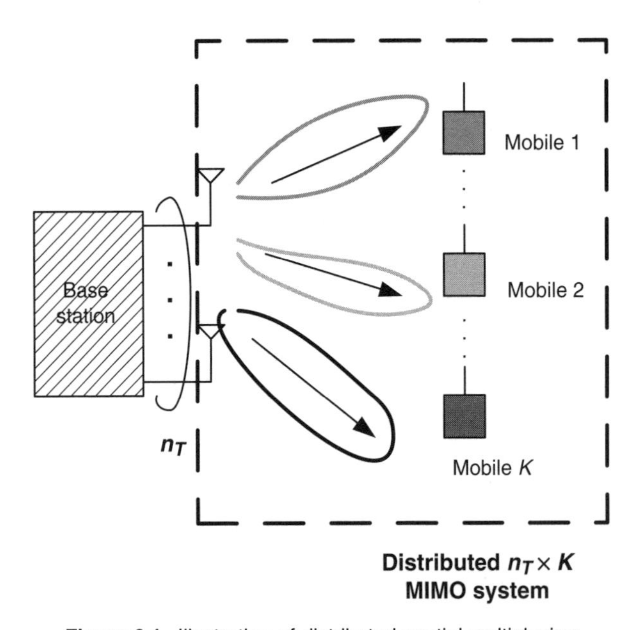

Figure 6.1. Illustration of distributed spatial multiplexing.

made by considering the instantaneous throughput achievable by the user. The case for multiple antennas has been demonstrated [75,76]. However, all the works cited are above based on a heuristic approach and the performance is evaluated through simulations. With the heuristic approach, we do not have design insight and do not know how far the performance of the algorithm is from the optimal. Tse and Hanly [132] proposed an analytical framework to optimize the reverse-link scheduling algorithm for SISO multiuser system with average power constraint based on an information-theoretic approach with respect to the utility function U_{thp}. It is shown that the optimal scheduling policy is to allocate resource to at most a single user (with the best channel condition) at a time. However, when multiple antennas are introduced to the multiuser system, the resource space is expanded to include the spatial dimension as well. Hence, the framework and the results do not generalize directly to multiuser multiple-antenna systems. Finally, it is worth mentioning that the advantage of scheduling with respect to network coverage is a relatively unexplored topic. An interesting work [25] considered the coverage performance of an uplink scheduling algorithm and showed that network coverage could also benefit from multiuser selection diversity through wireless scheduling.

In general, the cross-layer optimization problem is a complex problem involving both information theory (to model the physical layer) and queuing theory (to model the application level delay). In this chapter, we shall tackle the cross-layer optimization problem from the information-theoretic angle only. Specifically, we shall ignore the effect of source statistics and assume that all buffers have infinite size and always contain payloads to be transmitted once a resource is scheduled to the user. Hence, queueing theory is decoupled from information theory. In Chapter 11, we shall discuss the general problem of cross-layer design taking into account the queuing-theoretic angle as well.

This chapter is organized as follows. In Section 6.2, we discuss what we mean by system performance in multiuser systems. In Section 6.3, we outline the multiuser forward link channel model, the multiuser physical layer model, as well as the multiaccess control (MAC) layer model. In Section 6.4, we formulate the analytical design framework for spacetime scheduling problem with convex utility functions. In Section 6.5, low-complexity heuristics, namely, *greedy-based* and *genetic-based* algorithms, are introduced. In Section 6.6, we present numerical results to evaluate the performance of optimal and low-complexity spacetime schedulers. In Section 6.7, we extend the design framework for cross-layer scheduling with imperfect CSIT. Finally, we conclude with a brief summary of results in Section 6.8.

6.2 MULTIUSER SYSTEM PERFORMANCE

In multiuser systems with bursty source, it is less straightforward to define what is meant by *good system performance*, as this term can refer to many different aspects. For example, from a network perspective, the system capacity is definitely an important measure because it determines how efficient the network resource is utilized. On the other hand, from a user perspective, the fairness among users as well as quality of service (QoS) are also very important measures because they determine how smooth end-user applications are run on the wireless systems. Furthermore, network coverage is also a very important measure of system performance, especially during the initial deployment of multiuser systems.

When we have multiple antennas at the base station, the additional spatial degrees of freedom allow *spatial multiplexing gain* and/or *spatial diversity gain*. As we have introduced, the spatial multiplexing allows simultaneous transmission to multiple clients at the same time and frequency and therefore contributes to network capacity gain. On the other hand, the spatial diversity offers robustness to channel fading and contributes to the extension of network coverage. To simplify the picture and allow a sufficiently simple analytical model for cross-layer design, the system performance is defined in terms of a common convex utility function $U(\overline{R}_1, \ldots, \overline{R}_K)$, where $\overline{R}_K$ is the average data rate of user k. We shall first consider two common examples of utility functions in the subsections below. Afterward, we generalize the concept of *network capacity* and *network coverage* on the basis of system utility.

6.2.1 Examples of System Utility

We consider two common examples of system utilities related to the notion of *network capacity* below.

6.2.1.1 *Network Throughput.* One very important measure of system performance from the operator perspective is how efficient system resource (bandwidth, power, and spatial dimension) is utilized. Hence, a natural choice of a system utility function to reflect the resource utilization is given by

$$U_{\text{thp}}(\overline{R}_1, \ldots, \overline{R}_K) = \sum_{k=1}^{K} \overline{R}_k = \varepsilon\left[\sum_k r_k\right] \tag{6.1}$$

where $\overline{R}_k = \varepsilon[r_k]$ is the average throughput of user k and r_k is the instantaneous throughput of user k at any scheduling slot. Schedulers designed to optimize such utility function will result in the highest system capacity and are therefore called *maximal throughput schedulers*.

6.2.1.2 *Proportional Fairness.* While maximal throughput schedulers can result in optimal network resource utilization, it does not take into account the user's quality-of-service (QoS) perspective. For example, users with poor channel conditions will experience discrimination based on the maximal throughput scheduling policy and will suffer from *starvation*. This is obviously undesirable from the end-user perspective. Hence, another popular utility function used to strike a balance between system capacity and fairness among users is called *proportional fairness* (PF). A scheduler is called *proportional fair* [60] if it optimizes the utility function given by

$$U_{\mathrm{PF}}(\overline{R}_1, \ldots, \overline{R}_K) = \sum_{k=1}^{K} \log_2(\overline{R}_k) \tag{6.2}$$

It can be seen from the utility function in Equation (6.2) that there is a heavy penalty for terms with small $\overline{R}_k$ due to the concavity of the $\log_2(\,)$ function. Hence, to maximize the utility, the scheduler has to avoid the situation where some users obtain very small throughput $\overline{R}_k$. This is because for the same throughput difference Δ, the penalty of the low-throughput terms will be higher than the gain achieved by the high-throughput terms.

As we will see in later sections, optimization of the function in Equation (6.2) is not an easy problem because of the $\log_2(\varepsilon[r_k])$ terms. Hence, we shall try to simplify the utility for PF by the following lemma.

Lemma 6.1 (Proportional Fairness) A scheduler that maximizes $\sum_k \frac{r_k}{\overline{R}_k}$ will also maximize the utility function in Equation (6.2).

Proof From convex optimization theory, the utility function in Equation (6.2) is maximized if $dU_{\mathrm{PF}} \le 0$

$$dU_{\mathrm{PF}} = \varepsilon\left[\sum_k \frac{r_k' - r_k}{\overline{R}_k}\right] \le 0$$

for all r_k' where the expectation is taken with respect to all fading realizations across many scheduling slots. Hence, if a scheduler can maximize $\sum_k \frac{r_k}{\overline{R}_k}$ for every fading slot, then $dU_{\mathrm{PF}} \le 0$ and U_{PF} will also be maximized.

Hence, the equivalent utility function for proportional fairness is given by

$$U_{\mathrm{PF2}} = \varepsilon\left[\sum_{k=1}^{K} \frac{r_k}{\overline{R}_k}\right] \tag{6.3}$$

where $\overline{R}_k$ is the average throughput of user k, which depends on the heterogeneous level among users $[1, K]$. Because of the causality requirement on obtaining $\overline{R}_k$, we shall approximate $\overline{R}_k$ with the *moving-average throughput*. Letting t denote the tth scheduling instance, the moving-average throughput at the $(t + 1)$th scheduling instance, $\overline{R}_k(t + 1)$, is defined as

$$\overline{R}_k(t+1) = \left(1 - \frac{1}{t_c}\right)\overline{R}_k(t) + \frac{1}{t_c} r_k \tag{6.4}$$

where t_c is the *time constant* of the averaging window.

For a single-antenna system with TDMA constraint (selecting one user at a time), the PF scheduling involves selection of a user with the largest $\frac{r_k}{\overline{R}_k}$, which is equivalent to the original form of PF scheduler in Qualcomm's HDR [58]. However, this simple scheduling rule does not generalize to multiantenna systems easily.

6.2.2 Definition of Network Capacity

The network capacity is in general defined by an *average system utility function* as discussed in Section 6.2.1. In particular, we shall restrict the average system utility function, $U(\overline{R}_1, \ldots, \overline{R}_K)$, to be of the following form

$$U(\overline{R}_1, \ldots, \overline{R}_K) = \varepsilon[G(r_1, \ldots, r_K)] \tag{6.5}$$

where $G(r_1, \ldots, r_K)$ is the *instantaneous system utility* and r_k is the *instantaneous data rate*[1] of the kth user. Furthermore, for meaningful optimization, we constrain the instantaneous utility to satisfy

$$\frac{\partial G}{\partial r_k} > 0 \quad \forall r_k \geq 0 \tag{6.6}$$

6.2.3 Definition of Network Coverage

Assuming a sequence of N realizations of fading blocks, the instantaneous achievable data rate(s) of a scheduled mobile user(s) are random variables (functions of the specific fading realization) in that fading block. In conventional wireless systems where the scheduling is constrained to select one active user at any scheduling instance (fading block), *outage* is defined as the event that the instantaneous data rate of the scheduled user r is below a target data rate R_0. Let $N_{\text{outage}}(R_0)$ be the number of outage events over N realizations of fading blocks. The *outage probability* is defined as the likelihood of the outage event, $P_{\text{out}}(R_0) = \Pr[r \leq R_0] = \lim_{N\to\infty} N_{\text{outage}}(R_0)/N$. The traditional concept of network coverage is therefore defined as the maximum distance d between a mobile and a base station such that the outage probability of the scheduled user (at a target bit rate R_0) is below a specified target ε. That is, coverage is given by the maximum distance, $d(\varepsilon, R_0)$, such that

[1] *Instantaneous data rate* of a user refers to the data rate of a burst transmission at a particular fading block.

$$P_{\text{out}}(R_0) = \Pr[r(d) \le R_0] \le \varepsilon \tag{6.7}$$

where $r(d)$ denotes the instantaneous data rate of any user at a distance d from the base station. This is a commonly employed definition in cellular systems [117] as well as wireless systems with the scheduler constrained to select one user at a time [25].

However, when the base station has multiple transmit antennas, there are multiple spatial channels as a result of the additional degrees of freedom. This implies that a general multiple-antenna scheduler may select more than one active transmission at any scheduling slot (fading block). Hence, before we can discuss the design of schedulers optimized for network coverage, we must extend the definition of coverage to include multiple active user transmissions at a time.

Definition 6.1 (Utility-Based Outage) *Utility-based outage* is defined as the event that the instantaneous coverage utility $G_{\text{cov}}(r_1, \ldots, r_K)$ is below a target utility value G_0. The *outage probability* is defined as the probability of the occurrence of such event.

Definition 6.2 (Worst-Case Outage) Worst-case outage $P^*_{\text{out}}(G_0, d_0)$ is defined as the largest possible utility-based outage probability over all possiblemobile locations such that $\{\max(d_1, \ldots, d_K) \le d_0\}$:

$$P^*_{\text{out}}(G_0, d_0) = \max_{(d_1,\ldots,d_K):\max(d_1,\ldots,d_K)\le d_0} \Pr[G_{\text{cov}}(r_1,\ldots,r_K) \le G_0] \tag{6.8}$$

We observe that the *worst-case outage probability* in Definition 6.2 is achieved by equidistance of all the active mobiles as illustrated in Figure 6.2. This is summarized in the following lemma.

Lemma 6.2 (Worst-Case Outage) At any given target utility value G_0 and d_0, the worst-case outage probability $P^*_{\text{out}}(G_0, d_0) = \max_{(d_1,\ldots,d_K):\max(d_k)\le d_0} \Pr[G_{\text{cov}}(r_1, \ldots, r_K) \le G_0]$ is achieved by $d_k = d_0\ \forall k \in [1, K]$.

Proof Please refer to Appendix 6A.

Intuitively, because of the convexity of the instantaneous utility function $G(r_1, \ldots, r_K)$, the utility value will be the smallest (worst-case outage) if all the active users are equidistant from the base station at a distance of d_0. This is because otherwise, some users closer to the base station will contribute to a gain in the instantaneous utility.

As a result of Lemma 6.2, the *utility-based coverage* is defined by d_0.

Definition 6.3 (Utility-Based Coverage) *Utility-based coverage* is defined as the maximum distance d_0 such that the worst-case outage probability (at a specified target utility value G_0) is below a specified target ε. In other words, the utility-based coverage d^* is given by

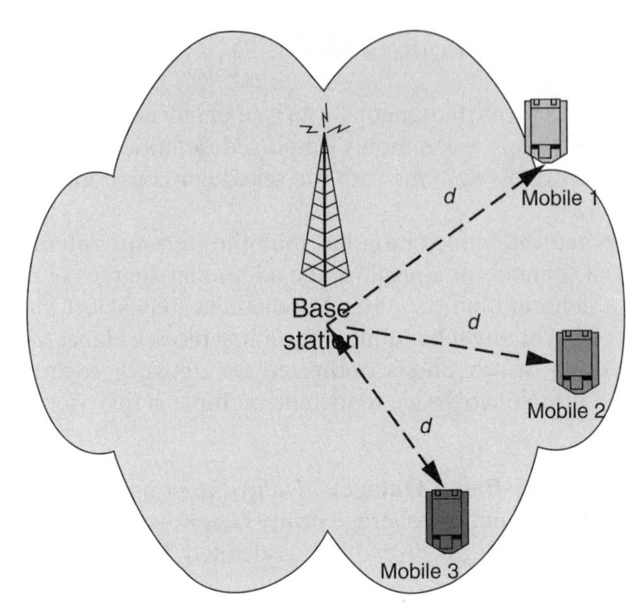

Figure 6.2. Illustration of worst-case outage.

$$d^* = \text{argmax}_{d_0} P^*_{\text{out}}(G_0, d_0) \quad \text{such that} \quad P^*_{\text{out}}(G_0, d_0) \le \varepsilon. \tag{6.9}$$

Note that utility-based coverage reduces to conventional coverage when the system utility is defined as $G(r_1, \ldots, r_K) = \Sigma_k r_k$ and the scheduler is constrained to select at most one active user at a time.

6.3 SYSTEM MODELS

Before cross-layer scheduling can be properly cast into an optimization problem, we need to develop some sufficiently simple analytical models for the multiuser channels and the multiuser physical layers as well as the MAC layer. Hence, we shall elaborate various mathematical models based on the information-theoretic approach. The motivation for applying information theory to model the physical layer performance is due to simplicity and generality of this layer. For instance, the data rate of a user is specified by Shannon's capacity; therefore, the physical layer model is decoupled and independent of specific implementation of the modulation and coding schemes[2]. On the other hand, to decouple the data source statistics from the system per-

[2]While Shannon's capacity is a theoretical data rate that a user can transmit reliably, it is a pretty good number with which to model the physical layer performance. This is because with advanced error correction code design such as the turbo code and LDPC code, one can come close to 0.2 dB of the SNR predicted by Shannon's theory.

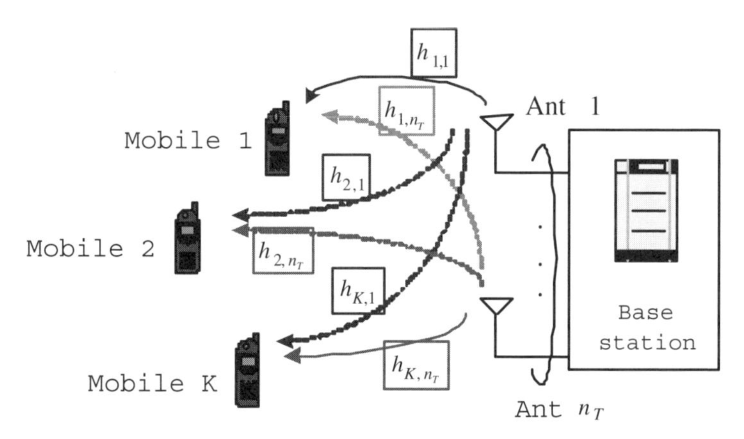

Figure 6.3. System model with multiantenna base station.

formance, we shall assume that source buffers are large in size so that they always contain source packets waiting to be transmitted. In other words, there are no empty scheduling slots due to insufficient source packets at the buffer. Hence, with these simple analytical models, one can obtain useful first-order insights from the cross-layer scheduling design framework.

6.3.1 Multiuser MIMO Channel Model

We consider a communication system with K mobile users having single receive antenna and a base station with n_T transmit antennas as shown in Figure 6.3. The microscopic channel fading between different users and different antennas is modeled as an i.i.d. complex Gaussian distribution with unit variance. In addition, it is assumed that the encoding and decoding frames are short bursts,[3] much shorter than the coherence time of the fading channel. Hence, we have quasistatic fading within each encoded frame.

Let Y_k be the received signal of the kth mobile. The $(K \times 1)$-dimensional vector of received signal $\mathbf{Y}$ at the K mobile stations is given by

$$\mathbf{Y} = \begin{bmatrix} Y_1 \\ \vdots \\ Y_K \end{bmatrix} = \begin{bmatrix} \sqrt{L_1}\mathbf{H}_1 \\ \vdots \\ \sqrt{L_K}\mathbf{H}_K \end{bmatrix} \mathbf{X} + \begin{bmatrix} Z_1 \\ \vdots \\ Z_K \end{bmatrix} \tag{6.10}$$

where $\mathbf{X}$ is the $n_T \times 1$ transmit symbol from the base station to the K mobiles, Z_k is the complex Gaussian noise with variance σ_z^2, $\mathbf{H}_k$ is the $(1 \times n_T)$-

[3]This is realistic because most high-speed wireless data systems operate in short burst mode. For example, in UMTS-HSDPA, a typical burst lasts for only 2 ms, which is much shorter than the coherence time of slow fading channels (at human pedestrian speed). Hence, the fading coefficients are highly correlated within a coding frame.

dimensional channel matrix between the n_T transmit antennas (at the base station) and the kth mobile, and L_k is the path gain between the base station and the kth mobile given by [117]

$$L_k(\text{dB}) = -20\alpha \log_{10}(d_k) + (G_t(\text{dB}) + G_r(\text{dB}) + 20\log_{10}\lambda - 20\log_{10}(4\pi)) \tag{6.11}$$

where d_k denotes the distance between the base station and the kth mobile, G_t and G_r respectively are the transmit and receive antenna gains, and λ is the wavelength of the carrier.

6.3.2 Multiuser Downlink Physical Layer Model

Before we can discuss the scheduling optimization problem, it is very important to define the physical layer model because different physical layer implementations will definitely affect the system-level performance. As mentioned in previous subsections, an information-theoretic approach is adopted in order to isolate the physical layer performance from specific implementation details (such as channel coding and modulation and multiple-access schemes like TDMA, FDMA, and CDMA). Specifically, the maximum achievable rate at the physical layer (with arbitrarily low error probability) is given by Shannon's capacity and is realized by random codebook and Gaussian modulation. In the following definitions, we shall first introduce some related definitions, followed by a description of the model.

Definition 6.4 (Achievable Rate Vector) A rate vector $(r_1, \ldots, r_K)$ is said to be "achievable" for the multiuser channel if there exists a multiuser code $(2^{Nr_1}, 2^{Nr_2}, \ldots, 2^{Nr_K}, N)$ with error probability $\lim_{N\to\infty} P_e(N) = 0$ where N is the length of the code.

Definition 6.5 (Capacity Region) The "capacity region" of the multiuser channel is the closure of the set of achievable rate vectors $(r_1, \ldots, r_K)$.

Theoretically, downlink transmission from the base station to the K mobiles falls under the class of broadcast channels in multiuser information theory [30]. For single antenna at the base station ($n_T = 1$), the multiuser systems belong to the class of *degraded broadcast channels* and the instantaneous capacity region $C_{n_T} = 1(h_1, \ldots, h_K)$ conditioned on the channel realizations $(h_1, \ldots, h_K)$ is given by

$$r_k \le \log_2\left(1 + \frac{|h_k|^2 p_k}{\sum_{j=1}^{K} p_j |h_k|^2 \mathbf{1}(|h_k| < |h_j|) + \sigma_z^2}\right) \quad \forall k \in [1, K] \tag{6.12}$$

where $\Sigma_{j=1}^{K} p_j = P_0$.

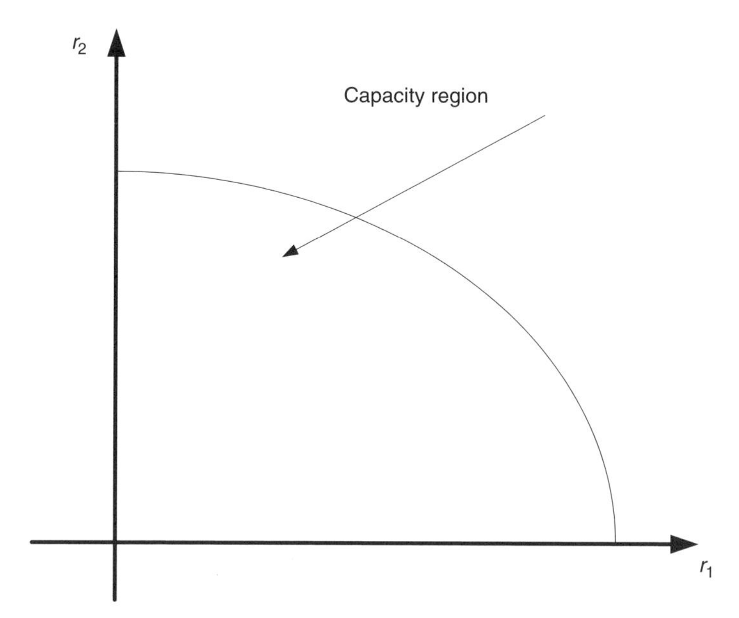

Figure 6.4. Capacity region of multiuser downlink channels with $n_T = 1$.

Figure 6.4 illustrates the capacity region of the Gaussian broadcast channel with $n_T = 1$ and $K = 2$.

Unfortunately, for multiantenna base station, the broadcast channel does not fall into the category of degraded broadcast channels and the process of determining the optimal capacity region is known to be very difficult [22,30]. Since mobile devices are battery-powered, we would also like to push all the processing power to the base station whenever possible. Furthermore, most communication devices such as wireless LAN (IEEE 802.11a/b) do not have multiuser processing capability at the receiving path. Hence, we adopt a simple zero-forcing approach,[4] called the *downlink orthogonal transmit beamforming* (OTBF) approach [83,139], as illustrated in Figure 6.5.

There are K streams of information data to K individual users at the base station transmitter, and they are channel-encoded independently. The vector $(K \times 1)$ of encoded symbols, $\mathbf{U} = [U_1, \ldots, U_K]$, are processed by the power control diagonal matrix $(K \times K)\mathbf{P} = \text{diag}(p_1, \ldots, p_K)$ followed by the beamforming matrix $(n_T \times K)$, $\mathbf{W} = [\mathbf{w}_1, \ldots, \mathbf{w}_K]$, where $\mathbf{w}_k$ is the $n_T \times 1$ complex beamforming weight of user k.

Hence, the transmitted vector of symbol $\mathbf{X}$ is given by

[4] While zero-forcing processing has been shown in Chapter 4 to be asymptotically optimal at high SNR, there are still some performance gaps at low SNR compared with MMSE processing in point-to-point MIMO links. However, in multiuser systems, the performance gaps of zero-forcing processing is negligible compared with MMSE processing at large K due to multiuser selection diversity.

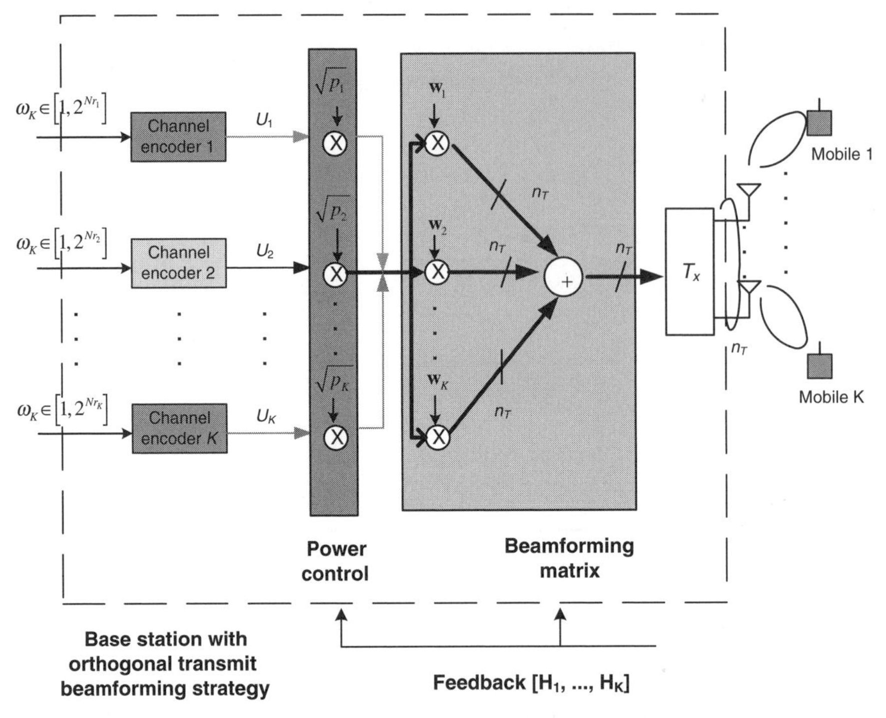

Figure 6.5. Orthogonal transmit beamforming (OTBF) strategy with multiple antennas.

$$\mathbf{X} = \mathbf{W}\sqrt{\mathbf{P}}\mathbf{U} = \sum_{k=1}^{K} \sqrt{p_k}\, U_k \mathbf{w}_k \tag{6.13}$$

where $p_k \geq 0$ is the average transmit power during the current scheduling instance for user k and $\varepsilon[|U_k|^2] = 1$. Since the encoding frame is short burst with quasistatic fading, no power adaptation within an encoding frame is required. At any scheduling slot, individual user(s) could be turned off by assigning $p_k = 0$. An *admitted user set*, $\mathcal{A} = \{k \in [1, K]: p_k > 0\}$, is defined as a set of selected users (users with nonzero allocated power) at any scheduling slot. The total transmit power out of the base station at any scheduling slot is constrained by P_0:

$$\sum_k p_k \leq P_0 \tag{6.14}$$

Calculation of OTBF Weights. Given an *admitted user set*, the transmit power $\{p_1, \ldots, p_K\}$ and a realization of the channel fading $\{\sqrt{L_1}\mathbf{h}_1, \ldots, \sqrt{L_K}\mathbf{h}_K\}$, the received signal of user k is given by

$$Y_k = \underbrace{\sqrt{p_k L_k}\mathbf{h}_k\mathbf{w}_k U_k}_{\text{Information}} + \underbrace{\sum_{j\in\mathcal{A}, j\neq k} \sqrt{p_j L_j}\mathbf{h}_k\mathbf{w}_j U_j + Z_k}_{\text{Multibeam interference}} \tag{6.15}$$

where the first term contains the desired signal and the middle term represents the *multibeam interference* due to simultaneous transmission of independent information streams. The following considerations apply:

Spatial Multiplexing. The level of spatial multiplexing is controlled by the *cardinality* of the admitted user set $|\mathcal{A}|$. The OTBF weight $\mathbf{w}_k$ is selected to satisfy

$$\mathbf{w}_k^*\mathbf{w}_k = 1 \quad \forall k \tag{6.16}$$

and the *orthogonality conditions*

$$\mathbf{h}_j\mathbf{w}_k = 0 \quad \forall j \in \mathcal{A}, j \neq k \tag{6.17}$$

where the subscript * denotes complex conjugate transpose. Note that when $p_k = 0$, the information stream for user k is turned off. In other words, the number of simultaneous transmissions is given by the cardinality of the admitted user set $\mathcal{A}$. Intuitively, we would like the transmit beam of user k to be orthogonal to the rest of the selected users in $\mathcal{A}$. Hence, the signal received by user k would consist of the desired signal only without the interference due to downlink signals to other users. Observe that there are $2n_T$ degrees of freedom in $\mathbf{w}_k$ and there are $2|\mathcal{A}| - 1$ equations from the constraints in Equations (6.16) and (6.17). Hence, we have

$$|\mathcal{A}| \leq n_T \tag{6.18}$$

This means that with n_T transmit antennas, the base station could support at most n_T spatial channels.

Spatial Diversity. The remaining degrees of freedom, $2(n_T - |\mathcal{A}|) + 1$, are utilized to realize the diversity gain to maximize the received SNR, $\mathbf{w}_k^*(\mathbf{h}_k^*\mathbf{h}_k)\mathbf{w}_k$.

Hence, given a certain admitted user set $\mathcal{A}$, the overall weight determination problem is given by Problem 6.1.

Problem 6.1 (Determination of OTBF Weights)

$$\mathbf{w}_k = \arg\max_{\mathbf{w}} \mathbf{w}^*\left(\mathbf{h}_h^*\mathbf{h}_k\right)\mathbf{w} \text{ subject to } \mathbf{w}^*\mathbf{w} = 1 \text{ and } \mathbf{h}_j\mathbf{w} = 0 \quad \forall_j \in \mathcal{A}, j \neq k \tag{6.19}$$

Please refer to Appendix 6B for solution of the weights.

Capacity Region of the OTBF. With the optimal beamforming weights $\{\mathbf{w}_k\}$, the *multibeam interference* becomes zero and there are $|\mathcal{A}|$ independent spatial channels.

The received signal for mobile user k is given by

$$Y_k = \sqrt{p_k L_k}\,\mathbf{h}_k \mathbf{w}_k U_k + Z_k \tag{6.20}$$

Hence, the maximumachievable date rate of the kth spatial channel during the fading block is given by the maximum mutual information between U_k and Y_k as

$$r_k \le \log_2\left(1 + \frac{p_k L_k |\mathbf{h}_k \mathbf{w}_k|^2}{\sigma_z^2}\right) \quad \forall k \in \mathcal{A} \tag{6.21}$$

In other words, the *conditional instantaneous capacity region* $C_{\text{OTBF}}(\mathbf{h}_1, \ldots, \mathbf{h}_K; \mathcal{A})$ of the multiuser systems with OTBF processing given the admitted user set $\mathcal{A}$ is given by Equation (6.21). For example, suppose $K = 3$, $\mathcal{A} = \{1, 2\}$ and $n_T = 2$. The conditional capacity region $C_{\text{OTBF}}(\mathbf{h}_1, \ldots, \mathbf{h}_K, \mathcal{A})$ is given by

$$C_{\text{OTBF}}(\mathbf{h}_1, \ldots, \mathbf{h}_K, \mathcal{A}) = \left\{(r_1, r_2, r_3) : r_1 \le \log_2\left(1 + \frac{p_1 L_1 |\mathbf{h}_1 \mathbf{w}_1|^2}{\sigma_z^2}\right), r_2 \le \log_2\left(1 + \frac{p_2 L_2 |\mathbf{h}_2 \mathbf{w}_2|^2}{\sigma_z^2}\right), r_3 = 0, p_1 + p_2 \le P_0\right\}$$

To evaluate the unconditional instantaneous capacity region, we note the following convexity property of the region. Given any two feasible rate vectors $\mathbf{r}_A$ and $\mathbf{r}_B$ in the capacity region, we have $\mathbf{r} = \lambda_A \mathbf{r}_A + \lambda_B \mathbf{r}_B$ belonging to the capacity region as well for any λ_A, λ_B satisfying $\lambda_A + \lambda_B = 1$. This is because the rate vector $\mathbf{r}$ can be achieved by *timesharing* between the two points $\mathbf{r}_A$ and $\mathbf{r}_B$. Hence, the instantaneous capacity region $C_{\text{OTBF}}(\mathbf{h}_1, \ldots, \mathbf{h}_K)$ is given by the *convex hull* of the union of $C_{\text{OTBF}}(\mathbf{h}_1, \ldots, \mathbf{h}_K; \mathcal{A})$ over all possible admitted user set combinations $\mathcal{A}$

$$C_{\text{OTBF}}(\mathbf{h}_1, \ldots, \mathbf{h}_K) = \overline{\bigcup_{\mathcal{A}:|\mathcal{A}| \le n_T} C_{\text{OTBF}}(\mathbf{h}_1, \ldots, \mathbf{h}_K; \mathcal{A})} \tag{6.22}$$

where $\overline{B}$ denotes the convex hull of the set B.

For example, when $K = 3$ and $n_T = 2$, the capacity region of the OTBF physical layer $C_{\text{OTBF}}(\mathbf{h}_1, \ldots, \mathbf{h}_K)$ is given by

$$C_{\text{OTBF}}(\mathbf{h}_1, \ldots, \mathbf{h}_K) = \overline{C_{\text{OTBF}}([1,2]) \cup C_{\text{OTBF}}([1,3]) \cup C_{\text{OTBF}}([2,3]) \cup C_{\text{OTBF}}([1]) \cup C_{\text{OTBF}}([2]) \cup C_{\text{OTBF}}([3])}$$

where $C_{\text{OTBF}}([1, 2])$ is the conditional capacity region given $\mathcal{A} = [1, 2]$.

Discussion. Observe that the data rate is a function of the received SNR, $\frac{p_k L_k |\mathbf{h}_k \mathbf{w}_k|^2}{\sigma_z^2}$. At *full spatial multiplexing* ($|\mathcal{A}| = n_T$), the implicit diversity order of $|\mathbf{h}_k\mathbf{w}_k|^2$ is 1 and therefore, the user is less robust with respect to fading. On the other hand, at *full spatial diversity* ($|\mathcal{A}| = 1$), the network capacity suffers since only one user is selected but the implicit diversity order of $|\mathbf{h}_k\mathbf{w}_k|^2$ is n_T (and hence, more robust to fading). Hence, a smooth tradeoff between network coverage and network capacity could be obtained by controlling the cardinality of the admitted user set $\mathcal{A}$.

6.3.3 MAC Layer Model

The scheduling algorithm in the MAC layer is responsible for the allocation of channel resource at every fading block. The system resource is partitioned into short frames as illustrated by Figure 6.6a. At the beginning of every frame, the base station estimates the channel matrix from the participating mobile users. As a result of short burst transmissions, the channel fading remains the same across the entire burst duration. The estimated CSI is passed to the scheduling algorithms in the MAC layer as illustrated in Figure 6.6b. The output of the scheduler consists of an admitted user set $\mathcal{A} = \{k \in [1, K] : p_k > 0\}$ (the set of user indices with nonzero power allocated at the current fading block), the corresponding power allocation $\{p_k\}$, and the *instantaneous rate* allocation $\{r_k\}$ of the selected users. The downlink payload is transmitted at the scheduled rate, and the rate is also broadcast on the downlink common channels to mobile users. Hence, the scheduling algorithm is an optimization problem with respect to an objective function.

6.4 CROSS-LAYER OPTIMIZATION WITH PERFECT CSIT

In this section, we shall discuss how to formulate the cross-layer scheduling design as an optimization problem. This is very important in order to obtain first-order insights into the nature of the design as well as the optimal performance as a performance upper bound. We first discuss two favors of the cross-layer scheduling design: the *capacity-optimized design* and the *coverage-optimized design*. Afterward, we shall illustrate that these two favors can be cast into a common optimization framework.

6.4.1 Capacity-Optimized Scheduler Design

The capacity-optimized scheduler design is equivalent to the following optimization problem.

Problem 6.2 (Capacity-Optimized Scheduling) Given any channel matrix realization, $\{\sqrt{L_1}\mathbf{h}_1, \ldots, \sqrt{L_K}\mathbf{h}_K\}$ choose the power allocation $\{p_k \geq 0 \ \forall k \in$

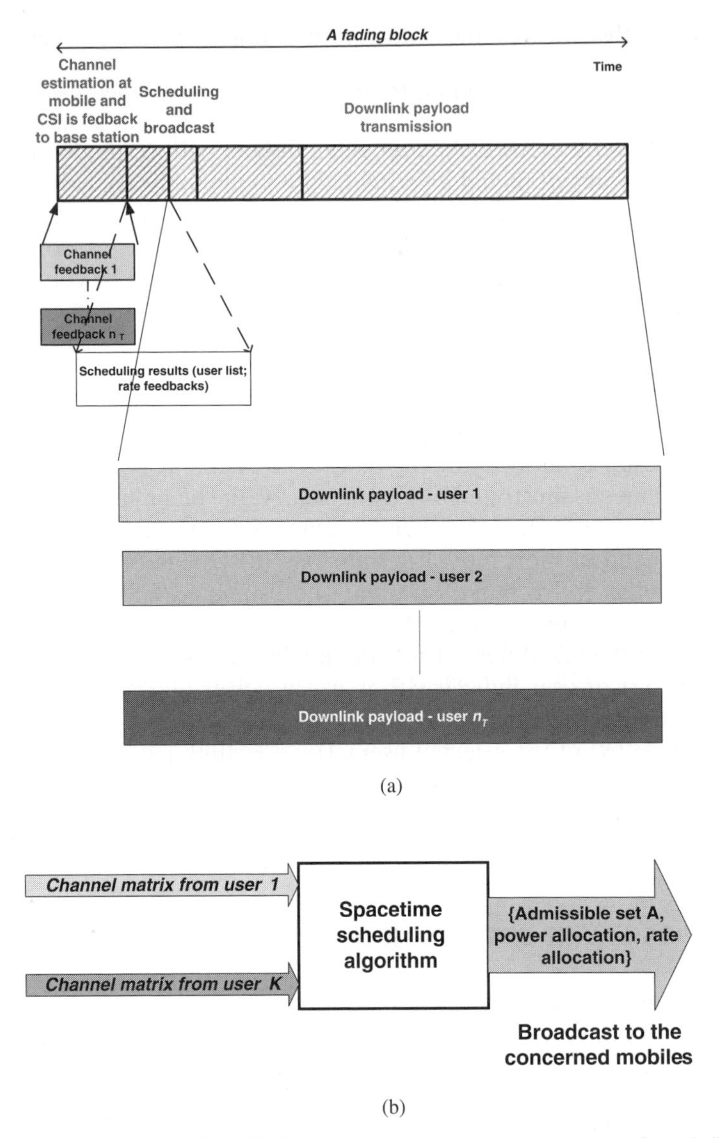

Figure 6.6. Block diagrams of the MAC layer: (a) timing of access and frame formal; (b) model of the scheduling layer.

$[1, K]\}$ to maximize the "capacity utility function", $G_{\text{cap}}(r_1, \ldots, r_K)$, such that $\Sigma_k p_k \leq P_0$ and $(r_1, \ldots, r_K) \in C$ where C denotes the "capacity region" of the multiuser physical layer.

We shall illustrate in Lemma 6.3 that at high SNR, the optimal admitted user set always has n_T users being selected. In other words, the available degrees of freedom should all be configured to exploit *spatial multiplexing* in order to maximize the *network capacity*.

Lemma 6.3 (Full Spatial Multiplexing at High SNR) The optimal solution in Problem 6.2 with $G_{cap} = \Sigma_k r_k$ and capacity region C_{OTBF} satisfies $|\mathcal{A}| = n_T$ when the SNR (P_0/σ_z^2) is high. Here, Adenotes the admitted user set of user indices with positive power allocation.

Proof Please refer to Appendix 6C.

6.4.2 Coverage-Optimized Scheduler Design

For the purpose of illustration, we consider the following user-centric *coverage utility*:

$$G_{cov} = \min_{k \in \mathcal{A}} r_k \tag{6.23}$$

where $\mathcal{A} = \{k : p_k > 0\}$ is the set of users with nonzero power allocation. Hence, coverage is defined with respect to the worst-case instantaneous data rate of all the selected users. The coverage-optimized scheduler design is equivalent to the following optimization problem.

Problem 6.3 (Coverage-Optimized Scheduling) Given any channel matrix realization of all mobile users, $\{\sqrt{L_1}\mathbf{h}_1, \ldots, \sqrt{L_K}\mathbf{h}_K\}$, select the power allocation $(p_1, \ldots, p_K)$ to minimize the "worst-case outage probability" P_{out}^* (defined in Lemma 6.2) such that $\Sigma_k p_k \leq P_0$ and $(r_1, \ldots, r_K) \in C$.

We shall illustrate in Lemma 6.4 that the optimal admitted user set always has one user being selected. In other words, the available degrees of freedom should all be utilized to exploit *spatial diversity* in order to minimize the outage probability.

Lemma 6.4 (Full Spatial Diversity) The optimal solution $(p_1, \ldots, p_K)$ in Problem 6.3 with $G_{cov}(r_1, \ldots, r_K) = \min_{k \in \mathcal{A}} r_k$ and the capacity region C_{OTBF} satisfies $|\mathcal{A}| = 1$, where $\mathcal{A} = \{k : p_k > 0\}$.

Proof Please refer to Appendix 6D.

6.4.3 Common Framework for Both Capacity-Optimized and Coverage-Optimized Scheduling

While both the network capacity and the network coverage are important performance measures, they are different optimization objectives, and therefore tradeoff is generally required in the scheduling optimization with respect to these two objectives. As a result of Lemma 6.3, we see that at high SNR, the capacity-optimized scheduler would fully utilize the degrees of freedom for *spatial multiplexing* ($|\mathcal{A}| = n_T$). On the other hand, from Lemma 6.4, the coverage-optimized scheduler would fully utilize the available degrees of freedom for *spatial diversity* ($|\mathcal{A}| = 1$). On the basis of these observations, we unify the *capacity-optimized* and *coverage-optimized* scheduling designs into a common optimization problem, $\mathcal{P}(Q, n_T)$, in the following theorem.

Theorem 6.1 (Common Scheduling Framework) Consider the optimization problem $\mathcal{P}(Q, n_T)$ in Problem 6.4.

Problem 6.4

$$\text{Select } (p_1, \ldots, p_K) \text{ to maximize } G_{\text{cap}}(r_1, \ldots, r_K) = \sum_k r_k \tag{6.24}$$

with constraints $\Sigma_k p_k \leq P_0$, $(r_1, \ldots, r_K) \in C$ and $|\mathcal{A}| \leq Q$ where $Q \in [1, n_T]$ is a control parameter on the level of spatial multiplexing and C is the capacity region of the corresponding physical layer.

When $Q = n_T$ and C corresponds to the OTBF capacity region, the solution of $\mathcal{P}(Q, n_T)$ optimizes the "capacity-optimized" scheduling problem with respect to $G_{\text{cap}} = \Sigma_k r_k$ in Problem 6.2. When $Q = 1$ and C corresponds to the OTBF capacity region, the solution of $\mathcal{P}(Q, n_T)$ optimizes the "coverage-optimized" scheduling problem with respect to P^*_{out} in Problem 6.3 and G_{cov} in (6.23).

Proof Please refer to Appendix 6E.

Hence, from Theorem 6.1, we can deduce that if the physical layer processing is OTBF at the base station, the optimization problem in $\mathcal{P}(Q, n_T)$ can result in a *capacity-optimized* scheduler design (when $Q = n_T$) as well as a *coverage-optimized* scheduler design (when $Q = 1$) by adjusting the cardinality parameter Q. Hence, for both design objectives, we can focus on solving a single optimization problem $\mathcal{P}(Q, n_T)$.

6.4.4 Optimal Solution—Single-Antenna Systems

For a single-antenna base station $n_T = 1$, the capacity region $C_{n_T=1}$ is given by the capacity region of degraded broadcast channels as illustrated in Equation (6.12). Without loss of generality, assume $|h_1| > |h_2| > \ldots > |h_K|$. For $G_{\text{cap}}(r_1, \ldots, r_K) = \Sigma_k \mu_k r_k$, where μ_k is a constant weight for user k, the cross-layer optimization problem [131] can be rewritten as follows.

Problem 6.5 (Cross-Layer Optimization—Single Antenna)

$$\max_{(p_1, \ldots, p_K)} L(p_1, \ldots, p_K; \lambda) \tag{6.25}$$

where $L = \left[\sum_k \log_2\left(1 + \frac{|h_k|^2 p_k}{|h_k|^2 \sum_{j<k} p_j + \sigma_z^2}\right) - \lambda \sum_k p_k\right]$ is the Lagrangian function and λ is the Lagrange multiplier for the constraint $\Sigma_k p_k = P_0$.

The solution of the problem is summarized in the following lemma.

Lemma 6.5 (Optimal Solution—Single Angenna) Define the *marginal utility functions*:

$$u_k(w) = \frac{\mu_k}{\sigma_z^2/|h_k|^2 + w} - \lambda \ln 2 \tag{6.26}$$

$$u^*(w) = \left[\max_k u_k(w)\right]^+ \tag{6.27}$$

and the sets

$$\mathcal{B}_k = \{w \in [0, \infty] : u_k(w) = u^*(w)\} \tag{6.28}$$

for all $k \in [1, K]$. The optimal value of the objective function in the optimization problem 15 is given by

$$L^* = \int_0^\infty u^*(w)dw \tag{6.29}$$

and is attained at the unique rate and power vector

$$r_k^*(\mu, \lambda) = \frac{1}{\ln 2} \int_{\mathcal{B}_k} \frac{1}{\sigma_z^2/|h_k|^2 + w} dw \quad \forall k \in [1, K] \tag{6.30}$$

$$p_k^* = \langle \mathcal{B}_k \rangle \tag{6.31}$$

where $\langle \mathcal{B} \rangle = \max_{w \in \mathcal{B}} - \min_{w \in \mathcal{B}}$.

Proof Please refer to Appendix 6F.

Note that from Lemma 6.5, we can deduce that the optimal cross-layer scheduling is given by the steps described below and the complexity is linear with respect to K:

- Set $w = 0$.
- Sort the user indices according to the descending order of $|h_k|$, i.e. $|h_1| > |h_2| > \ldots |h_K|$.
- For $k = 1$ to K, obtain the sets $B_k = \{w \in [0, w_0] : u_k(w) = u^*(w)\}$. Allocate the power to the kth user according to Equation (6.31).

In a special case when $G_{\text{cap}}(r_1, \ldots, r_K) = \Sigma_k r_k$, the optimal cross-layer scheduling design in Problem 6.5 will have the following result.

Lemma 6.6 (TDMA Scheduling) The optimal solution to Problem 6.5 with $G_{\text{cap}}(r_1, \ldots, r_K) = \Sigma_k r_k$ will always have $|\mathcal{A}| = 1$; that is, only one user will be allocated resource in every scheduling slot in order to maximize the total system capacity.

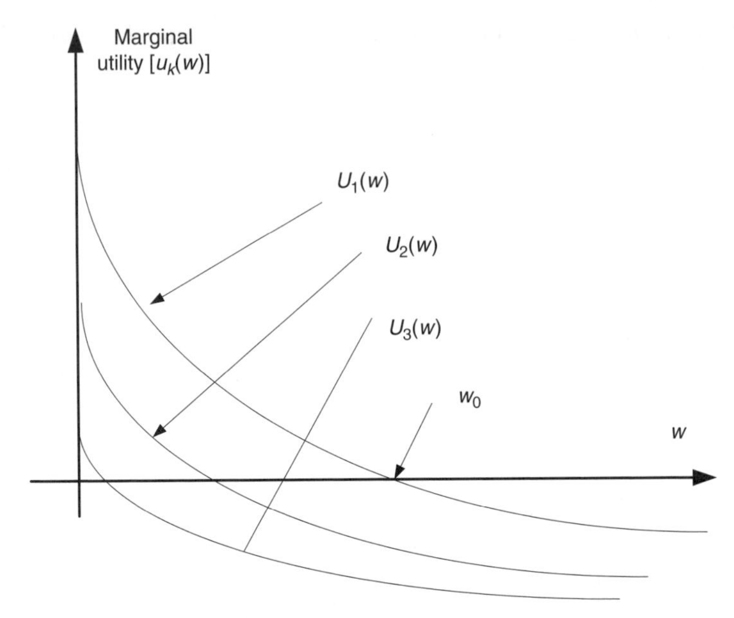

Figure 6.7. Illustration of the utility functions $u_1(w)$, $u_2(w)$, and $u_3(w)$ for single-antenna broadcast channels with $|h_1| > |h_2| > |h_3|$.

Proof When $G_{cap}(r_1, \ldots, r_K) = \Sigma_k r_k$, we have $\mu_k = 1$. Assume that we sort the user indices in the descending order of $|h_k|$. We have $u_1(w) > u_2(w) > \cdots > u_K(w)$ for all $w > 0$. In other words, the utilities $u_k(w)$ do not intersect with each other as illustrated in Figure 6.7. Hence, we have $u^*(w) = u_1(w)$ for all $w > 0$.

Let $u^*(w_0) = 0$. The optimal power allocation (as a result of Lemma 6.5) is given by $p_1^* = w_0$ *and* $p_k^* = 0$ for all $k \in [2, K]$. Hence, we have only one user with nonzero power allocation at each scheduling slot.

In other words, for single-antenna base stations, dynamic TDMA is the optimal strategy with respect to the sum-throughput system utility. Hence, while most of the commercial 3G systems (such as EV-DO, EV-DV and HSDPA) are CDMA-based to support voice users, they all use TDMA as the basic framework to support high-speed downlink data users. This will be elaborated further in Chapter 9 when we discuss commercial design examples in UMTS.

6.4.5 Optimal Solution—Multiple-Antenna Systems with OTBF Processing

On the other hand, for multiantenna base station with OTBF processing, the structure of the capacity region is a convex hull of the union of a number of *conditional capacity regions* over all possible combinations of admitted user sets. For the system utility functions we considered in this chapter, it can be shown that the optimizing rate vector $\mathbf{r}^*$ over the OTBF capacity region must belong to the vertex of one of the conditional capacity regions $C_{\mathrm{OTBF}}(\mathbf{h}_1, \ldots, \mathbf{h}_K; \mathcal{A})$ for some $\mathcal{A}$. The rate vector in the conditional capacity region $C_{\mathrm{OTBF}}(\mathbf{h}_1, \ldots, \mathbf{h}_K; \mathcal{A})$ has exactly $|\mathcal{A}|$ nonzero elements. Hence, it is more convenient to introduce a conditional variable, namely, the *admitted user set* $\mathcal{A}$, into the optimization problem. Given $\mathcal{A}$, we have $p_k = 0$ for all $k \notin \mathcal{A}$. In this way, the corresponding achievable rate vector can be expressed as a function of the admitted user set $\mathcal{A}$ directly from Equation (6.21). The modified optimization problem in $\mathcal{P}(Q, n_T)$ is given by Problem 6.6.

Problem 6.6 (Cross-Layer Optimization—Multiple Antennas)

$$\text{Select } \mathcal{A} \text{ and } \{p_k > 0 \ \forall k \in \mathcal{A}\} \text{ to maximize } G_{\mathrm{cap}}(r_1, \ldots, r_k) = \sum_k r_k \tag{6.32}$$

with constraints $\Sigma_k p_k \leq P_0$, $(r_1, \ldots, r_K) \in C_{\mathrm{OTBF}}(\mathcal{A})$ and $|\mathcal{A}| \leq Q$, where $Q \in [1, n_T]$ is a control parameter on the level of spatial multiplexing and $C(\mathcal{A})$ is the conditional capacity region of the corresponding physical layer.

The modified optimization problem now involves two sets of variables: the admitted user set $\mathcal{A}$ (which involves combinatorial search) and the optimal power allocation for the selected users $(p_1, \ldots, p_K)$ (which involves convex optimization). It can be solved by separating the admitted user set variable $\mathcal{A}$ from the continuous variables $(p_1, \ldots, p_K)$. The optimization flow is described below:

- Select an admitted user set $\mathcal{A}$ with $|\mathcal{A}| \leq n_T$.
- For every selected admitted user set $\mathcal{A}$, we compute the optimal power allocation $(p_1, \ldots, p_K)$ for those selected users. The optimal power allocation with respect to *capacity utility*, $G_{\mathrm{cap}} = \Sigma_k r_k$, is given by

$$p_k^* = \left(\frac{1}{(\ln 2)\lambda} - \frac{1}{|\mathbf{h}_k \mathbf{w}_k|^2} \right)^+ \tag{6.33}$$

  for all $k \in \mathcal{A}$ and λ is the Lagrange multiplier given by the solution of the equation $\Sigma_{k \in \mathcal{A}} p_k^* = P_0$. The proof of the optimal power allocation is shown in Appendix 6G.
- Compute the set of OTBF weights $\{\mathbf{w}_1, \ldots, \mathbf{w}_K\}$ according to the steps described in Section 6.3.2. Together with the computed optimal power allocation $(p_1, \ldots, p_K)$ in the previous step and the set of orthogonal weights, the instantaneous utility function is computed $G_{\mathrm{cap}}(r_1, \ldots, r_K)$. This represents the optimal utility function value given the selected admitted user set $\mathcal{A}$.

- Repeat the steps listed above with another feasible admitted user set $\mathcal{A}$ until all possible admitted user set combinations have been exhausted. The total search space is given by

$$\sum_{m=1}^{n_T} \binom{K}{m}$$

- Select the admitted user set $\mathcal{A}$ and the corresponding power allocation $(p_1, \ldots, p_K)$ that achieves the maximum utility value.

6.5 LOW-COMPLEXITY SCHEDULING ALGORITHMS

Since the scheduling process is a realtime process, the computational complexity of the cross-layer scheduling algorithm is very important. For single-antenna systems, the optimal scheduling strategy is TDMA-based and it is shown that there exists an optimal cross-layer scheduling with linear order of complexity with respect to the number of users K. Unfortunately, the computational complexity of the optimal algorithm exceeds the implementation limitation in most practical systems for moderate K and n_T. Therefore, it is necessary to look into some suboptimal algorithms with good complexity and performance tradeoff. Yet, the results obtained from the optimal algorithms can serve as a performance reference for comparison.

In this section, we shall introduce two low-complexity heuristic scheduling algorithms: the *greedy algorithm* and the *genetic algorithm*. For notation convenience, we introduce a binary vector $(\alpha_1, \ldots, \alpha_K)$ to represent the admitted user set $\mathcal{A}$ where $\alpha_k \in \{0,1\}$.

6.5.1 Suboptimal Solution A—Greedy Algorithm

The greedy algorithm is widely used in commercial systems such as 3G1X and Qualcomm HDR because of its simplicity. The algorithm is described below.

Algorithm 6.1 (Greedy Algorithm)

Step 1: For $k = 1$ to K, initialize $\alpha(k) = (0, \ldots, 1, \ldots, 0)$ (the only nonzero element is in the kth position). Calculate $\mathbf{r} = (0, \ldots, r_k, 0, \ldots 0)$ and $G^*_{\alpha,k} = G_{\text{cap}}(\alpha(k); \mathbf{p}(k))$ based on $\alpha(k)$ and the power vector $\mathbf{p}(k) = (0, \ldots, 0, P_0, 0, \ldots 0)$ (all zero entries except at the kth entry.).

Step 2: Sort the user indices according to the descending order of $\{G^*_{\alpha,k}\}$ calculated in step 1.

Step 3: The admitted user set is obtained by including the first n_T indices in the sorted list in step 2. The optimizing power vector is obtained from Equation (6.33).

Observe that the greedy algorithm complexity involves $K + O(n_T^2)$ function evaluations plus $O(K \log_2(K))$ sorting operations. In addition, it is the optimal algorithm when $n_T = 1$.

6.5.2 Suboptimal Solution B—Genetic Algorithm

As we shall illustrate in the next section, the performance gap between the greedy algorithms and the optimal algorithm is quite large for $n_T > 1$. Hence, this motivates the design of other suboptimal algorithms with better performance–complexity tradeoff. In this section, we shall apply genetic algorithm to the cross-layer scheduling problem.

The genetic algorithm is a family of computational models inspired by evolution and was first investigated by Holland [56]. These algorithms encode a potential solution to a specific problem on a simple *chromosome-like* data structure and apply recombination operators to this structure so as to preserve critical information. A *chromosome* is a string of bits that uniquely represents the independent variables of the optimization problem. Each chromosome is associated with a *fitness* value that indicates how good the chromosome is with respect to the optimization problem. The genetic algorithm is structured in such a way that chromosomes with a higher fitness value will have a higher chance of survival in the process of *evolution*. An implementation of a genetic algorithm begins with a population of random chromosomes. There are two slightly different approaches of the genetic algorithms regarding the population size. In the first approach, the population size remains unchanged throughout evolution. In the second approach, the population size changes during the evolution process. In this section, we shall adopt the first approach of maintaining a constant-size population. The main template of the genetic algorithm is illustrated below.

Algorithm 6.2 (Genetic Algorithm)

Step 1: *Initialization.* Initialize a population with N_p chromosomes [a chromosome is a sample of the optimizing variable $(\alpha_1, \ldots, \alpha_K)$, where $\alpha_k \in \{0, 1\}$]. These chromosomes are randomly picked, satisfying the constraint: $\Sigma_{k=1}^{K} \alpha_k \le n_T$.

Step 2: *Selection.* Construct an intermediate generation (which is a set of N_p chromosomes) based on the current population and the selection rule. The selection process we adopt is the *remainder stochastic sampling*. For each randomly selected chromosome in the current population, $\alpha(i) = (\alpha_1, \ldots, \alpha_K)$, we evaluate the *fitness* based on the utility function $G^*_{\alpha,i} = \max(p_1, \ldots, p_K)\, G_{\text{cap}}(r_1, \ldots, r_k)$ for $(r_1, \ldots, r_k) \in C$ and $\sum_k p_h \le p_0$. Let $\overline{G} = \frac{1}{N_p} \sum_i G_{\alpha,i}$ be the average fitness within the current population. The integer portion of $G^*_{\alpha,i}/\overline{G}$ indicates how many copies of that chromosome i are directly placed

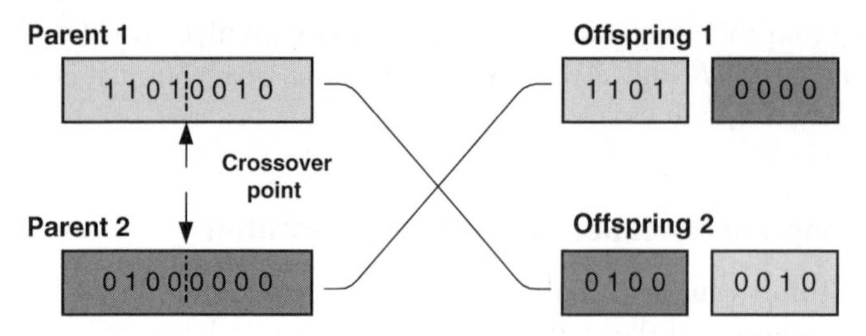

Figure 6.8. Illustration of crossover operation in genetic algorithm.

in the intermediate population. The fractional part of $G^*_{\alpha,i}/\overline{G}$ indicates the probability that an additional copy of that chromosome is placed in the intermediate population. For example, a chromosome with $G^*_{\alpha,i}/\overline{G} = 1.36$ places one copy in the intermediate generation and receives a 0.36 chance of placing a second copy. The selection process continues until all the N_p slots in the intermediate generation have been filled up. In this way, we see that chromosomes that are *fitter* will be allowed more chance to propagate into the next population.

Step 3: *Breeding.* The breeding process consists of two stages: the *crossover* and the *mutation*. Randomly select a pair of chromosomes in the intermediate population and recombine the two parents into two offspring according to crossover and mutation rules. The crossover operation is characterized by a *crossover probability* p_c. For every selected pairs of parents, there is p_c probability of performing crossover operation. By crossover, a randomly selected crossover point (between 1 and K) is selected for the pair of chromosomes. The two parents are split respectively in the crossover point selected, and the two offspring are obtained by crossing the fragments of the two parents as illustrated in Figure 6.8. For every bit in the chromosomes of the offspring, there is a p_m (mutation rate) chance of toggling the bit. This is called the *mutation operation*.

Throughout the remaining of this subsection, we shall dynamically adjust the mutation rate on the basis of the spread of the fitness in the current population. Specifically, the mutation rate of current generation is given by

$$p_m = \frac{1}{\beta_1 + \beta_2 \sigma_G / \overline{G}} \tag{6.34}$$

where σ_G is the standard derivation, $\overline{G}$ is the mean of the fitness of the current population (before selection), and β_1 and β_2 are two constants.[5]

These two processes introduce randomness into the intermediate generation so that the new population will be a combination of the best chromosome in the current population as well as some new random elements.

[5]Here, we set $\beta_1 = 1.2$ and $\beta_2 = 10$ for the maximal throughput and proportional fair utility functions.

Step 4: *Termination*. Replace the original population with the new population and repeat Steps 2 and 3 until the number of iterations reaches N_g. When forming a new population, it is ensured that the fitness chromosome in the current population will be saved and inserted into the next population, and all members of the next population are checked against the constraint $\Sigma_k \alpha_k \leq n_T$. If any chromosome violates this constraint, "0" is inserted into a randomly selected bit position in the violating chromosome until the constraint is satisfied.

The computation complexity of the genetic algorithm is bounded by $N_g \times N_p$ function evaluations. As will be illustrated in the next section, this represents enormous computational saving compared with the optimal algorithm.

6.6 RESULTS AND DISCUSSION

In this section, we shall discuss the performance of the *capacity-optimized* cross-layer scheduling algorithms for multiuser systems with multiple antennas at the base station. We first consider the system performance (network capacity) based on two commonly used system utilities: the *total throughput* and the *proportional fairness*. For each of the two system utilities, we compare the performance of the optimal cross-layer scheduling with respect to the greedy scheduling as well as the genetic-based scheduling. To highlight the contribution of *multiuser selection diversity*, we also compare the performance with respect to the *round-robin scheduler*, where the K users share the channel evenly in time irrespective of their channel matrices at every fading block. In addition, we shall discuss the effect of n_T on the overall network capacity.

Next, we shall discuss the performance of the *coverage-optimized* cross-layer scheduling. Similarly, we compare the coverage performance of the optimal algorithm versus the conventional round-robin, greedy scheduling as well as genetic-based scheduling.

Finally, we shall discuss the tradeoff between the capacity and coverage of multiuser systems. Note that in the simulations, each data point consists of 5000 realizations of channel fading. Channel fading of the K users are generated in accordance with an independent complex Gaussian distribution with unit variance. We assume 0 dB antenna gain in the transmit and receive antennas and 2 GHz carrier frequency. The data rate is expressed in terms of bits per second per hertz, and network coverage is expressed in terms of cell radius (m).

6.6.1 System Capacity Performance

Figure 6.9a–d illustrates the scheduling performance of the maximal throughput scheduler with respect to n_T and SNR. Mobile users are assumed to be homogeneous in terms of path loss and transmit power constraint. We observe

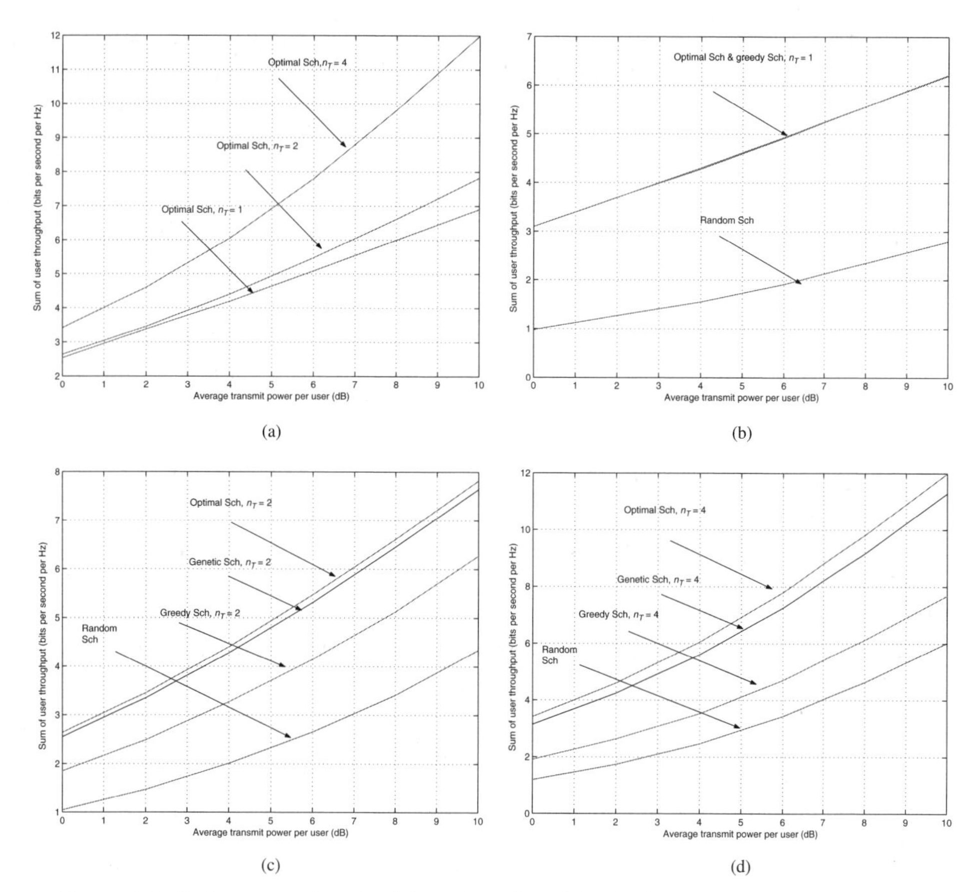

Figure 6.9. Performance of maximal throughput downlink schedulers versus SNR for multiple antenna systems: (a) optimal scheduler versus SNR, n_T = 1,2,4; (b) Max-thp schedulers versus SNR, n_T = 1; (c) Max-thp schedulers versus SNR, n_T = 2; (d) Max-thp schedulers versus SNR, n_T = 4.

that a significant gain in system capacity is achieved by increasing the number of transmit antennas n_T at high SNR, and this is due to the distributed spatial multiplexing. For instance, there are two factors affecting system performance. The first factor is the increased number of spatial channels as n_T increases, and this contributes to the increase in capacity. The second factor is *power splitting* between the increased spatial channels, and this contributes to a drop in the capacity of each spatial channel. As illustrated in the figures, the first factor wins for large SNR. For example, in Figure 6.9a, there are 14% and 71% total system capacity gain when comparing $n_T = 1$ at SNR = 10 dB with $n_T = 2$ and 4, respectively.

On the other hand, in Figure 6.9b, observe that the performance of the greedy algorithm coincides with the optimal scheduler when $n_T = 1$. However, in Figure 6.9c,d, there are 2.5 and 3.8 dB of SNR penalties between the per-

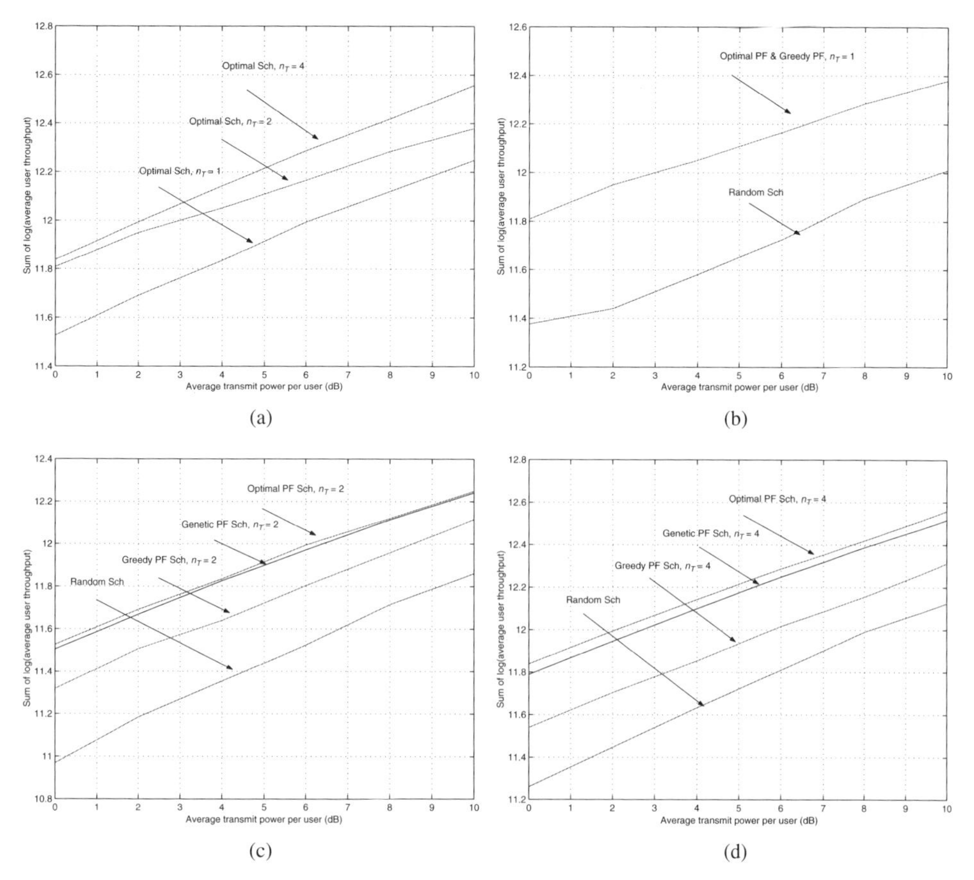

Figure 6.10. Performance of proportional fair downlink schedulers versus SNR for multiple-antenna systems: (a) optimal scheduler versus SNR, $n_T = 1,2,4$; (b) PF schedulers versus SNR, $n_T = 1$; (c) PF schedulers versus SNR, $n_T = 2$; (d) PF schedulers versus SNR, $n_T = 4$.

formance of the greedy and the optimal schedulers at $n_T = 2$ and 4, respectively. This motivates the application of genetic algorithm. We observe that the genetic algorithm has relatively small performance loss compared with the optimal scheduler.

6.6.2 Proportional Fairness Performance

Figure 6.10a–d illustrates the scheduling performance ($U_{\mathrm{PF}}(R_1, \ldots, R_K)$) of the proportional fair scheduler with respect to n_T and SNR. Mobile users are assumed to have heterogeneous path loss of variance 2 dB. The performance is represented by the utility function value. Similar to the maximal throughput utility function, significant gain is observed for $n_T = 2$ and $n_T = 4$. Similarly, the greedy algorithm achieves optimal performance only when $n_T = 1$. For large n_T, there is huge performance gap between the optimal and the greedy

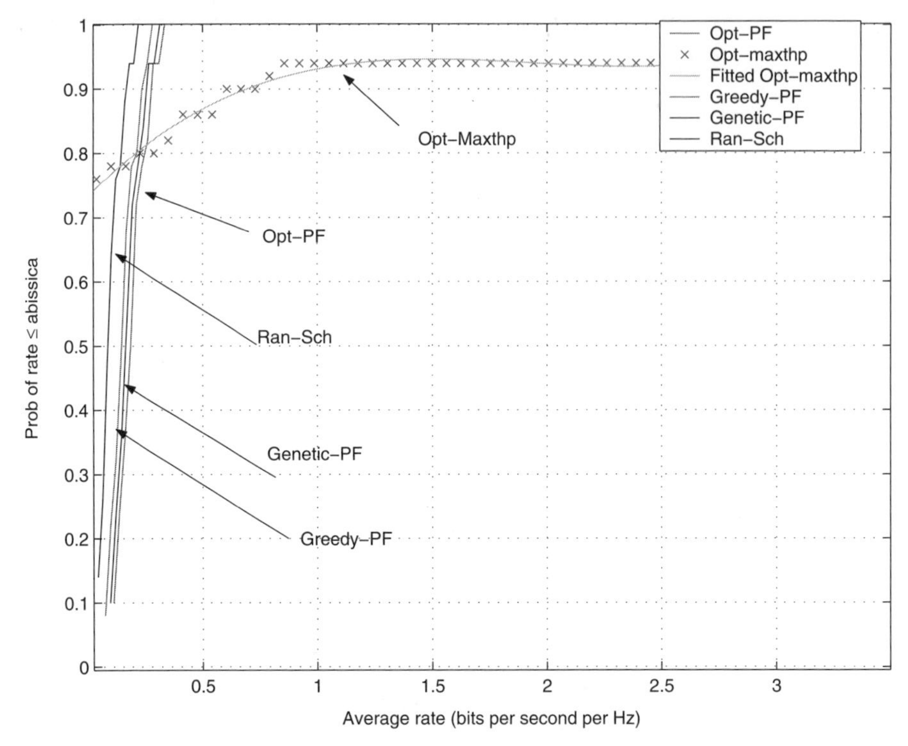

Figure 6.11. Illustration of fairness. Cumulative distribution function (cdf) of mobile data rate at SNR = 10 dB and $n_T = 2$.

algorithms. On the other hand, the genetic algorithm could fill up a majority of the performance gap at various n_T.

Another way to visualize the fairness dimension of the downlink scheduler performance is given in Figure 6.11, where the cumulative distribution function (cdf) of mobile throughput is plotted for $n_T = 2$ and SNR = 10 dB. Observe that while maximal throughput scheduler could achieve the highest total system capacity, the chance of mobile users achieving such a throughput is very low. On the other hand, for the PF scheduler, although the absolute maximum throughput achieved by any mobile user is smaller than the maximal throughput scheduler, mobiles have a higher chance of achieving a decent throughput. For example, over 80% of the time, mobiles with PF scheduling are able to achieve a data rate of at least 0.2, while mobiles of maximal throughput scheduler are able to achieve a data rate of 0. In addition, observe that the performance gap of the greedy algorithm against the optimal PF is very large.

6.6.3 Coverage Performance

Figure 6.12 illustrates the optimal scheduler performance in *coverage utility* $G_{\text{cov}}(r_1, \ldots, r_K) = \min_{k \in \mathcal{A}} r_k$ versus cell radius at various $Q = 1, 2, \ldots, 5$ and

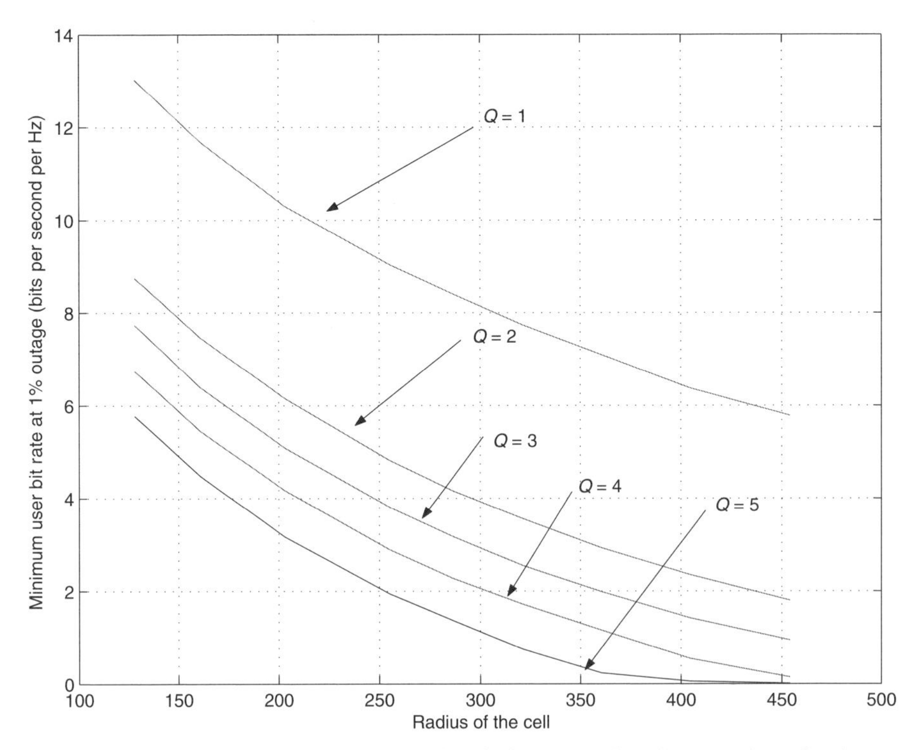

Figure 6.12. Minimum scheduled data rate (bps/Hz) versus cell radius at various $Q = 1, \ldots, 5$. Total transmit power of the base station is 30 dBm, path loss exponent is 4, and the total number of users is 10.

$n_T = 5$. The coverage utility we consider refers to the *minimum scheduled data rate* at 1% outage probability ($\in = 0.01$). Hence, a point (x, y) in the graph means that over 99% of the time, a scheduled user will be able to transmit at least y bps/Hz for system with the cell radius x. We observe that to achieve high coverage gain, the spacetime scheduler should exploit the spatial diversity ($Q = 1$) instead of spatial multiplexing ($Q = 5$). For example, there is a 16-fold area coverage gain at target network capacity of 5 bps/Hz when comparing $Q = 1$ to $Q = 5$.

6.6.4 Tradeoff between Capacity and Coverage

Figure 6.13 illustrates the tradeoff between network capacity and network coverage at SNR = 0 dB and 22 dB. The x axis represents the minimum scheduled user data rate (bps/Hz) [G_{cov}], and the y axis represents the total network capacity (bps/Hz) [G_{cap}]. As illustrated, by adjusting the parameter Q, the spacetime scheduler could be configured to optimize the coverage ($Q = 1$) or capacity ($Q = 5$) or a balance between the two dimensions.

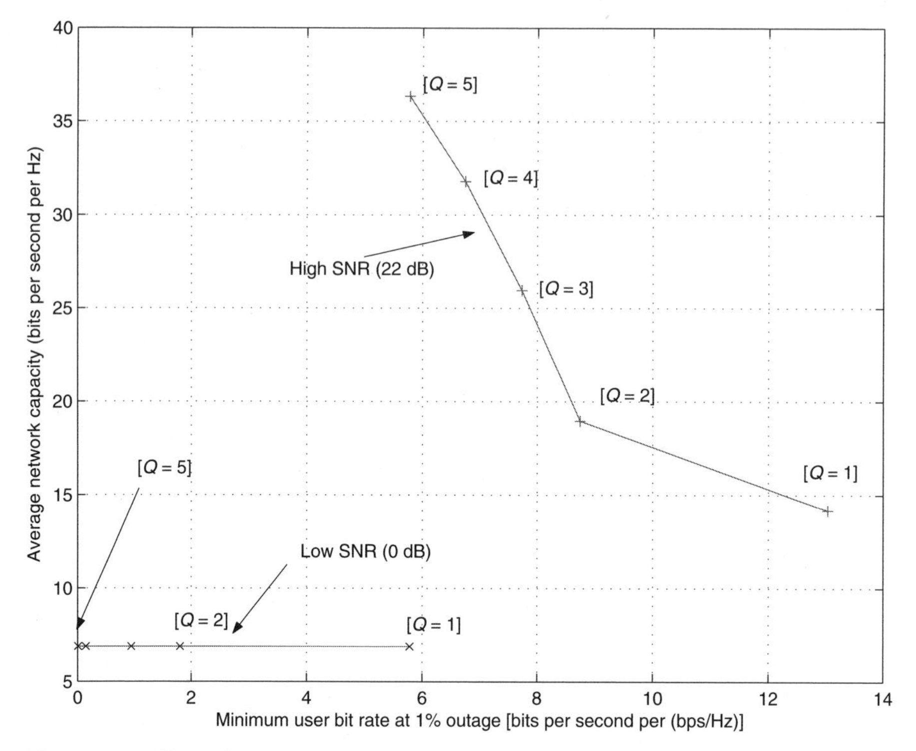

Figure 6.13. Tradeoff between the network capacity (sum of scheduled data rate) and the network coverage (minimum scheduled data rate) and at low and high SNRs.

6.6.5 Efficiency of Multiuser Diversity

Figure 6.14 illustrates the scheduling performance at SNR = 10 dB with respect to K. As K increases, the efficiency of multiuser selection diversity increases because at any scheduling instance, it is more likely to select user with good channel conditions. Yet, supporting a large K would induce a large signaling overhead for channel estimation at the base station.

As shown in the figure, the required number of users for efficient multiuser diversity is not too large. For instance, when $K > 10$, the gain of the multiuser diversity is quite significant already. Hence, the requirement on K is realistic for most practical systems.

6.6.6 Efficiency of Spatial Multiplexing

Figure 6.15 illustrates the network capacity (bps/Hz) versus cell radius of optimal scheduling at various $Q = 1, \ldots, 5$ and $n_T = 5$. As illustrated, the space-time scheduler should exploit the spatial multiplexing ($Q = 5$) at small radius instead of spatial diversity ($Q = 1$). For example, there is 1.7 times capacity gain of $Q = 5$ relative to $Q = 1$ at the cell radius of 260 m.

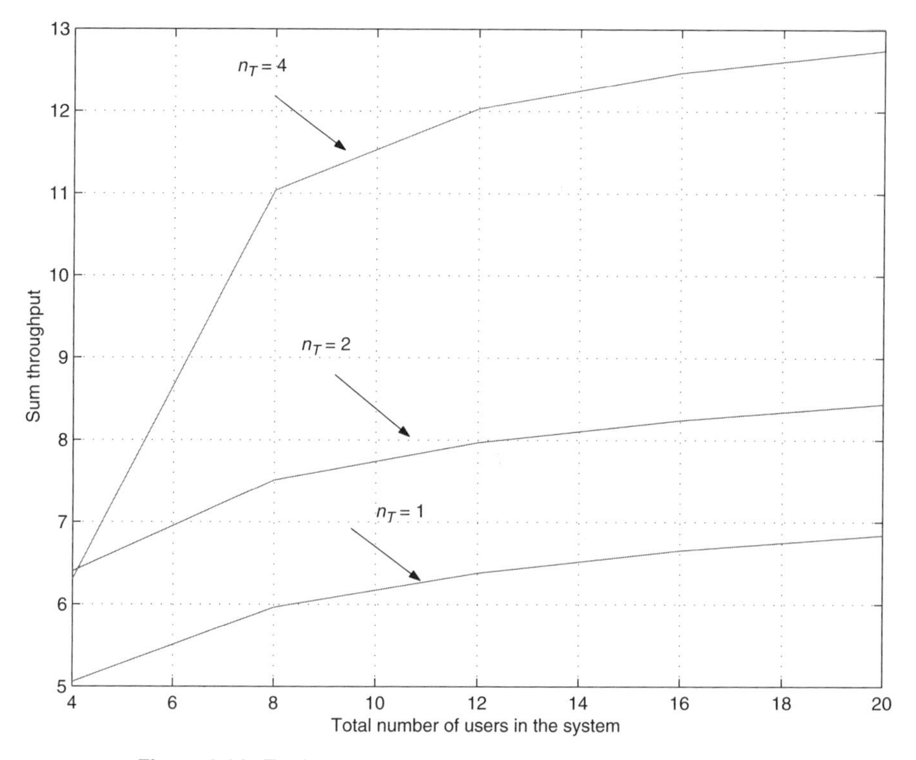

Figure 6.14. Total system capacity versus K and n_T at SNR = 10 dB.

6.6.7 Complexity Comparisons of Various Cross-Layer Algorithms

Table 6.1 compares the number of function evaluations of the optimal algorithm, the greedy algorithms, and the genetic algorithms at various Q and K. Observe that there are 8 times and 600 times saving in computation of genetic algorithm (compared with the optimal algorithm) when $(K, Q) = (10, 5)$ and $(20, 5)$, respectively. On the other hand, the genetic algorithm is about 5 times more complex than the greedy algorithm in both cases.

6.7 CROSS-LAYER SCHEDULING WITH IMPERFECT CSIT

As we have illustrated in the previous sections, cross-layer scheduling for base stations with multiple antennas achieves significant gains in network capacity and network coverage due to multiuser selection diversity over the spatial domain. To exploit the multiuser selection diversity, perfect knowledge of CSIT is assumed at the base station. When the base station has perfect knowledge of CSIT, the cross-layer scheduler selects a set of active users based on the CSI and allocates resource accordingly. The scheduled data rate of any

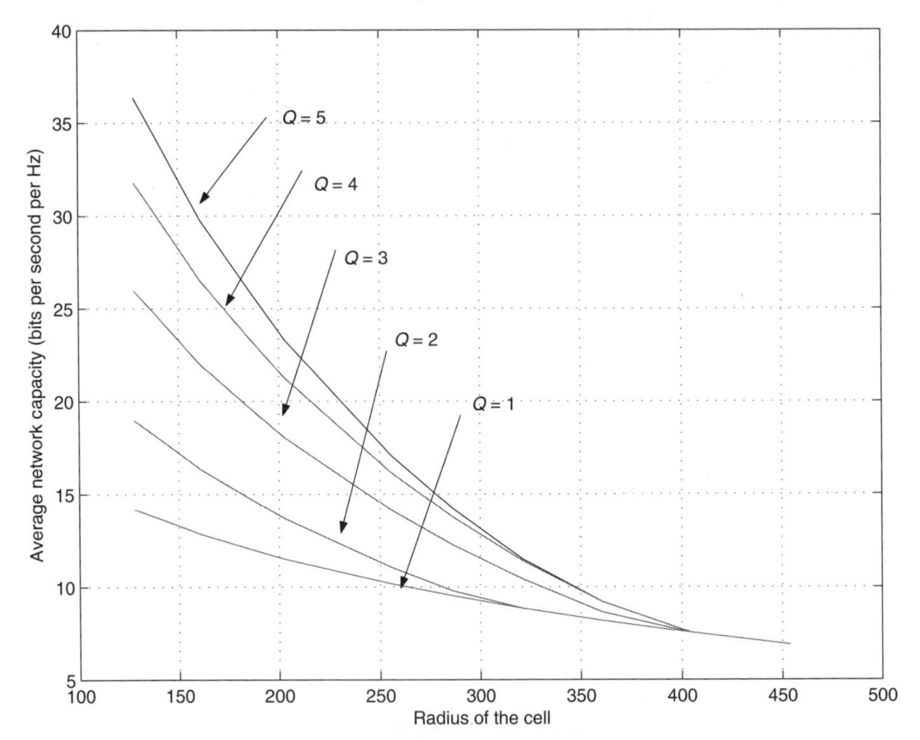

Figure 6.15. Total network capacity (bps/Hz) versus cell radius at various $Q = 1, \ldots, 5$ and $n_T = 5$. The transmit power at the base station is 30 dBm, path loss exponent is 4, and the total number of users is 10.

TABLE 6.1. Comparison of Computational Complexity (Number of Function Evaluations) of Greedy, Genetic, and Optimal Algorithms

(K, Q)	Greedy Algorithm	Genetic Algorithm	Optimal Algorithm
(10,2)	10	$10 \times 2 = 20$	55
(10,4)	10	$10 \times 5 = 50$	385
(10,5)	10	$10 \times 5 = 50$	385
(20,2)	20	$10 \times 5 = 50$	210
(20,4)	20	$20 \times 5 = 100$	3645
(20,5)	20	$20 \times 5 = 100$	60459

selected user is guaranteed to be lower than the *instantaneous channel capacity*, and the transmitted packet will be virtually error-free (with powerful error correction coding) in slow fading channels. However, when we have imperfect CSIT at the transmitter, the efficiency of multiuser scheduling is reduced because the wrong set of users may be selected for transmission (*misscheduling*). Moreover, the scheduled data rate of any active user may be larger than the instantaneous channel capacity (which is unknown to the transmitter), and this results in *packet transmission outage* even if powerful error correction code is applied. Conventional approach to tackle the imperfect CSIT is to design the scheduler assuming perfect CSIT and analyze the performance degradation due to the imperfect CSI. It is found that there is a significant degradation in the system performance due to imperfect CSIs [2]. However, this approach does not offer any design insight on what should be the optimal design and performance with imperfect CSIT as the optimal design can be quite different from that with perfect CSIT. In this section, we shall focus on the design of the cross-layer scheduler for the downlink of the multiuser multiantenna systems with imperfect knowledge of CSIT. In much the same way as in the previous sections, we consider a system with K users (each with single antenna) and a base station (with n_T antennas) operating in time-division duplexing (TDD) mode. Hence, the downlink CSIT can be estimated at the base station using uplink pilots due to channel reciprocity.[6] The cross-layer scheduling design with imperfect CSIT is formulated as an optimization problem, and the optimal scheduling solution is found to be computationally intensive. Therefore, two computationally effective suboptimal algorithms, namely, the *genetic algorithm* and *short-term SINR scaling*, are proposed and compared with the optimal performance.

6.7.1 Multiuser Physical Layer Model with Imperfect CSIT

In TDD mode, the base station can deduce downlink CSIT from the uplink pilots. In practice, the downlink pilot channel (from the base station) is usually allocated a higher power because it can be shared by the K users. Hence, the CSI estimation at the mobile terminals (CSIR) can be quite accurate, and we assume that the CSIR at the mobile station is perfect. On the other hand, on the uplink side, the base station estimate of the CSIT is based on the uplink pilots from the K users. Since the uplink pilots are dedicated pilots per user and cannot be shared, the power allocated in the uplink pilots are usually smaller and the CSIT obtained at the base station is likely to be imperfect. The estimated CSIT of all the K users at the base station, $\mathbf{H}^b$, is given by [70,87]

$$\mathbf{H}^b = \mathbf{H} + \Delta\mathbf{H}^b \tag{6.35}$$

where $\mathbf{H}$ is the actual CSIT, and $\Delta\mathbf{H}^b$ is the CSIT estimation error matrix with components given by i.i.d. Gaussian distribution variables (zero mean, variance $\sigma^2_{\Delta H}$). Assuming MMSE channel estimation, we have $\Sigma[\Delta\mathbf{H}^b\mathbf{H}^b] = 0$ by orthogonal principle.

[6] Strictly speaking, there are different biases in the uplink and downlink of TDD systems due to different RF gains in the transmit receive path. However, the biases are slowly time-varying relative to the channel fading and can be corrected on the basis of a slow calibration loop so that the instantaneous CSIT can be deduced from the uplink pilots.

For simplicity, we ignore the path loss terms and assume the channel fadings of the K users are i.i.d. The received signal at the kth mobile station can be expressed as

$$\begin{aligned} Y_k &= \underbrace{\sqrt{p_k}\,\mathbf{h}_k\mathbf{w}_kU_k}_{\text{Information}} + \underbrace{\sum_{j\neq k}\sqrt{p_j}\,\mathbf{h}_k\mathbf{w}_jU_j}_{\text{Multiuser interference}} + Z_k \\ &= \sqrt{p_k}\,\mathbf{h}_k\mathbf{w}_kU_k + \sum_{j\neq k}\sqrt{p_j}\,\mathbf{h}_k^b\mathbf{w}_jU_j - \sum_{j\neq k}\sqrt{p_j}\,\Delta\mathbf{h}_k^b\mathbf{w}_jU_j + Z_k \\ &= \sqrt{p_k}\,\mathbf{h}_k\mathbf{w}_kU_k \underbrace{-\sum_{j\neq k}\sqrt{p_j}\,\Delta\mathbf{h}_k^b\mathbf{w}_jU_j + Z_k}_{\text{Multiuser interference } V_k} \end{aligned} \tag{6.36}$$

where the term $\Sigma_{j\neq k}\sqrt{p_j}\mathbf{h}_k^b\mathbf{w}_jU_j$ is nulled out because of the ZF operation and V_k denotes the residual multiuser interference due to imperfect knowledge of CSIT. The maximum achievable data rate[7] of the kth user is given by the maximum mutual information between Y_k and U_k conditional on CSIR $\mathbf{h}_k$ and is expressed as follows:

$$\begin{aligned} C_k &= \max_{P(U_k)} I(U_k; Y_k|\mathbf{h}_k) \\ &= \max_{P(U_k)}\{H(Y_k|\mathbf{h}_k) - H(Y_k|U_k,\mathbf{h}_k)\} \\ &= \log_2\left(1 + \frac{p_k|\mathbf{h}_k\mathbf{w}_k|^2}{\sigma_z^2 + \sum_{j\neq k} p_j|\Delta\mathbf{h}_k\mathbf{w}_j|^2}\right) \end{aligned} \tag{6.37}$$

From (6.37), the instantaneous channel capacity is a function of $\mathbf{h}_k$ and $\Delta\mathbf{h}_k$ which are unknown to the base station. Hence, given any estimated CSIT $\mathbf{H}^b$, there is still uncertainty on the actual capacity C_k and *packet transmission outage* is possible when the scheduled data rate r_k exceeds the actual capacity C_k. To take into consideration of the packet outage, we define the *effective goodput* of the kth user as:

$$\rho_k = r_k\mathbf{1}[\mathbf{C_k} \geq \mathbf{r_k}] \tag{6.38}$$

where $\mathbf{1(E)}$ is the indicator function, which is equal to 1 if the event E is true and 0 otherwise.

6.7.2 System Utility Function

For illustration purposes, we consider the *average total goodput* as the measure of system performance and the optimization objective. The average total goodput of the multiuser multiantenna system is given by

[7]The *maximum data rate* here refers to the Shannon capacity based on random codebook and Gaussian constellation.

$$
\begin{aligned}
U_{\text{thp}}(\bar{\rho}_1, \ldots, \bar{\rho}_K) &= \varepsilon\left[\sum_{k=1}^{K} \rho_k\right] \\
&= \varepsilon_{\mathbf{h}^b}\left\{\sum_{k=1}^{K} r_k \varepsilon_{\mathbf{h}}\left[\mathbf{1}(r_k < C_k) \,|\mathbf{h}^b\right]\right\} \\
&= \varepsilon_{\mathbf{h}^b}\left\{\sum_{k=1}^{K} r_k \Pr\left[r_k < C_k \,|\mathbf{h}^b\right]\right\}
\end{aligned}
\tag{6.39}
$$

where $\bar{\rho}_k = \varepsilon[\rho_k]$ denotes the average effective throughput and $\varepsilon_{\mathbf{h}^b}[X]$ denotes the expectation of the random variable X w.r.t. $\mathbf{h}^b$. Define an *instantaneous total goodput* $G(r_1, \ldots, r_K)$ as

$$
\begin{aligned}
G(r_1, \ldots, r_K) &= \sum_{k=1}^{K} r_k \Pr\left[r_k < C_k \,|\mathbf{h}^b|\right] \\
&= \sum_{k=1}^{K} r_k \left[1 - P_{\text{out}}(r_k, \mathbf{h}^b)\right]
\end{aligned}
\tag{6.40}
$$

where $P_{\text{out}}(r_k, \mathbf{h}^b)$ is the conditional outage probability (conditioned on the CSIT $\mathbf{h}^b$) and the average total goodput is given by

$$
U_{\text{thp}}(\bar{\rho}_1, \ldots, \bar{\rho}_K) = \varepsilon_{h^b}[G(r_1, \ldots, r_K)]
$$

6.7.3 The Scheduling Problem Formulation

Since the scheduler determines the optimal $\{\mathcal{A}\}$, $\{p_k\}$, and $\{r_k\}$ once per CSIT realization, the cross-layer scheduling problem with imperfect CSIT can be summarized as the following optimization problem

Problem 6.7 (Cross-Layer Scheduling with Imperfect CSIT) Given any realization of the estimated CSIT for all mobile users $\{\mathbf{h}_1^b, \ldots, \mathbf{h}_K^b\}$, determine the "optimal admitted user set" $\mathcal{A}$, the "optimal power allocation" $\{p_k\}$ and the "optimal rate allocation" $\{r_k\}$, such that the instantaneously total goodput, $G(r_1, \ldots, r_K)$, is maximized subject to the total transmit power constraint $\Sigma_{k\in\mathcal{A}} p_k \leq P_0$ and the cardinality constraint $|\mathcal{A}| \leq n_T$.

6.7.4 The Optimal Scheduling Solution

The optimization variables in Problem 17 include $\{\mathcal{A}\}$, $\{p_k\}$ as well as $\{r_k\}$. Hence, the optimization problem involves a mixed convex optimization (w.r.t. $\{p_k\}$ and $\{r_k\}$) and combinatorial search (w.r.t. $\{\mathcal{A}\}$). Unlike the case with perfect CSIT where the optimal scheduled rates $\{r_k\}$ are simply given by the instantaneous channel capacities $\{C_k\}$, the selection of scheduled data rate with imperfect CSIT is quite tricky. For example, if the scheduled rate r_k is too large, the outage probability will be high and the overall average goodput is small (due to retransmissions). On the other hand, if the scheduled rate is too small, the outage probability will be small but the overall goodput will also be small (due to over-protection of packets). Hence, there exists an optimal choice of scheduled rate r_k. We shall elaborate the optimal scheduling solution w.r.t. the optimization problem (Problem 6.7) in the following two steps.

- *Step 1: Convex optimization on* $\{p_k\}$ *and* $\{r_k\}$. The constrained optimization problem (Problem 6.7) can be expressed using the Lagrangian approach. Given any admitted user set $\mathcal{A}$, the Lagrangian function is given by

$$L(\{p_k\},\{r_k\}) = \sum_{k=1}^{K} r_k (1 - P_{\text{out}}(r_k, \mathbf{h}^b)) - \mu \sum_{k=1}^{K} p_k \tag{6.41}$$

where μ is the Lagrange multiplier w.r.t. the transmit power constraint P_0. Let r_k^* and p_k^* be the zeros of the following equations:

$$\frac{\partial L}{\partial r_k} = 0 \tag{6.42}$$

$$\frac{\partial L}{\partial p_k} = 0 \tag{6.43}$$

From the KKT condition, the optimizing power allocations $\{p_k\}$ and rate allocations $\{r_k\}$ are given by $p_k = \max\{0, p_k^*\}$ and $r_k = \max\{0, r_k^*\}$.
- *Step 2: Combinatorial Search*. The system utility function $G(r_1, \ldots, r_K)$ is evaluated for every feasible admitted user set $\{\mathcal{A}\}$ satisfying the constraint $|\mathcal{A}| \le n_T$. The optimizing admitted user set is given by the one that achieves the maximum system utility. Similar to the case with perfect CSIT, the total search space is given by $\Sigma_{m=1}^{n_T} \binom{K}{m}$ which is quite large for moderate n_T and K.

Given any CSIT $\mathbf{h}_k^b$, the actual CSI $\mathbf{h}_k$ is complex Gaussian conditioned on the CSIT $\mathbf{h}_k^b$ with mean and co-variance given by

$$\varepsilon[\mathbf{h}_k | \mathbf{h}_k^b] = \mathbf{h}_k^b \tag{6.44}$$

and

$$\varepsilon\left[\mathbf{h}_k^* \mathbf{h}_k | \mathbf{h}_k^b\right] = \mathbf{h}_k^* \mathbf{h}_k + \sigma_{\Delta H}^2 \mathbf{I} \tag{6.45}$$

The conditional outage probability $P_{\text{out}}(r_k, \mathbf{h}^b)$ of the k-th user is given by:

$$P_{\text{out}}(r_k, \mathbf{h}^b) = \Pr[C_k < r_k | \mathbf{h}_k^b] \\ = \Pr[S_k < \Lambda_k | \mathbf{h}^b] \tag{6.46}$$

where

$$S_k = p_k |\mathbf{h}_k \mathbf{w}_k|^2 - \sum_{j \neq k} \zeta_j |\Delta \mathbf{h}_k \mathbf{w}_j|^2, \tag{6.47}$$

$\Lambda_k = (2^{r_k} - 1)\sigma_z^2$ and $\zeta_j = p_j \Lambda_k / \sigma_z^2$.

However, there is no closed form expression for the outage probability in (6.46). Moreover, conditioned on the CSIT $\mathbf{h}^b = \{\mathbf{h}_1^b, \ldots, \mathbf{h}_K^b\}$, the random variables $p_k|\mathbf{h}_k\mathbf{w}_k|^2$ and $p_j|\Delta\mathbf{h}_k\mathbf{w}_j|^2$ are mutually dependent and evaluation of the closed form cdf is not trivial. The complexity of the optimal scheduling solution is very high in both steps 1 and 2. In step 1, there is no closed-form expression for the optimizing rate r_k and the optimizing power p_k. In step 2, the combinatorial search space is very large (exponential in n_T). Hence, while the optimal solution allows us to obtain the performance upper bound, the solution itself is impractical for realtime implementations. In the following sections, we shall introduce a suboptimal but effective solution to simplify steps 1 and 2.

6.7.5 Suboptimal Solution

One way to simplify the computational requirement in step 1 is to give up the search for the optimal outage probability. Instead, for each given CSIT $\mathbf{h}^b$, the scheduled data rate is chosen such that the conditional outage probability $P_{\text{out}}(r_k, \mathbf{h}^b)$ is less than a specified target ε. For closed form solutions on r_k and p_k, we apply Gaussian approximation to S_k in the outage probability in (6.46) and the result is summarized below.

Lemma 6.7 (Gaussian Approximation) For sufficiently large $|A|$, the conditional outage probability can be approximated by

$$P_{\text{out}}(r_k, \mathbf{h}^b) \to 1 - Q\left(\frac{\Lambda_k - \mu_{S_k}}{\sigma_{S_k}}\right) \tag{6.48}$$

where Q(x) is the Gaussian Q-function,

$$\mu_{S_k} = p_k \sigma_{\Delta H}^2 + p_k |\mathbf{h}_k^b \mathbf{w}_k|^2 - \sum_{j \neq k} \zeta_j \sigma_{\Delta H}^2 \tag{6.49}$$

and

$$\delta_{S_k}^2 = p_k^2 \sigma_{\Delta H}^4 + 2 p_k^2 \sigma_{\Delta H}^2 |\mathbf{h}_k^b \mathbf{w}_k|^2 \\ \sigma_{\Delta H}^4 \left\| \sum_j \zeta_j \mathbf{w}_j \right\|^2 - 2\sigma_{\Delta H}^4 p_k \sum_j \zeta_j \left| \mathbf{w}_k^* w_j \right|^2 \tag{6.50}$$

Proof Note that conditioned on the CSIT $\{\mathbf{h}_1^b, \ldots, \mathbf{h}_K^b\}$, the random variables $p_k|\mathbf{h}_k\mathbf{w}_k|^2$ and $p_j|\Delta\mathbf{h}_k\mathbf{w}_j|^2$ are mutually dependent and the convergence of S_k to Gaussian in distribution is not trivial. Please refer to Appendix 6H for the proof.

To obtain a closed form solution for r_k and p_k, we consider a target outage probability given by ε. From (6.48), we have:

$$\frac{\Lambda_k - \mu_{S_k}}{\sigma_{S_k}} = \varphi_k \tag{6.51}$$

Since $\Lambda_k \geq 0$, we have the *short-term SINR scaling factor* given by

$$\varphi_k = \max\{-\mu_{S_k}/\sigma_S, Q^{-1}(1-\varepsilon)\}$$

From (6.51) and assuming $\Sigma_{j\neq k} p_j \approx P_0$ where P_0 is the total transmit power constraint, we have

$$\Lambda_k\left(1+\frac{\sigma_{\Delta H}^2 P_0}{\sigma_z^2}\right) - \sigma_{\Delta H}^2 p_k - p_k\left|\mathbf{h}_k^b \mathbf{w}_k\right|^2 = \sigma_{S_k}\varphi_k$$

and hence, the scheduled rate is given by:

$$r_k = \log_2\left(1+\frac{\varphi_k \sigma_{S_k} + p_k\left(\sigma_{\Delta H}^2 + \left|\mathbf{h}_k^b \mathbf{w}_k\right|^2\right)}{\sigma_z^2 + \sigma_{\Delta H}^2 P_0}\right). \tag{6.52}$$

For large n_T, the term $|\mathbf{h}_k^b\mathbf{w}_k|^2$ scales with n_T while the other terms in the variance expression in (6.50) does not scale with n_T. Hence, the variance in (6.50) can be approximated as

$$\sigma_{S_k} \approx \sqrt{2}\, p_k \sigma_{\Delta H}\left|\mathbf{h}_k^b \mathbf{w}_k\right|$$

The scheduled rate can therefore be approximated by

$$r_k = \log_2\left(1+\frac{p_k\left(\sqrt{2}\sigma_{\Delta H}\varphi_k\left|\mathbf{h}_k^b \mathbf{w}_k\right| + \sigma_{\Delta H}^2 + \left|\mathbf{h}_k^b \mathbf{w}_k\right|^2\right)}{\sigma_z^2 + \sigma_{\Delta H}^2 P_0}\right). \tag{6.53}$$

The optimal power allocation $\{p_k\}$ is obtained by optimizing the conditional average goodput $\Sigma_k r_k(1-\varepsilon)$ and is therefore give by:

$$p_k = \left(\frac{1}{\ln 2\mu} - \frac{\sigma_z^2 + \sigma_{\Delta H}^2 P_0}{\sqrt{2}\sigma_{\Delta H}\varphi k\left|\mathbf{h}_k^b \mathbf{w}_k\right| + \sigma_{\Delta H}^2 + \left|\mathbf{h}_k^b \mathbf{w}_k\right|^2}\right)^+ \tag{6.54}$$

for all $k \in \mathcal{A}$. Note that the power allocation strategy in (6.54) reduces to the conventional water-filling solution where $\sigma_{\Delta H} \to 0$ (perfect CSIT).

While the short-term SINR approach simplifies the step I, the combinatorial search in step II is still very complicated. Similar to the case with perfect CSIT, the combinatorial search in step II can be simplified with some sub-optimal approaches such as the greedy algorithm or genetic algorithm introduced in Section 6.5. We shall apply similar sub-optimal search algorithms as in the case of perfect CSIT, namely the *greedy* and the *genetic* algorithms, for the step II to determine the optimizing $\mathcal{A}$.

6.7.6 Numerical Results

Figure 6.16a illustrates the total network goodput versus n_T at various $\sigma^2_{\Delta \mathrm{H}}$ of the *ideal scheduler* (designed for perfect CSIT) in the presence of imperfect CSIT. At moderate CSIT errors, the total capacity becomes very small and is not able to scale up with n_T. In other words, the spatial multiplexing gains due to the multiantenna base station is heavily jeopardized by the packet outage due to imperfect CSIT. This indicates that cross-layer scheduling designed for perfect CSIT is not robust in the presence of imperfect CSIT, and it is very important for the cross-layer scheduling algorithm to match the imperfect CSIT.

Figure 6.16b illustrates the total network goodput versus n_T of the *robust scheduler* designed for imperfect CSIT as outlined in the previous sections. The network capacity actually scales with n_T and there is a graceful degradation of performance due to imperfect CSIT. This illustrates that the multiuser spatial multiplexing gain due to the multiantenna base station is still quite effective at small to moderate CSIT errors when there is a properly designed cross-layer scheduling algorithm.

Finally, Figure 6.17 illustrates the total network goodput versus n_T for different designs of the *robust scheduler*, namely, the optimal scheduler with optimal power and rate allocation, genetic scheduler with short-term SINR scaling, and greedy scheduler with short-term SINR scaling as well as round-robin scheduler, at various CSIT errors. Observe that the genetic scheduler with short-term SINR scaling achieves close-to-optimal performance at various CSIT errors $\sigma^2_{\Delta \mathrm{H}}$. Furthermore, there is a significant performance gap between the optimal and round-robin schedulers, and this illustrates the multiuser selection diversity gain in the presence of imperfect CSIT.

6.8 SUMMARY

In this chapter, we have considered the cross-layer design in multiuser systems with a base station (n_T transmit antenna) and K mobile stations (single receive antenna). We first defined the notion of system performance by introducing the concept of system utility and focused on the utility-based network capacity as well as the utility-based network coverage as performance measures.

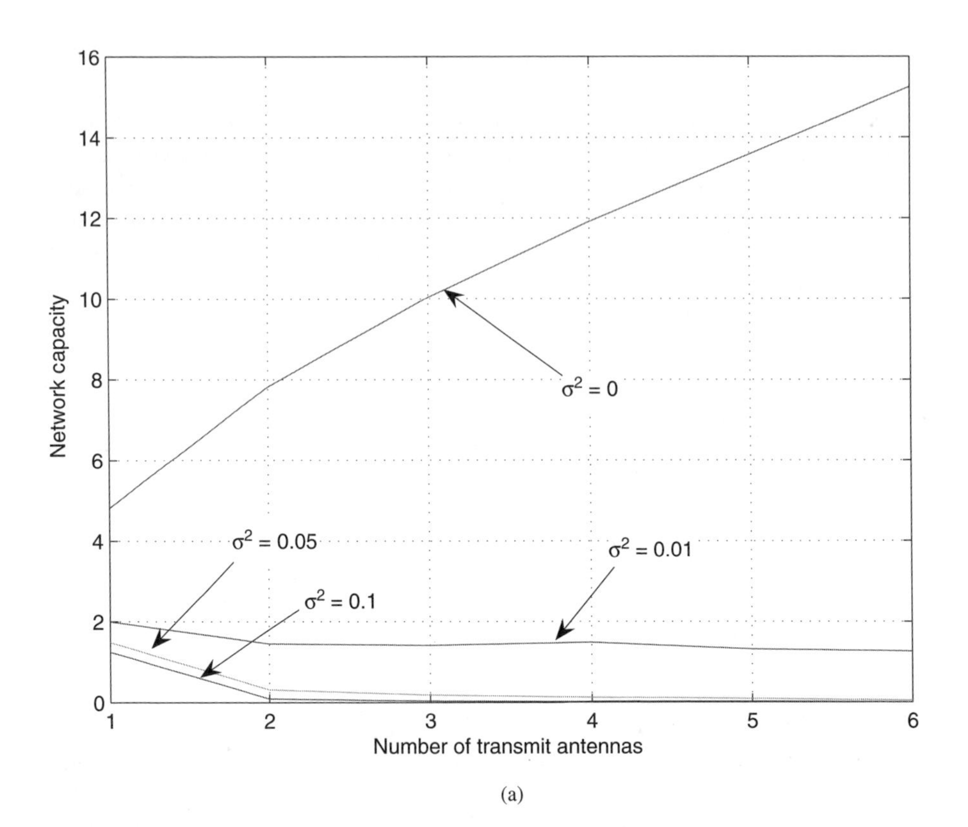

Network capacity
Number of transmit antennas
$\sigma^2 = 0$
$\sigma^2 = 0.05$
$\sigma^2 = 0.1$
$\sigma^2 = 0.01$

(a)

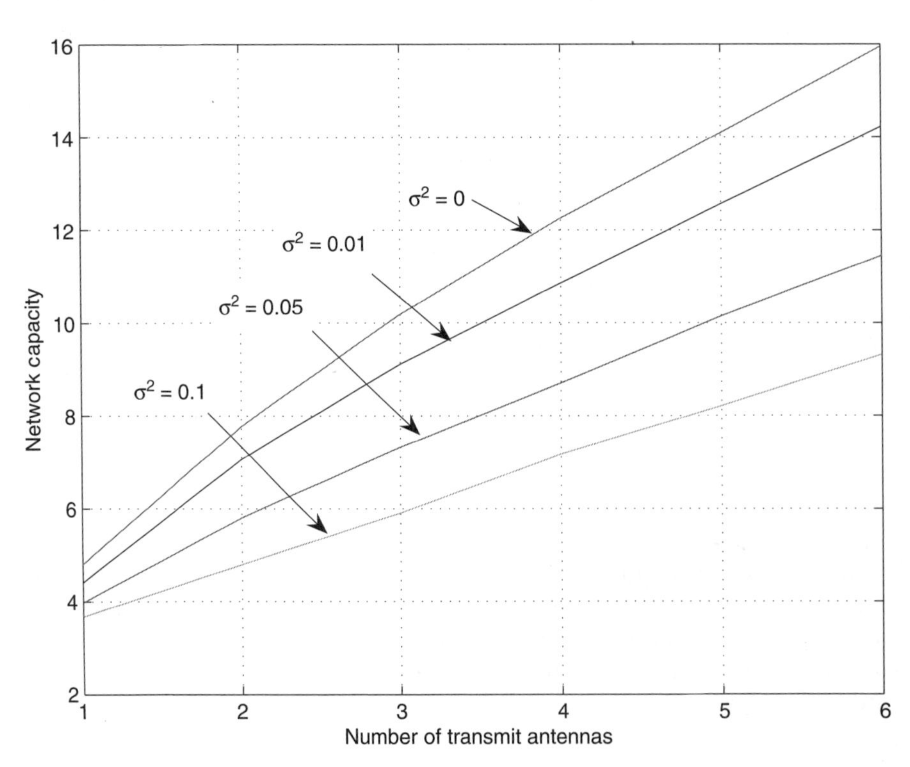

Network capacity
Number of transmit antennas
$\sigma^2 = 0$
$\sigma^2 = 0.01$
$\sigma^2 = 0.05$
$\sigma^2 = 0.1$

(b)

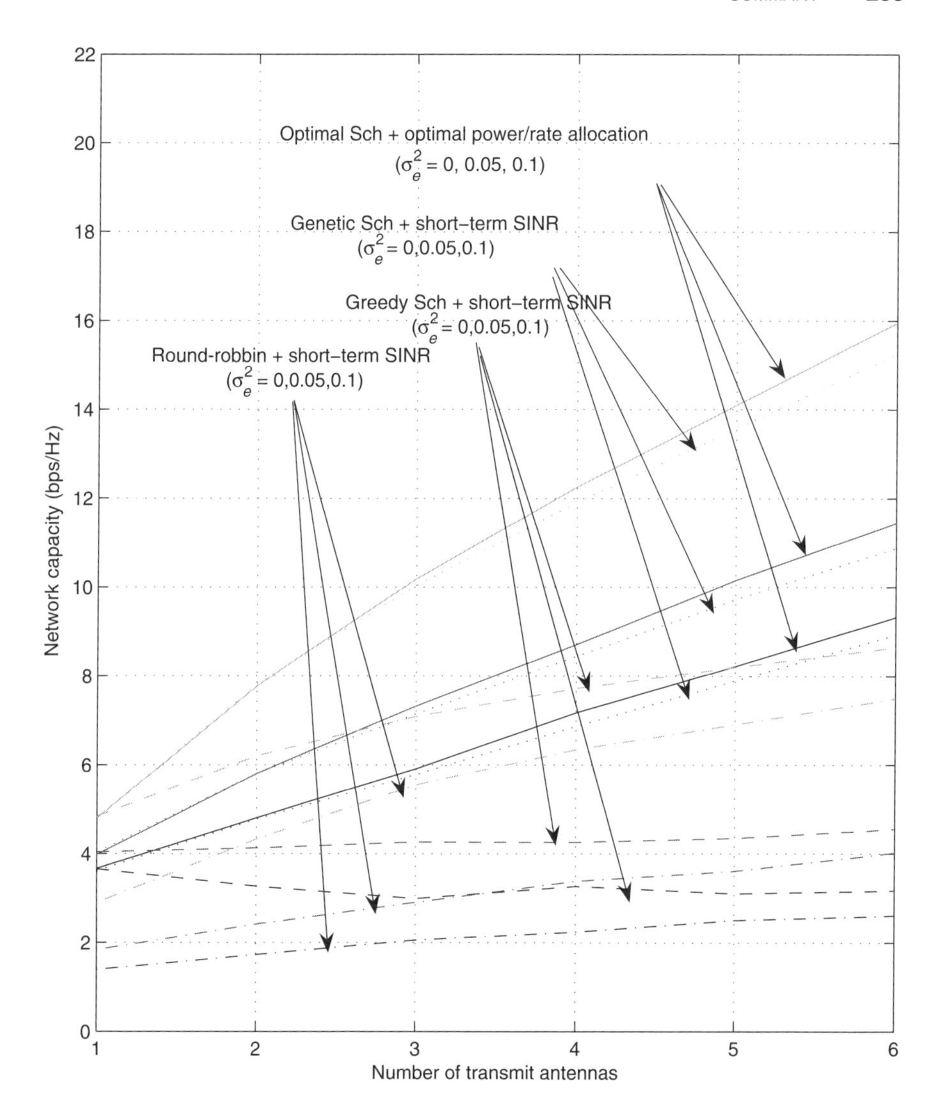

Figure 6.17. A comparison of the total network goodput (bps/Hz) versus n_T with imperfect CSIT on optimal scheduler with optimal power and rate allocation, genetic scheduler with short-term SINR scaling, greedy scheduler with short-term SINR scaling, and round-robin scheduler with short-term SINR scaling at $\sigma^2_{\Delta H}$ = (0, 0.05, 0.1), K = 10, and SNR = 10 dB.

Figure 6.16. Total network goodput (bps/Hz) versus n_T with imperfect CSIT at $\sigma^2_{\Delta H}$ = (0, 0.05, 0.1), K = 10, and SNR = 10 dB: (a) ideal scheduler—Scheduler designed for perfect CSIT; (b) robust scheduler—Scheduler matched to imperfect CSIT.

To tackle the challenging problem of cross-layer optimization, we decoupled the queuing theory (source statistics and delay performance) from the information theory (physical layer performance). Specifically, we assume the buffer size to be infinite and that the data users are delay-insensitive. Hence, there are always user payload packets available for transmission whenever resource is allocated to the user.

We have formulated the cross-layer design as an optimization problem. In order to obtain first-order design insights, we need a sufficiently simple physical layer model and the MAC layer model. For the multiuser physical layer, we model the performance using the concept of capacity region. While this is just a theoretical performance, advanced coding techniques such as turbo and LDPC codes can be shown to achieve the Shannon capacity to within 0.2 dB. On the other hand, the information-theoretic approach is independent of the specific implementations on the coding and modulation schemes and therefore, we can obtain useful and general design insight. For the single-antenna base station, the capacity region can be easily evaluated because the physical layer is modeled by a class of degraded broadcast channels. On the other hand, for the multiantenna base station, the capacity region is difficult to characterize, and therefore we adopt a linear processing constraint at the base station, namely, the orthogonal transmit beamforming (OTBF) approach, to obtain the corresponding capacity region.

The cross-layer design is therefore modeled as an optimization problem with two major focuses, namely, the *network capacity optimized* and *network coverage optimized* objectives. With multiple transmit antennas, we have extra degrees of freedom for *spatial multiplexing* or *spatial diversity*. Together with the *multiuser selection diversity* as a result of the scheduling, both the network capacity dimension and the network coverage dimension could benefit from these three factors. We propose an analytical framework and formulate the spacetime scheduler design as a mixed combinatorial and convex optimization problem. We found that both the *capacity-optimized* and *coverage-optimized* schedulers could be cast into a single optimization problem with an adjustable parameter $Q \leq n_T$. At $Q = 1$, the physical layer is configured to exploit spatial diversity and this optimizes the network coverage. At $Q = n_T$, the system exploits spatial multiplexing and this optimizes the network capacity.

For single-antenna systems, it is shown that the optimal cross-layer scheduling policy is TDMA-based. At any scheduling slot, only one active user will be allocated resource. However, this is no longer true when there are multiple antennas at the base station, due to the spatial degrees of freedom. While the optimal scheduling solution is found, the complexity is very high, and we have propose two computational efficient heuristic algorithms, namely, the greedy and genetic algorithms. The greedy algorithm is found to be optimal at $Q = 1$ (coverage optimized), but there is a huge performance gap from the optimal algorithm at $Q = n_T$ (capacity optimized). The genetic algorithm, on the other hand, can fill in the majority of the performance gap at a reasonable complexity.

Finally, we illustrate the design of cross-layer scheduling with imperfect CSIT for TDD systems. When the base station has perfect knowledge of CSIT, the cross-layer scheduler selects a set of active users based on the CSI and allocates resource accordingly. The scheduled data rate of any selected user is guaranteed to be lower than the *instantaneous channel capacity*, and the transmitted packet will be virtually error-free (with powerful error correction coding) in slow fading channels. However, when we have imperfect CSIT at the transmitter, the efficiency of multiuser scheduling is reduced because the wrong set of users may be selected for transmission (*misscheduling*). Moreover, the scheduled data rate of any active user may be larger than the instantaneous channel capacity (which is unknown to the transmitter), and this results in *packet transmission outage* even if powerful error correction code is applied. The cross-layer scheduling design is formulated as an optimization problem with the objective function given by the total average system goodput. Using the concept of *goodput*, the penalty of packet outage and retransmissions are taken into account. While the optimal scheduling solution is too complicated for practical realtime implementation, we proposed a suboptimal solution, namely, *short-term SINR scaling*, to simplify the determination of the scheduled rates and allocated powers. The performance of the suboptimal solution is compared with the optimal performance at various CSIT errors, and the performance is quite promising.

APPENDIX 6A: PROOF OF LEMMA 6.2

From Equation (6.21), r_k is a function of the path loss L_k and therefore is a function of the distance d_k. Without loss of generality, assume $d_0 = d_1 \geq d_2 \geq \ldots \geq d_K$. Since $r_k(d_k) \geq r_k(d_0)$ and $\partial G/\partial r_k \geq 0$, we have $G(r_1, \ldots, r_K) \geq G^*$, where $G^* = G(r_1(d_0), \ldots, r_K(d_0))$ is the utility value when all the active mobiles are equidistant from the base station. Since $\Pr[G \leq G_0] = \varepsilon[(1G \leq G_0)]$, where $1(A)$ is a binary indicator function, we have $1(G \leq G_0) \leq 1(G^* \leq G_0)$ for any fading realization. This implies $\varepsilon[1(G \leq G_0)] \leq \varepsilon[1(G^* \leq G_0)]$ or $P_{\text{out}} \leq P^*_{\text{out}}(G_0, d_0)$.

APPENDIX 6B: CALCULATION OF WEIGHTS

Without loss of generality, assume $\mathrm{A} = \{1, 2, \ldots, |\mathrm{A}|\}$. Since the channel fading of the K uses are iid, the interference space (with respect to use k is spanned by the columns of $\tilde{\mathrm{H}}_{\mathrm{k}} = [\mathbf{h}_1^* \mathbf{h}_2^* \ldots \mathbf{h}_{k-1}^* \mathbf{h}_{k+1}^* \ldots \mathbf{h}_{|A|}^*]$. Let $\mathbf{B}_{\mathcal{A},k}$ be a $n_T \times (n_T - |\mathcal{A}| + 1)$ matrix with columns given by the eigenvectors (corresponding to the zero eigenvalue) of $\tilde{\mathbf{H}} = [\mathbf{h}_1^* \mathbf{h}_2^* \ldots \mathbf{h}_{k-1}^* \mathbf{h}_{k+1}^* \ldots \mathbf{h}_{|A|}^*]$. In other words, columns of $\mathbf{B}_{\mathcal{A},k}$ span the $(n_T - |\mathcal{A}| + 1)$-dimensional *orthogonal subspace* of the $(|\mathcal{A}| - 1)$-dimensional *interference space* (spanned by $\tilde{\mathbf{H}}_k$). We have

$$\mathbf{h}_j \mathbf{B}_{\mathcal{A},k} \begin{cases} \neq \mathbf{0}, & j = k \\ = \mathbf{0}, & j \neq k, j \in \mathcal{A} \end{cases} \tag{6.53}$$

Observe that the constraint in Equation (6.17) is satisfied if and only if $\mathbf{w}_k$ belongs to the *orthogonal subspace* spanned by the columns of $\mathbf{B}_{\mathcal{A},k}$. Hence, $\mathbf{w}_k = \mathbf{B}_{\mathcal{A},k}\mathbf{u}$ for some $(n_T - |\mathcal{A}| + 1)$-dimensional complex column vector $\mathbf{u}$ and the optimization problem (Problem 6.1) reduces to the following.

Problem 6.8 Maximize $\mathbf{u}^*$ $(\mathbf{B}^*_{\mathcal{A},k}\mathbf{h}^*_k\mathbf{h}_k\mathbf{B}_{\mathcal{A},k})$ $\mathbf{u}$ subject to the constraint $\mathbf{u}^*$ $(\mathbf{B}^*_{\mathcal{A},k}\mathbf{B}_{\mathcal{A},k})$ $\mathbf{u} = 1$.

The optimal beamforming weight is given by the eigenvector of the largest eigenvalue of $(\mathbf{B}^*_{\mathcal{A},k}\mathbf{B}_{\mathcal{A},k})^{-1}$ $\mathbf{B}^*_{\mathcal{A},k}\mathbf{h}^*_k\mathbf{h}_k$ $\mathbf{B}_{\mathcal{A},k}$. (See Exercise 1 in this chapter) The solution involves matrix inversion and SVD. However, the size of the matrix involving inversion $\mathcal{B}^*_{\mathcal{A},k}\mathcal{B}_{\mathcal{A},k}$ is $(n_T - |\mathcal{A}| + 1) \times (n_T - |\mathcal{A}| + 1)$, which is usually quite small for moderate values of n_T (e.g., $n_T \leq 4$ for practical applications). Hence, the computational complexity of SVD and matrix inverse is limited.

APPENDIX 6C: PROOF OF LEMMA 6.3

Without loss of generality, assume that we normalize the noise variance $\sigma_z^2 = 1$. At high SNR, $P_0 \to \infty$, the optimal power allocation from Equation (6.33) when $|\mathcal{A}| = Q$ is given by

$$p_k^* |\mathbf{h}_k \mathbf{w}_k|^2 = \frac{P_0 |\mathbf{h}_k \mathbf{w}_k|^2}{Q} - 1 + \sum_{j \neq k} \frac{|\mathbf{h}_k \mathbf{w}_k|^2}{|\mathbf{h}_j \mathbf{w}_j|^2} \approx \frac{P_0 |\mathbf{h}_k \mathbf{w}_k|^2}{Q} - 1 \tag{6.54}$$

Hence, from Equation (6.21), the instantaneous capacity utility with $|\mathcal{A}| = Q$ is given by

$$C_{\text{cap}}(Q) = \sum_{k=1}^{Q} \log_2\left(\frac{P_0 |\mathbf{h}_k \mathbf{w}_k|^2}{Q}\right) = \sum_{k=1}^{Q} \log_2\left(\frac{P_0 \lambda_k^{(Q)}}{Q}\right)$$

where $\lambda_k^{(Q)}$ is the largest eigenvalue of $(\mathbf{B}^*_{\mathcal{A},k}\mathbf{B}_{\mathcal{A},k})^{-1}$ $\mathbf{B}^*_{\mathcal{A},k}\mathbf{h}^*_k\mathbf{h}_k\mathbf{B}_{\mathcal{A},k}$ as illustrated in Appendix 6A. Since

$$\begin{aligned} C_{\text{cap}}(Q) - C_{\text{cap}}(Q-m) &= m \log_2(P_0) + \sum_k \log_2(\lambda_k^{(Q)}) \\ &\quad - \sum_k \log_2(\lambda_k^{(Q-m)}) + Q \log_2(Q) - (Q-m)\log_2(Q-m) \\ &\geq 0 \text{ for large } P_0 \end{aligned}$$

we have $C_{\text{cap}}(Q) \geq C_{\text{cap}}(Q - m)$ for all $Q \leq n_T$ and $m \geq 1$. Therefore, the optimizing $\mathcal{A}$ has a cardinality of n_T.

APPENDIX 6D: PROOF OF LEMMA 6.4

To minimize the worst-case outage probability, one should select A such that the instantaneous $C_{cov}(r_1, \ldots, r_K)$ is optimized. Since $C_{cov} = \min_{k\in\mathcal{A}} r_k$, we should maximize the worst selected user data rate. From Equation (6.21), $r_k = \log_2(1 + P_0|\mathbf{h}_k\mathbf{w}_k|^2)$. As illustrated in Appendix 6A, $|\mathbf{h}_k\mathbf{w}_k|^2$ is given by the largest eigenvalue of $(\mathbf{B}^*_{\mathcal{A},k}\mathbf{B}_{\mathcal{A},k})^{-1}\ \mathbf{B}^*_{\mathcal{A},k}\mathbf{h}^*_k\mathbf{h}_k\mathbf{B}_{\mathcal{A},k}$. From Problem 6.8 in Appendix 6A, we could see that the eigenvalue above is larger if the cardinality of $\mathcal{A}$ is smaller because more degrees of freedom are allowed in **u** for optimizing $|\mathbf{h}_k\mathbf{w}_k|^2$. Hence, the largest possible $|\mathbf{h}_k\mathbf{w}_k|^2$ (and therefore the largest C_{cov}) is achieved by allocating all the power to a single user at any fading slot.

APPENDIX 6E: PROOF OF LEMMA 1.10

From Lemma 6.3, we see that G_{cap} is optimized when $Q = n_T$. Hence, the first part is trivial.

On the other hand, when $Q = 1$, $G_{cap}(r_1, \ldots, r_K) = (\Sigma_k r_k = r_{k_0} = G_{cov}(r_1, \ldots, r_K)$, where $k_0 \in \mathcal{A}$. Hence, the solution of Equation (6.24) optimizes $G_{cov}(r_1, \ldots, r_K)$ during any fading slot. From Equation (6.8) and Lemma 6.2, the worst-case outage probability at any given target G_0 is given by $P^*_{out} = \Pr[G_{cov}(r_1, \ldots, r_K) < G_0] = \varepsilon[1(G_{cov} < G_0)]$, where $1(A)$ is a binary indicator function that gives a value of 1 if the event A is true and a value of 0 otherwise. Define $G^*_{cov}(r_1, \ldots, r_K)$ as the instantaneous utility value with $\mathcal{A}^* = \{k_0\}$ and $(0, \ldots, 0, p^*_k, 0, \ldots, 0)$ as the scheduling and power allocation policy. We have $G_{cov} \leq G^*$. Hence, we have $1(G_{cov} < G_0) \geq 1(G^*_{cov} < G_0)$ for every realization and therefore, $\varepsilon[1(G_{cov} < G_0)] \geq \varepsilon[1(G^*_{cov} < G_0)]$. Hence, the minimum worst-case outage is attained using the scheduling policy $\mathcal{A}^* = \{k_0\}$ and power allocation policy $(0, \ldots, 0, p^*_{k_0}, 0, \ldots, 0)$. Hence, the solution at $Q = 1$ optimizes Problem 6.3.

APPENDIX 6F: PROOF OF LEMMA 6.5

The optimal value of the objective function L^* satisfies

$$
\begin{aligned}
L^* &= \sum_{k=1}^{K}(\mu_k r_k - \lambda p_k) \\
&= \sum_{k=1}^{K}\mu_k\left[\log_2\left(\sigma_z^2/|h_k|^2 + \sum_{j\leq k} p_j\right) - \log_2\left(\sigma_z^2/|h_k|^2 + \sum_{j\leq k-1} p_j\right)\right] \\
&\quad -\sum_k \lambda\left[\sum_{j\leq k} p_j - \sum_{j\leq k-1} p_j\right] \\
&= \left(\frac{\lambda}{\ln 2}\right)\sum_k \int_{\Sigma_{j\leq k-1}p_j}^{\Sigma_{j\leq k}p_j}\left(\frac{\mu_k}{\sigma_z^2/|h_k|^2 + w} - (\ln 2)\lambda\right)dw
\end{aligned}
$$

$$= \left(\frac{\lambda}{\ln 2}\right) \sum_k \int_{\Sigma_{j \le k-1} p_j}^{\Sigma_{j \le k} p_j} u_k(w) dw$$

$$\le \left(\frac{\lambda}{\ln 2}\right) \int_0^{\infty} u^*(w) dw$$

The remaining task is to show that the upper bound can be achieved. Note that $u_k(w)$ is a monotonic decreasing function of w. Also, $\lim_{w\to\infty} \max_k u_k(w) = -(\ln 2)\lambda < 0$. Hence, if $u^*(0) > 0$, then there exists a finite $w_0 > 0$ such that $u^*(w_0) = 0$. On the other hand, if $u^*(0) = 0$, define $w_0 = 0$ and the sets

$$\mathcal{B}_k = \{w \in [0, w_0] : u_k(w) = u^*(w)\}$$

that form a partition of $[0, w_0]$. Consider the rate and power allocations:

$$r_k^* \ (\mu, \lambda) = \frac{1}{\ln 2} \int_{\mathcal{B}_k} \frac{1}{\sigma_z^2 / |h_k|^2 + w} dw \quad \forall k \in [1, K]$$

$$p_k^* = \langle \mathcal{B}_k \rangle$$

It can be seen that $\Sigma_k \mu_k r_k^* - \lambda \Sigma_k p_k^* = \frac{1}{\ln 2} \int_0^{\infty} u^*(w) dw$, and therefore the upper bound is achieved. Hence, the rate vector $(r_1^*, \ldots, r_K^*)$ and the power vector $(p_1^*, \ldots, p_K^*)$ can optimize the Lagrangian function L. Finally, we need to illustrate that the optimizing rate vector $(r_1^*, \ldots, r_K^*)$ is in the capacity region given by Equation (6.12). Observe that any two utility functions $u_i(w)$ and $u_j(w)$ can intersect at most once at some w. Since $\mathcal{B}_k$ is the set of all w where $u_k(w)$ dominates over all other u_j, this implies that the sets $\mathcal{B}_k$ must be all contiguous, that is, single intervals. Using Equation (6.30) and the assumption $|h_1| > |h_2| > \ldots > |h_K|$, we have

$$\mathcal{B}_k = \left[\sum_{j<k} p_j^*, \sum_{j \le k} p_j^*\right]$$

and using Equation (6.30), I can write the rate as

$$r_k^* = \frac{1}{\ln 2} \int_{\Sigma_{j<k} p_j^*}^{\Sigma_{j \le k} p_j^*} \frac{1}{\sigma_z^2 / |h_k|^2 + w} dw$$

$$= \log_2 \left(1 + \frac{p_j^* |h_k|^2}{\sigma_z^2 + |h_k|^2 \Sigma_{j<k} p_j^*}\right)$$

Hence, the rate vector $(r_1^*, \ldots, r_K^*)$ belongs to the capacity region defined inEquation (6.12). This completes the proof.

APPENDIX 6G: OPTIMAL POWER ALLOCATION FOR NETWORK CAPACITY UTILITY

Since r_k is a function of weights $\mathbf{w}_1, \ldots, \mathbf{w}_K$ and $(p_1, \ldots, p_K)$ and since the weights are also the functions of the admitted user set $\mathcal{A}$, we could express the utility function $G_{\text{cap}}(r_1, \ldots, r_K)$ as $G_{\text{cap}}(r_1, \ldots, r_K) = G_{\text{cap}}(\mathcal{A}; p_1, \ldots, p_K)$. Given the channel matrix $\mathbf{H}$, let G^* denote the optimal utility function given by

$$G^* = \max_{\mathcal{A},(p_1,\ldots,p_K)} G_{\text{cap}}(\mathcal{A}; p_1, \ldots, p_K)$$

and

$$G^*_{\mathcal{A}} = \max_{(p_1,\ldots,p_K): \Sigma_{k\in\mathcal{A}} p_k \le P_0} G_{\text{cap}}(\mathcal{A}; p_1, \ldots, p_K).$$

Together with the constraints $\Sigma_{k\in\mathcal{A}} p_k \le P_0$ and $p_k \ge 0$, the optimization problem of $G^*_{\mathcal{A}}$ could be expressed as

$$\max_{(p_1,\ldots,p_K)} G_{\text{cap}}(p_1, \ldots, p_K; \mathcal{A}) - \lambda\left(\sum_{k\in\mathcal{A}} p_k - P_0\right) - \sum_k \mu_k p_k \qquad (6.55)$$

where λ and $(\mu_1, \ldots, \mu_K)$ are the Lagrange multipliers with respect to the constraints. According to the Kuhn–Tucker condition of optimality, we have

$$\lambda\left(\sum_{k\in\mathcal{A}} p_k - P_0\right) = 0 \qquad (6.56)$$

$$\mu_k p_k = 0 \qquad (6.57)$$

$$\frac{\partial G}{\partial p_k} - \lambda_{\alpha_k} - \mu_k = \frac{\partial G}{\partial r_k}\frac{\partial r_k}{\partial p_k} - \lambda - \mu_k = 0 \quad \forall k: \alpha_k = 1 \qquad (6.58)$$

Observe that since $G_{\text{cap}}(r_1, \ldots, r_K)$ is convex in $(r_1, \ldots, r_K)$ and r_k is convex in p_k, the Kuhn–Tucker condition is also the sufficient condition for global maximum.

From Equation (6.56), we can see that either $\lambda = 0$ or $\Sigma_{k\in\mathcal{A}} p_k = P_0$. Observe that the optimizing power vector $(p_1^*, \ldots, p_K^*)$ is nonzero. Without loss of generality, assume $p_k^* > 0$ and hence, $\mu_k = 0$. If $\lambda = 0$, then, from Equation (6.58), we have $\partial G/\partial p_k = 0$, which is not possible from Equations (6.6) and (6.21). Hence, we have $\lambda \ne 0$, and this gives (6.33).

APPENDIX 6H: ASYMPTOTIC DISTRIBUTION OF S_K

Let $S_k = p_k X_k - \Sigma_{j\neq k}\zeta_j Y_j$ where $X_k = |\mathbf{h}_k \mathbf{w}_k|^2$ and $Y_j = |\Delta\mathbf{h}_k \mathbf{w}_j|^2$. Since conditioned on the CSIT $\{\mathbf{h}_1^b, \ldots, \mathbf{h}_K^b\}$, $\{p_1, \ldots, p_K\}$ and $\{\mathbf{w}_1, \ldots, \mathbf{w}_K\}$ are constant and $\Delta\mathbf{h}_k$ is the only random variable. Hence, X_k and Y_j are mutually dependent on each other and we cannot apply central limit theorem as usual. We shall first establish the convergence in distribution.

For notation convenience, assume $|\mathcal{A}| - 1 = N$ and define $\tilde{S}_k = \dfrac{S_k - \mu_S}{N}$ where $\mu_S = \varepsilon[S_k|\mathbf{h}^b]$. Let $\Psi_{\tilde{S}}(jv)$ be the conditional characteristic functions of $\tilde{S}$ (conditioned on the CSIT $\mathbf{h}^b$), respectively. By Taylor series expansion, we have

$$\begin{aligned}\Psi_{\tilde{S}}(jv) &= 1 + jv\varepsilon[\tilde{S}_k|\mathbf{h}^b] - \frac{v^2}{2!}\varepsilon[\tilde{S}_k^2|\mathbf{h}^b] + \frac{jv^3}{3!}\varepsilon[\tilde{S}_k^3|\mathbf{h}^b] - \ldots \\ &= 1 - \frac{v^2}{2!}\varepsilon[\tilde{S}_k^2|\mathbf{h}^b] + R(v, N)\end{aligned} \tag{6.59}$$

where R(v, N) represents the remainder which depends on the third and higher order moments of $\tilde{S}_k$. Taking the log on both sides and using the series $\log(1 + x) \approx x - x^2/2 + \ldots$, we have

$$\begin{aligned}\log\Psi_{\tilde{S}}(jv) &= \log\left(1 - \frac{v^2}{2!}\varepsilon[\tilde{S}_k^2|\mathbf{h}^b] + R(v, N)\right) \\ &\approx -\frac{v^2}{2!}\varepsilon[\tilde{S}_k^2|\mathbf{h}^b] + R(v, N) - \frac{1}{2}\left(-\frac{v^2}{2!}\varepsilon[\tilde{S}_k^2|\mathbf{h}^b] + R(v, N)\right)^2 + \ldots\end{aligned} \tag{6.60}$$

The conditional mean of S_k is given by:

$$\begin{aligned}\mu_S &= p_k\varepsilon[X_k|\mathbf{h}^b] - \sum_{m\neq k}\zeta_m\varepsilon[Y_m|\mathbf{h}^b] = p_k|\mathbf{h}_k^b\mathbf{w}_k|^2 + p_k\sigma_{\Delta H}^2 - \sum_n \zeta_m\sigma_{\Delta H}^2 \\ &\approx p_k|\mathbf{h}_k^b\mathbf{w}_k|^2 + p_k\sigma_{\Delta H}^2 - \sigma_{\Delta H}^2\, P_0\Lambda_k/\sigma_z^2\end{aligned}$$

Since $\Sigma_m p_m \leq P_0$ and $\|\mathbf{w}_m\|^2 = 1$, the term $|\mathbf{h}_k^b\mathbf{w}_k|^2$ is of the order N. Hence, we have $\frac{1}{N}\mu S$ approaches a constant independent of N.

Define $\mu_m = \varepsilon[Y_m|\mathbf{h}^b]$, $\mu_X = \varepsilon[X_k|\mathbf{h}^b]$. The conditional variance of S_k is given by:

$$\begin{aligned}\sigma_S^2 &= p_k^2\varepsilon[(X_k - \mu_X)^2|\mathbf{h}^b] + \sum_m\sum_n \zeta_m\zeta_n[\varepsilon[Y_mY_n|\mathbf{h}^b] - \mu_m\mu_n] \\ &\quad - 2p_k\sum_m \zeta_m[\varepsilon[X_kY_m|\mathbf{h}^b] - \mu_X\mu_m] \\ &= p_k^2\sigma_{\Delta H}^4 + 2p_k^2\sigma_{\Delta H}^2|\mathbf{h}_k^b\mathbf{w}_k|^2 +\end{aligned}$$

$$= p_k^2\sigma_{\Delta H}^4 + 2p_k^2\sigma_{\Delta H}^2|\mathbf{h}_k^b\mathbf{w}_k|^2 +$$

$$= \sum_m\sum_n \zeta_m\zeta_n\sigma_{\Delta H}^4\left|\mathbf{w}_m^*\mathbf{w}_n\right|^2 - 2\sigma_{\Delta H}^4 p_k\sum_m \zeta_m\left|\mathbf{w}_k^*\mathbf{w}_n\right|^2$$

$$= p_k^2\sigma_{\Delta H}^4 + 2p_k^2\sigma_{\Delta H}^2|\mathbf{h}_k^b\mathbf{w}_k|^2 + \sigma_{\Delta H}^4\left\|\sum_m \zeta_m\mathbf{w}_m\right\|^2 - 2\sigma_{\Delta H}^4 p_k\sum_m \zeta_m\left|\mathbf{w}_k^*\mathbf{w}_m\right|^2 \quad (6.61)$$

Since $\Sigma_m p_m \leq P_0$ and $\|\mathbf{w}_m\|^2 = 1$, the term $|\mathbf{h}_k^b\mathbf{w}_k|^2$ is the order N. Similarly, the terms $|\mathbf{w}_k^*\mathbf{w}_m|^2$ and $\|\Sigma_m\zeta_m\mathbf{w}_m\|^2$ are bounded by a constant independent of N. Hence, the variance $\sigma_{\tilde{S}}^2$ can be expressed as $\sigma_{\tilde{S}}^2 \approx O(N)$. In other words, the second order moment of $\tilde{S}_k$ is given by

$$\varepsilon[\tilde{S}_k^2|\mathbf{h}^b] = \frac{1}{N^2}\varepsilon\left[(S_k - \mu_S)^2|\mathbf{h}^b\right] \approx O(1/N)$$

Similarly, due to the fact that $\varepsilon[\mathbf{h}_k\Delta h_k^*|\mathbf{h}^b] = -\sigma^2\mathbf{I}$, it can be shown that the m-th order moment of $\tilde{S}_k$ given by

$$\varepsilon[\tilde{S}_k^m|\mathbf{h}^b] = \frac{1}{N^m}\varepsilon\left[(S_k - \mu_S)^m|\mathbf{h}^b\right] \approx O\left(1/N^{m-1}\right)$$

The remainder term $R(v, N)$ in (6.60) is of the order $1/N^2$ which drops faster than $\sigma_{\tilde{S}}^2$. Hence, the characteristic function $\Psi_{\tilde{S}}(jv)$ converges to

$$\Psi_{\tilde{S}}(jv) \rightarrow e^{\frac{-v^2}{2}\sigma_{\tilde{S}}^2}$$

as N increases. In other words, $\Phi_{\tilde{S}}$ converges to the characteristic function of a Gaussian random variable with zero mean and variance $\sigma_{\tilde{S}}^2$. Hence, V_k approach Gaussian distribution as N increases with mean and variance given by

$$\varepsilon[V_k|\mathbf{h}^b] = \varepsilon\left[p_k\mathbf{w}_k^*\mathbf{h}_k^*\mathbf{h}_k\mathbf{w}_k|\mathbf{h}_k^b\right] - \sum_{j\neq k}\zeta_j\varepsilon\left[\mathbf{w}_j^*\Delta\mathbf{h}_k^*\Delta\mathbf{h}_k\mathbf{w}_j|\mathbf{h}_k^b\right]$$

$$= p_k\left(|\mathbf{h}_k^b\mathbf{w}_k|^2 + \sigma_{\Delta H}^2\right) - \sum_{j\neq k}\zeta_j\sigma_{\Delta H}^2.$$

and

$$\mathrm{Var}(V_k|\mathbf{h}^b) = p_k^2\sigma_{\Delta H}^4 + 2p_k^2\sigma_{\Delta H}^2|\mathbf{h}_k^b\mathbf{w}_k|^2$$

$$+ \sigma_{\Delta H}^4\left\|\sum_m \zeta_m\mathbf{w}_m\right\|^2 - 2\sigma_{\Delta H}^4 p_k\sum_m \zeta_m\left|\mathbf{w}_k^*\mathbf{w}_m\right|^2.$$

The outage probability is given by:

$$P_{\text{out}}(r_k, \mathbf{h}^b) = \Pr[V_k < \Lambda_k | \mathbf{h}^b] = 1 - Q\left(\frac{\Lambda_k - \mu_{V_k}}{\sigma_{V_k}}\right)$$

EXERCISES

1. *Zero forcing beam-forming matrix in multi-user MISO systems* Consider a multi-users MISO system adapting zero forcing approach for downlink interference cancellation. Suppose there are K users and the channel is under quasi-static fading. Let X be the $n_T \times 1$ dimenional transmitted symbol given by:

$$X = W\sqrt{P}U$$

where U is the channel encoded symbols (having dimension $K \times 1$), P is a $K \times K$ diagonal matrix for power control and $W = [w_1, \ldots, w_K]$ is the $n_T \times K$ spatial multiplexing matrix as illustrated in figure 6.5. The received signal of user k, Y_k, is equal to

$$Y_k = \sqrt{p_k}\, h_k w_k U_k + \sum_{j \in \mathcal{A}, j \neq k} \sqrt{p_j}\, h_k w_j U_j + Z_k$$

where $\{h_1, \ldots, h_K\}$ is a realization of the $1 \times n_T$ dimensional channel fading vectors of the K users and $\mathcal{A}$ is the admitted user set. The beam-forming weights $\{w_1, \ldots, w_K\}$ are selected according to the zero forcing rule in (6.16) and (6.17).

(a) Prove that the maximum size of the admitted user set $\mathcal{A}$ is n_T.

(b) Given an admitted user set $\mathcal{A} = \{1, 2, \ldots, 10\}$, let $n_T \geq 10$ and I be the signal space spanned by the columns of $\tilde{H}_k = [h_1^* h_2^* \ldots h_{k-1}^* h_{k+1}^* \ldots h_{10}]$. Show that the dimension of the space I is 9 with probability 1.

(c) Let $B_{A,k}$ be the matrix with columns that span the null space of the interference space I. Express $B_{A,k}$ in terms of $\tilde{H}_k$ and what is the dimension of the null space.

(d) Show that the solution of the zero-forcing weight selection in (6.16) and (6.17) always can be expressed as $w_k = B_{A,k} u$.

(e) Show that the optimal SNR $|h_k w_k|^2$ is given by the largest eigen-value of $(B_{\mathcal{A},k}^* B_{\mathcal{A},k})^{-1} B_{\mathcal{A},k}^* \mathrm{h}_k^* \mathrm{h}_k B_{\mathcal{A},k}$.

2. *Cross Layer Scheduling with imperfect CSIT* Recall from the text that the imperfect CSIT H^b is defined to be

$$H^b = H + \Delta H^b$$

where ΔH^b is modeled as the zero-mean complex Gaussian CSIT error independent of H. Prove that given any CSIT h_k^b for the k-th user, the conditional distribution of the actual CSI h_k is complex Gaussian conditioned on the CSIT estimate h_k^b with mean and variance given by

$$\varepsilon[h_k|h_k^b] = \frac{1}{1+\sigma_{\Delta H}^2} h_k^b$$

$$Var[h_k|h_k^b] = \frac{\sigma_{\Delta H}^2}{1+\sigma_{\Delta H}^2} I$$

PART 2

DESIGN EXAMPLES AND APPLICATIONS

7

REVIEW OF MAC LAYER DESIGN FOR WIRELESS SYSTEMS

7.1 OVERVIEW

Multiple-access control (MAC) involves the allocation of communication bandwidth resources to multiple participating devices. The central issue is the "separation" of users at the receiver in order to achieve effective multi-point-to-point communication. From an information-theoretic point of view an interesting analytical goal is to determine the maximum capacity of a multiple-access channel (as discussed in previous chapters), but in practice there are many other engineering issues to be tackled. For one thing, a pertinent practical issue is how the participating devices should organize their transmissions, under the constraint that a practical receiver cannot afford to be of a very high complexity (e.g., a MMSE receiver may not be practicable in some systems of a certain cost). On the other hand, we have to design effective means to maintain the CSIT knowledge up-to-date at the base station. In addition, there is finite error probability associated with packet transmissions, which is ignored in the information-theoretic model. In this chapter, we focus on the MAC layer design with practical considerations such as receiver complexity and effect of packet error probability as well as the cost of maintaining CSIT knowledge.

In practice, we want to allow users to transmit data simultaneously such that their transmissions can be separated at the receiver. However, such transmission simultaneity can be manifested in time, in space, in frequency, or in all of these domains. Which form of simultaneity is preferable depends on the cost and the application of the system. Different choices of transmission simultaneity lead to different user separation schemes (i.e., different methods to

Channel-Adaptive Technologies and Cross-Layer Designs for Wireless Systems with Multiple Antennas: Theory and Applications. By V. K. N. Lau and Y.-K. R. Kwok
ISBN 0-471-64865-5

provide orthogonality). For example, in a time-division multiple-access (TDMA) system, users transmit data simultaneously from the frequency-domain perspective but are separated in the time domain. The reverse is true for a frequency-division multiple-access (FDMA) system. In a code-division multiple-access (CDMA) system, users transmit data simultaneously in both time and frequency domains, but are separated in the "code" domain. Practical systems are usually "synchronized" in that the simultaneous transmissions are structured in "rounds," which are more commonly called *frames*.

A further concern in a MAC protocol is to decide which users are "allowed" to participate in each simultaneous transmission. Specifically, the MAC protocol may need to restrict the number of simultaneous transmissions in order to provide service to each user with acceptable quality. This issue usually can be further divided into two problems: *admission control* and *scheduling*. The first problem concerns determining which group of users should be allowed to participate continuously in the system. The second problem concerns the ordering of the participants both within each round and across different rounds.

In a practical setting, the MAC protocol layer is at layer two as shown in Figure 7.1. With the engineering concerns mentioned above, the MAC protocol layer is commonly divided into two sublayers: *request collection* and *request scheduling*. The request collection sublayer, which subsumes an admission control process, deals with the mechanism of getting transmission requests from the participating users. It should be noted that even if a user is admitted into the system, it may not always have data to send. On the other hand, a busy user may have much more data to send than an average user. In other words, "requests" are usually *heterogeneous* in that an audio request may entail asking for just one unit of resources (e.g., one TDMA timeslot) while a video

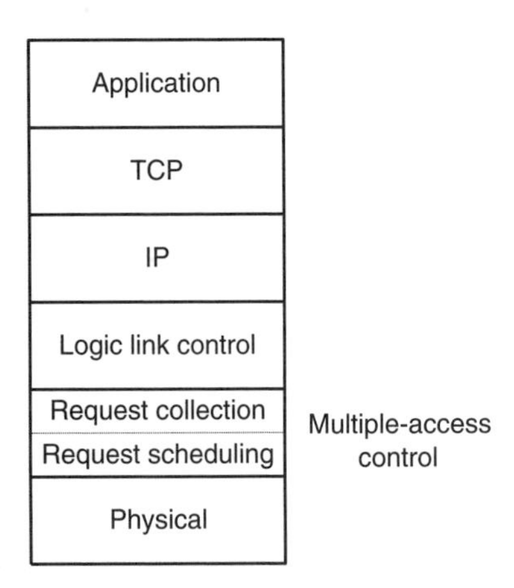

Figure 7.1. A simple practical protocol stack in which the MAC layer, divided into the request collection sublayer and the request scheduling sublayer, is situated at level 2.

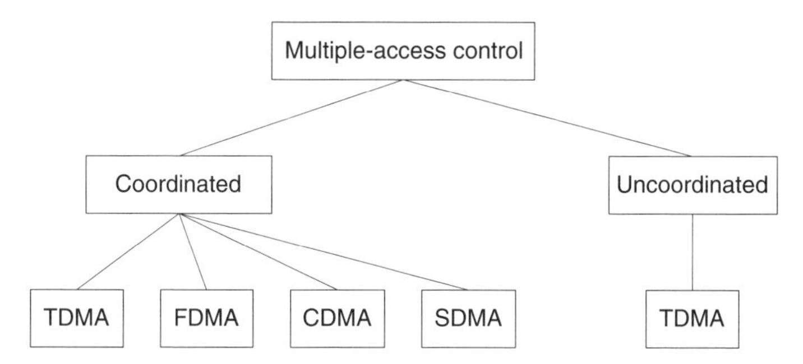

Figure 7.2. A simple taxonomy of different practical user separation methods classified as coordinated and uncoordinated approaches.

request may need a large number of resource units. Thus, the MAC protocol cannot assume to have a regular fixed requesting pattern from the participating users, as in the simplest form of a TDMA or FDMA communication system.

On the other hand, the request scheduling sublayer deals with the mechanism of selecting possibly a subset of collected requests for servicing and ordering such selected requests in the next round.

Many methods have been proposed to implement a practical MAC protocol with this two-sublayer structure. Indeed, Figure 7.2 shows a simple classification of practical MAC protocols, mainly from the "separation" perspective. Early MAC protocols are usually "uncoordinated" in the sense that there is no interaction among the participating devices to achieve the separation goal; that is, a participating device may just transmit data at will without regard to the activities of other devices. For instance, the famous ALOHA protocol, which will be discussed in Section 7.2.1, is an uncoordinated TDMA protocol.

Here, we use the word "coordination" in a broad sense—we consider a system as coordinated if the participating devices follow some "rules" with regard to other devices' activities, even though there is no centralized "coordinator." Thus, an "uncoordinated" system is not well defined in the frequency domain, in the spreading code domain, or in the space domain (e.g., sectoring of a cell). In Sections 7.2 and 7.3, we provide brief reviews of TDMA and CDMA MAC protocols, respectively.

Indeed, as detailed in the subsequent sections in this chapter, some MAC protocols rely purely on distributed coordination without using a centralized arbitration agent. However, as will be evident in our subsequent discussions, more coordination is usually better in a practical system. Most notably, in order to satisfy several conflicting goals pertinent to a useful communication system, a judicious centralized scheduler is needed. We discuss these scheduling issues in Section 7.4. In Sections 7.5 and 7.6, we describe in detail two practical jointly adaptive design examples for TDMA and CDMA systems, respectively.

7.2 TDMA-BASED MAC LAYER DESIGN

In this section, we review four well-known TDMA-based MAC protocl designs, which illustrate the mechanisms of coordination (and the lack thereof).

7.2.1 ALOHA

The ALOHA protocol [6] works by allowing each participating device (e.g., a wireless terminal in our context) to "grab" the shared transmission medium. The distributed access of the medium is completely uncoordinated in the strict sense because each device just sends a packet at will. Indeed, there is no synchronization among the devices. There is no room for scheduling in this fully distributed random contention protocol. Figure 7.3 illustrates the ALOHA protocol. Assuming a Poisson distribution for the packet arrival process, the ALOHA protocol gives a theoretical throughput of $1/2e \approx 0.18$, which is fairly low by all accounts.

7.2.2 Slotted ALOHA

A simple yet effective enhancement of the ALOHA protocol is achieved by adding synchronization—the devices attempt to "grab" the medium only at the beginning of each regular time interval, which we refer to as a *timeslot*. Figure 7.4 illustrates the slotted ALOHA protocol. Again with a Poisson packet arrival, the slotted ALOHA protocol gives a theoretical throughput of $1/e \approx 0.36$, which is still too low to be considered bandwidth-efficient.

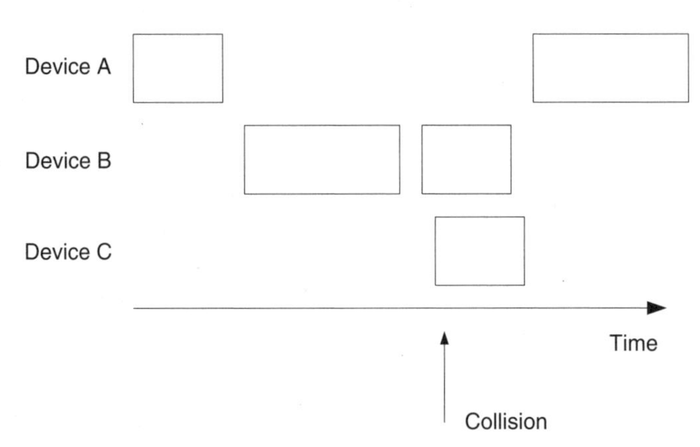

Figure 7.3. Conceptual illustration of the ALOHA protocol—devices B and C collide when they attempt to transmit at about the same time.

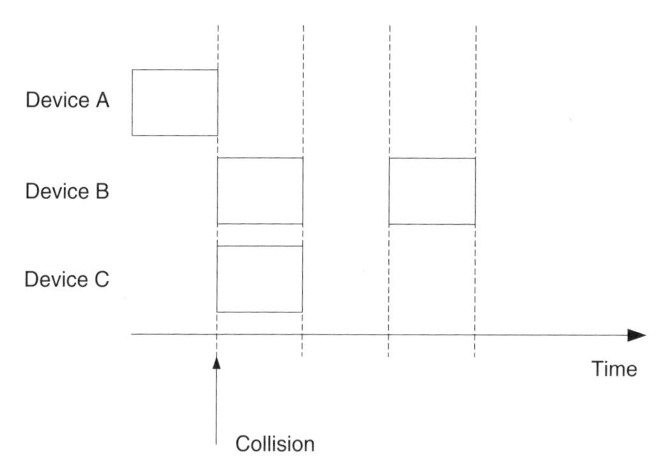

Figure 7.4. Conceptual illustration of the slotted ALOHA protocol—devices B and C collide when they attempt to transmit at exactly the same timeslot.

7.2.3 PRMA

The design of the *packet reservation multiple-access* (PRMA) protocol [48] further improves the efficiency by adding some more coordination among the devices. Specifically, all the participating devices obey the "reservation" rule—once a device has successfully grabbed a timeslot, the device automatically "reserves" the same slot in subsequent frames until the device voluntarily gives it up. Here, the "reservation" process requires strict cooperation among devices such that no device would attempt to transmit in a reserved timeslot. Figure 7.5 illustrates the PRMA protocol.

7.2.4 TDMA with Request Subframe

TDMA protocols with request subframes represent a class of MAC techniques that improve the efficiency of the reservation process in terms of resource usage [115]. As we can see in Figure 7.5, a collision in a timeslot leads to the loss of the entire slot. Although arguably the "capture" effect [67] can rescue some of the collided slots, the loss of bandwidth due to reservation collisions is still considerable. Here, the crux of the problem is that to achieve the purpose of reservation, we do not need a large timeslot. Thus, a natural improvement is to segregate the TDMA timeframe into two parts: request subframe and traffic subframe, as illustrated in Figure 7.6. In the request subframe, devices contend for the tiny request timeslots using slotted ALOHA.

Theoretically, if we make the number of request timeslots equal to that of the traffic timeslots, then we can impose a one-to-one mapping of request slots to traffic slots. Thus, a device that successfully transmits a request in the request timeslot i can then grab the corresponding traffic timeslot i in the traffic subframe.

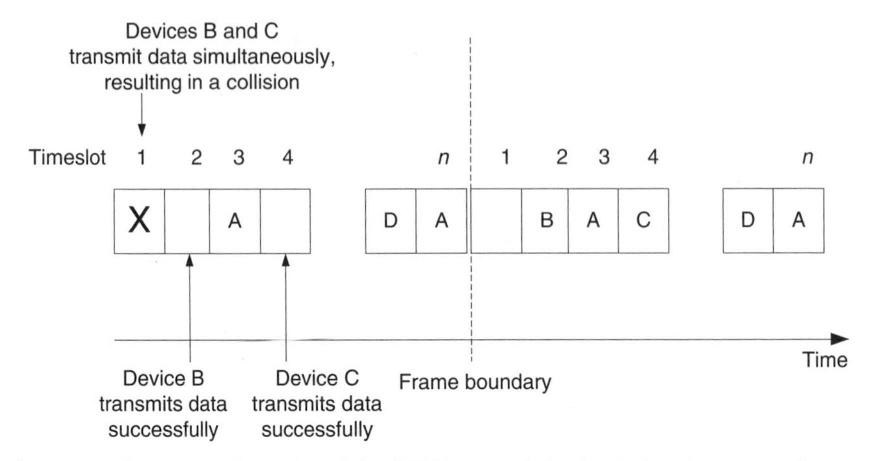

Figure 7.5. Conceptual illustration of the PRMA protocol. Device A already reserves timeslots 3 and n, while device D already reserves timeslot $n-1$. Blank boxes denote unoccupied slots. In the first frame, devices B and C successfully transmit their data, and, as such, reserve timeslots 2 and 4, respectively.

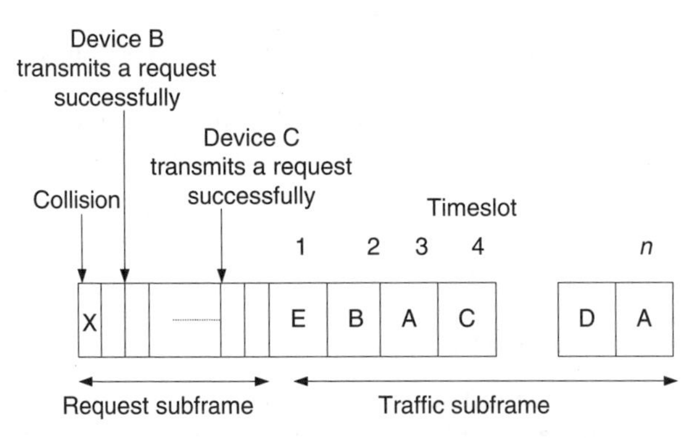

Figure 7.6. Conceptual illustration of a TDMA protocol with request subframe—devices B and C, having successfully transmitted requests in the request subframe, are assigned (via a downlink FDD or TDD frame) to transmit data in timeslots 2 and 4, respectively.

However, to make more efficient use of the request subframe, the number of request timeslots is usually significantly larger than that of the traffic timeslots. As such, an important implication of the design is that in general we need a centralized controller to "collect" the requests in the request subframe. The centralized controller, manifested as a base station or an access point, then assigns the traffic timeslots to the requests according to some criteria. In Section 7.4, we shall elaborate the issues related to the underlying scheduling algorithm that governs this traffic timeslot assignment process.

7.3 CDMA-BASED MAC LAYER DESIGN

In a CDMA wireless communication system, a MAC protocol needs to work under two systemwide constraints: (1) the downlink link power budget (in dB) and (2) the uplink interference limit [79]. Without loss of generality, we shall illustrate the design using terminologies in CDMA2000 systems. There are typically two types of bandwidth resource in a CDMA system: the *fundamental channels* (FCHs), which are intended for low-bit-rate voice services; and the *supplemental channels* (SCHs), which are used for high-bit-rate data services. Each of these "channels" is manifested as orthogonal spreading codes assigned by the centralized controller (i.e., the base station).

Let us consider that there are n high-data-rate transmission requests. In cell k, suppose that the power required to support a single SCH of request j is τ_{jk}, which can be measured at the mobile device, and its value depends on the current channel condition of the user. If the current downlink power level is Γ_k and the power budget is Γ_{max}, then we have the following constraint

$$\Gamma_k + \sum_{j=1}^{n} m_j \tau_{jk} \leq \Gamma_{max} \tag{7.1}$$

where m_j denotes the number of SCHs to be allocated to user j.

For the uplink, we need to consider two different cases: a cell k in soft handoff with the mobile device j (e.g., the host cell) and a cell k' not in soft handoff with the device j (e.g., as depicted in Figure 7.7, the data user is in soft handoff with cells 0, 1, and 2, which constitute the *active set*, but not with cell 3). In the former case, the extra interference λ_{jk} introduced by device j is given by

$$\lambda_{jk} = m_j \mu_{jk} \tag{7.2}$$

where μ_{jk} denotes the uplink received power at the base station k for one SCH and its value depends on the channel condition (which controls the processing gain required; detailed later), current total received power Λ_k, and the received bit energy:interference ratio. Here, for simplicity we use m_j again, meaning that the user request is for both uplink and downlink and is symmetric; that is, the same number of SCHs will be allocated to the user for both the uplink and downlink. Of course, this is not necessarily true in practice, and such an assumption is presented for illustration purposes. Nevertheless, the physical implication of this assumption deserves a bit more elaboration here. While previously many people considered high-data-rate services to be asymmetric (i.e., downlink is much more demanding than the uplink, as in the Web surfing situation), we believe that this assumption may not be appropriate for wireless Internet devices because a quickly emerging hot application, called Multimedia Messaging Services (an extension of the very successful Short Message Service), requires the user mobile device to be capable of transmit-

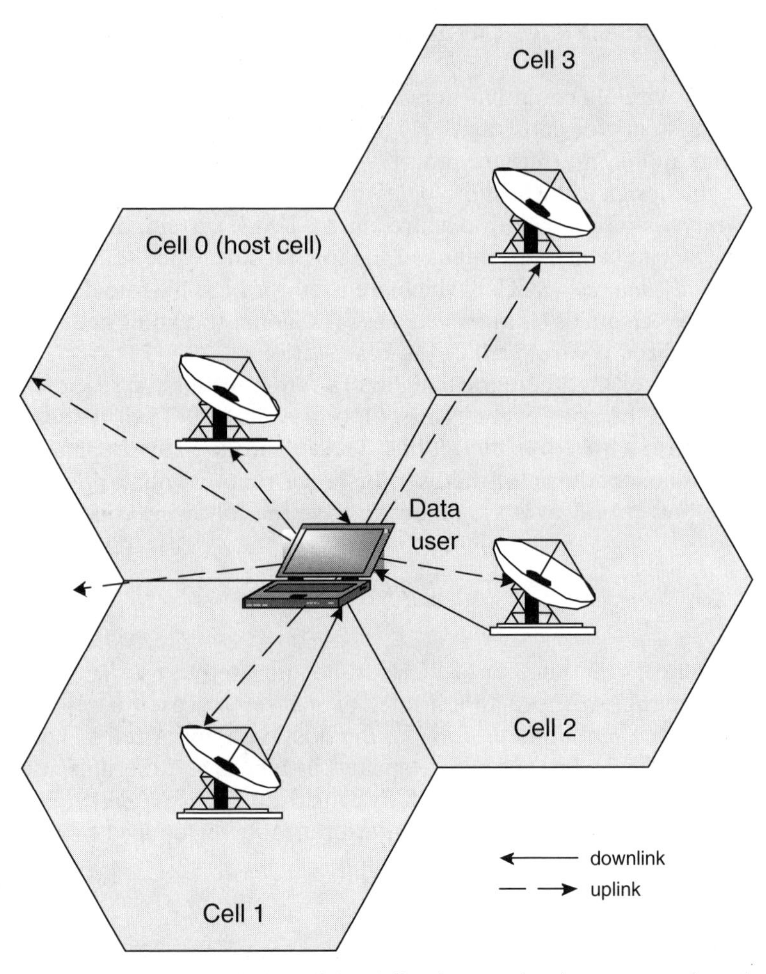

Figure 7.7. A scenario in which the soft-handoff active set of a data user consists of cells 0, 1, and 2 but not cell 3.

ting high-bandwidth data (e.g., a video clip) to the peer devices through the base station. Hence, high data rate is needed on the uplink as well.

Note that it is infeasible to obtain measurements of the received bit energy : interference ratio and thus, we need to make use of the uplink pilot strength p_{jk}^{UL}, and the transmit power ratio of the data channel and the pilot channel, denoted by ϕ_j, of request j. We have

$$\mu_{jk} = \phi_j p_{jk}^{\mathrm{UL}} \Lambda_k \tag{7.3}$$

Thus, for a cell k in soft handoff with user j, the extra interference introduced is

$$\lambda_{jk} = m_j \phi_j p_{jk}^{\mathrm{UL}} \Lambda_k. \tag{7.4}$$

For a cell k' not in soft handoff, the major difference is that it cannot obtain the uplink pilot strength p_{jk}^{UL}. However, it is practical to accurately estimate its value using the downlink pilot strength using the relative path loss (given by the ratio of the downlink pilot strength of the cell k' and a soft handoff cell k) as follows:

$$p_{jk'}^{\mathrm{UL}} = p_{jk}^{\mathrm{UL}} \frac{p_{jk'}^{\mathrm{DL}}}{p_{jk}^{\mathrm{DL}}} \tag{7.5}$$

Thus, the extra interference $\lambda_{jk'}$ is given by

$$\lambda_{jk'} = m_j \phi_j p_{jk}^{\mathrm{UL}} \frac{p_{jk'}^{\mathrm{DL}}}{p_{jk}^{\mathrm{DL}}} \Lambda_{k'}. \tag{7.6}$$

In summary, we have the following constraint for the uplink:

$$\Lambda_k + \sum_{j=1}^{n} \lambda_{jk} \leq \Lambda_{\max} \tag{7.7}$$

where λ_{jk} is computed using Equation (7.4) if k is in soft handoff with user j, and using Equation (7.6) for other cells.

We can see that we need a judicious scheduling algorithm to properly allocate the bandwidth resources in a CDMA system in order to optimize the efficiency of the system.

7.4 ISSUES IN REQUEST SCHEDULING

As we can see from the brief review of TDMA and CDMA MAC techniques above, we need to use a scheduler to arbitrate the requests from the participating devices (i.e., devices that have pending packets to be sent) so as to achieve a more efficient utilization of the bandwidth resources as well as to provide a higher quality of service (QoS) to the devices. However, these two goals are often in conflict with each other. Thus, in the literature, there is a plethora of scheduling techniques for striking a proper balance between them.

Specifically, a good scheduling algorithm in the MAC protocol needs to consider the following factors:

1. *System Throughput.* The most obvious goal is to achieve a maximum overall system throughput; that is, it is always desirable to deliver the highest throughput achievable with a given dimensioning of the system. Another advantage of using system throughput as a performance metric is that it is

easily quantifiable—usually expressed as a sum of all the mean data rates. However, although maximizing the overall system throughput is an easy target, in both theoretical analysis and real-life practice, it is usually subject to several other system or user constraints such as fairness and QoS as described below.

2. *Channel Quality*. The quality of the wireless channel for each device is inevitably different and is time-varying because each device exhibits different mobility and experiences different signal propagation conditions. A device under a better channel condition suffers fewer packet losses compared with one under a poor channel condition. Thus, from the system's perspective, it is wise to allocate more resources (e.g., more timeslots) to a device with a better channel quality because such a device delivers a higher data throughput. However, this is in conflict with the goal of providing fairness as described below. A related problem is how the scheduler can determine the channel conditions of individual users. A feasible method is to require the device to report the channel conditions to the scheduler from time to time using some feedback control channels. A critical issue is then about the accuracy and timeliness of the channel condition information. This is because outdated channel condition information can obviously lead to erroneous resource allocation.

3. *Fairness*. Strictly favoring devices with good channel conditions is obviously unacceptable to those devices suffering from poor channel conditions because, more often than not, all devices are of equal status in the system (e.g., the users pay the same service subscription fee). Thus, the scheduler needs to provide some sort of fairness. A critical question is then how fairness should be quantified. However, there is still no universally accepted notion of fairness. Different considerations lead to different definitions of fairness. Another issue is that even if we have a well-defined notion of fairness, it is not easy to map it to some metric that is quantifiable from an application perspective.

4. *Application QoS Requirements*. The application running on the wireless device may demand some specific QoS requirements. For example, a videophone application running on a multimedia cellular phone may specify a minimum video frame rate, a minimum response time, and a maximum delay jitter. Thus, given these requirements, the scheduler may need to adjust the strategy of maximizing system throughput or providing a high level of fairness. Indeed, even if a device is suffering from a poor channel condition, the scheduler may still need to allocate many resources to it because the device has very urgent packets to send.

5. *Admission Control*. Another responsibility of a scheduler is to determine whether a new device is to be admitted into the system given the current traffic demand conditions. Specifically, since the requests of many devices are recurrent in nature (e.g., video or audio services that need regular allocation of resources across different transmission frames), a scheduler has to accurately judge whether a new user should be admitted into the system because such a user would inevitably aggravate the contention among the existing users.

7.5 DESIGN EXAMPLE—A JOINTLY ADAPTIVE TDMA MAC PROTOCOL

In this section, we describe in detail a design example on jointly adaptive TDMA MAC protocols. We first present an overview of the issues involved in an advanced TDMA protocol. We then discuss the various design considerations of our example, which is called *synergistic channel-adaptive multiple access*.

7.5.1 Overview of TDMA MAC Protocol Design

Figure 7.8 depicts a partial taxonomy of TDMA-based MAC protocols. In general there are two types of TDMA-based protocols: pure TDMA and dynamic TDMA (DTDMA). In a *pure* TDMA protocol, the slot assignment, which is determined statically, is fixed throughout the communication process without regard to the actual requirements of the users. Thus, even if a user does not have any information packet to send, the assigned timeslot is still occupied and therefore is wasted. In view of the poor utilization and inflexibility, pure TDMA protocols are not used in practical wireless networks because of the dynamic user population involved.

Dynamic (DTDMA) protocols can be classified as TDD (time-division duplex)- or FDD (frequency-division duplex)-based. A TDD-based protocol can exploit the asymmetry between the uplink (heavily loaded) and downlink (lightly loaded) communications in that the bandwidth for downlink communication can be converted for uplink communication when mobile traffic load is high. However, a TDD-based system suffers from the long and irregular delay in both the uplink and downlink communications. A FDD-based scheme, by contrast, can exploit the almost instantaneous acknowledgment (i.e., short propagation delay) in both uplink and downlink transmissions, and as such,

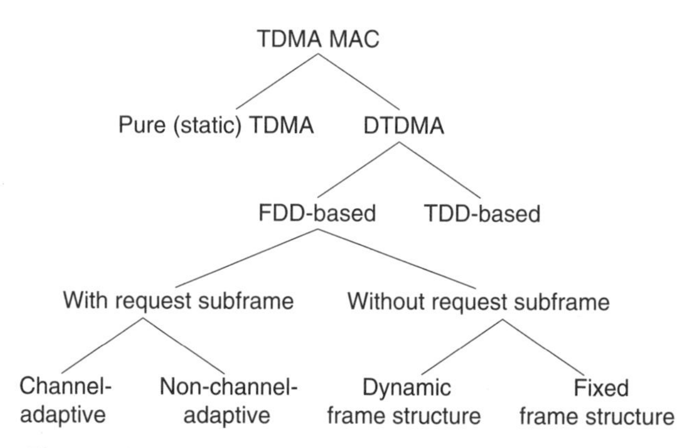

Figure 7.8. A partial taxonomy of TDMA-based MAC protocols.

the interface design is also simpler. We shall focus on FDD schemes only in this section.

Time on the FDD-based uplink channel is divided into a contiguous sequence of TDMA frames, which are subdivided into request slots and information slots. The information slots can be classified into CBR (constant-bit-rate), VBR (variable-bit-rate), and ABR (available-bit-rate) slots.

MAC protocols are classified into two categories: dynamic frame structure and fixed frame structure. In the former category, the frame structure, in terms of either frame length or frame format, is varied over time. The objective is to optimize the bandwidth usage for subframe request and information transmission.

Finally, in the fixed-frame-structure category, some more recently proposed protocols use a variable-throughput frame supported by a channel-adaptive physical layer. Such channel-adaptive protocols are motivated by the exploitation of the channel error characteristics to enhance performance. We describe in detail one such protocol in the next section.

7.5.2 Synergistic Channel-Adaptive Multiple Access

The SCAMA (synergistic channel-adaptive multiple access) [78] is based on exploiting the synergy between two protocol layers instead of strictly following the traditional information hiding protocol design paradigm. Figure 7.9 highlights the differences in the designs among a traditional fixed-throughput MAC protocol (denoted generically as DTDMA/FR) and two variable-throughput protocols (DTDMA/VR and SCAMA).

7.5.2.1 Wireless Channel Model. The wireless communication environment considered here is the reverse-link situation of a wireless system where a number of mobile terminals contend to transmit packets to a wireless server. As discussed in Chapter 1, the wireless link between a mobile terminal and the server is characterized by two components: the microscopic fading component and the macroscopic shadowing component. *Microscopic fading* is caused by the superposition of multipath components and is therefore fluctuating in a fast manner (on the order of a few milliseconds). *Macroscopic shadowing* is caused by terrain configuration or obstacles and is fluctuating in a relatively slow manner (on the order of 1–2 s). Let $c(t)$ be the combined channel fading given by

$$c(t) = c_l(t)c_s(t)$$

where $c_l(t)$ and $c_s(t)$ are the long-term macroscopic and short-term microscopic fading components, respectively. Both $c_s(t)$ and $c_l(t)$ are random processes with coherent time on the order of a few milliseconds and seconds, respectively.

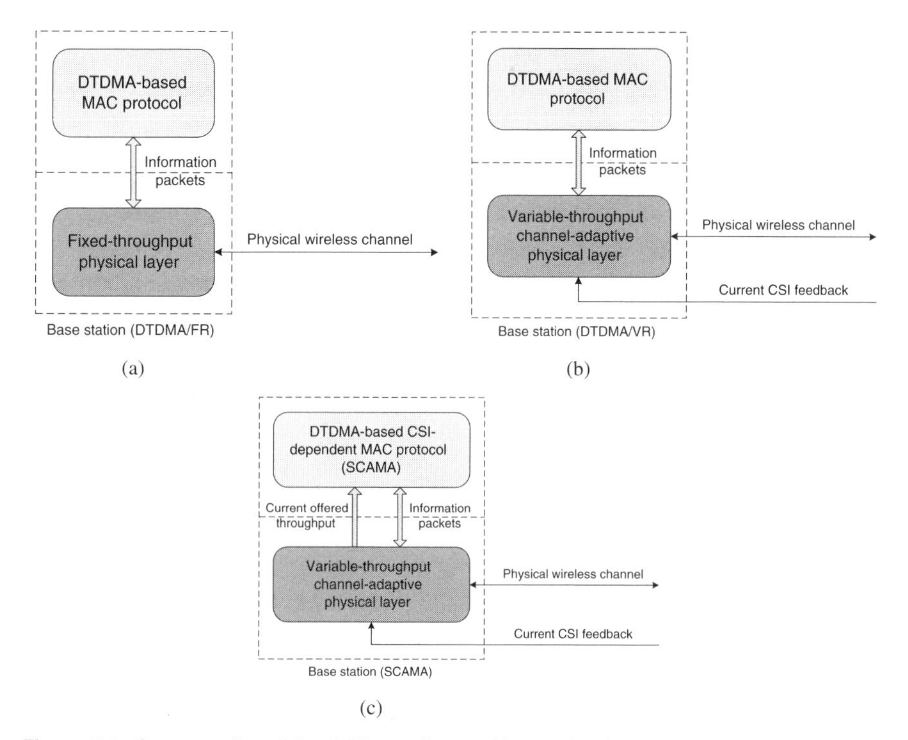

Figure 7.9. Conceptual models of different forms of interaction between the physical and MAC layers: (a) DTDMA/FR (fixed-rate channel encoder); (b) DTDMA/VR (variable-rate channel encoder); (c) SCAMA (adaptive channel encoder with MAC interaction).

Short-Term Microscopic Fading. Without loss of generality, we assume $\varepsilon[c_s^2(t)] = 1$. The probability distribution of $c_s(t)$ follows the Rayleigh distribution, which is given by

$$f_{c_s}(c_s) = c_s \exp\left(-\frac{c_s^2}{2}\right)$$

We assume the maximum mobile speed to be 20 km/h and hence, the Doppler spread $f_d \approx 25$ Hz. It follows that the *coherent time* T_c is approximately given by

$$T_c \approx \frac{1}{5 f_d} \tag{7.8}$$

which is about 8 ms.

Long-Term Macroscopic Shadowing. The long-term fading component $c_l(t)$ is also referred to as the *local mean*, which, as shown by field test measurement, obeys the *lognormal* distribution, $f_{c_l}(c_l)$

$$f_{c_l}(c_l) = \frac{4.34}{\sqrt{2\pi}\sigma_l c_l} \exp\left(-\frac{(c_l(\text{dB}) - m_l)^2}{2\sigma_l^2}\right)$$

where m_l, σ_l are the mean (in dB) and the variance of the lognormal distribution, namely, $c_l(\text{dB}) = 20\log(c_l)$. Since $c_l(t)$ is caused by terrain configuration and obstacles, the random process fluctuates over a much longer timescale. Again, from field test results, the order of timespan for $c_l(t)$ is about one second. Since mobile terminals are scattered geographically across the cell and are moving independently of each other, we assume that the channel fading experienced by each mobile terminal is independent of the fading of the other channels.

7.5.2.2 Variable-Throughput Channel Coding. Redundancy is incorporated in the information packet for error protection. To exploit the time-varying nature of the wireless channel, a variable-rate channel-adaptive physical layer is employed as illustrated in Figure 7.10a. Channel state information (CSI), $c(t)$, which is estimated[1] at the receiver, is fed back to the transmitter via a low-capacity *feedback channel.* Based on the CSI, the level of redundancy and the modulation constellation applied to the information packets are adjusted accordingly by choosing a suitable transmission mode.[2] Thus, the instantaneous throughput is varied according to the instantaneous channel state.

In this design example, a six-mode variable-rate adaptive bit-interleaved trellis-coded modulation scheme (ABICM) is employed [82]. Transmission modes with *normalized throughput*[3] varying from $\frac{1}{2}$ to 5 are available depending on the channel conditions. For realtime sources such as CBR or VBR, the physical layer employs a variable-throughput forward error correction (FEC) code. For a nonrealtime source such as ABR, the physical layer employs a variable-throughput error correction code embedded with error detection and retransmission.

Information packets per user are transmitted in the assigned traffic slots of a TDMA frame. Since the coherence time of short-term fading is around 8 ms, the CSI remains approximately constant within a traffic slot duration. Hence, all the transmitted symbols of the traffic slot (per user) share the same transmission mode, which is determined by the current CSI level. Specifically, transmission mode q is chosen if the feedback CSI, $\hat{c}$, falls within the *adaptation thresholds*, (ζ_{q-1}, ζ_q). Here, the operation and the performance of the ABICM scheme are determined by the set of adaptation thresholds $\{\zeta_0, \zeta_1, \ldots\}$.

In addition, the ABICM scheme is operated in the *constant BER* mode [82]; that is, the adaptation thresholds are set optimally to maintain a target trans-

[1] In this design example, we assume that CSI is estimated using the pilot symbol approach [82].
[2] *Transmission mode* refers to the combination of channel encoding rate and modulation constellation level.
[3] *Normalized throughput* refers to the number of information bits carried per modulation symbol.

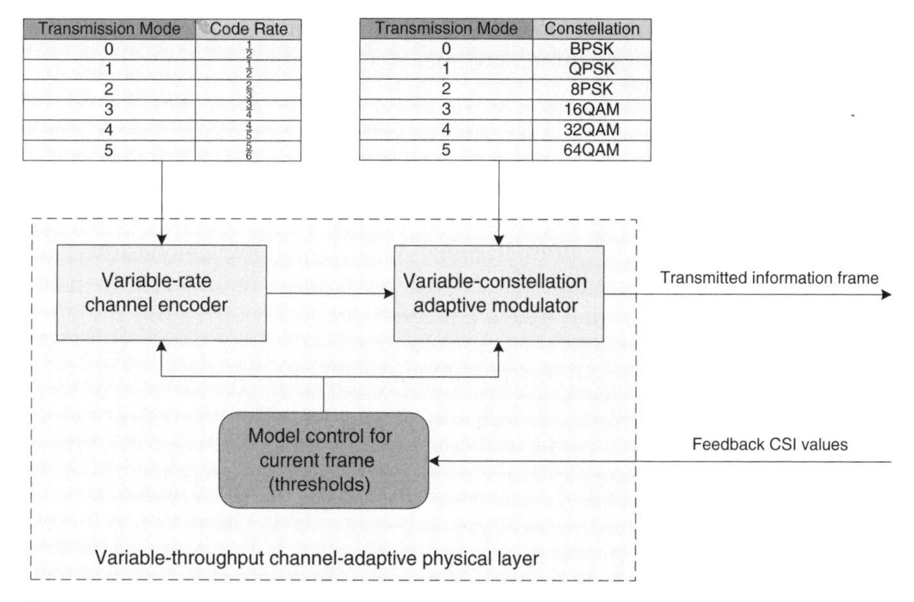

Figure 7.10. A conceptual block diagram of the variable-throughput channel-adaptive physical layer.

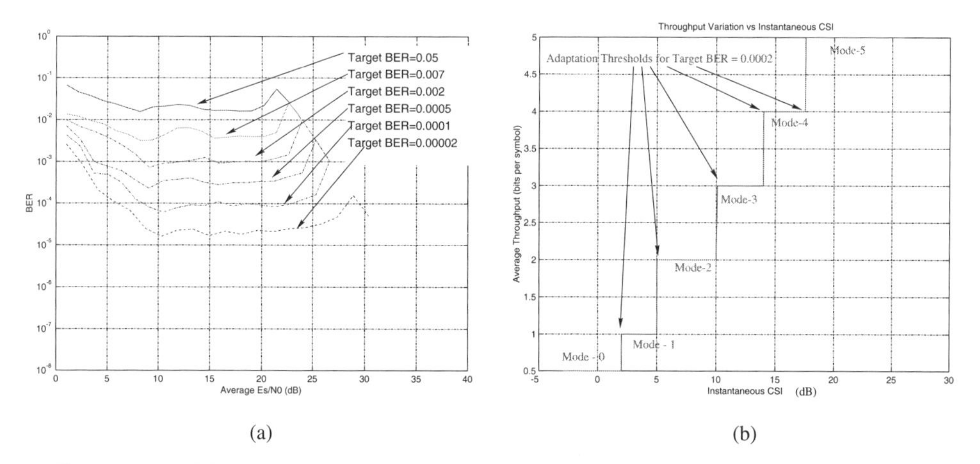

Figure 7.11. BER and throughput of ABICM scheme: (a) instantaneous BER and the adaptation range; (b) instantaneous throughput versus CSI.

mission error level over a range of CSI values. When the channel condition is good, a higher mode can be used and the system can enjoy a higher throughput. On the other hand, when the channel condition is poor, a lower transmission mode is used to maintain the target error level at the expense of a lower transmission throughput. The concept of constant BER operation is illustrated in Figure 7.11a.

Given the above mentioned considerations about the channel state, the instantaneous throughput offered to the MAC layer, ρ, is also varying and is therefore a function of the CSI $c(t)$ and the target BER P_b, denoted by $\rho = f_\rho(c(t), P_b)$. Figure 7.11b illustrates the variation of ρ with respect to the CSI at a given BER $P_b = 2 \times 10^{-4}$.

As indicated earlier, for most of the aforementioned schemes, bandwidth allocation in the MAC layer for a certain traffic type is based essentially only on a first-come first-served strategy. Some of them do not even consider CBR and VBR separately, that is, without regard to the different QoS requirements. For those MAC schemes that treat CBR, VBR, and ABR separately, no mechanism is incorporated to adjust the priority between classes in a flexible way.

Most importantly, these previous protocols do not take into account the CSI in the bandwidth allocation process. However, in a wireless communication system in which burst errors due to fading are inevitable, CSI is a critical factor in achieving a higher overall utilization of the precious bandwidth in the system.

7.5.2.3 Frame Structure. Figure 7.12 shows the uplink and the downlink frame structures of the SCAMA protocol. To incorporate the channel-adaptive feature of the SCAMA MAC protocol, the TDMA frames for the uplink and downlink are divided into subframes as follows.

In the uplink, a frame is divided into three subframes as illustrated in Figure 7.12a: the *request subframe*, the *traffic subframe*, and the *reporting subframe*. Specifically, there are N_r minislots in the request subframe for CBR, VBR, and ABR request contention. Note that an ABR user is not allowed to make reservation in the sense that even if an ABR user is granted traffic slot(s) in the current frame, it has to contend again in the next frame for the remaining data packets. On the other hand, CBR and VBR users can reserve slots in succeeding frames. Specifically, a CBR or VBR user who successfully makes a transmission request in one of the N_r minislots does not need to contend again in the next period, and the request will be automatically generated in the MAC layer until the current burst ends. There are N_i information slots in the traffic subframe for the transmission of CBR, VBR, or ABR packets. Finally, there are N_b minislots in the reporting subframe. The functions of the three subframes will be elaborated in detail later in Section 7.5.2.4. Here, the frame duration is 2.5 ms. Such a short frame duration has the advantage of shorter delay and is practicable in wideband systems.

A downlink FDD frame is similarly partitioned into four subframes: the *acknowledgment subframe*, the *polling subframe*, the *traffic subframe*, and the *announcement subframe*. The frame duration is also 2.5 ms, and the numbers of slots in the subframes are given by N_r, N_b, N_i, and N_b, respectively. The functionality and operation of each subframe are described in Section 7.5.2.4. Note that in both uplink and downlink, variable-throughput adaptive-channel coding and modulation are applied to traffic slots only. For the minislots of the other subframes, traditional QPSK modulation is applied.

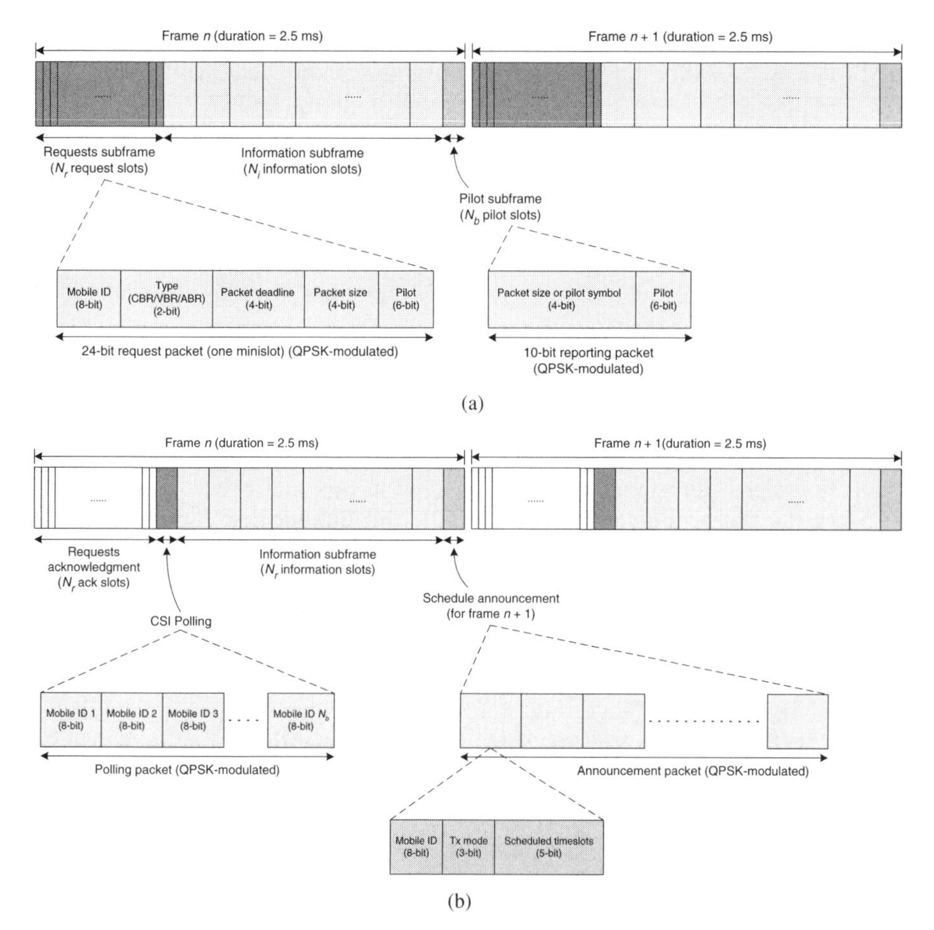

Figure 7.12. Uplink (a) and downlink (b) frame structures of the SCAMA protocol for CBR, VBR, and ABR users.

7.5.2.4 Protocol Operations. The operation of the SCAMA protocol is divided into two phases: the *request phase* and *transmission phase*. In the request phase, mobile terminals that have packets to transmit will send a request packet in one of the N_r request minislots governed by the respective permission probability. The request packet is short (24 bits[4]) and occupies only a minislot as illustrated in Figure 7.12a. It contains the mobile terminal ID, request type (CBR, VBR, or ABR), data deadline, number of information data packets desired to transmit, and pilot symbols for CSI estimation.

If more than one mobile terminal sends request packets in the same request minislot, collision occurs and all the request packets are lost if capture effect

[4]The 24-bit packet size includes the payload only. Other common header and trailer bits, such as guard bits and CRC bits, are not shown here for brevity.

is not considered.[5] After each request minislot, an acknowledgment packet will be broadcast from the base station through the acknowledgment minislot in the downlink frame as illustrated in Figure 7.13a. The acknowledgment packet contains only the successful request packet ID. Mobile terminals that fail to receive an acknowledgment will retransmit the request packet in the next request minislot, which is again governed by the permission probability. On the other hand, successfully acknowledged users will wait for announcement on the allocation schedule of the traffic slots from the base station.

Unlike the traditional MAC protocols, the base station will collect all requests in the current request phase as well as the *backlog requests* from the previous frames before allocation of traffic slots. All the requests will be assigned priorities according to the deadline, the CSI, and the service type (CBR, VBR, or ABR), as well as the waiting time of the request (i.e., the number of elapsed frames since the request is acknowledged). The time slot allocation algorithm is conceptually depicted in Figure 7.13b.

Since the physical layer offers a variable throughput that is dependent on the CSI, the rationale behind the SCAMA MAC protocol is to give higher priority to the mobile terminals that are in better channel conditions in the bandwidth allocation process. The motivation of this strategy is that a user with better channel condition, with the support of the variable-rate channel encoder, can enjoy a larger throughput and therefore, can use the system bandwidth more effectively. Nevertheless, for fairness, information slots should also be allocated to mobile terminals that are approaching their deadlines, despite their possibly worse channel states; otherwise, the queued information packets will be dropped.

7.5.2.5 Priority Function for Slot Allocation. In general, a priority function for efficient slot allocation should satisfy the following goals:

- Give priority to requests with high CSI value
- Maintain priority (i.e., prevent priority inversion) between different classes (CBR, VBR, or ABR)
- Maintain fairness (delay jitter) within each class

The SCAMA protocol employs a general priority function that provides a flexible balance of these conflicting goals. Furthermore, the slot allocation mechanism is also very flexible for incorporating other types of allocation algorithms such as deficit round robin (DRR), weighted fair queueing (WFQ), and class-based queueing (CBQ). Specifically, the *priority metric* of the ith request (which may be a new request or a backlog request), μ_i, is given by the following equation:

[5] If capture is considered, the request with the highest signal energy may be successfully received.

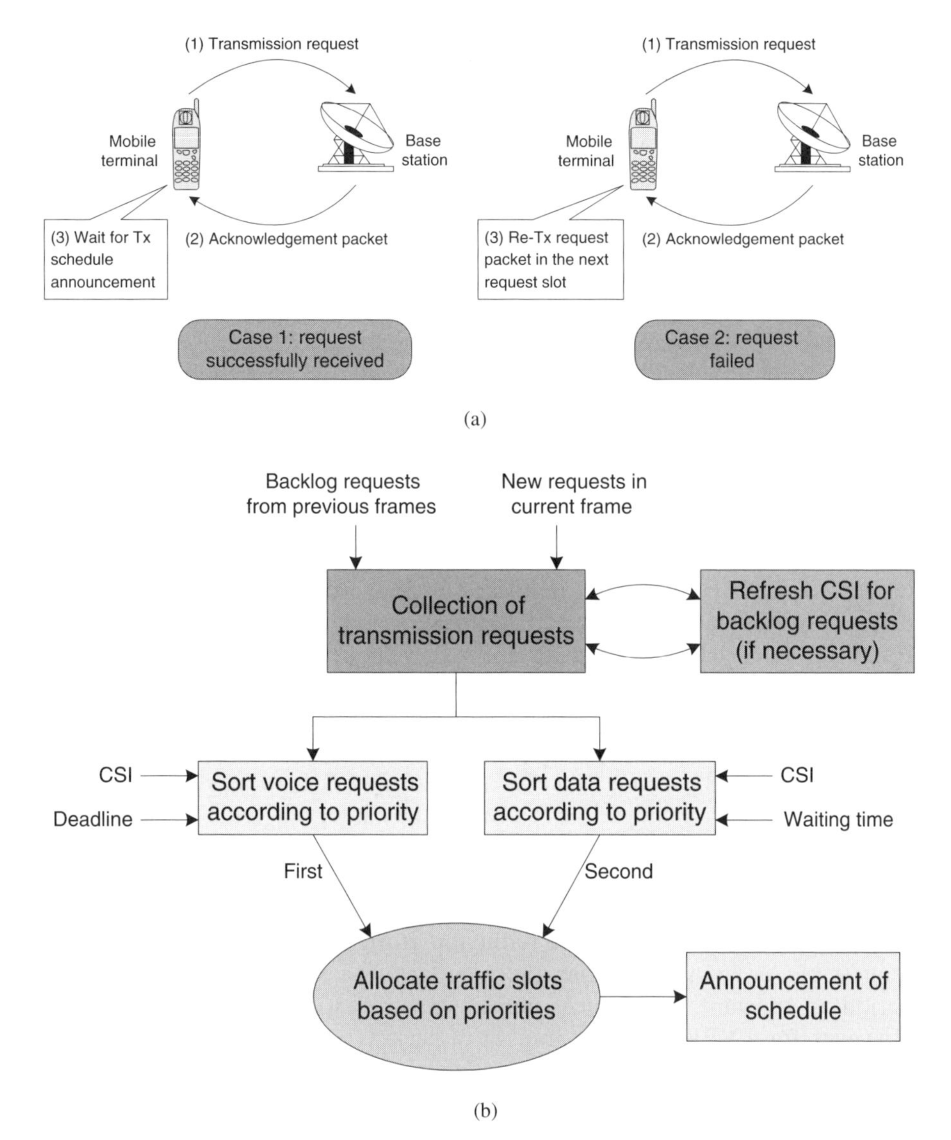

Figure 7.13. Key operations of the SCAMA protocol: (a) operation of request and acknowledgment; (b) timeslot allocation.

$$\mu_i = \begin{cases} f_\rho(\mathrm{CSI}^{(i)}) + \lambda_{\mathrm{CBR}}\left(T_d^{(i)}\right)^{-\beta_{\mathrm{CBR}}} + \Delta_{\mathrm{CBR}} & \text{for CBR request} \\ f_\rho(\mathrm{CSI}^{(i)}) + \lambda_{\mathrm{VBR}}\left(T_d^{(i)}\right)^{-\beta_{\mathrm{VBR}}} + \Delta_{\mathrm{VBR}} & \text{for VBR request} \\ f_\rho(\mathrm{CSI}^{(i)}) + \lambda_{\mathrm{ABR}}\left(T_w^{(i)}\right)^{-\beta_{\mathrm{ABR}}} & \text{for ABR request} \end{cases} \tag{7.9}$$

where $T_d^{(i)}$, $T_w^{(i)}$, λ_{CBR}, λ_{VBR}, λ_{ABR}, β_{CBR}, β_{VBR}, β_{ABR}, Δ_{CBR}, and Δ_{VBR} are the deadline, the waiting time, the *forgetting factors* of the CBR, VBR, and ABR requests, and the *priority offsets* assigned to the CBR and VBR users, respectively. From Equation (7.9), the first term is intended to enforce a higher priority for requests with a higher throughput. The objective of the second term is to maintain fairness[6] within each of the service classes. Finally, the last term is responsible for maintaining priority between different classes. The balance between the three goals could be easily adjusted by tuning λ and Δ.

Thus, in the allocation phase, traffic slots are allocated to service requests according to the sorted priority metrics. If there are insufficient traffic slots to serve all requests, remaining requests are queued and reconsidered in the next frame.[7] After the request phase, the results of traffic slot allocation will be broadcast in the *announcement subframe* of the downlink frame. The announcement packet contains the traffic slot allocation schedule as well as the transmission mode as illustrated in Figure 7.12b. Mobile terminals will then start to transmit information packets on the allocated traffic slot(s).

7.5.2.6 Handling Heterogeneous User Requirements. The SCAMA protocol is reservation-based for CBR and VBR users only. As mentioned earlier, for a ABR user, even if traffic slots have been assigned for its successfully acknowledged request, the allocation is meant only for the current frame and the ABR user needs to initiate another request for any remaining data packets. By contrast, for a CBR or VBR user, when traffic slots have been assigned for its successfully acknowledged request, additional requests will be *automatically generated* by the base station (hence, reservation) periodically at 20 ms (by taking voice as an example CBR source) and 40 ms (by taking video as an example VBR source) time intervals for CBR and VBR respectively.

Thus, the CBR or VBR user no longer needs to contend for request minislots in the current talkspurt. By avoiding unnecessary requests, the advantage of this reservation strategy is the reduction of the contention collisions in the request phase. For a CBR user, the number of packets generated per CBR period is constant and hence, this basic reservation scheme works fine. However, for a VBR user, the number of packets generated per VBR period is a random variable and hence, this information needs to be updated per VBR period in order to make proper reservation for VBR. The mechanism for VBR updating is illustrated in the following section.

7.5.2.7 CSI Determination. On the other hand, a critical component in the SCAMA MAC protocol is determination of the current CSI for each user. As mentioned earlier, we assume that the coherence time for short-term fading is around 8 ms, while the frame duration is only 2.5 ms. Thus, the CSI remains approximately constant for at least three frames. For a new request, known

[6] The second term will be large for requests with an urgent deadline or long waiting time.

[7] If the deadline for a remaining request has expired, this request will not be queued anymore and the information packet at the mobile terminal will be dropped.

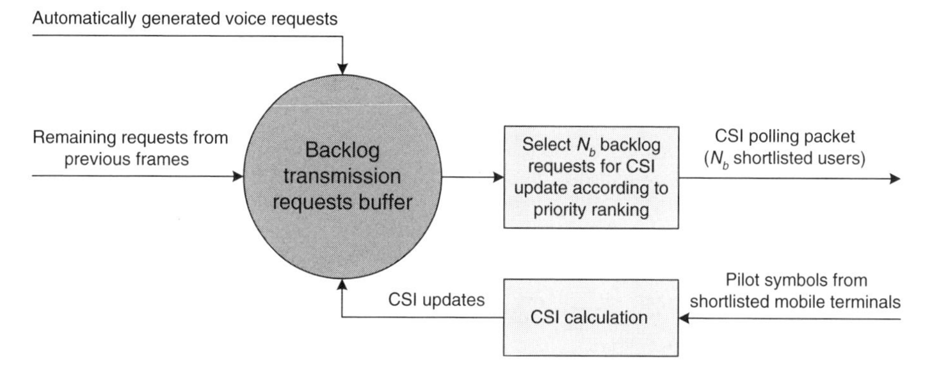

Figure 7.14. Information (CSI or VBR demands) updating mechanism for backlog requests.

pilot symbols are embedded in the request packets so that the current CSI can be estimated at the base station, and this estimated CSI is valid for the next few frame duration. However, for a backlog request, the estimated CSI value obtained previously during a previous request phase may be obsolete and thus, a mechanism is needed to obtain updated CSI.

Both the VBR reservation and the CSI update rely on a special updating procedure as illustrated in Figure 7.14.

At the beginning of each frame, the base station shortlists N_b backlog requests (those with the CSI values expired or those requiring VBR reservation update) according to their priorities. A *polling packet* is then broadcast to the mobile terminals in the *polling subframe*. The CSI polling packet contains the mobile terminal IDs that are shortlisted by the base station. The structure of the polling packet is shown in Figure 7.12b.

Mobile terminals listed in the polling packet respond at the appropriate reporting minislots according to the order specified in the polling packet. If the shortlisted request is of CBR type with CSI value expired, the mobile station will transmit pilot symbols in the reporting minislot. Otherwise, the mobile station will transmit a VBR demand update packet as shown in Figure 7.12a. The VBR demand update packet contains the number of VBR packets generated in the current frame as well as pilot symbols for CSI updating.

Thus, the base station could update the VBR reservation requests as well as backlog request CSI values (which are valid for at least two consecutive frames). The estimated CSI value is used to determine the transmission mode in the physical layer as well as the priority of the request in the MAC layer.

With the preceding design considerations, the SCAMA protocol is outlined below in pseudocode format. Simulation results reported in Reference 78 show that the SCAMA protocol is effective for handling heterogeneous QoS requirements by exploiting the channel adaptiveness.

1: **loop**
2: contention phase {requests are acknowledged but no slot is assigned}
3: {CSI of each request is recorded}
4: merge the new requests to the request queue
5: sort the request queue according to the request priority
6: assign information slots according to the request queue order
7: remaining unassigned requests are queued
8: **end loop**

7.6 DESIGN EXAMPLE—A JOINTLY ADAPTIVE CDMA MAC PROTOCOL

In this section, we describe in detail a design example on jointly adaptive CDMA MAC protocols. We first describe how channel adaptation can be realized in a CDMA network. We then present various algorithms for optimized rate allocation and request scheduling.

7.6.1 Channel Adaptation in CDMA Systems

In Section 7.3, we describe the basic framework of MAC protocol design for CDMA systems. The key quantities, τ_{jk} (power required for a downlink SCH) and μ_{jk} (interference of an uplink SCH), are critically affected by the channel conditions. Indeed, again by using the channel state information (CSI), which is the signal-to-interference ratio (SIR) as measured by the mobile device and reported to the base station (i.e., the feedback CSI as indicated in Figure 7.15), τ_{jk} and μ_{jk} can be carefully selected to maximize the achievable data rate for a certain specified bit error rate (BER). In other words, the same data rate can be achieved by using possibly smaller values of τ_{jk} and μ_{jk}.

In this CDMA design example, we use a an eight-mode VTAOC (variable-throughput adaptive orthogonal coded modulation) module [81]. The available instantaneous physical layer throughput is defined as the number of information bits carried per modulation symbol. It ranges from $\frac{12}{2}$ to $\frac{19}{2}$, and its value depends on depending on which path of convolutional encoding and Walsh–Haadamard mapping the data bitstream is to traverse as governed by the CSI. Specifically, transmission mode q is chosen from the current information bit if the CSI falls within the *adaptation thresholds* (ζ_{q-1}, ζ_q). The operation and the performance of the VTAOC scheme is determined by the set of adaptation thresholds $\{\zeta_0, \zeta_1, \ldots, \zeta_M\}$.

Similar to Section 7.5, the VTAOC scheme operates in the constant BER mode [81], in which the adaptation thresholds are set optimally to maintain a target transmission error level over a range of CSI values as illustrated earlier in Figure 7.11a. When the channel condition is good, a higher transmission mode could be used and the system can enjoy a higher physical layer throughput at a given power level. On the other hand, when the channel condition is bad, a lower transmission mode is used to maintain the target error level at

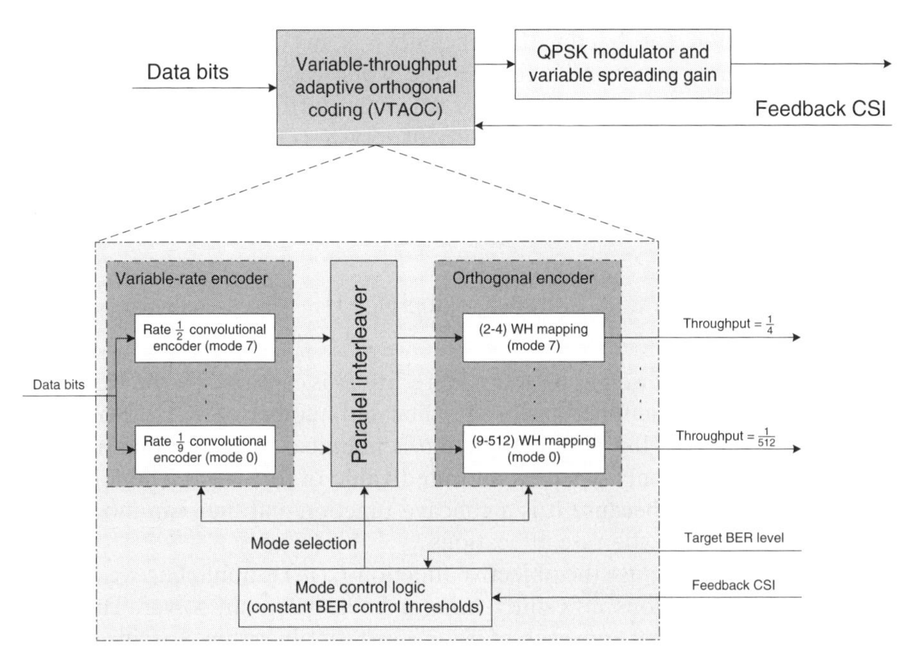

Figure 7.15. Block diagram of the variable-throughput adaptive physical layer for a CDMA system.

the expense of a lower transmission throughput. Essentially, with reference to the BER curves (against CSI) and the adaptation thresholds (as shown earlier in Figure 7.11b), the values of τ_{jk} and μ_{jk} can be proportionately scaled down or up to allocate a certain number of SCHs to a user.

In the following, we shall consider five different approaches: In 7.6.2, we consider the *throughput optimal approach* and the *near-optimal utility-based approach*. In 7.6.3, we consider the *channel-adaptive fair heuristic*. In 7.6.4. we consider the *longest queue first* and *the largest delay first*.

7.6.2 Optimal and Near-Optimal Approaches

We consider a throughput optimal approach in which the following objective function of maximizing the aggregate rate is used:

$$J_T(\vec{m}) = \sum_{j=1}^{n} m_j \tag{7.10}$$

Thus, using $J_T(\vec{m})$ together with the power and interference constraints [i.e., Equations (7.1) and (7.7)] constitutes a linear programming problem. Using a linear program solver,[8] the optimal rate vector can be computed such that the

[8] In our study, we use AMPL/CPLEX [42] as the linear programming solver.

system utilization (or bandwidth efficiency) is maximized. However, this approach requires exponential time in the worst case with time complexity $O(n^3 n!)$ [103] and thus may not be practicable. Furthermore, this approach may not necessarily give the best quality of service (QoS) to the users.

Because maximizing the total throughput may not always be desirable from the user's QoS point of view, we also consider an approach that optimizes the aggregate utility

$$J_U(\vec{m}) = \sum_{j=1}^{n} w_j \log(m_j + 1) \tag{7.11}$$

where w_j is a constant weight to user j. Here, utility is captured by the weighted sum of the logarithmic functions of the allocated number of SCHs. Note that the term 1 inside the logarithmic function is intended to ensure that a user j receives zero utility only when its allocated value of m_j is zero. The logarithmic function is used because it is a concave function and thus can model the diminishing return of the rate allocations.

Unfortunately, because the objective function $J_U(\vec{m})$ is nonlinear, we cannot obtain optimal solutions efficiently using a linear program solver. Thus, we resort to a near-optimal approach of using a genetic algorithm. Specifically, we encode each m_j as a 4-bit number[9] and then concatenate all the m_j values as a single binary string.[10]

Using such encoding, a population of randomly generated strings, each representing a different allocation, is used. Note that in our study, the population size, denoted by N_p is 2^{20}, using about 5 Mbytes of memory. As a binary encoding is used, we can apply the two genetic operators, crossover and mutation, with their usual definitions. For crossover, two strings are randomly selected from the population. A random position (the crossover point) within the bit string is chosen, and then the two strings exchange their parts at the crossover point. For mutation, a string is randomly selected and a randomly selected bit is flipped.

The invocations of crossover and mutation are governed by the crossover and mutation rates, respectively. With reference to the genetic algorithm literature [35], we set crossover probability as 0.1 and mutation probability as 0.02. The genetic algorithm is iterative in nature, and each iteration is called a *generation*.[11] In each generation, the crossover and mutation operators are applied N_p times. At the end of a generation, the *fitness* value of each string is computed.

We define *fitness* as the value of $J_U(\vec{m})$, and an invalid string (i.e., one that violates the power and interference constraints) is assigned a fitness of zero.

[9] The maximum value of m_j is only 16, but this is legitimate in practice because for a reasonable number of background load in a cell, the number SCHs that a user can obtain is less than 10.
[10] Again in practice, the number of high-data-rate requests is less than 10, and thus, the string will be around 40 bits long.
[11] In our study, the number of generations is 1000.

Using the fitness values, a selection process can be applied in that after eliminating all the zero-fitness strings from the population, new strings are generated by duplicating the remaining strings in a proportionate manner—the fittest string gets the largest share of duplication, and so on. A new generation can begin using the new population.

The time complexity of the genetic algorithm is $O(nN_p)$ because each evaluation of the fitness requires $O(n)$ time. With carefully crafted implementation, the genetic algorithm can be quite efficient in practice. The merit of a genetic approach is that it can generate near-optimal solutions [35].

7.6.3 Approaches that Maintain Fairness

The fairness concept has its roots in the design of the packet service disciplines at a router in which different sessions of packets are contending for a single outgoing link [149]. Assuming that the packets are infinitesimally divisible (i.e., the fluid model), a fair policy can be implemented by using the generalized processor sharing (GPS) approach [108]; in a round-robin manner, each packet session gets an infinitesimally small share of transmission time on the outgoing link. The shares may be weighted by the sessions' requested rates, which are determined when they are admitted in the system.

However, because in practice packets are not infinitesimally divisible and are of variable sizes, the GPS approach cannot be used in actual implementation. Thus, a variety of different approaches is devised to approximate the behavior of the GPS approach [149]. These approaches differ in time complexity and analytical performance in terms of delay.

For wireless networks, there have also been a number of more recent attempts in designing fair allocation policies. The major difference between the rate allocation problem for wireline networks and that for wireless networks is that in a wireless network, the channel quality is time-varying and location-dependent. Thus, it is possible that the channel quality of a user can be so poor that data cannot be successfully transmitted. Thus, the wireline policies cannot be directly applied in a wireless setting.

The general approach of tackling the problem is that a wireline policy is used as a reference system in which the channel is assumed to be error-free. The allocation for each user is then computed in this ideal system. In the real system, for each user we try to allocate the computed rate share to the user as far as possible subject to the constraint that the user's channel condition can support such an allocation. If a user's channel condition happens to be poor for one or more rounds of allocations, the user will then be a *lagging* user because a smaller amount of rate share will be obtained as compared to the one computed in the reference system.

On the other hand, the "surplus" rate allocations (because the users with poor channel quality cannot transmit) are shared proportionately by the users with better channel conditions. Thus, these users get more rate shares as compared to those computed in the reference system, and are called *leading* users.

The existing algorithms for fair wireless rate allocation are based mostly on this general principle. However, because these algorithms are designed using a very simplified channel model—a two-state model (either good or bad), they are not suitable for use in our model, in which the channel is accurately simulated, taking into account both fast fading and shadowing effects.

In our study, we devise a channel-adaptive fair rate allocation policy using a priority metric Q_i that is simple to implement:

$$Q_i = \gamma_i e^{-\beta\Delta_i} \tag{7.12}$$

Here, γ_i is the SIR (i.e., the channel state) of user i, Δ_i is the leading amount in terms of the actual number of information bits transmitted, and β is a scaling factor for balancing the effects of γ_i and Δ_i. We use the STFQ (start-time fair queueing) [149], which is an efficient variant of the GPS approach, as the reference allocation algorithm. Thus, the allocation policy works by using the following "filling" procedure:

1. Sort the requests in the descending order of Q_i.
2. Allocate one SCH to the first request and check the constraints. If the constraints are satisfied, repeat this step with the next request; otherwise, undo the allocation and stop.
3. Update the Q_i of all requests. Go back to the first step.

The time complexity of this above channel-adaptive fair policy is $O(\mathcal{M}n \log n)$, where $\mathcal{M}$ is the largest possible value of m_j. As mentioned before, in practice the value of $\mathcal{M}$ is quite small (less than 10) and thus, the channel-adaptive fair policy is quite fast.

7.6.4 User-Oriented Heuristics

The schemes described above are based on a more global view of the system: maximizing system throughput, maximizing aggregate utility, and maintaining systemwide user fairness. We also consider two heuristics that are more user-oriented. The first one is called *longest queue first* (LQF), which gives a higher priority to a user that has a larger backlog of packets to be sent. This heuristic is useful for mobile devices with only a limited amount of buffer space. Indeed, variants of LQF were found to be highly effective in one study [63].

The second one is *largest delay first* (LDF), which gives a higher priority to a user that has its head of line packets delayed by the longest amount of time. However, it is important to note that unlike the four approaches described earlier, these user-oriented heuristics do not make use of the channel adaptation mechanism in that users are not selected on the basis of their relative channel conditions. Thus, although a channel-adaptive physical layer (e.g., the VTAOC scheme) is still incorporated in the model, there is no synergy between the rate allocation policy and the physical layer.

TABLE 7.1. Physical Layer Parameters

Parameter	Value
Path loss exponent	4
Lognormal shadowing variance	8 dB
Channel bandwidth	5 MHz
Cell radius	800 m
Chip rate	7.7328 Mcps
Modulation	QPSK modulation with quadrature spreading
FCH rate	14.4 kbps
SCH rate	14.4 kbps–1.8432 Mbps
Processing gain (FCH)	512
Processing gain (SCH)	[512–4]
BER (FCH)	10^{-2}
BER (SCH)	10^{-4}
E_s/I_0 (FCH)	7 dB
E_s/I_0 (SCH)	13 dB
Frame duration	20 ms
Burst duration	5 frames
Average adjacent cell load	50%, 75%

7.7 PERFORMANCE COMPARISONS

Under our system model, the six rate allocation schemes described above are evaluated by simulations. We use a seven-cell environment: a center hexagonal cell with six surrounding cells. We focus on the center cell and model the surrounding cells as background load. However, when a user is situated within the soft-handoff region (which is assumed to be at a distance of 0.9 × cell radius from the base station), three neighboring cells will participate in soft handoff. In the simulation of the physical layer, we employ a path loss exponent of 4 (i.e., signals are attenuated as d^{-4}, where d is the distance of propagation). We model the Rayleigh fast fading and lognormal shadowing (with variance of 8 dB) environments. Power control is incorporated also. The other physical layer parameters are listed in Table 7.1.

We use a simple model for the mobility of the users. Each user (voice or data) selects a random starting position in the center cell, which is uniformly distributed over the cell. The direction of motion is also randomly selected. The motion is rectilinear, and a random inward direction is selected again if the user hits the boundary of the center cell. This is done to maintain the same number of users in the system throughout the whole simulation period. The speed of motion is assumed to be constant (at 35 m/h). A voice source is assumed to be continuously toggling between talkspurt and silence states. The duration of a talkspurt and a silence period are assumed to be exponentially distributed with mean 1.0 and 1.35 s, respectively. We assume that a talkspurt and a silence period start only at a frame boundary. For a high-data-rate source, we model it as generating large data files (e.g., videoclip files for multimedia messaging services). The arrival time of the file data generated by a mobile device is assumed to be exponentially distributed with mean equal to 1 s. The data size is also assumed to be exponentially distributed with mean equal to 10 kbytes. Again we assume that the data packets are generated at a frame boundary. Each test case (with a certain combination of number of voice users and number of data users) is run for 1000 s of simulation time and is repeated 10 times with different random-number seeds to obtain the average results.

We consider two aspects of performance: systemwide and user QoS. For the systemwide aspect, we measure the capacity, coverage, data request admission probability, and voice outage probability. For the user QoS aspect, we measure the average delay and average throughput. We describe these performance metrics in more detail below.

- We quantify the *capacity* of the system as the total number of SCHs available. This is an important performance parameter for the downlink. Obviously, the capacity is limited by the power and interference constraints. Because of the power level required and because the interference generated by a user critically depends on that person's distance from the base station (e.g., path loss), the capacity of the system depends highly on the geographic distribution of the users.
- *Coverage* is a performance metric closely related to capacity, but it indicates performance from another angle. Specifically, coverage can be defined as the fraction of cell area in which a data user can be served. Again, this is also an important performance parameter for the downlink. Obviously, the coverage area depends on how many SCHs are to be allocated to the data user. For example, suppose that a data user is to be allocated only 1 SCH, then he/she may successfully get allocated even if he/she is situated at the boundary of the cell. On the other hand, if the user wants to get a large number of SCHs, then that may be successful only if the user is situated near the base station.
- *Data request admission probability* is defined as the fraction of cases where a data request successfully is allocated a positive number of SCHs (note that m_j may end up to be zero in the throughput optimal approach). This metric is an important performance parameter for the uplink. It should be noted that this metric also reflects the level of QoS that the system can offer to a user with respect to how often a user's data "call" can successfully be made.
- *Voice outage probability* is an important parameter in that admitting the requests of data users should not disturb the QoS of the existing voice users (in other words, voice service is treated as the default service that should always be available). Again, this is an important performance parameter for the uplink. The outage probability is computed by calculating the fraction of frames that a voice packet cannot be successfully transmitted because the received E_s/I_0 falls below the threshold (i.e.,

7 dB). Because of the isochronous nature of voice, such a packet is useless (i.e., need not be retransmitted), and the quality of the voice service has been disrupted.

- Each data packet is *timestamped* when it is generated. Thus, the delay experienced by a packet can be calculated by subtracting this timestamp from the receipt time of the packet at the base station. Average delay is obtained by taking the mean over all delay values within the whole duration of a test case. Average throughput is computed in a similar fashion.

Figure 7.16 shows the capacity results for adjacent-cell load of 50% and 75%, which represent moderate background load and heavy background load, respectively. These results are obtained using 5 data users and 25 voice users. In the figure, we use the following abbreviations for the six rate allocation schemes: ROPT (throughput optimal using a linear program solver), UTIL (utility optimal using a genetic algorithm), CAF (channel-adaptive fair heuristic), PF (proportional fair heuristic), LQF (longest-queue-first heuristic), and LDF (largest-delay-first heuristic). As can be seen, the capacity of the system drops significantly when the data users are midway between the base station and the cell boundary. This is because the power-level requirements increase significantly as the user gets farther away from the base station. At the cell boundary, the situation is even worse, not only because of the distance effect but also soft-handoff load. Indeed, at the cell boundary, we find that there is at most one high-data-rate user who can be served (should he/she be selected) because the capacity is just around 10 SCHs. The performance ranking of the six schemes is as follows: ROPT, UTIL, CAF, PF, LQF, and LDF. In fact, the performance of LQF and LDF are very similar. ROPT clearly excels in the capacity aspect because ROPT inherently optimizes the power and interference "usage" to allocate rates to users. UTIL optimizes the logarithm of rate and thus is slightly inferior to ROPT. CAF and PF perform similarly. With a heavy background load (i.e., 75% adjacent-cell load), the performance of the system becomes remarkably worse. This is because the higher load in the surrounding cells introduces a much higher interference to the users at midcell area as well as the cell boundary.

Figure 7.17 shows the coverage results (again with 5 data users and 25 voice users, and for adjacent-cell loads of 50% and 75%). We can see that the coverage of ROPT is very much higher than in all the other schemes, including even UTIL. This is because, except for ROPT, all schemes do not pay particular attention to maximizing the number of SCHs that can be allocated with a high efficiency of utilizing the power budget. UTIL is a distant second, and the other four schemes are much worse. A scrutiny of the simulation trace reveals that all schemes, except ROPT, need to cater for one or more user-oriented constraints: utility (diminishing return), fairness, and delay. Thus, the power budget is almost never utilized to the maximum possible extent. When the background load is higher (i.e., 75% of adjacent cell load), the performance of LQF and LDF are remarkably worse. This can be explained by the

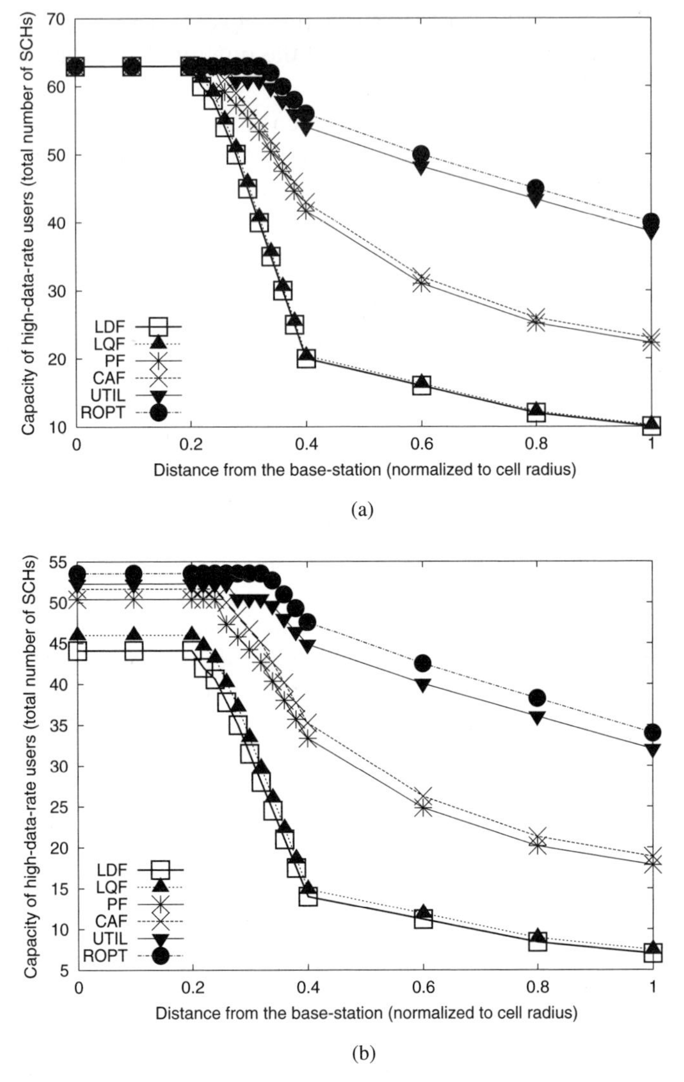

Figure 7.16. Number of SCHs available on the downlink with varying distances from the base station, with adjacent-cell loads of (a) 50% and (b) 75%.

observation that these schemes make allocation "mistakes" more frequently (i.e., allocating SCHs to a user with a poor channel condition or who is far away from the base station) under a tight power budget.

Figure 7.18 shows the uplink admission probability results (5 data users and 25 voice users; adjacent-cell load 50% and 75%). We can see that, similar to the situation of the downlink capacity, the admission probability drops significantly when the data users are midway between the base station and the cell boundary. It is interesting to note that, except for ROPT and UTIL, the admis-

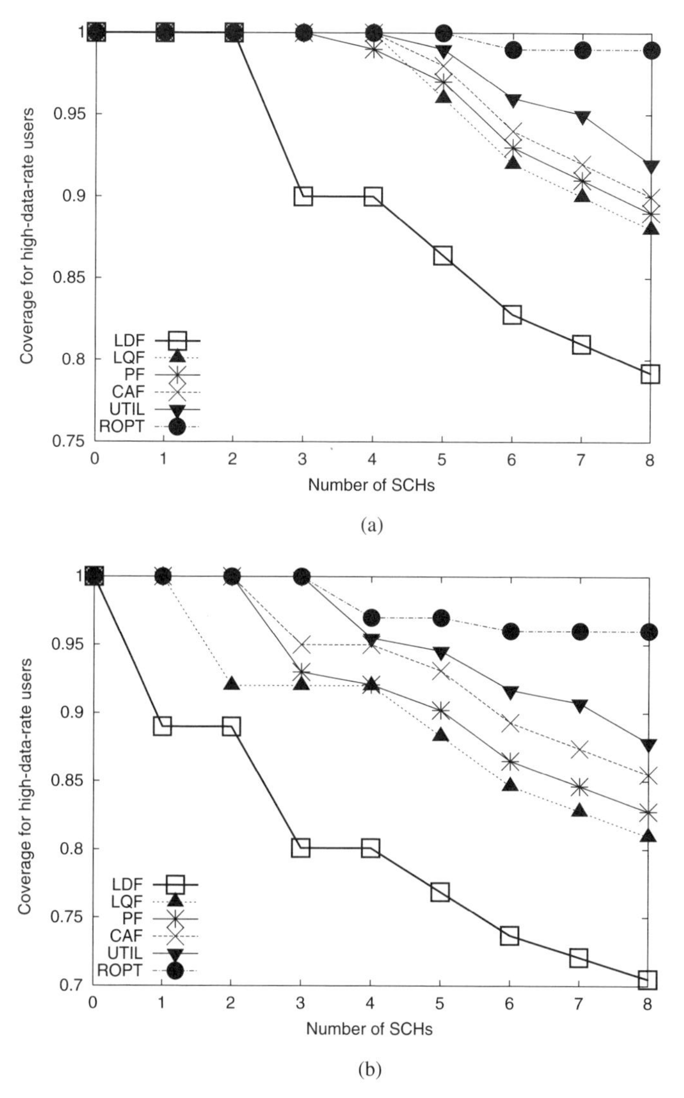

Figure 7.17. Downlink link coverage area as a function of the number of SCHs allocated, with adjacent-cell loads of (a) 50% and (b) 75%.

sion probabilities of all schemes are not close to 1 even when the data users are very near the base station. This is because in CAF, PF, LQF, and LDF, a data user with a very good position and (hence) very good channel condition does not necessarily get allocated. For example, if it happens that a severely lagging user encounters a good channel state, an even better user (in terms of channel state) may not be allocated any SCH because the latter may be leading by a large margin. The results for a heavy background load environment show similar trends.

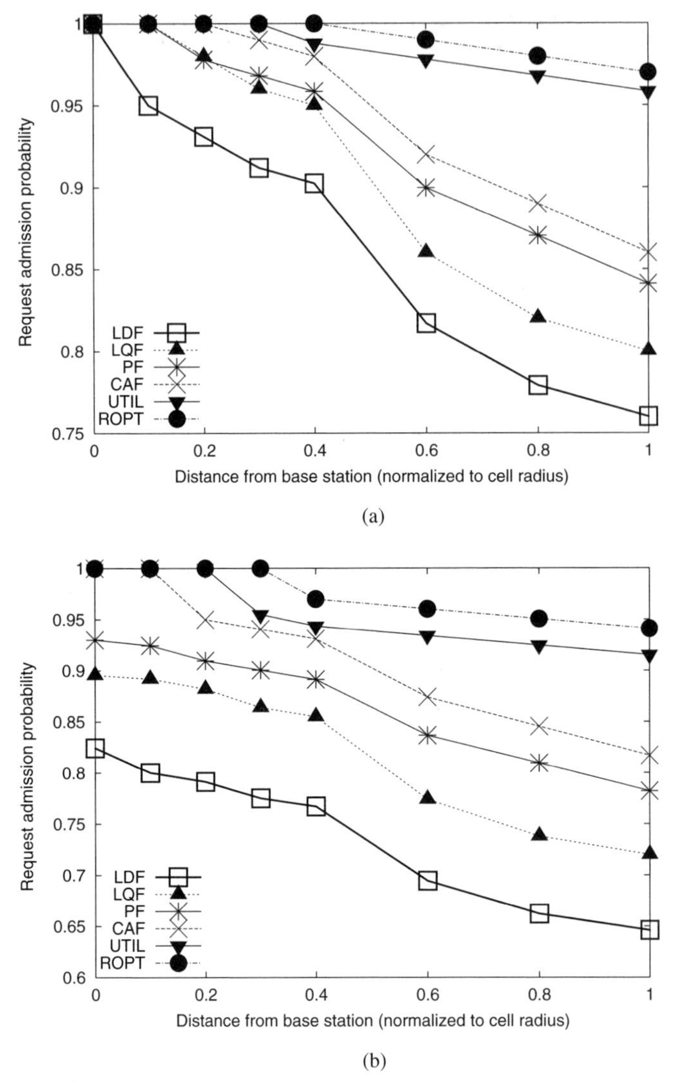

Figure 7.18. Data request admission probabilities on the uplink, with adjacent-cell loads of (a) 50% and (b) 75%.

Figure 7.19 shows the uplink outage probabilities of the voice users. These results are obtained by using 5 data users and varying number of voice users (from 15 to 33). We can see that even for ROPT, the voice user's QoS becomes unacceptable (cannot get service in 5% of the time) when there are about 30 voice users in the system. For other rate allocation schemes, the situation is even worse. The reason for this phenomenon is that the interference levels introduced by the data users and, more importantly, by the peer voice users, are quite high and the system capacity is reached most of the time. Thus, a

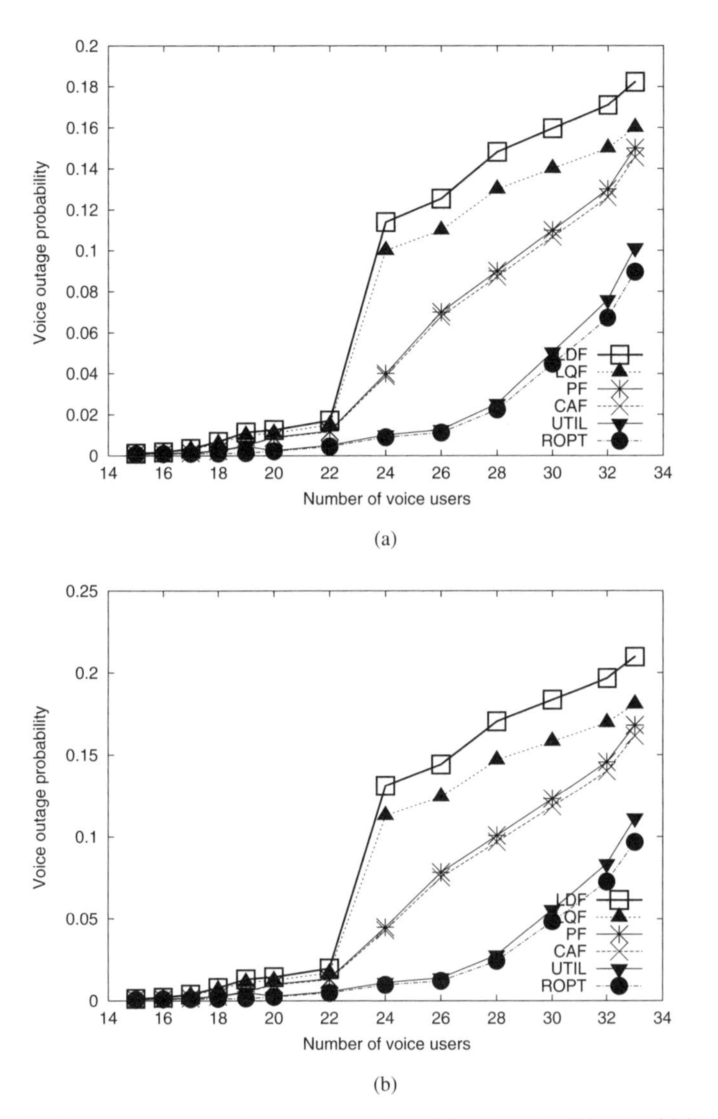

Figure 7.19. Voice outage probabilities on the uplink, with adjacent-cell loads of (a) 50% and (b) 75%.

stable operating point of our simulated system should be around 20 voice users with 4–6 data users.

Now, let us consider the user QoS aspect of the performance, in terms of average data delay and average data throughput. We use 20 voice users and vary the number of data users from 3 to 7. Moreover, we use a heavy background load of 75%. These results are shown in Figure 7.20. As expected, the average throughput of ROPT is much more higher than in the other schemes. UTIL is obviously worse than ROPT but not by a large margin. The per-

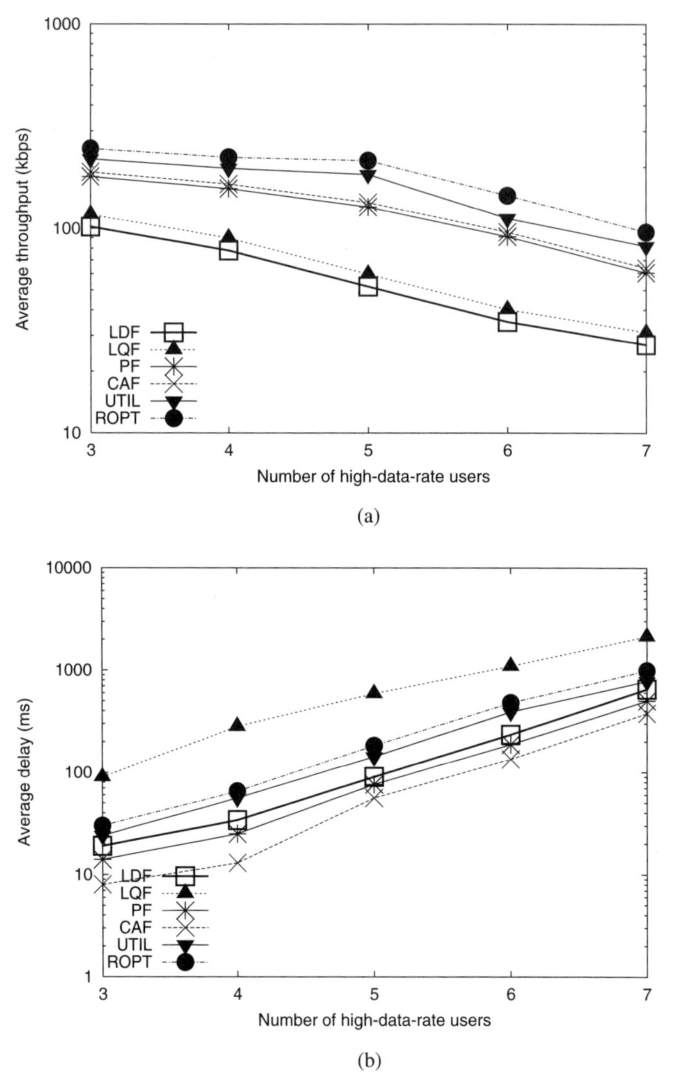

Figure 7.20. Average and throughput (a) and average delay of data requests with 20 voice users and adjacent-cell load of 75%.

formance levels of CAF and PF are very similar. Finally, the performance levels of LQF and LDF are also very close. These results by and large concur with the observations that we have seen from the systemwide results (coverage, capacity, etc.). However, it is interesting to note that for the average delay, the ranking of the schemes is quite different: CAF and PF are among the best, followed by LDF, UTIL, ROPT, and finally LQF. This phenomenon was intriguing and thus, we looked at the simulation traces very carefully. We find that the ROPT and UTIL schemes only focus on rate as they are so designed, often generating a rate allocation vector $\vec{m}$ with many zeros: effectively, some data users do not get allocation for several burst durations. A large delay thus results for such users. On the other hand, while CAF and PF do not explicitly cater for the delay metric, the action of trying to balance the service shares among the users has the effect of controlling the delay also. Of course, for many cases, the absolute values of the delays are not as low as those best cases achieved by LDF. However, the LDF scheme seldom can allocate a high rate for the data users and thus, the gain in choosing largest delay users is frequently offset by the loss in rate. Thus, LDF does not perform well overall.

7.8 SUMMARY

The major goal of this chapter is to motivate the reader to appreciate the importance of various practical issues in dealing with the multiuser communication problem. A practical MAC protocol is needed for such multiuser communication situations and is usually structured as two complementary components: request collection and request scheduling. The former component entails receiving users' requests in some way according to the user separation method used in the system, whereas the latter concerns ordering the requests so as to achieve a balanced optimization of several conflicting goals, including maximizing system throughput and providing fairness. Scheduling is a complicated and challenging problem. To illustrate the issues involved in a jointly adaptive MAC protocol design, we have described in detail two examples for TDMA and CDMA systems, respectively. In Chapter 8, we proceed to discuss the various designs of several interesting packet scheduling algorithms.

EXERCISES

7.1 What is the major goal of a MAC protocol? Under what communication environments do we need a MAC protocol?

7.2 Describe two major concerns in the design of a MAC protocol.

7.3 Describe the "coordination" achieved in the following MAC protocols: (a) slotted ALOHA; (b) PRMA; (c) TDMA with request subframe.

7.4 Describe the two major constraints that need to be handled in a CDMA MAC protocol.

7.5 Describe four major issues pertaining to the design of a MAC scheduling algorithm.

7.6 Explain why it may not be a good idea to have a variable-length frame structure in a TDMA protocol.

7.7 What would be the adverse effects of an inaccurate CSI in a channel-adaptive TDMA protocol?

7.8 Why are reservations provided for CBR and VBR services?

7.9 What are the major constraints of using a fair queueing algorithm in a CDMA system?

7.10 What are the differences between the following objective functions: (a) maximum aggregate rate and (b) proportional fair?

PROBLEM

1. *ALOHA Channels* ALOHA systems employed the simplest solution to solve the multiplexing problem by allowing every user to transmit packets whenever they want. The problem of concern with this scheme is that packets from different users may overlap and how many users can such a system serve?

This question lead us to compute the capacity of ALOHA channels. Assuming the packets come to the channel in a point poisson process with rate λ and each packet last for a time τ. The traffic introduced to the channel is

$$G = \lambda\tau$$

and the throughput of the system is the amount of packets that do not overlap with other packets. Define $\lambda' < \lambda$ to be the rate of occurrence that a packet do not overlap and the throughput S is given by

$$S = \lambda'\tau$$

(a) prove that the probability that a packet do not overlap with other packets is e^{-2G}.
hint: There is no overlap of packets if a packet do not start before τ and after τ seconds of a given packet.

(b) prove that $S = Ge^{-2G}$ and show that it is upper bounded by $\frac{1}{2e}$.

2. *Slotted ALOHA Systems* In order to improve the capacity of the ALOHA channels, a variation of the scheme is to allow users only to transmit packets at several time slots. If packets overlap, they completely overlap, not partially. Assuming users work independently to other users. Define G_i to be the probability that user i transmit packets at the slot; G to be the normalized traffic of all users

$$G = \sum_{i}^{N} G_i$$

where N is the total number of users. Denote S_i to be the probability that user i's packet get through correctly with no interference from other users.

(a) write S_i a function of G_i

(b) Assuming identical users, $S_i = \frac{S}{N}$ and $G_i = \frac{G}{N}$ and prove that $S = G$ $(1 - \frac{G}{N})^{N-1}$ where S is the normalized throughput.

(c) prove that $S = Ge^{-G}$ and is upper bounded by $\frac{1}{e}$. what is the gain in capacity in slotted ALOHA comparing to the traditional ALOHA scheme?

8

CHANNEL-ADAPTIVE WIRELESS FAIR SCHEDULING

8.1 OVERVIEW

In Chapter 7, we introduced the role of scheduling in aMAC protocol—to judiciously allocate the precious bandwidth resources to the collected requests from the request collection sublayer so as to optimize the system performance while satisfying the users' QoS requirements.

Fair queueing algorithms [149], which are the major techniques for packet scheduling, have been extensively studied in wireline networks for providing QoS guarantees in routers to connections among end hosts. More recently, as wireless networks proliferate, researchers have also devoted much effort in extending the fair queueing techniques for applications in a wireless environment. Usually, a major assumption made in these wireless fair scheduling techniques is that the channel model is rather unrealistic; the channel is either in a "good" state, in which a session (defined as an active data connection) can transmit using full bandwidth, or in a "bad" state, in which a session cannot transmit any data.

As illustrated in Chapters 6 and 7, multiuser system capacity can be enhanced significantly through cross-layer scheduling so as to exploit the time-varying nature of the wireless channels. To facilitate the multiuser selection diversity gain, both the physical layer and the MAC layer have to be jointly adaptive to the changing channel states. In the physical layer, the effective throughput is adaptive by choosing an appropriate level of FEC (forward error correction) and modulation level according to the instantaneous CSI. In the MAC layer, priority is given to the users with good channel states because these users can enjoy a higher instantaneous throughput using the adaptive physical layer. On the basis of the jointly adaptive approach, the overall system

Channel-Adaptive Technologies and Cross-Layer Designs for Wireless Systems with Multiple Antennas: Theory and Applications. By V. K. N. Lau and Y.-K. R. Kwok
ISBN 0-471-64865-5

capacity will be enhanced. However, a question remains as to what a scheduler should do in order to maintain the fairness among the sessions in the system that, more often than not, are in a "not so good" channel state. In this chapter, we focus on the notion of *fairness* in wireless systems using some design examples.

In the next section, we first provide a discussion on the scheduling model and use a motivating example to illustrate the difficulty of the scheduling problem. In Section 8.3, we describe a general design framework for a practical wireless packet scheduler. In Section 8.4, we provide a detailed discussion of the different fairness notions in wireline and wireless networks and demonstrate that a new fairness notion is needed in order to cater for the multilevel channel qualities. We then present a detailed qualitative analysis of several representative wireless fair queueing techniques in Section 8.5. The objective of the analysis is to illustrate the distinctive features in the existing approaches. Numerical examples are described in Section 8.6. Finally, channel-adaptive scheduling for CDMA systems are presented in Section 8.7.

8.2 SCHEDULING MODEL

In this section, we first describe the classical fluid fair queueing (FFQ) approach, which is based on which other scheduling algorithms are developed. We then describe the service model to be supported by a scheduler. Using a motivating example, we illustrate the difficulties involved in designing a wireless packet scheduling algorithm.

8.2.1 Fluid Fair Queueing

Fluid fair queueing (FFQ), which is also known as generalized processor sharing (GPS), is a theoretical model in providing fair sharing of channel bandwidth. Specifically, for a set of *flows* (defined as packets having the same source–destination pair) with prenegotiated service rates $r_1, r_2, \ldots, r_N$, each flow i will receive an actual bandwidth share of $\frac{r_i}{\Sigma_i ri} C$, where C is the channel bandwidth.

The theoretical good features of FFQ include bounded delay and guaranteed minimum throughput. However, from a practical perspective the major drawback of FFQ is that it is based on the unrealistic assumptions that (1) the server can serve all flows simultaneously and (2) each traffic flow is infinitesimally divisible. In practice, the atomic unit of a transmission of each flow is a single packet, the size of which can be several hundreds of bytes.

8.2.2 Service Model

FFQ provides full separation between flows, where the minimum guarantees provided for a flow are unaffected by the behavior of other flows. However,

FFQ assumes that the channel is error-free, or at the very least, that errors are not location-dependent (i.e., all backlogged flows have the ability to transmit at a given time, or none of the flows can; here, a flow is called *backlogged* if it has packets pending to be transmitted). Specifically, FFQ is neither fair nor able to provide minimum throughput bounds in the presence of location-dependent channel errors.

In order to capture the behavior of flows in a wireless environment while bearing the constraints of the channel in mind, the *error-free service* of a flow is used as the service that it would have received at the same time instant if all channels had been error-free, given identical offered load. A flow is then said to be *leading* if it has received channel allocation in excess of its error-free service. On the other hand, a flow is said to be *lagging* if it has received channel allocation less than its error-free service. A flow that is neither leading nor lagging is said to be *in sync*.

Ideally, an efficient packet scheduling algorithm should support a fair service model with the following properties:

1. Short-term fairness among in-sync backlogged flows that perceive an error-free channel
2. Short-term throughtput bounds for flows with an error-free channel
3. Channel-conditioned delay bounds for packets
4. Long-term fairness among backlogged flows with bounded channel errors
5. Long-term throughtput bounds for all flows with bounded channel errors
6. Support for both delay-sensitive and error-sensitive data flows
7. (Optional) Optimization of the schedulable region by decoupling the delay and bandwidth requirements of flows

8.2.3 A Motivating Example

In FFQ, each flow i is given a weight r_i (the negotiated service share), and for any time interval $[t_1, t_2]$ during which there is no change in the set of backlogged flows, the channel capacity granted to each flow i, denoted by $W_i(t_1, t_2)$, satisfies the following property:

$$\left|\frac{W_i(t_1,t_2)}{r_i} - \frac{W_j(t_1,t_2)}{r_j}\right| = 0 \tag{8.1}$$

Now, consider three backlogged flows during the time interval [0, 2] with $r_1 = r_2 = r_3$. Flow 1 and flow 2 have error-free channels, while flow 3 perceives a channel error during the time interval [0,1). If the scheduler is aware of the channel state of the flows, then it will not consider flow 3 during [0,1). Thus, by applying Equation (8.1) over the time periods [0,1) and [1,2], the following allocations result: $W_1[0,1) = W_2[0,1) = 0.5$, $W_1[1,2] = W_2[1,2] = W_3[1,2] = \frac{1}{3}$.

However, over the time interval [0,2], the allocations are $W_1[0,2] = W_2[0,2] = \frac{5}{6}$, while $W_3[0,2] = \frac{1}{3}$. Obviously, the overall allocation is unfair to flow 3.

This example illustrates that it is difficult to define fairness and design a suitable scheduler. The problem is that wireless fair queueing must distinguish between a nonbacklogged flow (for which no compensation is provided in fair queueing) from a backlogged flow that perceives an error channel. However, compensating for the latter will void the separation property of the fair queueing.

8.3 DESIGN FRAMEWORK

In general, the behavior of a wireless packet scheduler can be structured as consisting of the following components: (1) error-free service model, (2) lead–lag model, (3) compensation model, and (4) slot and packet queues. They are elaborated in the following text.

8.3.1 Error-Free Service Model

The error-free service provides a reference for how much service a flow should receive in an ideal error-free channel environment. Typically, the error-free service is some packetized approximation of FFQ and a commonly used algorithm is the weighted fair queueing (WFQ). In WFQ, each flow i in a set of flows is allocated a weight r_i. The kth packet p_i^k of flow i is assigned a start-time tag $S(p_i^k)$ and a finish-time tag $F(p_i^k)$ according to the following algorithm:

- $S(p_i^k) = \max\left\{V(A(p_i^k)), S(p_i^{k-1}) + \frac{L_i^{k-1}}{r_i}\right\}$, where L_i^k is the length of the kth packet of the flow i, $A(p_i^k)$ is the arrival time of the packet, and $V(t)$ is the virtual time at time instant t.
- $F(p_i^k) = S(p_i^k) + \frac{L_i^k}{r_i}$.
- $\frac{dV}{dt} = \frac{C}{\sum_i r_i}$.
- At each time, the packet with the minimum finish tag (i.e., the packet whose last bit would complete transmission first among all backlogged packets in the fluid model) is transmitted.

8.3.2 Lead–Lag Model

There are two approaches in this model:

1. The lag of a flow is computed as the difference between the error-free service and real service received by the flow. In this case, a flow that falls

behind its error-free service is compensated irrespective of whether its lost service was utilized by other flows.

2. The lag of a flow is computed to be the number of slots[1] allocated to the flow during which it could not transmit as a result of channel errors, but another backlogged flow that had no channel error transmitted in its place and increased its lead. In this case, the lag of a flow is incremented on a lost slot only if another flow that took this slot is prepared to relinquish a slot in the future.

8.3.3 Compensation Model

The purpose of the compensation component is to enable the lagging flows to reclaim service lost as a result of channel errors, and to cause the leading flows to relinquish excess service received in the past. There are several possible compensation models for leading and lagging flows.

- *No Explicit Compensation.* A lagging flow is not compensated explicitly. Rather, the scheduling proceeds according to the error-free service except that a flow perceiving channel error is skipped. As long as the load offered to the scheduler is stable (e.g., input traffic into each scheduler is policed), this approach provides long-term fairness among flows with bounded channel errors.
- *Flow with Maximum Lag Preferentially Allocated the Channel.* There are two variants to this compensation model: (1) the flow with the maximum lag is granted access to the channel whenever it can transmit (there is no explicit punishment of leading flows), and (2) the scheduler grants channel access to the flow with the minimum finish tag that perceives an error-free channel. This mechanism explicitly maintains the precedence of lagging and leading flows, but in-sync flows may also be affected by compensation of lagging flows.
- *Leading and Lagging Flows Swapping Slots.* There are several variants to this compensation model. When a leading flow is allocated a slot, it decides whether to relinquish or retain the slot according to one of three heuristics: (1) a leading flow always gives up its slots, (2) a leading flow gives up a constant fraction of its slots (i.e., the compensation is linear), and (3) a leading flow gives up a varying fraction of its slots, where the fraction of slots relinquished decreases exponentially as the size of the lead decreases.

  When a leading flow relinquishes a slot, a lagging flow is picked up according to one of three heuristics: (1) the lagging flow with the minimum finish-time tag, (2) the lagging flow with the maximum lag, and (3) a lagging flow from a weighted round-robin allocation of lagging flows where the weight of a flow is its lag. The heuristic for the leading flow to

[1] We assume that the resources to be allocated are timeslots in a frame.

relinquish its slots determines how gracefully leading flows degrade, while the heuristic for the lagging flow chosen for compensation determines how fairly lagging flows make up their lag.

- *Bandwidth Reserved for Compensation.* A fraction of channel bandwidth is statically reserved for compensation by creating a "compensation flow" and scheduling it in the error-free service along with other flows. A lagging flow reclaims additional channel access from the slots allocated to the compensation flow.

8.3.4 Slot Queue and Packet Queue

Typically, packets are tagged as soon as they arrive in wireline fair queueing algorithms. This works well if there is no channel error; thus, a scheduled packet will never be lost. However, in a wireless channel, a lost packet may need to be retransmitted for an error-sensitive flow. Retagging the packet after a transmission loss will cause it to join the end of the flow queue and thus cause packets to be delivered out of order.

Fundamentally, a separation between "when to send the next packet" and "which packet to send next" is required. The first question should be answered by the scheduler, while the second question is really a flow-specific decision and should be beyond the scope of the scheduler. In order to decouple the answers to these two questions, one additional level of abstraction can be used in order to decouple "slots" (the unit of channel allocation) from "packets" (the unit of data transmission).

When a packet arrives in the queue of a flow, a corresponding slot is generated in the slot queue of the flow and is tagged according to the wireless fair queueing algorithms. At each time, the scheduler determines which slot will get access to the channel, and the head-of-line packet in the corresponding flow queue is then transmitted. The number of slots in the slot queue at any time is exactly the same as the number of packets in the flow queue. While this description applies to the case of fixed packet sizes, the same concept can be extended to variable packet sizes as well.

8.4 FAIRNESS NOTIONS

8.4.1 Effort Fair and Outcome Fair

In a broad sense, fairness can be defined with respect to two aspects: effort and outcome [24]. Intuitively, a policy is called *effort fair* if the allocation of services to different sessions is fair, without regard to the actual amount of data successfully delivered by the sessions using the allocated services. Informally, "fair" means that a session gets the service amount that it deserves to get. On the other hand, a policy is called *outcome fair* if the actual realized data throughput among the sessions is fair.

Effort Fair. A scheduler is fair if the bandwidth (e.g., the amount of timeslots) that the system allocates to different sessions is proportional to the different service shares. Mathematically, this means that the difference between the normalized services, that the system allocates to any two sessions i and j is bounded as follows

$$\left| \frac{S_i(t_1, t_2)}{r_i} - \frac{S_j(t_1, t_2)}{r_j} \right| < \varepsilon \tag{8.2}$$

where $S_i(t_1, t_2)$ denotes the allocated service of a certain session i during time interval $(t_1, t_2]$, r_i is the requested service share, and ε is a finite constant. Such a fair scheduler can be considered as effort fair [33] in that the scheduler guarantees only that the effort expended on the sessions is fair, without regard to the actual throughput achieved by the different sessions.

Outcome Fair. A scheduler is fair if the difference between the normalized amount of realized throughput of any two sessions i and j is bounded as follows

$$\left| \frac{T_i(t_1, t_2)}{r_i} - \frac{T_j(t_1, t_2)}{r_j} \right| < \varepsilon \tag{8.3}$$

where $T_i(t_1, t_2)$ denotes the actual throughput session that i achieves during the time interval $(t_1, t_2]$. Such a fair scheduler can be considered as outcome fair [96] in that the scheduler tries to provide a fair actual performance achieved by the sessions (rather than the "nominal" performance as in the effort fair definition discussed above).

In a TDMA system, "effort" is the number of timeslots allocated, while "outcome" is the actual data throughput using the allocated timeslots. Note that a "variable actual throughput" is manifested by the fact that some data may be lost as a result of poor channel conditions, thus inducing retransmissions; or, in adaptive FEC schemes such as ABICM (discussed in Chapter 7), the amount of data protection may vary according to the channel conditions.

8.4.2 Fairness Notions for Wireline Networks

In wireline networks, the classical fairness notion is based on the generalized processor sharing (GPS) concept [108]. In GPS, the notion of fairness is defined in the following manner. Let $G_i(t_1, t_2)$ denote the throughput of session i in a given time interval $(t_1, t_2]$, where $i \in F$ and F is the set of all backlogged sessions (i.e., sessions having data pending to be sent) in the system. A scheduler is fair if and only if, for any $j \in F$ such that both session i and session j are continuously backlogged in $(t_1, t_2]$, we have

$$\frac{G_i(t_1, t_2)}{G_j(t_1, t_2)} \geq \frac{r_i}{r_j} \tag{8.4}$$

where r_i and r_j are the allocated rates of the two sessions.

Because of the high computational complexity involved in keeping track of the GPS fairness, there are many other improved variants [149] such as D-EDD, FFQ, VC, WFQ, WF2Q, and SCFQ (see Zhang [149] for a detailed survey). For all these fairness notions, one important point to note is that in a wireline network, effort fair and outcome fair are equivalent because the channel (link) state is constant.

While a fairness notion such as GPS (or its variants) works well in a wireline network, it is unsuitable for a wireless environment in which the channel quality of different sessions may vary considerably as a result of different shadowing and fading effects [109]. Specifically, using such a fairness measure, the scheduler will try to allocate the same throughput levels to different sessions.

However, this is not efficient from a resource utilization point of view because those sessions suffering from deep fading (i.e., channel quality is not good) will not be able to utilize the timeslots efficiently (e.g., data loss may occur more frequently). Essentially, in a wireless network, effort is not necessarily equal to outcome. A more intelligent method is to allow the sessions having better channel states to proceed first.

8.4.3 Fairness Notions for Wireless Networks

8.4.3.1 Overview. Much research has more recently been undertaken on devising new algorithms for fair queueing in wireless networks. Many algorithms have been proposed [24]. For an excellent discussion of wireless fairness notions and a useful unified framework, the reader is referred to the research papers by Lu et al. [96].

The general idea of wireless scheduling algorithms is as follows. The scheduler simulates an error-free system running a wireline packet scheduling algorithm when the sessions are in good channel state (at which the effective throughput is maximum). When the session that is scheduled to transmit data encounters a bad channel state, it will give up the transmit opportunity to other error-free sessions (i.e., in good channel state). These error-free sessions will give their transmit rights back to the error session to compensate when it escapes from a bad channel state.

Thus, essentially, the scheduler tries to swap the allocated timeslots between error-free sessions and error-prone sessions when sessions encounter error. Here, the goal is to hide the short-term channel error burst from the end users. The system maintains long-term fairness at the expense of instantaneous fairness between sessions.

In the literature, there are many notable scheduling algorithms designed for wireless networks: WPS (wireless packet scheduling) [96], IWFQ (idealized wireless fair queueing algorithm) [96], CIFQ (channel-condition-independent fair queueing) [106], SBFA (server-based fairness algorithm) [116], CS-WFQ (channel-state-independent wireless fair queueing) [89], ELF (effort-limited fairness) [33], PF (proportional fairness) [60], and WFS (wireless fair service)

[102]. A scrutiny of these current scheduling algorithms for wireless networks reveals that in most of these algorithms, there is one common major feature: the channel model is a two-state (good or bad) model.

In general, these previous algorithms work well in that they schedule the error-free sessions to transmit data while leaving the error sessions (in a bad channel state) waiting until their channel states become good again. Thus, to maintain fairness, it suffices to guarantee that the error sessions can catch up (i.e., get back the missing service share) within a bounded period of time. Unfortunately, usually nothing can be said about the behavior and the time bound of the error period.

Furthermore, the key assumption, which we believe is the major drawback, is that a session in a bad channel state can transmit nothing. This is undeniably an oversimplification because channel-adaptive and variable-rate physical layer protocols are commonly sought to combat the time-varying nature of wireless channels. Algorithms that use such a simplified assumption include CIFQ, IWFQ, SBFA, and WFQ.

On the other hand, the more practical algorithms, such as ELF, CS-WFQ (which uses a principle similar to that in ELF), PF, and CAFQ, allow sessions to transmit packets even though the sessions are in a poor channel state (hence, effort is very likely not equal to outcome). To illustrate the different design philosophies of these existing algorithms, we describe CIFQ, ELF, and PF in detail below.

8.4.3.2 Channel-Condition-Independent Fair (CIF). Ng et al. [106], define the notion of channel-condition-independent fair (CIF) in the following manner. To achieve CIF, a packet fair queueing algorithm should provide

- Delay and throughput guarantees for error-free sessions
- Long-term fairness for error sessions
- Short-term fairness for error-free sessions
- Graceful degradation for sessions that have received excess service

The long-term fairness for error sessions can be interpreted as a *lagging* session (a session that cannot realize its requested service rate) that enjoys an error-free channel after some time will be guaranteed to catch up with a certain time bound. The short-term fairness for error-free sessions can be interpreted as that between any two error-free sessions that are in the same status (both are leading, or both are satisfied, or both are lagging), the difference in the normalized amount of service they receive is bounded for some short time interval [106].

This notion of fairness is adapted to the wireless environment as it takes the nonperfect channel states (i.e., burst errors occur) into consideration. It allows the scheduler to delay the service of a session if it does not have a good channel state, as long as it can catch up in the long run. We can see that using

CIF, because the scheduler does not allow a session having "intermediate" (not so good or not so bad) channel states to transmit, effort is equivalent to outcome. However, this scheduling philosophy is unsuitable in a practical environment, where we can expect that a significant portion of sessions will have intermediate channel states. This can be further clarified in the following simple example.

To ease the discussion, let us use a simplified channel quality model in which there are five possible channel states: A, B, C, D, and E. Accordingly, assume that the effective throughput that can be achieved in different channel states are as follows: in channel state A, 100% of the maximum bandwidth can be realized; in channel state B, 75%; in channel state C, 50%; in channel state D, 25%; and in channel state E, 0%. Suppose that there is a mobile device currently in channel state B; what should we do to this session when we apply the CIFQ scheduler?

If we treat only channel state A as "good" in a two-state model and leave all other states as "bad," the session will receive no service. However, this is obviously inefficient and unfair because in fact the session can still transmit some (albeit a smaller amount of) data. If we treat channel state B as a "good" channel state and let the session transmit as usual, some bandwidth will be wasted because there is discrepancy between the effort and the outcome.

In CIFQ, in the short term, the system gives priority to the sessions that have a good channel state while all the other error sessions (with possibly different channel states) are simply treated as the same—not allowed to transmit. This is in fact unfair among sessions and may increase the average delay of the sessions. Specifically, the CIFQ algorithm assumes a rather ideal wireless environment in which sessions have a high probability of having a good channel state and a lagging session can be guaranteed to catch up only when it has an error-free channel afterward.

However, in reality, the time period during which a session can make full use of the bandwidth (i.e, in channel state A) is usually limited; in other words, a session can spend most of the time in channel states between the perfect (channel state A) and the worst (channel state E). An efficient scheduling algorithm should provide fairness and performance guarantee even while the channel states of the sessions are varying among different quality levels.

Moreover, the CIF notion is still not comprehensive enough in that the fairness among sessions suffering from intermediate channel states is not properly handled. We believe that, from the end user's point of view, the behavior (i.e., fairness and performance) of the sessions suffering from bad channel states is of utmost importance because it can indicate the worst-case service quality that the session can possibly get in the wireless network. This motivates a new notion of fairness as detailed in Section 8.4.4.

8.4.3.3 Effort-Limited Fairness (ELF). Using ELF [33], outcome fair is maintained among sessions unless a session has a channel state poorer than a predefined threshold. Among the sessions with channel states higher than the threshold, the normalized amount of timeslots allocated can be quite different. In order to maintain outcome fair, the scheduler allocates more timeslots (i.e., exerts more effort) on a session with a very poor channel state. Thus, outcome fair is maintained at the expense of the system throughput.

However, in order to avoid the pathological case that the poor session wastes too much of the system throughput (i.e., despite significant effort, the outcome is still not enough), a "power factor" is used to control the amount of effort exerted on such extremely unlucky sessions. In order to maintain outcome fair, the ELF approach cannot avoid wasting some bandwidth so as to achieve a fair distribution of realized throughput to the sessions with poor channel conditions.

8.4.3.4 Proportional Fairness (PF). Designed for HDR (high-data-rate) services in CDMA systems, *proportional fair* [60] (PF) is considered to be a simple yet effective fairness notion. Specifically, based on the utility based concept[2] introduced by Kelly [66], HDR (downlink) scheduling is performed in a TDMA manner (i.e., only one user is selected for high-data-rate transmission in each burst session).

At the scheduling time t, suppose that a session i has an average realized throughput $H_i(\tau)$ over a past time window of length τ (i.e., from time $t - \tau$ to t), and the real throughput that can be achieved by session i at time t is $\lambda_i(t)$. A scheduler is said to achieve proportional fair if it selects for transmission the session with the largest value of

$$\frac{\lambda_i(t)}{H_j(\tau)} \tag{8.5}$$

Furthermore, the notion of proportional fairness has the nice property that a proportional fair allocation cannot be replaced by any other arbitrary allocation that does not lead to a reduction in the aggregate fractional rate change.

It should be noted that a proportional fair scheduler heuristically tries to balance the services of the sessions in terms of outcome, while implicitly maximizing the system throughput in a greedy manner. Obviously, the proportional fairness notion is a purely outcome fairness metric. Thus, while such a metric is simple to use, proportional fairness does not guarantee fairness in a strict sense. For example, consider the situation where a session has experienced a prolonged period of poor channel states and hence has a small value of $H_i(\tau)$. It may not get service even though its channel condition improves [e.g., with a moderately large value of $\lambda_i(t)$] if there is a "dominant" session that has a very good channel state [i.e., a very large value $\lambda_i(t)$]. Furthermore,

[2] As illustrated in Chapter 6, *proportional fair utility* is defined as a logarithmic function of the rate allocated to a user. Because of the convex nature of the logarithmic function, diminishing return is modeled.

the delay experienced by sessions can also be very high in a proportional fair system.

8.4.4 Channel-Adaptive Fairness

There is a relatively new notion of fairness called *channel-adaptive fairness* (CAF). Specifically, a scheduler is channel-adaptive fair if in the short term the difference between the normalized throughput (normalized with respect to the channel capacity) of any two backlogged sessions i and j is bounded as follows

$$\left|\frac{T_i(t_1,t_2)}{r_i f(\Phi_i)} - \frac{T_j(t_1,t_2)}{r_j f(\Phi_j)}\right| < \varepsilon \tag{8.6}$$

where Φ_i denotes the channel state (e.g., one of the five classes A, B, C, D, and E), and $f(\Phi_i) = M(\Phi_i)^\eta$, in which $M(\Phi_i)$ is the *effective throughput factor* $(0 \leq M(\Phi_i) \leq 1)$. The effective throughput factor is channel-state-dependent. For example, $M(\Phi_i) = 0.75$ if Φ_i is channel state B. Here, η is a *punish factor*, which is a positive number. Thus, in our definition of fairness, the throughput that a session receives will be proportional to its channel quality and outcome fair is maintained among all sessions in the long term.

This fairness notion considers explicitly different channel states. Unlike the CIFQ algorithm that prevents the sessions without excellent channel state from transmitting, and unlike the ELF algorithm that distributes the normalized amount of service inversely proportional to their channel states, a CAF scheduler provides transmission opportunities to all sessions that do not suffer from the worst channel state in the short term, and at the same time, it punishes the sessions without good channel states to a different extent. Furthermore, unlike the proportional fair scheduler, using the CAF scheduler does not necessarily schedule the session with the best channel condition to transmit first. With the channel-adaptive fairness, we can formalize a new fair queueing algorithm that is explained in detail in the following section.

The punish factor η can help us decide between making use of bandwidth more efficiently and treating every session more fairly. When a larger value of η is used, we punish the nonperfect channel state sessions that transmit packets more seriously and prevent them from wasting too much bandwidth. In effect, the bandwidth is used more efficiently, the average delay of the total system is decreased, and the throughput is increased. However, if there is a session that is more unlucky than the others and has a higher probability of having a bad channel state, its average delay and throughput may be very bad because it is punished seriously and prevented from occupying the bandwidth.

When a smaller punish factor is used, this kind of unlucky session will be punished only moderately and both the average delay and throughput of these sessions will be reduced. But as they have more chance to access the bandwidth and hence incur a larger wastage of bandwidth, the total throughput and

TABLE 8.1. Qualitative Comparison of Fairness Notions

Fairness	Short-Term	Long-Term
CIF	Short-term fairness is maintained only among sessions with perfect channel states; neither outcome fair nor effort fair is considered for sessions with "not so good" channel states	Outcome fair, provided the sessions are under homogeneous error characteristics in the long run
ELF	Outcome fair is maintained among sessions with channel states better than a predefined threshold	Not precisely defined
PF	Short-term fairness is not precisely maintained	Not precisely defined
CAF	Short-term fairness (normalized by channel states) is maintained among all the sessions unless the session has the worst channel state; a compromise is achieved in attaining outcome fair and efficient bandwidth usage	Outcome fair, provided the sessions are under homogeneous error characteristics in the long run

average delay of the system will be adversely affected. Thus, the punish factor can be used to tune the utilization of system resources.

8.4.5 Comparison with Other Fairness Notions

The comparison among different existing fairness notions is summarized in Table 8.1.

8.5 WIRELESS PACKET SCHEDULING ALGORITHMS

8.5.1 Idealized Wireless Fair Queueing (IWFQ)

IWFQ uses WFQ for its error-free service. Each arriving packet is tagged as in WFQ, and the service tag for a flow is set to the finish tag of its head-of-line packet. Among the flows that can transmit, namely, backlogged flows with an error-free channel, the flow with the least service tag is picked and the head-of-line packet is transmitted. IWFQ also simulates error-free service for identical arrivals.

The lead of a leading flow is the difference between the service tag of the flow and that of the flow in the error-free simulation, upper-bounded by a per-flow parameter. The lag of a lagging flow is the difference between the service tag of the flow in the error-free simulation and that of the flow in the real system, upper-bounded by Br_i, where B is the scheduler parameter and r_i is the normalized weight of the flow (i.e., $\Sigma r_i = 1$).

The compensation model implicitly favors channel access for lagging flows. Since the precedence of tags is maintained, a lagging flow has a low service tag and captures the channel whenever it perceives an error-free channel. Among lagging flows with clean channels, the flow with the lowest tag gets to transmit until either it perceives a dirty channel or its finish tag is greater than that of some other flows with clean channels. This compensation model guarantees that lagging flows will catch up (compensate for) their lag but may starve out leading flows in the short term.

8.5.2 Channel-Condition-Independent Fair Queueing (CIFQ)

CIFQ uses start-time fair queueing (STFQ) as the error-free service. STFQ is an approximation of WFQ that eliminates the $\frac{dV}{dt}$ computation complexity by setting $V(t)$ to the start tag of the transmitting packet. As in IWFQ, CIFQ simulates an error-free service. The lag of a flow is the difference in service between the error-free service and the real service (i.e., lead is negative lag).

A flow is considered to be "active" if it is either leading or backlogged. The error-free service is applied among all active flows. If a backlogged leading flow is allocated a slot, it relinquishes the slot with a probability of α (a system parameter). If an nonbacklogged leading flow is allocated a slot, it relinquishes the slot. A relinquished slot is allocated to the lagging flow with the maximum normalized lag.

Lagging flows receive additional service only when leading flows relinquish slots. These relinquished slots are given to the lagging flow with the maximum normalized lag where the normalization is done using the rate weight of a flow. As a result of this compensation policy, in-sync flows are not disturbed if lagging flows can receive the additional service, and leading flows degrade their service gracefully. However, in pathological cases, a lagging flow may capture the channel as in IWFQ and starve out other flows.

8.5.3 Server-Based Fairness Approach (SBFA)

SBFA provides a generic framework for adapting different service disciplines to the wireless domain, although the properties satisfied by the service discipline in the wireline domain may not be translated to the wireless domain. SBFA reserves a fraction of the channel bandwidth statically for compensation by specifying a virtual compensation flow. If a backlogged flow is allocated a slot but cannot transmit because of channel errors, it enqueues a slot request in the compensation flow.

The error-free service serves the compensation flow along with the other packet flows. When the compensation flow is allocated a slot, it hands over the slot to the flow to which its head-of-line slot request belongs. SBFA does not have the concept of a leading flow. The lag of a lagging flow is the number of slot requests in the compensation flow.

Since compensation flow is treated like any other flow by the error-free service, in-sync flows are not affected. However, when there are also slots in the compensation flow, its bandwidth is shared by all flows perceiving excellent channels at that time instant. Thus, in-sync flows receive excess service in this scenario. Lagging flows share the compensation flow, and hence the rate of aggregate compensation received is statically bounded by the reserved share of the compensation flow. Head-of-line blocking of the compensation flow is not prevented.

Since the lead of a leading flow is not monitored, leading flows do not give up their lead. SBFA is fundamentally different from the other aforementioned algorithms because it statically reserves a fraction of the channel for compensation. Therefore, all the bounds supported by SBFA are only with respect to the remaining fraction of the channel bandwidth. The performance of SBFA is sensitive to the statically reserved fraction.

8.5.4 Wireless Fair Service (WFS)

WFS uses an enhanced version of WFQ in order to support delay–bandwidth decoupling. In WFS, each flow is allocated two parameters: a rate weight r_i and a delay weight ϕ_i. The start-time tag of a packet is computed as in WFQ. However, computation of the finish-time tag is based on ϕ_i rather than r_i, specifically, $F(p_i^k) = S(p_i^k) + \frac{L_i^k}{\phi_i}$. The service tag of a flow is the finish tag of its head-of-line packet.

At a time t with virtual time $V(t)$, WFS transmits the flow with the minimum service tag and a clean channel subject to the constraint that the start tag of the head-of-line packet for the flow must be less than $V(t) + \rho$, where ρ is a lookahead parameter of the scheduler. If $\rho = \infty$, the error-free service is earliest-deadline-first. If $\rho = \infty$ and $r_i = \phi_i$, the error-free service is WFQ. If $\rho = 0$ and $r_i = \phi_i$, the error-free service is WF^2Q. Decoupling the delay and the rate weights allows for delay–bandwidth decoupling.

If a backlogged flow perceiving an error channel is allocated the channel, its lag is increased only if there is another flow that can transmit in its place and increase its lead (or reduce its lag). Both lead and lag are bounded by per-flow parameters. In effect, the lag of a flow reflects the number of slots that the flow is entitled to make up in the future while the lead of a flow reflects the number of slots it must relinquish in the future.

A leading flow with a lead of l and a lead bound of l_{max} relinquishes a fraction l/l_{max} of the slots allocated to it by error-free service. This leads to an exponential reduction in the number of slots relinquished as a function of the lead of the flow and implies that a leading flow asymptotically relinquishes all its lead.

8.5.5 Utility-Based Opportunistic Scheduling

Liu et al. [91] proposed a novel scheduling technique for maximizing resource utilization while controlling the levels of fairness among users. An interesting feature of their approach is that they do not use any formal notion of fairness. Instead, they define a utility function for a user such that optimizing the value of this utility function would ensure that no individual user would be denied access to the wireless channel.

Liu et al. [91], used a stochastic model is for capturing the time-varying and channel-condition-dependent performance of each user. Specifically, the level of performance of user i at time k is denoted as U_i^k. Here, a salient feature of such modeling is that U_i^k is just an abstraction of possibly a complex function of physical layer performance metrics such as throughput, power consumption, and FEC overhead. Such details are not explicitly considered so that the design of the scheduling can focus on the maximization based on the U_i^k. The only constraint is that U_i^k are additive and comparable.

Let r_i denote the time fraction assigned to user i. This time fraction is a long-term guaranteed service share of user i. Here, we have $\Sigma_{i=1}^N r_i = 1$, where N is the total number of users. Thus, at timeslot k, the scheduler selects user i to transmit if

$$U_i + v_i^* \geq U_j + v_j^* \quad \forall j \tag{8.7}$$

where v_i^* is an offset used for satisfying the time fraction assignment constraint for user i. This offset is then used to provide fairness implicitly in this model. Note that if $v_i^* = 0$ for all i, then the scheduling algorithm reduces to a utility-maximizing (e.g., throughput-maximizing) greedy algorithm.

A key issue is then how to obtain and keep track of the values v_i^* for all i. Specifically, the ultimate goal is that the probability of user i gets scheduled $= r_i$ is achieved. If the probability is larger than r_i, then v_i^* needs to be reduced. On the other hand, if the probability is smaller than r_i, then v_i^* needs to be increased. Liu et al. [91] employ the stochastic approximation algorithm

$$v_i^{k+1} = v_i^k - \frac{1}{k}(\delta_i - r_i) \tag{8.8}$$

where $\delta_i = 1$ if user i gets scheduled at timeslot k.

The CAFQ technique is formalized by the following three algorithms.

8.5.6 Channel-Adaptive Fair Queueing

CAFQ [143] has the following features:

- A new notion of fairness is employed.
- Contrary to CIFQ, graceful degradation is not ensured to help the lagging session more efficiently.

Algorithm 1 Receive

```
 1: INPUT: VC (virtual compensation flow), p (packet), and A (set of backlogged flows);
 2: if queue[ID] ∉ A before receiving p then
 3:    ID.lag = 0;
 4:    ID.V initialize;
 5:    ID.C initialize;
 6:    if (ID.lag > 0) AND (ID.channel == 1) AND (ID ∉ VC queue) then
 7:       Add ID to the VC queue;
 8:    end if
 9:    A = A + {ID};
10: end if
11: Add p to queue[ID];
```

- A *punish factor* is used to decide how seriously the scheduler punishes a nonperfect channel state session that transmit packets (thus, the notion of "punishment" is defined with respect to the goal of maximizing overall system throughput).
- A virtual compensation session is incorporated to help the lagging sessions catch up.

Fairness should be maintained in that as long as a session can transmit some data, it should be provided with some chance to transmit. At the same time, QoS should also be met. However, from the system manager's viewpoint, it is hard to meet these two sometimes conflicting goals with a limited bandwidth and channels that have time-varying quality. This is because whenever a session without an excellent channel state is allowed to transmit, there part of the bandwidth will be wasted, and the wasted bandwidth can never be replenished.

It should be noted that this is very different from the idea of swapping sessions that are error-free and error-prone as in existing scheduling algorithms such as CIFQ. When an error-free session takes the opportunity of an error-prone sessions, it will relinquish the service when the error-prone one is in a good channel state.

Indeed, if abundant bandwidth is available or the channel state is most likely to be excellent, we should maintain the graceful degradation and prevent the leading sessions from starving. However, in a realistic system in which the channel is usually not so good, we cannot expect to achieve perfect allocations, but rather we should meet the sessions QoS first. Thus, in the CAFQ algorithm, graceful degradation is not implemented and the rationale is to compensate the lagging sessions as soon as possible so as to quickly resume a higher throughput and reduce the delay.

Algorithm 2 Schedule

```
1: i = min{V_i | i ∈ A};
2: j = min{C_i | i ∈ A};
3: k = min{F_i | i ∈ A};
4: if VC queue is empty then
5:   n = –1;
6: else
7:   n = ID at the head of the VC queue;
8: end if
9: if (i == VC) AND (n ≠ –1) then
10:  Send(n, i);
11: else
12:  if j ≠ –1 then
13:    Send(j, i);
14:  else
15:    if k ≠ –1 then
16:      Send(k, i);
17:    else
18:      Send(0, –3); {No packet to send}
19:    end if
20:  end if
21: end if
```

Algorithm 3 Schedule(*j*, *i*)

```
1: if i == –3 then
2: Schedule in the next frame; {No packet to send}
3: else
4: Transmit the packet at head of queue[j];
5: if queue[j] is empty then
6:   A = A – {j};
7: end if
8: if i == VC then
9:   VC.V update;
10:  j.lag = j.lag-packetlength;
11:  if j.lag just becomes non-lagging then
12:    initialize j.F;
13:  end if
14:  if j continues to be non-lagging then
15:    update j.F;
16:  end if
17:  if j continues to be lagging then
18:    update j.C;
19:  end if
20:  if ((j just becomes non-lagging) OR (j ∉ A)) AND (j ∈ VC) then
21:    Delete j from VC;
22:  end if
23: else
24:  i.V update;
25:  if i == j then
26:    if j continues to be lagging then
27:      Update j.C;
```

```
28:         else
29:            Update j.F;
30:         end if
31:         if (j ∈ VC) AND (j ∉ A) then
32:            Delete j from VC;
33:         end if
34:         else
35:            i.lag = i.lag + packetlength;
36:            j.lag = j.lag - packetlength;
37:            if j continues to be lagging then
38:            Update j.C;
39:         end if
40:         if j continues to be non-lagging then
41:            Update j.F;
42: end if
43: if j just becomes non-lagging then
44: Initialize j.F;
45: end if
46: if i just becomes lagging then
47: Initialize i.C;
48: end if
49: if (j ⇐ VC) AND ((j just becomes non-lagging) OR (j ∉ A)) then
50: Delete j from VC;
51: end if
52: if (i.channel == 1) AND (i ⇐ A) AND (i just becomes lagging) then
53: Add i to VC;
54: end if
55: end if
56: end if
57: end if
```

As in existing algorithms, CAFQ is associated with an error-free system to account for the service lost or gained by a session as a result of errors. A session is classified as leading or nonleading depending on the difference of the service that it received between the error-free system and the real one. A session is leading if it has received more service in the real system than in the error-free one, while it is nonleading if it has received less or the same amount.

CAFQ simulates SFQ (start-time fair queueing) [149] in the error-free system for reasons of simplicity because it is hard to schedule according to the finish times of the packet in the wireless environment. In the SFQ, when packet k of session i arrives, it is stamped with a virtual start time $S(P_{i_k})$ given by

$$S(P_{i_k}) \leftarrow \max\{V((A(P_{i_k})), F(P_{i_{k-1}}))\} \tag{8.9}$$

$$F(P_{i_k}) \leftarrow S(P_{i_k}) + \frac{l_{i_k}}{r_i} \tag{8.10}$$

where P_{i_k} is the kth packet of session i, $F(P_{i_k})$ is the virtual finish time of packet P_{i_k}, $V(A(P_{i_k}))$ is the virtual clock of the system at the arrival time $A(P_{i_k})$ of the packet, r_i is the preallocated service share of session i, and l_{i_k} is the length of the packet.

The virtual time of the packets is initialized to zero. In the error-free system, a session i is selected in the increasing order of the virtual start times among sessions that are backlogged. Since it is possible that the packet of another session instead of session i will be transmitted in the real system, a session's virtual time keeps track only of the normalized service received by the session in the error-free system.

Another parameter, Δ, is used to keep track of the difference of the service that a session received in the real system and in the error-free one. The Δ of a session is initialized to zero. A session is nonleading if Δ is greater than or equal to zero, while it is leading if Δ is less than zero.

In CAFQ, fairness is maintained in two aspects. In the short term, CAF is maintained among the leading sessions and nonleading sessions separately unless the sessions have the worst channel state (cannot transmit). In the long term, outcome fair is ensured with the help of a virtual compensation session.

8.5.6.1 Short-Term Fairness. CAFQ uses two parameters, N and L, to implement the channel-adaptive fairness in the short term. N_i keeps track of the normalized amount of services received by session i, which is proportional to its channel state function when it is nonleading. When a session i becomes nonleading and does not suffer from the worst channel state, N_i will be initialized as follows

$$\max\left\{N_i, \min_{k \in \Psi}\{N_k | \text{lag}_k \geq 0\}\right\} \tag{8.11}$$

where Ψ denotes a set of sessions that are backlogged, and for a nonleading session chosen to transmit packets in the real system, N_i is updated as follows

$$N_i \leftarrow N_i + \frac{l_i}{r_i f(\Phi_i)} \tag{8.12}$$

and L_i is defined in a similar way. Here, L_i keeps track of the normalized amount of services received by session i that is proportional to its channel state function when it is leading. When a session i becomes leading and no longer suffers from the worst channel state, L_i will get initialized in a way analogous to that expressed in Equation (8.11).

In the real system, selection is made among the nonleading sessions first. The session with the minimum N_i will be selected, and the packet at the head of the waiting queue of this session will be transmitted and N_i will be updated accordingly. If there is no such kind of session that is nonleading and backlogged, the system will select from the leading ones in the increasing order of the sessions' L_i, and then L_i will be updated accordingly.

If all sessions are not backlogged (a very unlikely situation in a mobile computing system with a reasonable number of active users), dummy packets will be sent. If the session j selected in the real system is not the one chosen in the error-free one and it is i that is selected in the error-free system, the Δ of i and j will both be updated: $\Delta_i \leftarrow \Delta_i + l_i, \Delta_j \leftarrow \Delta_j - l_j$; otherwise, Δ will not be changed.

When a session with a comparatively bad channel state transmits a packet, either N_i or L_i will increase more rapidly than will a session with a better channel state. As the punish factor changes, we can decide how serious we should punish a session that does not have a perfect channel and transmits packets. The larger the punish factor is, the more serious we punish the unlucky sessions.

Let us consider a simple example. Suppose that session i has channel state 0.75 (i.e., class B), session j has channel state 0.25 (i.e., class D), and both are nonleading and have the same service rate, and all packets are of the same length l. If the punish factor is 1, then after both of them transmit one packet, N_i increases by $1.33 \times \frac{l}{r_i}$ and N_j increases by $4 \times \frac{l}{r_j}$. The reason why N_j increases much more than N_i is that j has a much poorer channel state than does i while both of them get one packet transmitted. After that, j has less chance of transmitting because of the large N_j value, so what the system does is give j the chance to transmit but punishes it because it wastes bandwidth. If we change the punish factor to 0.5, then N_i increases by $1.15 \times \frac{l}{r_i}$ after i transmits one packet and N_j increases by $2 \times \frac{l}{r_j}$ after that. So, as expected, j is punished only moderately as the punish factor decreases.

8.5.6.2 Long-Term Fairness.

Nonetheless, there is still one issue to be considered. Although the sessions that do not have excellent channel states but get packets transmitted are punished, they are given some chance to transmit, and part of the bandwidth of the system is wasted and can never be compensated. Because the scheduler will not schedule a leading session to transmit if there is a lagging one that is backlogged and is not in the worst channel state (i.e., state E), the scheduler will not conserve the effort of the system as most of the other scheduling algorithms do.

Thus, we assign a service share to a virtual compensation session to help in the long term. This preallocated service share is used to help the lagging ones with perfect channel state because only when a session has a perfect channel state, can it get compensation most efficiently. When a lagging session exits from nonperfect channel states, its session ID will be queued in the virtual compensation session. Sessions that are queued in the virtual compensation session are in the decreasing order of their Δ. So we give bonus service to the lagging sessions if it has a perfect channel state, and the session that lags the most will get it first so that it can be assisted in catching up and thus, long-term outcome fair can be maintained.

The design of the virtual compensation session is quite different from that in SBFA. In SBFA, only sessions with perfect channel states will be scheduled while sessions with bad channel states will get their transmission opportunities back only from the preallocated bandwidth. Thus, compensation is implemented by using the preallocated bandwidth instead of using the excess services as in CIFQ and IWFQ. In CAFQ, however, compensation is not implemented purely by such swapping because some of the bandwidth wasted by the poor sessions may not be recovered. Thus, the virtual compensation session in CAFQ is for compensating the wasted bandwidth such that outcome fair can be maintained in the long term.

In the error-free system, a session i is selected among all the backlogged sessions and the virtual compensation session in the increasing order of the virtual time. If it is the virtual compensation session that is selected and there is session ID waiting in the queue, the session with the ID at the head of the virtual compensation queue will be scheduled to transmit in the real system. The Δ of this session will be decreased as $\Delta_i \leftarrow \Delta_i - l_i$. If it is not the virtual compensation session that is selected or there is no session ID waiting in the queue, the system will select a session to transmit in the real system from the nonleading ones according to N_i and then from the leading ones according to L_i if there is no nonleading one to take the service as we have mentioned above.

The leading sessions may possibly be starved because we always select from the nonleading ones first. The reason why graceful degradation is not maintained in CAFQ as in CIFQ and WFS is that it is better to maximize the performance of the nonleading sessions first. One of the most important goals of the scheduling algorithms is to meet the QoS of the sessions, so CAFQ will not help a session that has achieved the same amount of service as it should have in the error-free system if there are lagging sessions whose QoS has not been fulfilled.

8.6 NUMERICAL EXAMPLES

8.6.1 Parameters

The following assumptions are made in the example scenarios:

- All the packets are of the same length of 1000 bits.
- The bandwidth of the system is 4 Mbps.
- The punish factor η is set to be 1 if not otherwise stated.
- Each mobile device moves in a straight line with a speed of 20 km/h in random directions within a field of 200×200 m (a new random direction is generated when the device hits the boundary); the fast fading and shadowing effects are thus computed on the basis of the instantaneous geographic locations.
- The parameter α of CIFQ [106] is set to be 0.1.
- In CS-WFQ the thresholds for all sessions are $\frac{1}{3}$, so when a session has a channel state poorer than state C (which corresponds to $\frac{1}{3}$ as elaborated below), it will not be scheduled.

TABLE 8.2. Channel States and Error Modes[a]

Channel State/Error Mode	*A*	*B*	*C*	*D*
1	0.44	0.49	0.054	0.016
2	0.38	0.48	0.12	0.02
3	0.31	0.44	0.19	0.06
4	0.24	0.40	0.27	0.09
5	0.17	0.33	0.34	0.16

[a]The values in each row indicate the corresponding steady-state probabilities at states *A*, *B*, *C*, and *D* respectively.

To facilitate understanding the scenarios, we employ a slightly simpler physical layer model based on the Markov chain [153]. Using such a discrete Markov model with four states (A, B, C, D) allows us to trace the operations of the algorithms easily. Let the effective throughputs of the physical layer in channel states A, B, C, D be $1, \frac{2}{3}, \frac{1}{3}$, and 0, respectively. The better the channel quality is, the higher the effective throughput is. In other words, the sessions that are in channel states A, B, and C can make use of 100%, $\frac{2}{3}$, and $\frac{1}{3}$ of the allocated bandwidth but the sessions under channel state D cannot transmit any data effectively.

We identify five kinds of error modes as they differ in the steady-state probability shown in Table 8.2. We would like to emphasize that such a hypothetical error model is used merely for illustrative purposes. As we can see, error mode 1 has the best overall channel state while error mode 5 has the worst overall channel state. In each simulation run, the sessions begin with different channel states selected at random.

8.6.2 Scenario 1

We simulate CIFQ and CAFQ with the environment as follows. The service shares of the two sessions are both 0.5 in CIFQ. The virtual compensation session has a service rate of 0.1, and the session rates are 0.45. Both of the sessions are continuously backlogged. Session 1 has an error-free channel state all along [i.e., $M(\Phi_1)$ is always equal to 1], while the channel state of session 0 changes periodically as follows $(k = 0,1,2,3)$:

- $M(\Phi_0) = 1$ when $8k \le t \le 8k + 2$
- $M(\Phi_0) = \frac{2}{3}$ when $8k + 2 \le t \le 8k + 4$
- $M(\Phi_0) = \frac{1}{3}$ when $8k + 4 \le t \le 8k + 6$
- $M(\Phi_0) = 0$ when $8k + 6 \le t \le 8k + 8$

We keep track of the difference between the expected service and the actual service of these two sessions (denoted by β) in the two algorithms. The result is shown in Figure 8.1.

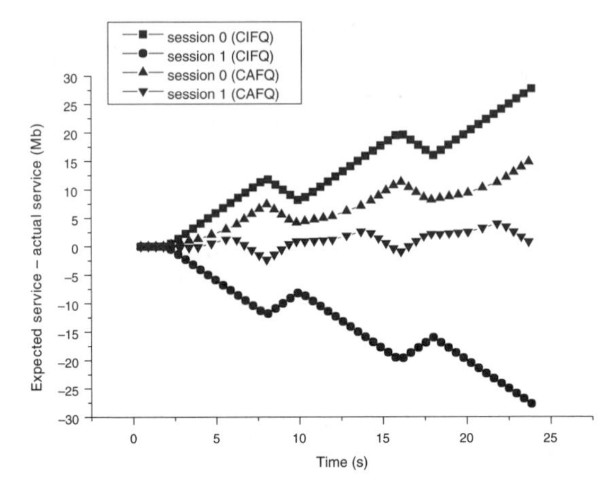

Figure 8.1. Variations of Δ in CIFQ and CAFQ.

For CIFQ, β of session 0 increases at the rate of 2 Mbps while β of session 1 decreases at the same rate during the period when session 0 does not have an excellent channel state. This is because session 1 occupies the bandwidth to transmit when session 0 cannot transmit because of its channel state. After session 0 has a perfect channel state, it will be compensated and β will decreases at the rate of 1.8 Mbps. At the same time, β of session 1 will increase at the rate of 1.8 Mbps. Using the simple periodic error pattern, it suffices to show that when the session has a comparatively high probability of not having a perfect channel state, it is possible that its β grows very fast. So session 0 is seriously lagging when session 1 is seriously leading. It is because CIFQ tries to make use of the bandwidth greedily, and will not save session 0 at the expense of wasting bandwidth.

The CAFQ algorithm works better because it always allows session 0 to transmit provided the session is not under the worst channel state (i.e., state D). Meanwhile, session 0 is also punished (as governed by η) for transmitting data under an error-prone channel (hence inducing losses and wasting bandwidth). So session 1 will not lead as much as it will do in CIFQ; nor will session 0 lag that much. The only drawback is that session 1 sometimes is affected as its β exceeds 0 occasionally. But the amount of β that exceeds is not large and is soon recovered. This is because we always try to help the nonleading one first and the leading one will be ignored until it gets lagged. In summary, we find that

- The long-term fairness of CIFQ may not be maintained when there is a comparatively high probability that a session will have a nonperfect channel state.
- CAFQ helps the nonleading session more efficiently.

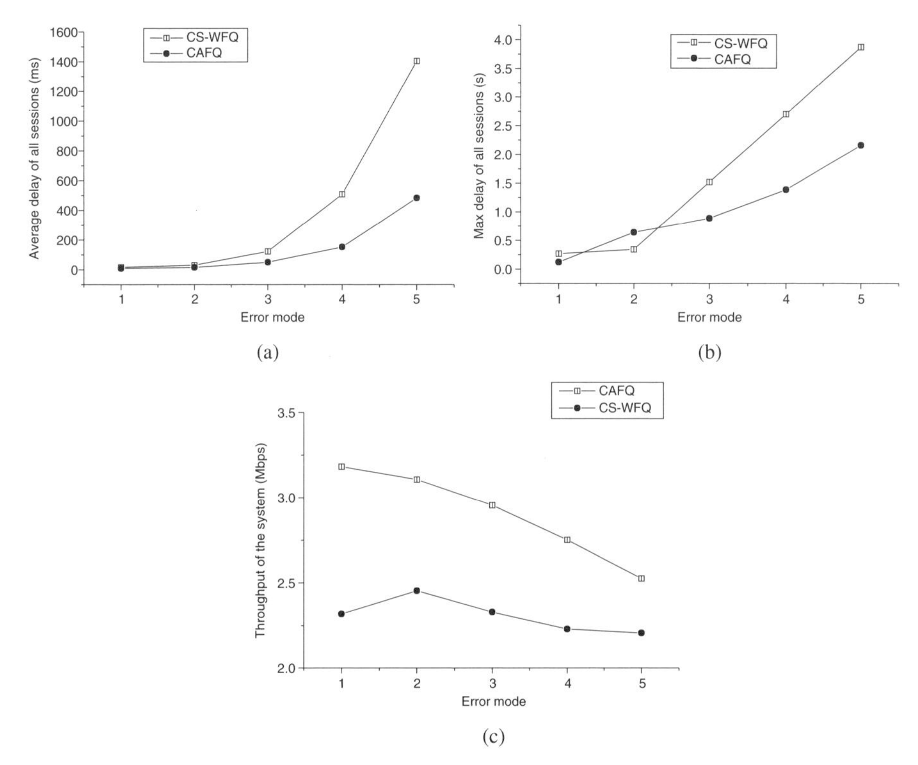

Figure 8.2. Comparison between CS-WFQ and CAFQ: (a) average delay; (b) maximum delay; (c) throughput.

8.6.3 Scenario 2

In this scenerio, the simulation time is 500s and we computed the average result over 10 simulation runs. We simulate CS-WFQ and CAFQ under five kinds of error modes. There are three sessions in the system. Their preallocated service rates are 0.25, 0.25, and 0.5 in CS-WFQ. The virtual compensation session has the rate of 0.2, and the other sessions rates are 0.2, 0.2, and 0.4 in CAFQ. The data source of the sessions are Poisson sources with arrival rates of 0.8, 0.8, and 1.6 Mbps. All the sessions in the system have the same kind of error mode in each simulation run.

We calculate the average delay, maximum delay of all the sessions, and the system throughput in CAFQ and CS-WFQ when the error mode changes. The results are shown in Figure 8.2.

As can be seen, the average delay and maximum delay increase, and the throughput decreases as the overall channel state worsens in both CS-WFQ and CAFQ. However, the rate of increase (decrease) is slower in CAFQ, and the average delays and the maximum delays are always smaller in that with CAFQ than in that with CS-WFQ. In addition, the throughputs in CAFQ are always higher than that in CS-WFQ. This is because CS-WFQ wastes the bandwidth seriously by maintaining outcome fair within the effort limit.

Thus, fewer packets can be transmitted in a given time period, and packets need to wait for a longer time before they are transmitted so that the average delays and maximum delays grow. On the contrary, CAFQ allows the session without excellent channel state to transmit at the same time that it punishes that session, so this session has a chance to transmit but less so if it has a worse channel state.

In the short term, CAF is maintained and thus the sessions' need for maintaining outcome fairness is handled. At the same time, efficient utilization of bandwidth is also achieved such that the precious bandwidth is not wasted in order to desperately maintain outcome fair. In the long run, the virtual compensation session helps the session that lags most seriously and has perfect channel state. This helps reduce the average and maximum delays.

8.6.4 Scenario 3

Simulation time and bandwidth of the system are the same as in scenario 2. There are three sessions in the system with rates 0.2, 0.2, and 0.4, and the sources are Poisson sources with arrival rates 1, 1, and 2 Mbps, respectively. The channel error model is a simple four-state Markov chain. The steady-state probability of both session 0 and session 2 is error mode 1, and the channel state probability distribution of session 1 follows that of error mode 5 as defined in Table 8.2. We keep track of the throughput and the maximum delay of the sessions. The results are shown in Figure 8.3.

When the punish factor increases, the system punishes the session that experiences a poor channel condition but uses significant bandwidth. Thus, when a session has a poor channel state, it has less chance of getting its packets transmitted. As the punish factor increases, the packets are more likely to get backlogged in the queue, and that is the major cause of the increase in maximum delay as shown in Figure 8.3a. Furthermore, sessions with poor channel conditions have less chance to waste bandwidth (i.e., inability to deliver the desired outcome given the effort allocated) and thus, the system throughput improves as the punish factor increases as illustrated in Figure 8.3b.

In Figure 8.3c, we can see that the throughputs of sessions 0 and 2 increase as the punish factor increases. More importantly, we can see that the two curves are very close to each other, demonstrating that CAFQ treats sessions with similar channel conditions in a fair manner. Note that under CIFQ, the throughputs of the three sessions are depicted as horizontal lines independent of the punish factor.

8.6.5 Scenario 4

It is also interesting to examine the performance of CIFQ and CS-WFQ when the channel-adaptive fairness is incorporated.

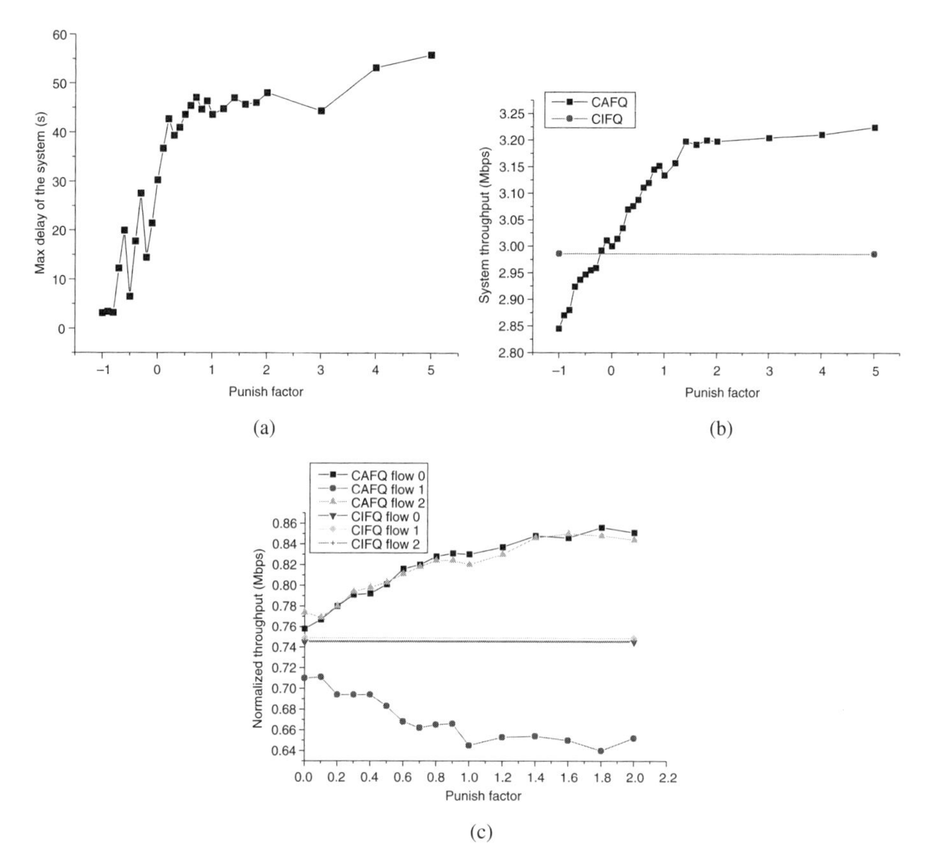

Figure 8.3. Performance of CAFQ and CIFQ with various values of punish factor: (a) maximum delay; (b) system throughput; (c) normalized throughput.

There are three sessions in the system. Their preallocated service rates are 0.2, 0.2, and 0.4, respectively. The preallocated service share for the virtual compensation session is 0.2. The sources are the Poisson sources with arrival rates 0.8, 0.8, and 1.6 Mbps. The sessions channel model is a four-state Markov chain. The steady-state probability of session 1 is as error mode 1, and the remaining sessions are as error mode 2. We calculated the throughput and the average delay of the system assuming that the system runs the original CIFQ or CS-WFQ algorithm. The modified algorithms (with channel-adaptive fairness incorporated), called *CIFQ** and *CS-WFQ**, are also tested in the same environment. The results are shown in Figure 8.4.

We simulate CIFQ* twice when the punish factors are 3 and 5. It is found that the throughput and the average delay are improved more if the punish factor is greater. This is because we punish the sessions that access the bandwidth by improving C_i or F_i more if they experience worse channel states. So the bandwidth is protected from being wasted by the unlucky sessions with worse channel states and thus, the throughput and the average delay of the system are improved.

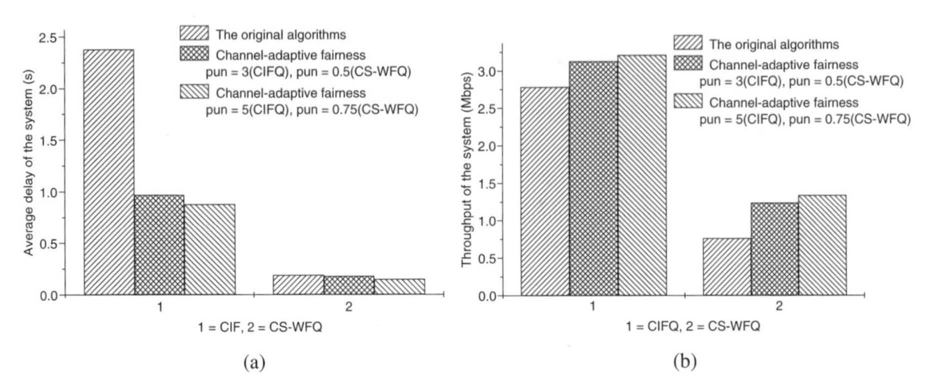

Figure 8.4. Performance of CIFQ and CS-WFQ with CAF incorporated: (a) average delay; (b) average throughput.

We find a similar situation in CS-WFQ*. As session 1 always perceives a better channel than does session 2, let us consider a time instant when session 1 has a perfect channel state while session 2 can make use of only one-third of the allocated bandwidth. In CS-WFQ, the bandwidth ratio is $\frac{1}{3}$. Although CS-WFQ can ensure the same normalized amount of throughput at the devices, it is unfair to session 1. Session 1 has the same service share as session 2 but is allocated much less bandwidth than is session 2, so in fact session 1 is punished by the system because session 2 has a poor channel state.

In CS-WFQ*, if the punish factor is 0.5, the bandwidth is as follows: $(\frac{1}{3})^{0.5} = 0.577 > \frac{1}{3}$. In comparison with the situation in the rate proportional GPS, it improved the throughput at the device of session 2 as it tried to leave more effort with session 2 at the base station. Although it wastes part of the bandwidth, it helps session 2 have a better QoS, which is nearer to the predetermined one despite the bad channel state. Comparing with the situation in CS-WFQ, it reduces the wasting of bandwidth seriously at the same time that it helps session 2.

When the punish factor is 0.75, the bandwidth ratio is now $(\frac{1}{3})^{0.75} = 0.76 > 0.577 > \frac{1}{3}$. In comparison with the situation when the punish factor is 0.5, less bandwidth is granted to session 2 to avoid wasting the bandwidth, although it still allocates more bandwidth to help session 2 in comparison with GPS. So the throughput and average delay of the system are further improved. The price to be paid is that when the punish factor is greater, the session with the worse channel state achieves less throughput at the device than its preallocated share.

8.7 CHANNEL-ADAPTIVE SCHEDULING FOR CDMA SYSTEMS

In this section, we describe how to apply the CAFQ approach to packet scheduling in CDMA systems. We first describe the modeling of code allocation in multicode CDMA systems. We then discuss in detail how the CAFQ approach can be used.

8.7.1 Code Allocation in Multicode CDMA Systems

Liu et al. [92] proposed a unified bandwidth-on-demand fair-sharing platform together with the maximum capacity power allocation (MCPA) criterion. The principle of MCPA is that the maximum capacity power allocation is realized when the power assigned to each code among multirate mobile terminals is minimized while fulfilling the target signal-to-interference ratio (SIR) at the base station. From the results of the maximum capacity power allocation, the system capacity (i.e., the number of basic rate mobile terminals that the system can accommodate) of a multicode CDMA system is defined as

$$S = \frac{\gamma_0 + 1.5G}{\gamma_0} \tag{8.13}$$

where γ_0 is the target SIR at the base station and G is the spread-spectrum processing gain. If a mobile terminal is assigned to transmit at a higher rate (e.g., m times the basic rate), it will occupy the following amount of the system capacity:

$$S_m = \frac{m \times (\gamma_0 + 1.5G)}{m \times \gamma_0 + 1.5G} \tag{8.14}$$

Thus, the remaining system capacity S' after this mobile terminal is scheduled is given by

$$S' = S - S_m \tag{8.15}$$

After this mobile terminal gets its packet transmitted, it will return this part of system capacity back. Most importantly, for a system with remaining capacity S' after some of the capacity has been assigned, the maximum multirate $m_{\max}$ at which a mobile terminal can be permitted to transmit is given by

$$m_{\max} = \left\lfloor \frac{1.5G \dfrac{(S' - \mathcal{D})}{\gamma_0}}{\left\lceil \dfrac{\gamma_0 + 1.5G}{\gamma_0} - S' - \mathcal{D} \right\rceil} \right\rfloor \tag{8.16}$$

where $\mathcal{D}$ is the system capacity reservation factor.

8.7.2 CAFQ for CDMA Systems

In CDMA networks, multiple mobile terminals can transmit at the same time. Using a multicode approach, multiple packets from a certain mobile terminal can transmit at the same time. However, as governed by the power and interference budgets, the number of simultaneous transmissions of packets is limited. We use the multicode CDMA framework proposed by Liu et al. [92] (i.e., equations (17) to (22) in that article [92]) and the method of calculating the optimal power allocation given the code/rate allocation.

In applying the CAFQ algorithm [142] in a multicode TD-CDMA system, we define *resource* as the number of codes that the base station can support or the number of channels in which the base station can transmit packets at the same time given the target SINR value. Assuming perfect power control, the resource of the cell can be calculated as given by the following equation [92]

$$N = \frac{G}{\gamma_b} + 1 \tag{8.17}$$

where γ_b is the required bit-energy-to-noise-density ratio at the base station. Packets from the same mobile terminal can be delivered simultaneously to reduce the interference level of the system, but the number is limited by the BER (bit error rate) requirement of their traffic types. For a certain mobile terminal transmitting packets of service type T, the allowed number of code channels is bounded by $\mathcal{M}_T$. For voice service mobile terminals, only one packet needs to be transmitted at a time. However, for data service mobile terminals, several packets can be transmitted together depending on the channel conditions and the interference/power budgets.

At the beginning of a frame, the system resource is initialized as governed by Equation (8.17), so there is full resource available to schedule. Then the session with the highest priority according to CAFQ is selected to receive service and the number of codes allocated to this session is determined using Equation (8.16). After the scheduling, the remaining resource is updated accordingly.

For example, suppose that there is resource remaining and session i is selected according to CAFQ; the system will allocate C_i codes to session i, where C_i is the smaller one between the remaining resource C and $\mathcal{M}_T$ if session i is service type T. After the scheduling decision, the remaining resource will be decreased as $C \rightarrow C - C_i$. After the packet at the head of the session i gets transmitted, the remaining resource will be increased as $C \rightarrow C + C_i$. When it is not the end of the simulation and there is no remaining resource, it will wait until a certain session finishes transmitting and returns the resource.

When a session is selected, the system will compare the effective service share $\hat{r}_i$ of this session and the remaining resource C. The smaller one is the number of codes assigned to it. Hence, $C_i \rightarrow \min(\hat{r}_i, C)$. The lagging and nonlagging definitions; the updating of N_i, L_i and Δ; and the virtual compensation session are manipulated as described in Section 8.5.6.1. For a lagging session chosen to transmit packets in the real system, N_i is updated as follows

$$N_i \rightarrow N_i + \frac{C_i l_i}{r_i f(\Phi_i)} \tag{8.18}$$

where C_i is the number of codes assigned to session i. When a session i becomes nonlagging and no longer suffers from the worst channel state, L_i is initialized in a manner analogous to that for N_i.

When there is remaining system resource, the algorithm works as follows. When it is a normal session that is selected in the simulated reference system (round-robin runs in the reference system when $\mathcal{M}_T \neq 1$ because the system allocates the codes according to service shares in the real system), a backlogged lagging session with the smallest N_i will be selected in the real system to get service. It will allocate C_i codes to mobile terminal i, and after this, the resource will be decreased by C_i.

If there is no such kind of lagging session available, the service chance goes to the nonlagging session with the least L_i. If all sessions are not backlogged, scheduling will be done in the next frame. If the session i selected in the real system is not the session j that is selected in the simulated error-free one, the Δ values of i and j will both be updated as $\Delta_i \rightarrow \Delta_i - C_i \times l_i$ and $\Delta_j \rightarrow \Delta_j + r_j \times l_j$, respectively. Otherwise, the Δ values will not be changed.

When it is the virtual compensation session that is selected in the reference system and there is a session ID waiting in the queue, the session with the ID at the head of the virtual compensation queue will be scheduled to transmit in the real system. The Δ value of this session i will be decreased as $\Delta_i \rightarrow \Delta_i - C_i \times l_i$. Session i must be the one that is currently having a perfect channel state and lags most seriously.

8.8 SUMMARY

In this chapter, we have presented a qualitative and quantitative analysis of different fair queueing scheduling algorithms in wireless networks. Because of the time-varying nature of the wireless channel in a practical situation, burst errors are the norm rather than an exception. Thus, we believe that a good scheduling algorithm should take into consideration and exploit the variations of channel conditions among the mobile devices. In this regard, we describe a new notion of fairness in which a scheduler is fair with respect to the throughput normalized by the channel capacity. Using this new fairness definition, we describe a new scheduling algorithm called *CAFQ* (channel-adaptive fair queueing). In the numerical results, we have seen that the CAFQ algorithm can balance the frequently conflicting goals of maintaining fair service and maximizing overall system throughput.

There are several possible avenues of further research. The CAFQ algorithm is a centralized approach. It would be more practicable if we could devise a distributed implementation such that both the base station and the mobile devices contribute in the scheduling process, in a manner similar to the algorithm suggested by Kelly [66]. Furthermore, aided by the information-theoretic understanding of the throughput capacity of a multiaccess fading channel, it would be interesting to study the theoretical behaviors of the CAFQ algorithm for scheduling multimedia data transmission in a CDMA system that involves several more system dimensions such as interference, power control, and soft handoff.

EXERCISES

8.1 Explain *effort fair* and *outcome fair*.

8.2 Explain why compensation is needed in a wireless packet scheduling scheme.

8.3 With compensation, how does wireless scheduling deviate from the idealized GPS algorithm?

8.4 Explain the differences between short-term fairness and long-term fairness.

8.5 Discuss the effects of a multilevel channel state wireless medium on the scheduling policy.

PROBLEM

1. *Weighted Fair Queueing and Packet-by-Packet Generalized Processor Sharing* Weighted Fair Queueing (WFQ), also known as Generalized Processor Sharing (GPS), was designed as a scheduling algorithm for wire line system. It assumed the sessions work to infinitely divisible and the system can support all the sessions simultaneously. Every session shares the bandwidth proportionally to the priorities preset to the sessions which are then served in a Round Robin manner. PGPS, being a packetised version of GPS, transmit packet which comes first and has the highest priority. In order words, only 1 packet can depart at any time instant.

Suppose there are 2 users with priorities ϕ_1 and ϕ_2. Decide the work progress of GPS if the packets of users 1 and 2 arrive as shown in figure 1, three packets of user 1 being backlogged at time 0 and 1 packet from user 2 backlogged at time 0 and some more packets arrive later.

Assume the system capacity is of unit 1.

(a) Compute the work progress of GPS if $\phi_1 = \phi_2$.

(b) Compute the work progress for GPS if $\phi_1 = 2\phi_2$.

(c) Compute the work progress of PGPS if $\phi_1 = \phi_2$.

(d) Compute the work progress for PGPS if $\phi_1 = 2\phi_2$.

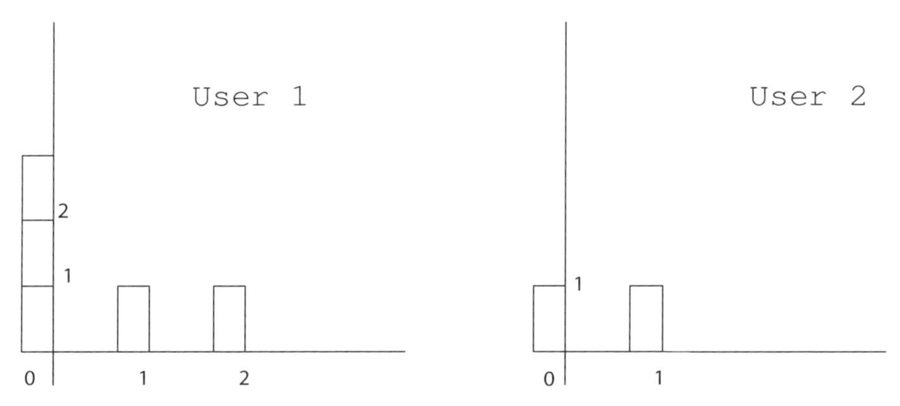

Figure 1. Packets arrivals for GPS and PGPS.

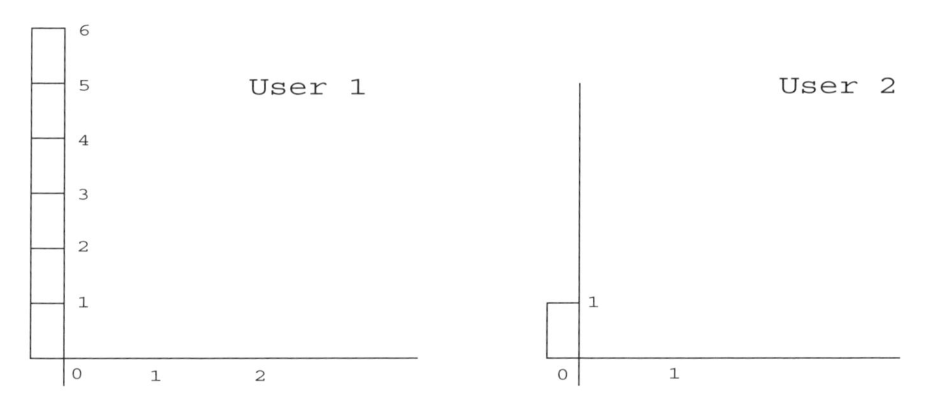

Figure 2. Packets arrivals in the comparisons for GPS, PGPS and WF^2Q.

2. *Worst case Weighted Fair Queueing* PGPS is claimed to have a performance no worse than a packet with maximum packet size compared to GPS system. However, packets can depart at PGPS faster than GPS even if they started later in PGPS. And that would lead to discrepancy between GPS and PGPS and alternate high and low throughput in the PGPS. WF^2Q is designed as a better approximation of GPS and the performance is proved to be the same as GPS asymptotically. In this question, we are going to take a look at GPS, PGPS and WF^2Q and understand their relationship.

In WF^2Q, the server considers packets that have started receiving service in GPS. Or formally,

$$p_i^k \,|\, b_{i,GPS}^k \le \tau \le b_{i,PGPS}^k$$

where $b_{i,PGPS}^k$ is the start time of service of packet p_i^k in GPS system whereas $b_{i,PGPS}^k$ is the start time of service of packet p_i^k in PGPS system.

Using the arrival of packets describing in figure 1, do part (a) and (b). Then using 2, repeat doing part (a) and (b). Is there any difference in these 2 examples in terms of packets' delays?

(a) Compute the work progress of PGPS if $\phi_1 = 2\phi_2$.

(b) Compute the work progress for WF^2Q if $\phi_1 = 2\phi_2$.

3. *Virtual Time Implementation of PGPS* Virtual Time function is designed to track the work done in GPS system and this function can provide a good guide to the scheduler which packet to transmit first. When a packet comes, it is stamped with its virtual finish time and the packet with the samllest virtual finish time departs first.

Denote the k-th packet of session i to be p_i^k, the priority values of session i to be ϕ_i, the arrival time to be $A(p_i^k)$, the virtual start time and finish time to be S_i^k and F_i^k respectively. L_i^k is the length of the k-th packet and ϕ_i is the priority values of session i. The virtual time function V is defined as follows,

$$V(t_{i-1} + \tau) = v(t_{i-1}) + \frac{\tau}{\phi_i}, \tau leq t_i - t_{i-1}$$

$$S_i^k = \max(V(A(p_i^k)), S_i^{k-1})$$

$$F_i^k = S_i^k + \frac{L_i^k}{\phi_i}$$

(a) Explain the physical meaning of the priority values and how it affects the virtual time function.

(b) Explain the formulation of the virtual time function. Why does this scheduling rule lead to proportional fairness?

(c) Using figure 1 as an example, Update the virtual time function for both users and decide which packet should depart first. Justify the result. Assume each packet is of unit length and the users have equal priority.

9

PACKET-SWITCHED DATA SERVICES IN W-CDMA SYSTEMS

9.1 OVERVIEW

Engineers and practitioners have always been trying hard to utilize the elegant results from wireless communication research to produce high-quality systems that can meet the ever-increasing demands from consumers. Third-generation (3G) mobile communication systems are the most recent prominent examples. Specifically, in 3G systems, channel-adaptive technologies are used to further boost the bandwidth efficiency of the wireless spectra. As detailed in this chapter, high-performance channel-adaptive scheduling techniques are implemented in 3G systems to enable high-quality services to meet the heterogeneous demands of various diverse mobile applications such as videophone, multimedia messaging, and traditional voice services. We shall focus on the Universal Mobile Telecommunications System (UMTS) as an illustration.

UMTS, which is a wideband CDMA (WCDMA) standard, can manifest in two different versions in pairing uplink and downlink: FDD (frequency-division duplex) and TDD (time-division duplex). In the FDD mode (the paired spectrum mode due to the need of two 5-MHz-wide frequency bands for uplink and downlink), each physical channel consists of a unique code. In the TDD mode (the unpaired spectrum mode, so called because it needs only one 5-MHz-wide frequency band), each physical channel consists of a unique code and unique timeslots in a frame. Note that the chip rate in UMTS is 3.84 Mcps and the frame duration is 10 ms. The 10-ms frame is further divided into 15 slots. Thus, we have a total of 2560 chips per slot, leading to a symbol rate of 2560 symbols per slot. With a spreading factor ranging from 4 to 256 for an FDD uplink and from 4 to 512 for an FDD downlink, the channel symbol rate

Channel-Adaptive Technologies and Cross-Layer Designs for Wireless Systems with Multiple Antennas: Theory and Applications. By V. K. N. Lau and Y.-K. R. Kwok
ISBN 0-471-64865-5

ranges from 7500 symbols per second to 960 kilosymbols per second (ksym/s). On the other hand, the spreading factor in TDD mode ranges from 1 to 16 and thus, the channel symbol rate ranges from 240 ksym/s to 3.84 Msym/s.

UMTS can support both circuit-switched connections (e.g., for conventional voice services) and packet-switched connections. Indeed, packet-switched connections are envisioned as the major driving force in the 3G market because such connections enable on-demand variable-rate application services (e.g., multimedia messaging, videophone, and location-based services) and are well suited for bursty sources.

The scheduling actions in the early UMTS standard—Rel 99—are located at the network (RNC). As will be elaborated in this chapter, the packet data scheduling in UMTS—Rel 99 is based on a *macroscopic timescale* where the resources allocated to a *data bearer* are dynamically adjusted according to the source buffer status. For example, if there are 10 packet-switched data users in the cell, there will be a 10-packet-switched data-bearer setup and the radio resource assigned to these 10 data bearers will be dynamically adaptive to the individual buffer status. On one hand, when the buffer has plenty of packets, high-data-rate dedicated physical traffic channels will be set up for the data bearer. On the other hand, when the buffer is empty, the dedicated physical traffic channels associated with the data bearer will be released. As discussed in Chapter 6, this approach does not exploit the multiuser selection diversity; therefore, the bandwidth efficiency is low. Nevertheless, the performance of the packet data services in UMTS Rel 99 is already far superior to that in GPRS or EDGE systems because of the faster data rates in the dedicated physical channels.

To exploit the advantages of channel adaptation and multiuser selection diversity, the high-speed downlink packet access (HSDPA) scheme is incorporated in a more recent version of the UMTS standard—Rel 5. The HSDPA system adopts a *microscopic scheduling* approach on the radio resource (timeslots/codes) for its serving active users over very short duty cycles (up to 2 ms). To facilitate the microscopic scheduling, the scheduling algorithm resides at the base station instead of the RNC to minimize the potential delay in the execution of scheduling. The base station obtains the instantaneous channel quality estimates from the mobiles (UEs) and selects one mobile (UE) to transmit on the high-data-rate shared-traffic channel at the current timeslot, based on a scheduling algorithm (which factors in system throughput, QoS requirements, fairness, or a combination of these). This is fundamentally different from the Rel 99 approach. For example, if there are 10 packet-switched data users in the cell, there will be one high-data-rate traffic channel shared dynamically between these 10 users in the HSDPA systems. In fact, it is shown that the microscopic scheduling approach in the HDSPA systems is theoretically optimal because of the multiuser selection diversity.

In Section 9.2, we first briefly introduce the key features of the UMTS architecture. For the fine details and complete specifications, the reader is referred to the literature for some excellent texts and research articles:

[14,32,46,50,57,74,107,141]. In Section 9.3, we discuss the design of the *macroscopic scheduling* mechanism in the UMTS Rel 99. In Section 9.4, we focus on the *microscopic scheduling* mechanism in the HDSPA systems.

9.2 UMTS ARCHITECTURE

A high-level view of the UMTS architecture is shown in Figure 9.1. We can see that the UMTS architecture is divided into three main components: user equipment (UE), UTRAN (UMTS terrestial radio access network), and core network (CN). Following the GSM architecture, UE is a mobile device consisting of a subscriber identification module (SIM) and a wireless transceiver with the appropriate air interface equipment (WCDMA in this case). UTRAN is responsible for handling all radio-related functionalities. CN, adopted from the GSM architecture for easy migration of technology, is responsible for switching and routing calls or data connections to and from external networks such as ISDN, PSTN, and the Internet.

In the UTRAN, the radio resource management functions such as the overload control, admission control, code allocation, outer-loop power control, soft handover, and dynamic bearer reconfigurations are reside primarily in the RNC. Node B is responsible for the physical layer processing over the air such as channel coding, interleaving, rate adaptation, and spreading. Manifested as a base station (BS), a node B also handles inner-loop power control and diversity combining for softer handover between different sectors of node B. In

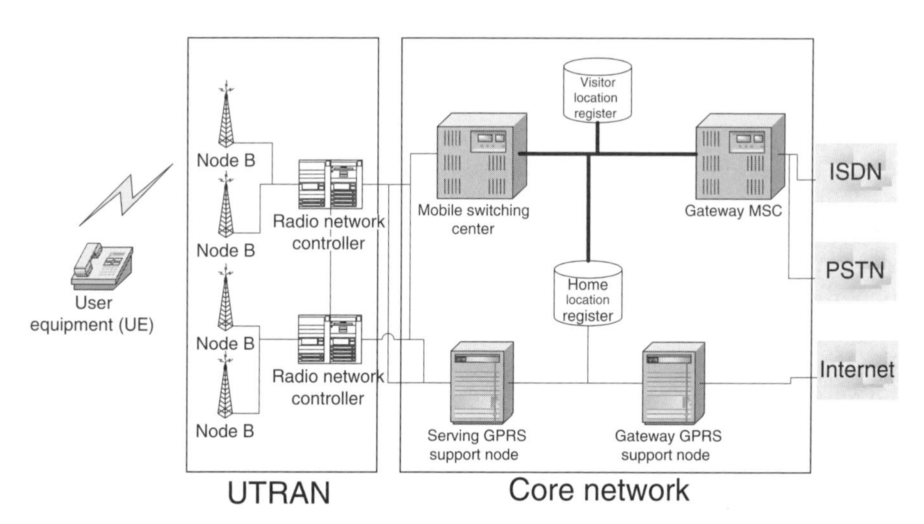

Figure 9.1. A high-level view of the UMTS architecture, composed of the user equipment (UE), radio network controller (RNC), node B, home location register (HLR),visitor location register (VLR), mobile switching center (MSC), gateway MSC, serving GPRS support node (SGSN), and gateway GPRS support node (GGSN).

Rel 5, it is also responsible for "fast scheduling," as detailed in Section 9.4. Each RNC is designated to handle multiple node Bs, and there can be multiple RNCs in a UTRAN. Node B and RNC are connected using fixed-line transmission facilities such as ATM, and the connection is defined as the Iub interface. The RNC may also be interconnected with each other, and the connection is defined as the Iur interface. Besides offering higher physical bit rate over the air, the UMTS system is differentiated from the 2.5G systems (such as GPRS) in the quality-of-service (QoS) dimension as well. For instance, UMTS offers four QoS classes over the wireless interface, and the QoS differentiation is enforced at the RNC as well.

In the CN, the primary role is to support the switching of circuit-switched connections as well as packet-switched connections. Unlike the GSM or IS-95, UMTS offers both circuit-switched services and packet-switched services. However, there is an implicit challenge in the design of the core network in UMTS. On one hand, the target public network of the circuit switched service is essentially the Public Switched Telephone Network (PSTN), which is based on circuit switching architecture. On the other hand, the target public network of the packet-switched service is essentially the Internet, which is based on the packet-switching architecture. Hence, a fundamental issue on the core network design of UMTS system is whether it should be circuit-switched or packet-switched. In Rel 99, an *overlay approach* is adopted. Basically, the core network has two layers: the circuit-switched layer and the packet-switched layer. In the circuit-switched layer, it consists of the mobile switching center (MSC), which is the core element for mobility management, authentication, and switching of users' calls. The CN is connected to the PSTN via the gateway MSC (GMSC) over the PCM-64 payload interface and SS7 signaling interface, respectively. In the packet-switched layer, it consists of serving GPRS support node (SGSN) and gateway GPRS support node (GGSN). Both the SGSN and GGSN are IP routers with additional functionality such as mobility management, authentication, and data encryption for packet-switched services. The CN is connected to the public Internet via the GGSN, which routes the IP packets between the UMTS CN and the Internet. In addition to the switching elements in the CN, there are distributed databases to support user mobility and supplementary services such as call forwarding, namely, the home location register (HLR) and the visiting location register (VLR). HLR is a database for storing the master copy of a user's service profile, which includes the user's service specifications and dynamic status information such as call forwarding details. VLR is a database for storing a visiting user's service profile.

UMTS is an evolving standard. The abovementioned architecture is the basic structure as specified in UMTS standard Rel 99. More recent releases are Rel 4 and Rel 5 (as such, Rel 99 is retrospectively called "Rel 3"). The major addition in Rel 4 is the specification of using ATM for QoS (quality-of-service) control inside the CN. On the other hand, in Rel 5, an all-IP-based network is specified for use inside the CN. In other words, a single IP-based core network is proposed in the UMTS—Rel 5 to interface with both the PSTN and the public Internet. While the interface to the public Internet is straightforward, the interface of the circuit-switched voice service to the PSTN is done via VoIP (Voice over Internet Protocol) gateway. In addition, an IP multimedia subsystem (IMS) is added to interface with the GGSN in the CN. In the IMS, there are three components, namely the media resource function (MRF), call session control function (CSCF), and media gateway control function (MGCF). MRF is responsible for controlling media stream resources and the multiplexing of different media streams. CSCF is the gateway interfacing the GGSN and the IMS. CSCF is also used as a firewall separating the CN from other operators' network. MCCF is responsible for handling protocol conversions. We shall not elaborate further on the core network operation as this is beyond the scope of this chapter.

9.2.1 Radio Interface

The radio interface protocol architecture is shown in Figure 9.2. The protocol stack can be partitioned into two vertical planes: the control plane and the user plane. The control plane is responsible for the signaling, coordination, and control functions in call setup (both circuit switched and packet switched), call release, radio resource management, and session management as well as mobility management. The user plane is responsible for the user applications such as voice, videostreaming, and TCP/IP. In practice, the protocol layers above the physical layer are usually implemented in firmware and the proto-

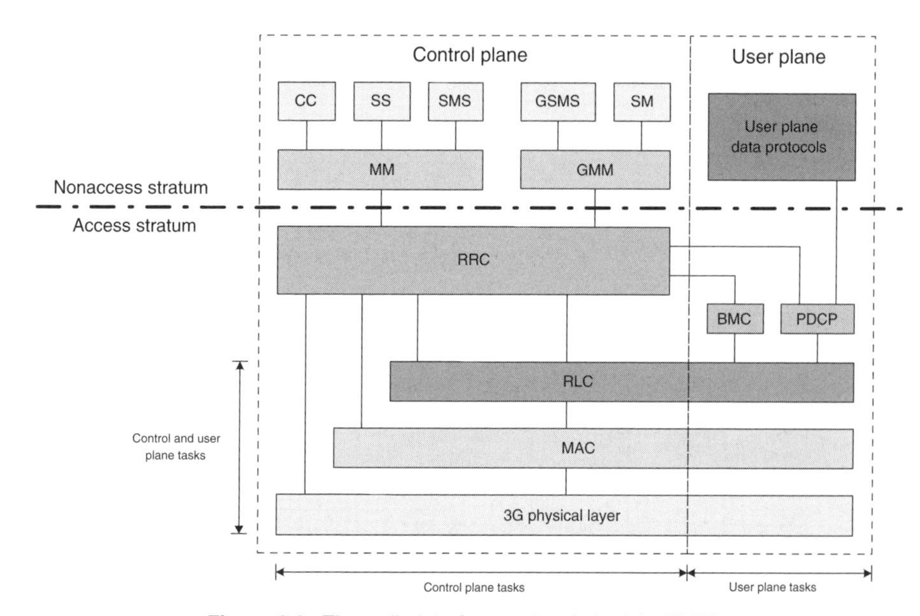

Figure 9.2. The radio interface protocol stack in UMTS.

col layers in both planes coexist as concurrent threads or processes in the mobile phone.

The upper protocol layers in both the user plane and the control plane are all supported by a common suite of lower layers, namely, the radio link control (RLC) layer, the medium access control (MAC) layer, and the physical layer (PHY). The RLC layer is responsible for offering *logical channels* to the upper-layer applications for both planes. For example, packet retransmission, segmentation, and reassembly are the primary services offered by the RLC layer. The MAC layer, is responsible for the multiplexing between the *transport channels* (offered by the physical layer) and the logical channels (offered to the RLC layer). The PHY layer is responsible for offering various *transport channels* to the MAC layer, based on various physical channels. It is responsible for the modulation, coding, spreading, power control, and multiplexing of packets from various transport channels into physical channels. Note that the RLC layer and the MAC layer terminate at the RNC while the PHY layer terminates at node B.

In the user plane, user-specific applications such as TCP/IP are interfaced to the logical channels offered by the RLC layer via the packet data convergent protocol (PDCP), which is responsible mainly for IP header compression. The PDCP layer terminates at the RNC; therefore, the IP header is regenerated at the RNC and delivered to the SGSN as a regular IP packet as illustrated in Figure 9.3. The IP layer in the user-specific application terminates at the GGSN (transparent to the UTRAN), and the TCP layer is transparent to the UMTS network and the Internet.

In the control plane, there are various upper-layer signaling protocols, namely, the radio resource manangement (RRC), the *mobility management* (MM), and the session management (SM). The RRC layer, which terminates at the RNC, is responsible for the radio-dependent resource management such

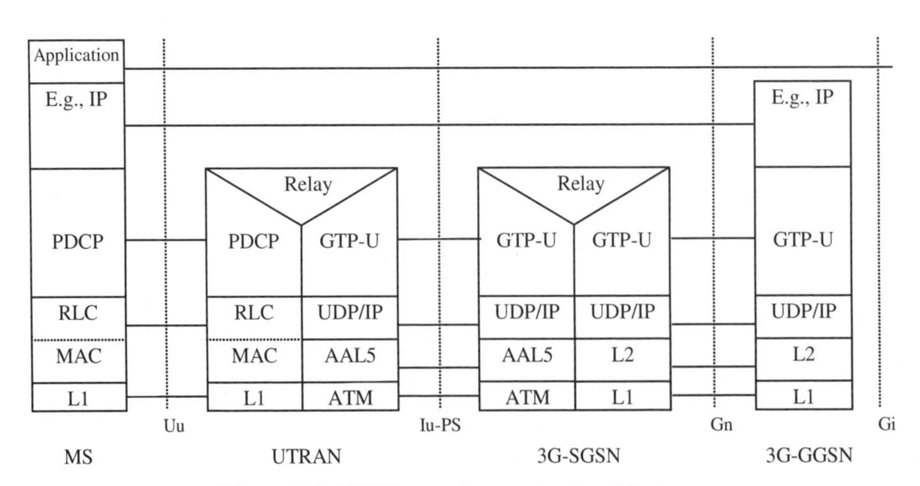

Figure 9.3. UMTS user plane protocol architecture.

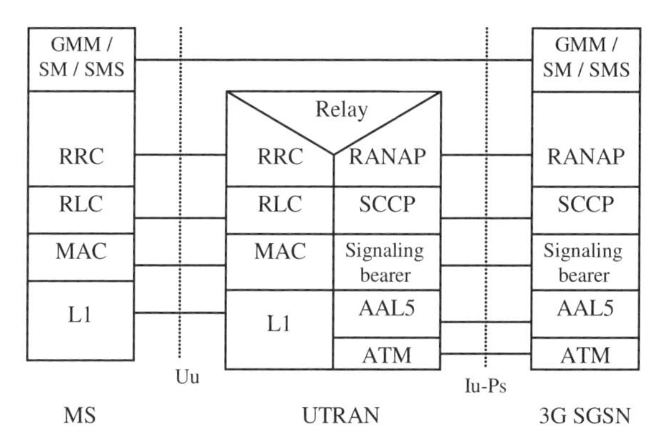

Figure 9.4. UMTS control plane protocol architecture.

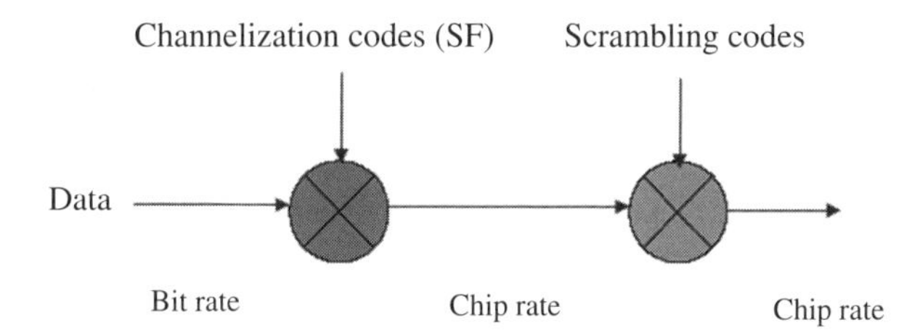

Figure 9.5. Channelization code and scrambling code in UMTS.

as outer-loop power control, soft handoff, admission control, overload control, and dynamic bearer reconfiguration. The MM and SM layers terminate at the core network and are responsible for the location area/routing area update as well as QoS management of the established sessions. This is illustrated in Figure 9.4. We shall elaborate the functions of the PHY layer, MAC layer, RLC layer, and RRC layer in the following sections.

9.2.2 PHY Layer

The PHY layer of UMTS is based on wideband CDMA technology. Specifically, the user separation is based on hybrid *deterministic CDMA*[1] and *random CDMA* approaches. In the downlink, each physical channel is spread using a channelization code and a scrambling code as illustrated in Figure 9.5. The *channelization code* is used for the actual spreading operation (bandwidth expansion) and is targeted for the separation of intracell users, whereas the *scrambling code* is used for the suppression of intercell interference. The downlink scrambling code is unique per cell. In the uplink, the physical channels

[1] *Deterministic CDMA* refers to the case where the spreading codes between different users are completely orthogonal.

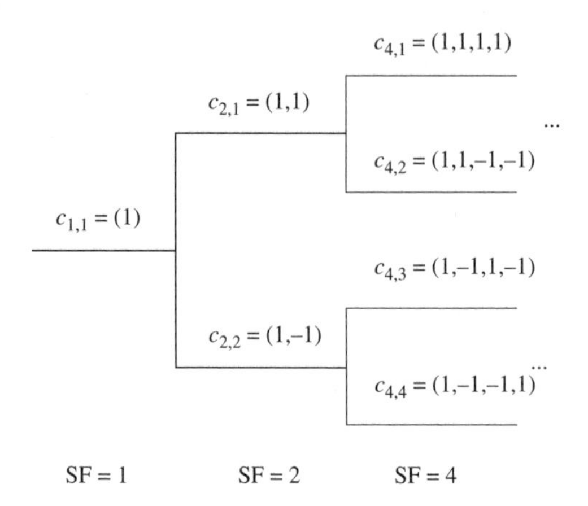

Figure 9.6. Orthogonal variable-spreading-factor (OVSF) codes.

from the mobiles are spread by both the channelization code and the scrambling code. However, the channelization code, which is used for the actual spreading operation (bandwidth expansion), is targeted to separate different physical channels from the same mobile, whereas the scrambling code is used to suppress interference between different users (intracell as well as intercell). Unlike the downlink case, each mobile is assigned a unique scrambling code (by the RNC).

To support various bit rate in the physical channels, the variable spreading factor is adopted in the design of channelization codes in UMTS. The spreading factor ranges from 4 to 512 and 4 to 256 in the FDD downlink and uplink systems, respectively. The smaller the spreading factor of the channelization code, the higher the physical bit rate is. Unlike the IS-95 systems where the capacity bottleneck is usually in the uplink direction (interference-limited), the UMTS system might be code-limited in the downlink direction. This is because the orthogonal channelization codes are designed in accordance with the *orthogonal variable-spreading-factor* (OVSF) codes as illustrated in Figure 9.6. On one hand, if all the users are low-data-rate users or voice users using channelization codes with a spreading factor of 128, the downlink can support at most 128 orthogonal channels. However, if there is one high-data-rate user using a physical channel with a spreading factor of 4 ($c_{4,1}$), all the other codes (of higher spreading factor) derived from code $c_{4,1}$ cannot be used in order to maintain mutual orthogonality between code channels. Hence, a single high-data-rate user consumes one-fourth of the voice capacity in the cell. If a UMTS cell has a few high-data-rate users, the number of voice channels will be quite limited because of the shortage of orthogonal codes. In other words, the system may be *code-limited*. This situation is aggregated according to the manner in which packet-switched data connections are serviced in UMTS Rel 99. For instance, if there are 10 packet-switched data users, there might be 10 dedicated physical channels set up. One solution is to deploy the secondary scrambling code in the base station at the expense of increased mutual interference between code channels in the two scrambling code domains. In UMTS Rel 5 (HSDPA), the problem will be reduced because only one high-speed shared physical channel will be set up to service the various packet data users.

In UMTS systems, channels can be organized into three layers: the physical channels, the transport channels, and the logical channels. A *physical channel* is characterized by a channelization code and a scrambling code based on WCDMA technology. A *transport channel* defines how and with which type of characteristics the data are transferred in the physical layer and serves as the interface between the MAC layer and the PHY layer. A logical channel is a virtual connection for disseminating a certain type of information (which can be control information from the control plane or data applications from the user plane), and it is the interface between the MAC layer and the RLC layer. Figure 9.7 shows the relationships among the three layers of channels in UMTS.

The UMTS logical channels are briefly described below:

- *Common control channel* (CCCH)—a channel that is used for both downlink and uplink to carry control information between the UE and the RNC before any dedicated control channel is set up
- *Dedicated control channel* (DCCH)—a bidirectional dedicated channel carrying control information between a specific UE and the RNC
- *Dedicated traffic channel* (DTCH)—a bidirectional channel between a specific UE and the CN carrying traffic data (user plane)
- *Common traffic channel* (CTCH)—a downlink channel carrying traffic data to a designated group of UEs
- *Broadcast control channel* (BCCH)—a common downlink channel carrying all the necessary system or cell-specific information, such as the random access codes or timeslots, to all mobiles in the cell
- *Paging control channel* (PCCH)—a common downlink channel carrying paging or notification messages from the network to a specific UE, that is, when the network wants to initiate communication with the terminal

These logical channels are multiplexed or encapsulated by the transport channels in the MAC layer. Transport channels can be divided into three categories: common transport channels, dedicated transport channels, and shared transport channels. *Common transport channels* are shared by all users in the cell and are sometimes referred to as *overhead channels* because they seldom not contribute to the transmission of user data.[2] Shared transport channels are shared by a designated group of users only. Usually, there is no closed-loop

[2] In fact, RACH and FACH can be used to carry short bursts of user data.

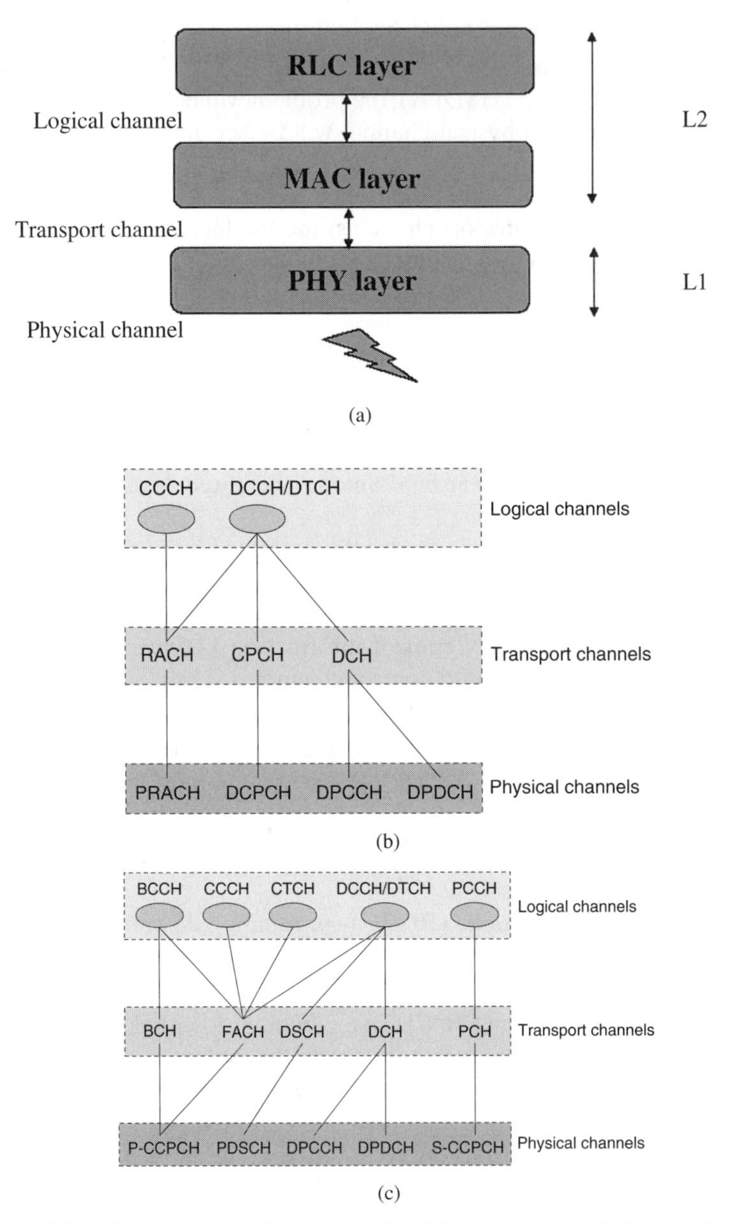

Figure 9.7. Three layers of channels (a) and uplink (b) and downlink (c) channels in UMTS.

power control or soft handoff in both common transport channels and shared transport channels. Dedicated transport channels are set up for a specific user only and support closed-loop power control and soft handoff. Note that a dedicated logical channel is not neccessarily mapped to a dedicated transport channel in the MAC layer. This is further elaborated in Section 9.3. A list of transport channels are elaborated as follows:

- *Broadcast channel* (BCH)—a downlink common transport channel supporting the BCCH on the downlink.
- *Paging channel* (PCH)—a downlink common transport channel supporting thePCCH on the downlink.
- *Forward access channel* (FACH)—a downlink common transport channel supporting several logical channels such as the CCCH, CTCH, DCCH, and DTCH. The FACH channel can be used to carry signaling information or a small amount of user data to a UE within a cell. A cell can have several FACH channels, but the primary one must have low data rate in order to be received by all users. Inband signaling is used in the FACH to indicate for which user the data are intended.
- *Random access channel* (RACH)—an uplink common transport channel supporting the CCCH and DCCH to carry control information or supporting DTCH to carry small amount of bursty data.
- *Common packet channel* (CPCH)—an uplink commontransport channel supporting the DCCH to carry control information or DTCH to carry a small amount of bursty data.
- *Downlink shared channel* (DSCH)—a downlink shared transport channel supporting DCCH or DTCH.
- *Dedicated channel* (DCH)—a dedicated downlink–uplink transport channel supporting DCCH and DTCH between the network and a specific UE.

The data carried by the transport channels are delivered by the MAC layer to the physical layer in the basic units of *transport blocks* once every *transmission-time interval* (TTI) (10, 20, 40, or 80 ms). The transport format of each transport block is identified by the transport format indicator (TFI), which is used in the interlayer communication between the MAC layer and the PHY layer. Several transport channels can be multiplexed together into a single code composite transport channel (CCTrCh) in the physical layer as illustrated in Figure 9.8. Cyclic redundancy check (CRC) and channel encoding are performed on a per-transport channel basis. The encoded bitstream from a transport channel is zero-padded, interleaved (over the TTI), and partitioned into a number of radio frames. For example, a transport channel with TTI = 40 ms is partitioned into four radio frames after channel encoding and CRC. Various transport channels—each possibly having different TTIs, channel coding rates, and data size—are multiplexed together to form a single CCTrCh based on a *rate matching algorithm* on a 10-ms frame-by-frame basis. The CCTrCh is further mapped into a number of physical channels and transmits over the air using CDMA.

In fact, UMTS protocol architecture allows QoS specification per transport channel because each transport channel can have different channel encoding, TTI, block size, and FER target as well as rate matching attributes.[3] This is a

[3] The rate matching attribute is a parameter defined per transport channel indicating the priority of bit puncturing or repetition in the CCTrCh multiplexing.

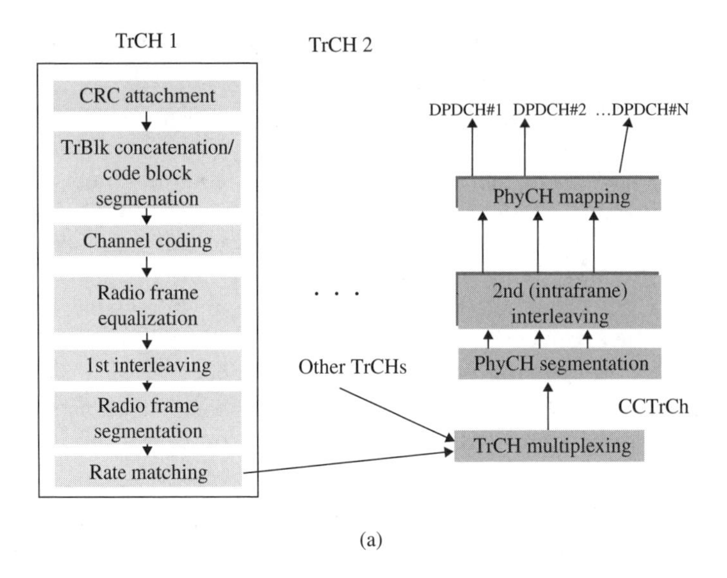

(a)

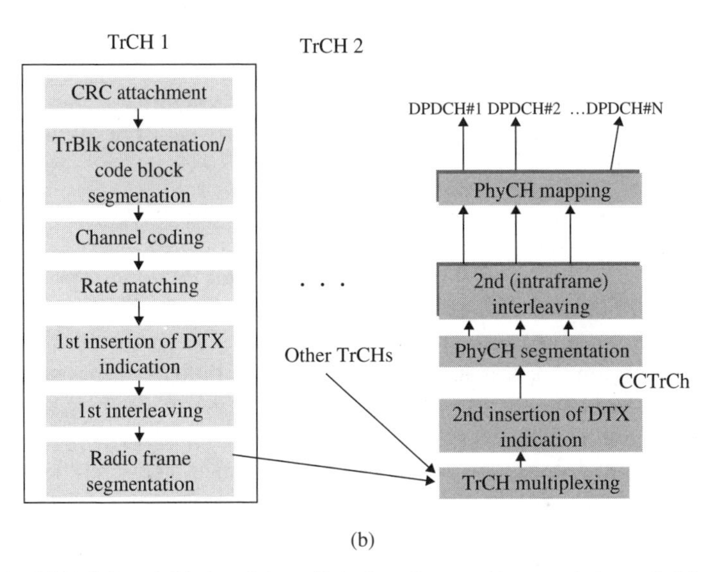

(b)

Figure 9.8. (a) Uplink and (b) downlink multiplexing of several transport channels into a single CCTrCh in the physical layer.

unique feature that cannot be found in other competing standards such as 3G1X. Similar to the transport channels, there are also three types of physical channels: the *common channels*, the *dedicated channels*, and the *shared channels*:

- *Common pilot channel* (CPICH)—a downlink common channel with a constant spreading factor of 256 scrambled by the *cell-specific* primary

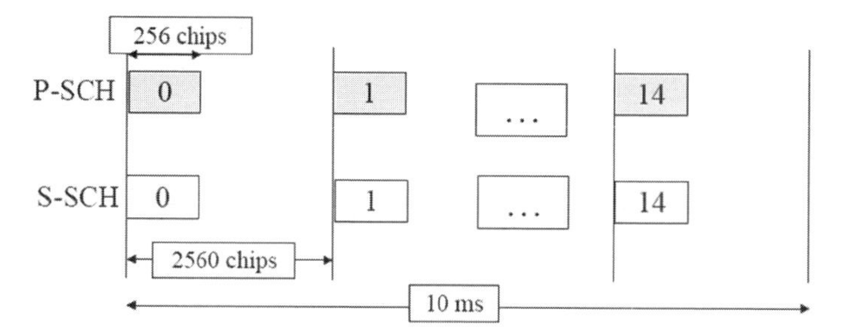

Figure 9.9. Structure of P-SCH and S-SCH for cell search.

scrambling code. This is used as a common reference for channel estimation and timing acquisition at the mobile.

- *Synchronization channel* (SCH)—a common physical channel used for cell searching by the mobiles. There are two subchannels, namely, the primary SCH and secondary SCH as illustrated in Figure 9.9. Both the P-SCH and S-SCH are sent only during the first 256 chips of each timeslot and are transmitted in parallel. Please see References 1 and 3 for details on the cell search operation.
- *Primary common control physical channel* (P-CCPCH)—a common-physical channel supporting the BCH transport channel, carrying cell broadcast information. It has a fixed data rate of 30 kbps and a fixed spreading factor of 256. Since the P-CCPCH is time-multiplexed with the SCH, the first 256 chips in every timeslot are not transmitted and hence, the effective data rate is reduced to 27 kbps. There is no closed-loop power control in this channel, and it needs to be demodulated by all terminals in the cell.
- *Secondary common control physical channel* (S-CCPCH)—a common physical channel used to support the PCH and FACH (transport channels) in the downlink. A fixed spreading factor of 256 is used, and variable data rates can be supported by discontinuous transmission (DTX) and rate matching. Similar to the P-CCPCH, there is no closed-loop power control in S-CCPCH.
- *Physical random access channel* (PRACH)—a common physical channel supporting the RACH transport channel. It is a contention-based uplink channel that can be used to carry control information or a small amount of user data from the mobile to the network. Similarly, no closed-loop power control is supported.
- *Physical common packet channel* (PCPCH)—a common physical channel supporting the CPCH transport channel for carrying bursty data traffic in the uplink. However, unlike the PRACH, this channel can be reserved for several frames, and fast power control is supported.
- *Physical downlink shared channel* (PDSCH)—a shared physical channel supporting the DSCH transport channel on the downlink. It is used to

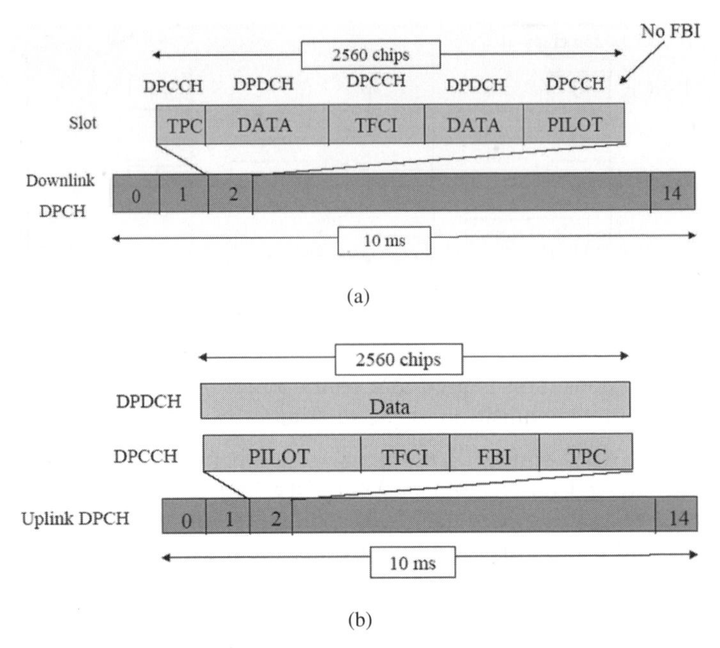

Figure 9.10. Multiplexing of DPDCH and DPCCH in downlink (a) and uplink (b).

carry control information or bursty data and is shared by several designated users. It supports the use of fast power control, but there is no soft handoff for this channel. In addition, it supports a variable spreading factor on a frame-by-frame basis so that the bit rate is adaptive to the physical channel condition. The PDSCH must always be associated with a downlink DPCH to indicate which terminal should decode the information in the PDSCH and the associated spreading code. Since the information of the PDSCH is associated with a downlink DPCH, the PDSCH frame is not started before three timeslots after the end of the associated downlink DPCH.

- *Dedicated physical channel* (DPCH)—a dedicated physical channel in both uplink and downlink used to support the DCH between the RNC and a specific mobile terminal. Two subchannels constitute the DPCH: the *Dedicated physical data channel* (DPDCH) and the *Dedicated physical control channel* (DPCCH). The DPDCH carries the data part of the DCH, whereas the DPCCH carries the physical layer control information such as the fast power control commands, dedicated pilots, and the transport format combination indicator (TFCI).[4] In the downlink, the DPDCH and the DPCCH are time-multiplexed and use normal QPSK modulation as illustrated in Figure 9.10a. The spreading factor of the DPCH does not

[4]TFCI is used at the receiving side to demultiplex data from a single CCTrCh into a number of transport channels on a frame-by-frame basis.

Spreading Factor	Channel Symbol Rate (kbps)	Channel Bit Rate (kbps)	DPDCH Channel Bit Rate Range (kbps)	Max. User Data Rate with $\frac{1}{2}$-Rate Coding (approx.)
512	7.5	15	3–6	1–3 kbps
256	15	30	12–24	6–12 kbps
128	30	60	42–51	20–24 kbps
64	60	120	90	45 kbps
32	120	240	210	105 kbps
16	240	480	432	215 kbps
8	480	960	912	456 kbps
4	960	1920	1872	936 kbps
4, with 3 parallel codes	2880	5760	5616	2.3 Mbps

(a)

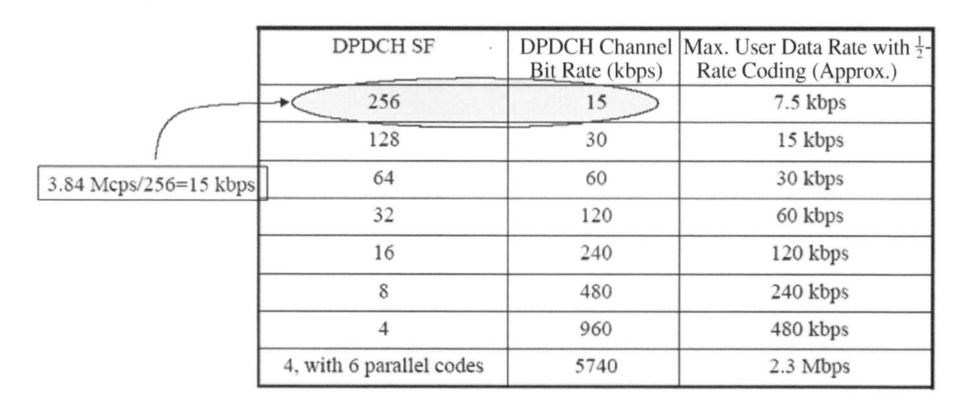

DPDCH SF	DPDCH Channel Bit Rate (kbps)	Max. User Data Rate with $\frac{1}{2}$-Rate Coding (Approx.)
256	15	7.5 kbps
128	30	15 kbps
64	60	30 kbps
32	120	60 kbps
16	240	120 kbps
8	480	240 kbps
4	960	480 kbps
4, with 6 parallel codes	5740	2.3 Mbps

(b)

Figure 9.11. Tables of spreading factor and bit rates in uplink (a) and downlink (b) DPCH.

vary on a frame-by-frame basis, and the data rate is varied by the rate matching operation by bit puncturing, bit repetition, or discontinuous transmission (DTX). In the uplink, the DPDCH and the DPCCH are I-Q multiplexed with different channelization codes c_D and c_C, respectively, as illustrated in Figure 9.10b. The DPCCH always has a fixed spreading factor of 256, whereas the DPDCH has a variable spreading factor on a frame-by-frame basis (4–256).

Figure 9.11 tabulates two sets of possible spreading factors and the associated bit rates of DPCH in the downlink and uplink.

Transport Channel	Physical Channel
(UL/DL) Dedicated channel **DCH**	Dedicated physical data channel **DPDCH** Dedicated physical control channel **DPCCH**
(UL) Random access channel **RACH**	Physical random access channel **PRACH**
(UL) Common packet channel **CPCH**	Physical common packet channel **PCPCH**
(DL) Broadcast channel **BCH**	Primary common control physical channel **P-CCPCH**
(DL) Forward access channel **FACH** (DL) Paging channel **PCH**	Secondary common control physical channel **S-CCPCH**
(DL) Downlink shared channel **DSCH**	Physical downlink shared channel **PDSCH**
Signaling physical channels	Synchronisation channel **SCH**
	Common pilot channel **CPICH**
	Acquisition indication channel **AICH**
	Paging indication channel **PICH**
	CPCH Status indication channel **CSICH**
	Collision detection/channel assignment indicator channel **CD/CA-ICH**

Figure 9.12. A summary of transport channel mappings in UMTS.

Figure 9.12 summarizes the mapping of transport channels to physical channels in UMTS systems.

9.2.3 MAC Layer

The MAC layer in the UMTS protocol stack is a very important component in that it serves many purposes:

- Mapping logical channels to transport channels
- Choosing suitable transport format for each transport channel based on the instantaneous source data rate
- Managing priority in a set of data traffic flows within each UE
- Managing priority among different UEs using dynamic scheduling techniques
- Identifying specific UEs on the common transport channels
- Marshaling of data PDUs between transport block sets and the underlying physical layer dedicated channels
- Measuring and monitoring of traffic volume
- Ciphering for transparent mode RLC
- Choosing access service class for RACH and CPCH sessions

As shown in Figure 9.13, the MAC layer can be divided into four major components: MAC-d, MAC-c/sh, MAC-b, and MAC-hs. MAC-d is responsible

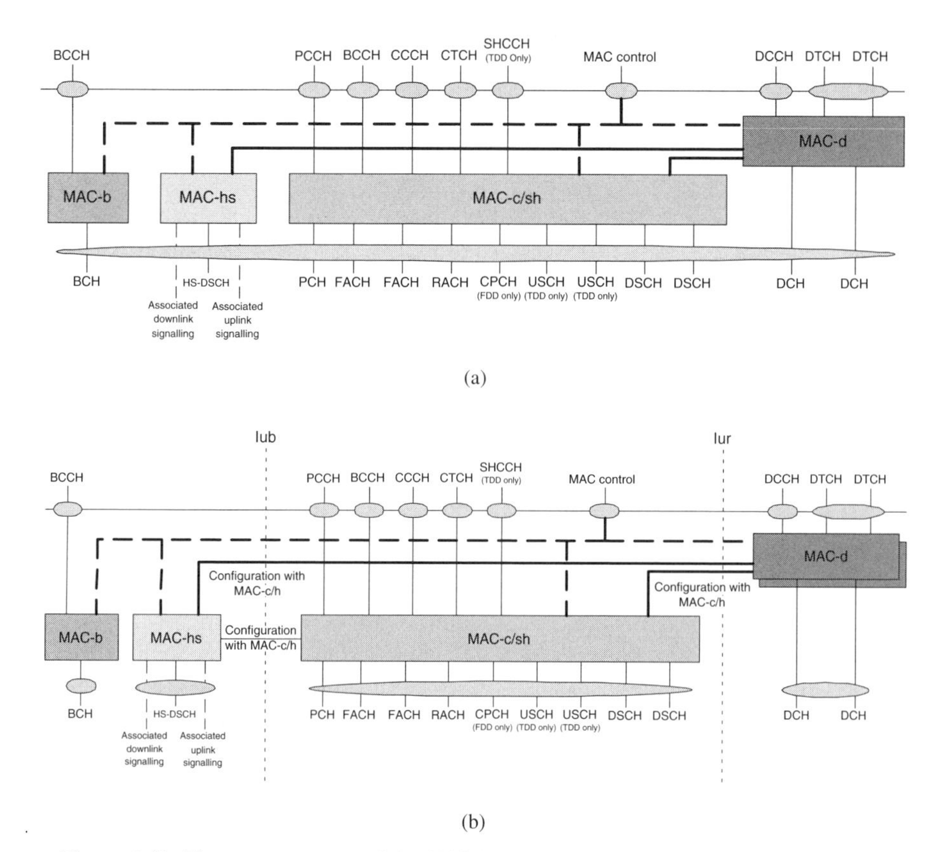

Figure 9.13. Three components of the MAC layer in each UE (a) and in (b) the UTRAN.

for handling dedicated channels and resides at the serving RNC (S-RNC).[5] MAC-c/sh is responsible for handling the common and shared channels. MAC-b is responsible for handling the broadcast channel. BothMAC-c and MAC-b reside at the controlling RNC (C-RNC).[6] MAC-hs is a high-speed extension of the MAC layer in HSDPA and is responsible for the high-speed DSCH transport channel. Unlike the other MAC components, the MAC-hs resides at node B to facilitate fast scheduling.

9.2.4 RLC Layer

The major role of the RLC layer is to provide segmentation and retransmission functions for the logical channels offered by the MAC layer; it resides at

[5]The S-RNC is the anchor point with respect to a UE when the call is set up. As the mobile moves to other node Bs, the S-RNC will not be changed unless the S-RNC relocation procedure has been executed. With respect to a UE, the S-RNC is unique.
[6]Each node B has a unique RNC, and this is the C-RNC.

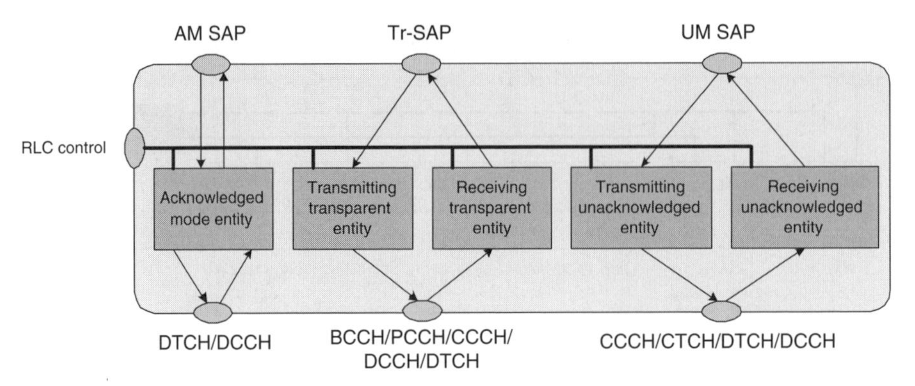

Figure 9.14. The architecture of the RLC layer.

the S-RNC. There are three different operation modes: transparent mode (Tr), unacknowledged mode (UM), and acknowledged mode (AM). The architecture of the RLC layer is shown in Figure 9.14. We can see that Tr and UM modes are characterized as unidirectional, and AM mode is featured as bidirectional.

The Tr mode offers basic service data unit (SDU) transfer, segmentation and reassembly, and SDU discard and does not add any protocol overhead to higher-layer SDUs. Specifically, corrupted SDUs are either discarded or marked as erroneous. Segmentation and reassembly are optional features in the Tr mode. If segmentation is not configured in the RLC layer, the SDU from the upper layers is used as the protocol data unit (PDU) to the MAC layer. On the other hand, if segmentation is configured, the packet length of the SDU received from the upper layers must be an integer multiple of the PDU length so that segmentation is done without any RLC overhead. This mode is useful for realtime applications, such as streaming or conversational applications, where low-overhead logical channel is needed.

The UM mode offers concatenation, padding, ciphering, and sequence number checking in addition to the services offered by the Tr mode. It does not offer retransmission function and as such, data delivery is not guaranteed. Consequently, corrupted data PDUs are either marked or discarded depending on the system configuration. Unlike the Tr mode, RLC protocol overhead is added in the UM mode to facilitate concatenation, padding, ciphering, and sequence number checking. For instance, the UM mode is used for certain RRC signaling procedures, whereas only a unidirectional logical channel is required and there are acknowledgment and retransmission procedures defined in the RRC signaling already.

In the AM mode, an ARQ (automatic repeat request) mechanism is employed for error correction. Figure 9.15 shows the components and mechanisms involved in theAM mode. We can see that the transmitting party of the AM-RLC entity receives RLC SDUs from upper layers via the AM-SAP. These RLC SDUs are then segmented (in case of a large SDU) into fixed-

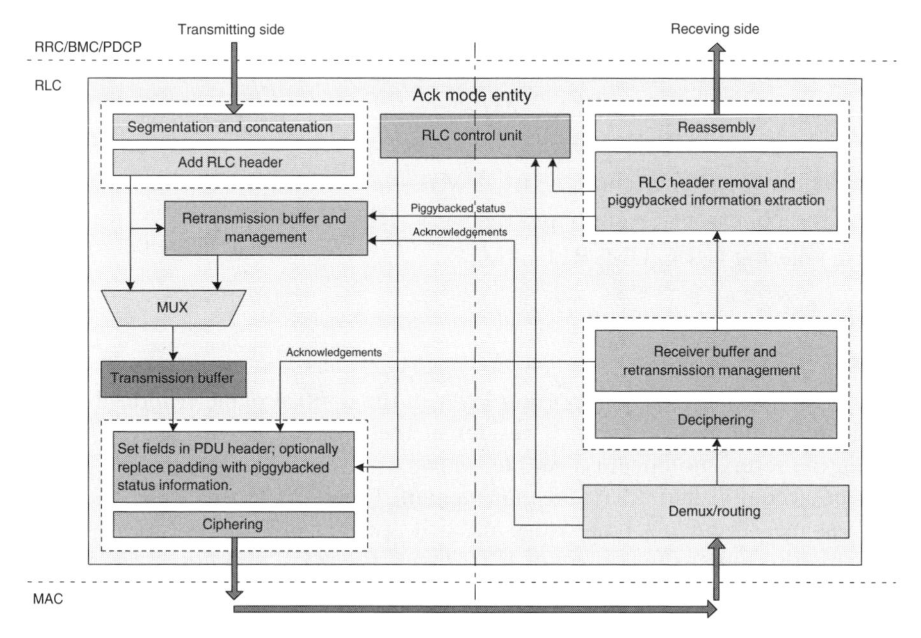

Figure 9.15. Illustration of the components and mechanisms involved in the RLC-acknowledged mode.

length AMD PDUs. Alternatively, the SDUs may be concatenated into an AMD PDU if the SDUs are smaller than the AMD PDU length. The AMD PDUs buffered in the retransmission buffer are either deleted or retransmitted. The MUX multiplexes the AMD PDUs from the retransmission buffer that need to be retransmitted. Ciphering is performed (if configured), followed by sending the encrypted PDUs to either one or two DCCH or DTCH logical channels.

On the other hand, the receiving party of the AM-RLC entity receives AMD and control PDUs via the specific logical channels. Afterward, the AMD PDUs are routed to the deciphering unit followed by delivery to the receiving buffer. Reassembly is performed in that the AMD PDUs accumulate in the buffer until a complete RLC SDU is received. The receiver then acknowledges successful reception (or requests retransmission of missing AMD PDUs).

In summary, the RLC layer provides the following services:

- Segmentation and reassembly of higher-layer PDUs into/from smaller RLC payload units (RLC-Tr, RLC-UM, RLC-AM)
- Assembling by concatenation of RLC SDUs (RLC-UM, RLC-AM)
- Padding (RLC-UM, RLC-AM)
- Transfer of user data (RLC-Tr, RLC-UM, RLC-AM)

- Error correction (RLC-AM)
- In-order delivery of higher-layer PDUs (RLC-AM)
- Duplicate detection (RLC-AM)
- Flow control (RLC-AM)
- Sequence number check (RLC-UM)
- Protocol error detection and recovery (RLC-AM)
- Ciphering (RLC-UM, RLC-AM)

9.2.5 RRC Layer

The RRC layer is a very important layer in the control plane responsible for radio resource management. It resides at the S-RNC, and using RRC messages, the setup, modification, and teardown of layers 2 and 1 protocol entities can be accomplished. Mobility management functions in the UTRAN level such as measurements, handovers, and cell updates with respect to a specific UE are also under the control of RRC layer. The RRC architecture is shown in Figure 9.16. We can see that there are four major functional entities:

- The *dedicated control function entity* (DCFE), which is responsible for handling all functions and signaling for one particular UE. It relies mostly on AM mode RLC

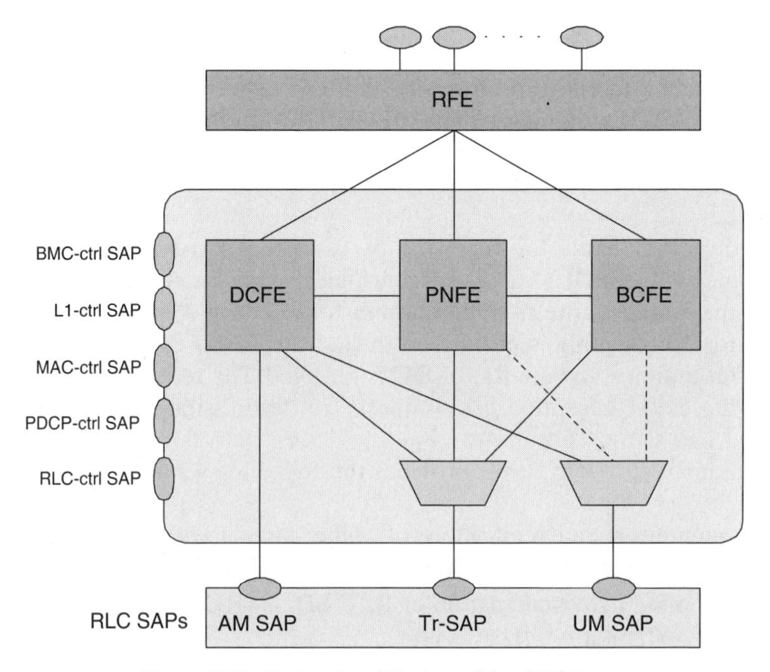

Figure 9.16. Protocol architecture of the RRC layer.

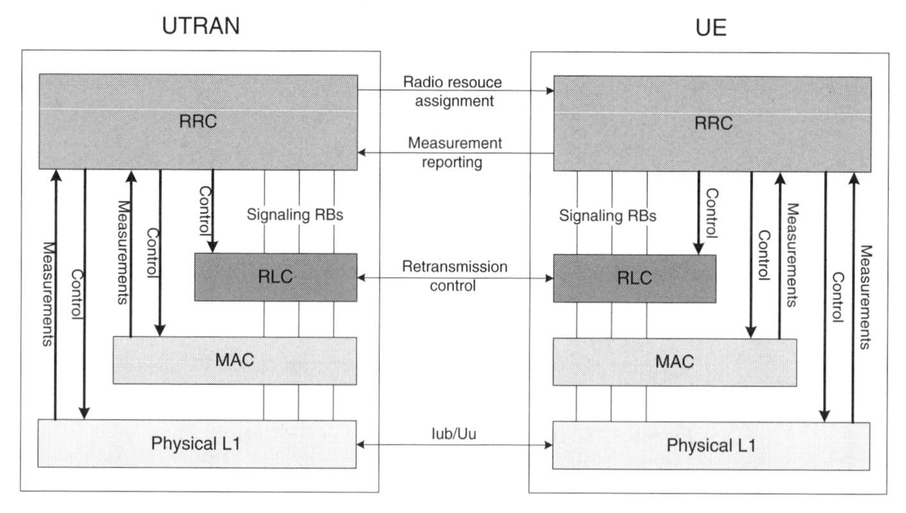

Figure 9.17. Interactions between the RRC layer and other protocol components.

- The *paging and notification control function entity* (PNFE), which is responsible for handling the paging of idle-mode UEs; employs mainly the PCCH logical channel
- The *broadcast control function entity* (BCFE), which is responsible for handling the system information broadcasting; utilizes either BCCH or FACH logical channels
- The *routing function entity* (RFE), which is responsible for the routing of higher-layer messages

RRC interacts closely with each layer in the UMTS protocol stack so as to provide control information, and in turn receive measurement feedback from these layers. This is illustrated in Figure 9.17. Specifically, RRC provides the following functions:

- Cell broadcast service (CBS) control
- Initial cell selection and cell reselection
- Paging
- Broadcast of information
- Establishment, maintenance, and release of an RRC connection between a UE and the UTRAN
- Assignment, reconfiguration, and release of radio resources for the RRC connection
- Control of requested QoS
- UE measurement reporting and control of the reporting
- RRC message integrity protection

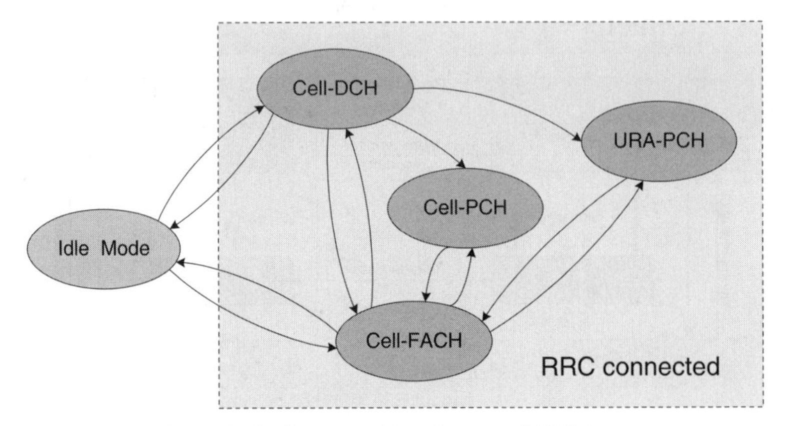

Figure 9.18. State transition diagram of RRC layer.

- Arbitration of radio resources on uplink DCH
- Slow dynamic channel allocation (DCA) in TDD mode
- Timing advance in TDD mode
- RRC connection mobility functions
- Outer-loop power control
- Control of ciphering functions

Figure 9.18 shows the transition mechanism of a UE's RRC states. After the UE is powered on, it chooses a suitable cell with the strongest pilot signal strength for service through monitoring the common downlink pilot channel. After the UE acquires the pilot channel through cell search procedure, the UE is able to receive system information and cell broadcast messages from the BCH channel. At this moment, the RRC layer at the UE is at the idle state. The UE stays in the idle mode until it sends a request to the UTRAN to establish a dedicated control channel (DCCH) with the RRC peer at the S-RNC. When this happens, the RRC layer is moved from the idle state to the *connected state*. Note that setting up a DCCH does not necessarily imply the setting up of dedicated transport channels and dedicated physical channels.

When a logical connection (DCCH) is set up between the UE and the S-RNC, both the RRC peers in the UE and the S-RNC will be at the connected state. However, there are four substates in the connected state: the *cell-DCH* state, the *cell-FACH* state, the *cell-PCH* state, and the *URA-PCH* state. In the *cell-DCH* state, a dedicated physical channel is allocated to the UE for the logical DCCH and DTCH (or radio bearer). In addition, the S-RNC recognizes the location of the UE down to the cell level. The UE carries out measurements and transmits measurement reports according to downlink measurement control information from the RNC. This state can be entered directly from the idle state during call setup or from the *cell-FACH* state (as triggered by traffic volume change).

In the *cell-FACH* state, the UE is not allocated any dedicated physical channel. The DCCH and DTCH logical channels are supported by the RACH, CPCH, and FACH transport channels. In this state, the UE is also capable of listening to the BCH for acquiring system information as well as performing cell reselections. Similar to the *cell-DCH* state, the S-RNC recognizes the location of the UE down to the cell level. Hence, after a reselection process is performed, the UE needs to send a *cell update* message to notify the S-RNC about its new location. This state can be entered directly from the idle state or from the *cell-DCH* state (as triggered by traffic volume change).

In the *cell-PCH* state, the UE is still recognized on a cell level but it can be reached only via the PCH. The major feature of this state is that the battery consumption is less than that in the *cell-FACH* state because monitoring of the paging channel includes a discontinuous reception (DRX) capability. The UE also obtains system information on the BCH. If the UE performs a cell reselection, it has to move to the *cell-FACH* state first so as to carry out the *cell update* procedure.

Finally, in the *URA-PCH* state, it is very similar to the *cell-PCH* state except that the UE needs to execute the *cell update* procedure only if the UE detects a change in the *UTRAN registration area* (URA). The UE obtains the URA of the current cell from the BCH, and a cluster of cells will share the same URA. This arrangement is to reduce the uplink signaling loading to the system due to *cell updating*. The UE in this state is basically in a dormant mode, and the DCCH cannot be used in this state. The UE has to move to *cell-FACH* state to reactivate the DCCH for any potential activities.

We shall illustrate the important role of the RRC states in the dynamic bearer reconfiguration of packet-switched connections in the next section.

9.3 PACKET-SWITCHED CONNECTIONS IN UMTS (Rel 99)

As variable-rate bursty data services are the major new incentives for using 3G wireless systems, great efforts have been exerted to design efficient algorithms to enable such services. However, provisioning of packet data services efficiently is much more challenging than that of the constant-bit-rate voice services because of the inherent heterogeneity in the users' requirements and their associated resource conditions (signal-to-noise ratio, channel quality, data urgency, etc.).

Specifically, there are four basic types of traffic classes supported by UMTS:

- *Conversational class*—realtime symmetric traffic, mostly voice services
- *Streaming class*—realtime asymmetric traffic, usually from the network to the user, (e.g., video streaming)

- *Interactive class*—nonrealtime packet traffic, high integrity (e.g., Web surfing)
- *Background class*—nonrealtime packet traffic, delay-insensitive (e.g., emails)

Judicious radio resource management is needed to satisfy users' QoS requirements while maximizing the system's performance and utilization. In UMTS Rel 99, the scheduler resides at the S-RNC and is responsible for the dynamic allocation of radio resources, in terms of time durations, codes, and power levels, in the radio bearers set up for the packet data users over a macroscopic timescale. As will be discussed below, the scheduler needs to decide on what types of channels to use for different UEs in a dynamic manner according to the system loading and the user buffer levels. In the following paragraphs, we shall focus on the packet-switched data connection setup and dynamic reconfiguration incorporated in the Rel 99 standard of UMTS.

9.3.1 Radio Bearer for Packet-Switched Users

In the UMTS systems, the *radio bearer* is defined as a logical channel between the UE and the RNC to carry data payload (from the user plane) or signaling data (from the control plane). In other words, the radio bearer is the logical channel offered by the RLC layer to the upper layers. If the radio bearer is used to carry signaling data, it is called the *signaling radio bearer*. Otherwise, it is called the *data radio bearer*.

While the data radio bearer is a logical channel for the user plane, it can be mapped into three types of transport channels: the *common transport channels* (CPCH, RACH, FACH), the *dedicated transport channels* (DCH), and the *shared transport channels* (DSCH). Common transport channels can be used to carry user packet data. In a typical deployment scenario, there is usually only one or just a handful of RACH and FACH channels per sector. A distinctive merit of using common transport channels for user data is that the setup time is minimal because these channels are already set up. However, there is no closed-loop power control as well as soft handoff. Instead, the cell reselection mechanism is employed when the UE moves to the next cell. Hence, this is most suitable for transmitting small and bursty user packets.

Dedicated transport channels are bidirectional in nature and support fast power control and soft handoff. Radio-link-level performance is thus greatly enhanced. However, the setup time for dedicated channels is usually quite long, and this introduces setup overhead. In addition, the dedicated transport channels will consume more code space in the downlink. Hence, it is most suitable for transmitting a large volume of user packets.

The shared transport channel (DSCH) operates in a time-division scheduling manner in that a single orthogonal code is shared among many users in the time domain. Compared with the dedicated transport channels, the DSCH can save code consumption in the downlink as well as reduce the associated

TrCh	DCH	RACH	FACH	CPCH	DSCH
TrCH type	Dedicated	Common	Common	Common	Shared
Applicable UE state	CELL_DCH	CELL_FACH	CELL_FACH	CELL_FACH	CELL_FACH
Direction	Both	Uplink	Downlink	Uplink	Downlink
Code usage	According to maximum bit rate	Fixed code allocations in a cell	Fixed code allocations in a cell	Fixed code allocations in a cell	Fixed code allocations in a cell
Power control	Fast closed-loop	Open-loop	Open-loop	Fast closed-loop	Fast closed-loop
SHO support	Yes	No	No	No	No
Target data traffic volume	Medium or high	Small	Small	Small or medium	Medium or high
Suitability for bursty data	Poor	Good	Good	Good	Good
Setup time	High	Low	Low	Low	Low
Relative radio performance	High	Low	Low	Medium	Medium or high

Figure 9.19. A summary of transport channel mappings to support data radio bearer in UMTS.

setup time. Hence, it is suitable for transmitting large-volume or bursty user packets. In fact, in HSDPA, a high-speed version of the DSCH (HS-DSCH) is created to support fast scheduling of packet data users. A shortcoming of the DSCH is that it does not support soft handoff.

Figure 9.19 summarizes the possible mappings of various transport channels for supporting data radio bearers.

9.3.2 Setup of Packet-Switched Connection

In order to set up a packet-switched connection for a data user, the UE has to first establish a DCCH (based on either DCH or FACH/RACH). This is initiated by the UE sending a *RRC connection request* message to the S-RNC via the RACH channel (which is mapped to the CCCH). The corresponding RRC state change can be from the idle state to the cell-DCH state or from the idle state to the cell-FACH state. Once the RRC connection is successfully set up, the next step is to establish a data radio bearer. Specifically, a *radio bearer setup* message is sent from the S-RNC to the UE over the DCCH (already established RRC connection). Figure 9.20 illustrates an example of packet-switched data connection setup with low traffic volume. The RRC state of the UE is changed to the *cell-FACH* after the DCCH is set up. Next, the data radio bearer (RB1) is also set up in accordance with the FACH transport channel.

Unlike that of the circuit-switched connection, the resource requirement of a packet-switched connection has to be dynamically adjusted because of the bursty nature of the source. This is called the *dynamic bearer reconfiguration* or scheduling over macroscopic timescale, and the decision is made by the RRC layer at the S-RNC. To facilitate the S-RNC to perform dynamic bearer reconfiguration, the RRC layer has to monitor the instantaneous system

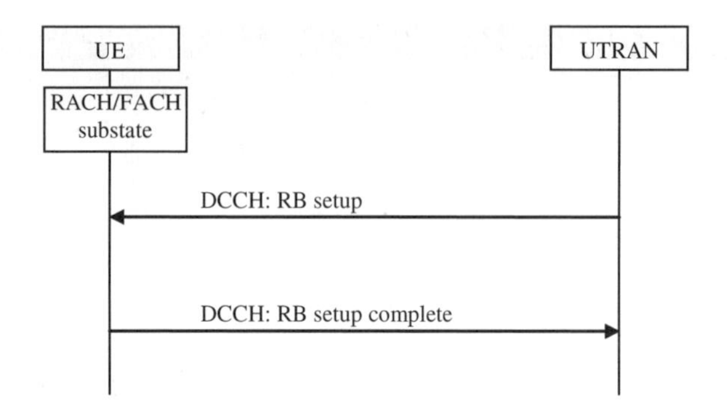

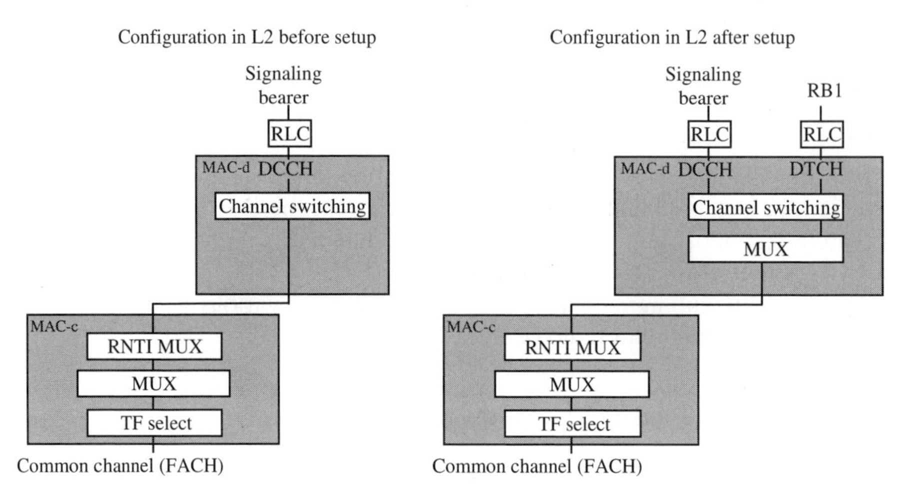

Figure 9.20. An example of packet-switched data connection setup in UMTS with low traffic volume.

loading (from the MAC layer) as well as the instantaneous traffic volume (from the RLC layer). In the downlink, these measurement can be obtained directly from the MAC layer and the RLC layer of the S-RNC. In the uplink, the traffic volume measurement is performed at the UE RLC layer per instruction of the S-RNC.[7] This is illustrated in Figure 9.21.

Suppose that the downlink traffic volume increases such that the buffer level at the RLC rises. The S-RNC may trigger a *transport channel reconfiguration* as illustrated in Figure 9.22. Dedicated transport channels (DCH1 and DCH2) are set up to support the existing DCCH and the radio bearer (RB1) in the reconfiguration in response to the rise in the downlink traffic volume.

[7] To trigger uplink traffic volume measurement, the S-RNC can send a *traffic volume measurement* request to the UE, and the UE will respond by sending a *measurement report* to the S-RNC. The traffic volume measurement can also be configured as periodic or event-driven.

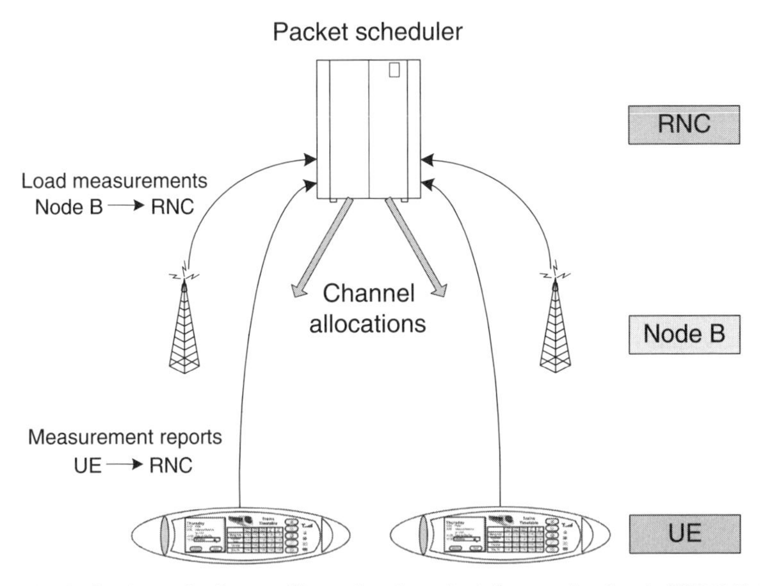

Figure 9.21. A schematic diagram illustrating the scheduling mechanism in UMTS Rel 99.

In this case, the RRC state changes from the *cell-FACH* state to the *cell-DCH* state because dedicated transport channels are set up after reconfiguration.

On the other hand, the dedicated transport channels may be released and replaced by the common transport channels to support the DCCH and RB1 in the example when the traffic volume drops. This transition is usually triggered by the *inactivity timer*, which times the idle period when the buffer in the RLC layer is empty. When this happens, the RRC state will change from the *cell-DCH* state back to *cell-FACH* state. Figure 9.23 summarizes the RRC state changes in the lifetime of a packet-switched data connection.

9.3.3 Scheduling Algorithms

The scheduling algorithm for UMTS Rel 99 can be done along two different resource dimensions, namely, the code dimension and the time dimension:

1. *Code-Division Scheduling.* Different competing users are allocated code channels simultaneously so that the system capacity is shared among all users. The data rate allocated to each user is usually quite low, but the channel can be occupied for a longer period of time. If more users are admitted (on the basis of certain admission control criteria), the data rate for each user will be reduced. The major advantages are lower interference level and longer uplink range, due to the lower data rate allocated to each user. The major disadvantages are longer application-level delay, due to low data rate and lower total throughput. Code-division scheduling is more suitable for dedicated transport channels. Since the scheduling algorithm for packet-switched data

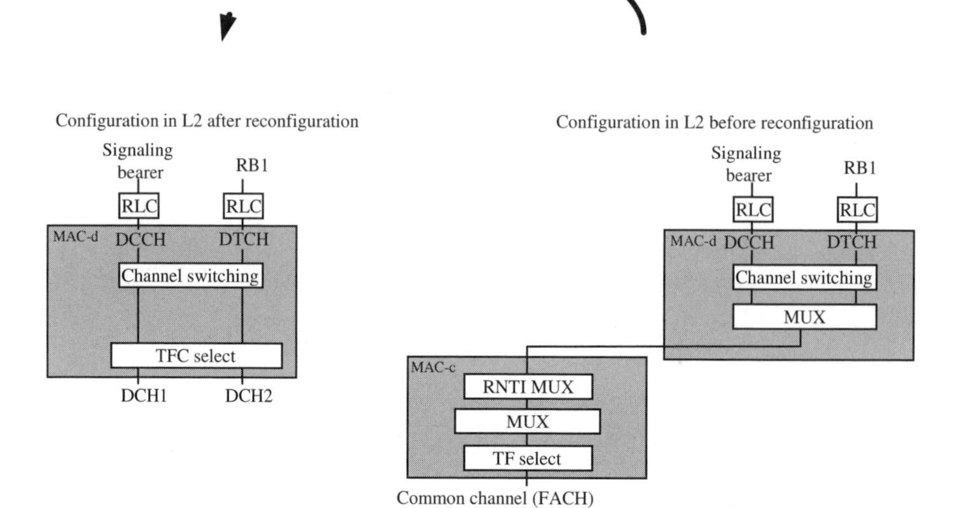

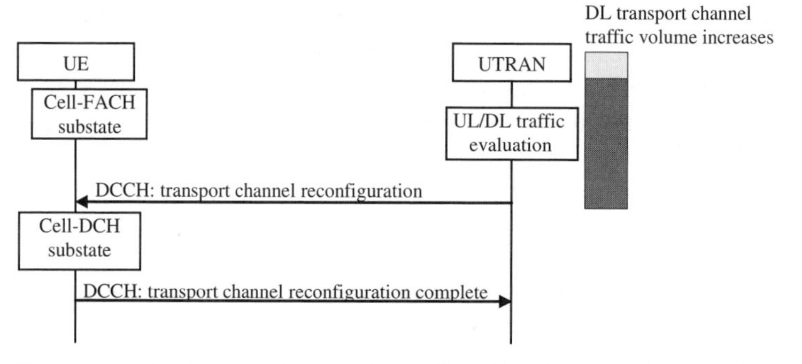

Figure 9.22. An example of transport channel reconfiguration due to an increase of downlink traffic volume.

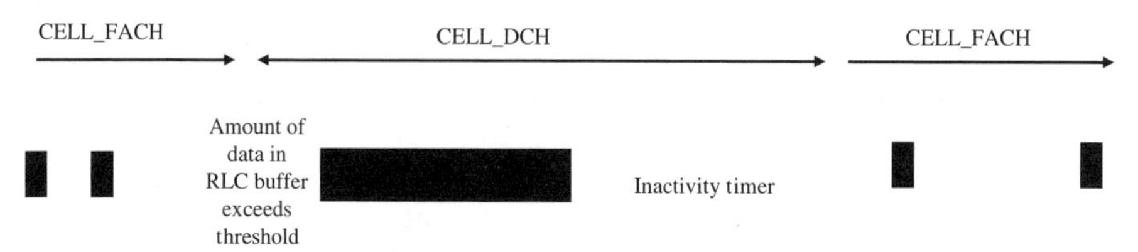

Figure 9.23. An example of RRC state changes in the lifetime of a packet-switched data connection.

users in UMTS Rel 99 resides in the S-RNC and is done over a macroscopic timescale (e.g. 500 ms), it cannot exploit the microscopic variation of the fading channel as we discussed in Chapter 6. Hence, code-division scheduling cannot exploit the multiuser selection diversity, and the capacity of the user is given by the *average capacity* rather than the *peak capacity* over fading channels.

2. *Time-Division Scheduling.* A shared transport channel is set up and shared by a group of packet data users in a time-division manner. At any given time, a single user is assigned to use the shared channel and all the available capacity is then allocated to the selected user. Consequently, a much higher data rate (maximum rate) can be achieved. From our discussions in Chapter 6, this approach is actually capacity-optimal. However, in UMTS Rel 99, the scheduling algorithm resides at the S-RNC and fails to exploit the fast variation of the channel state. Therefore, the efficiency of multiuser selection diversity as a result of the time-division scheduling policy is limited. Yet, time-division scheduling over DSCH channels is more efficient compared to the code-division scheduling over DCH channels.

In both code-division scheduling and time-division scheduling, the physical data rate allocated to a user can be adaptive with respect to the channel condition. For instance, a user closer to the base station requires less transmission power and can get a higher bit rate whereas a user on the cell edge requires higher transmission power and can get a lower bit rate. This adaptation is done at the MAC layer on a frame-by-frame basis.

9.3.4 Performance

To evaluate the efficacy of a scheduling technique for the UMTS, we need to consider both link-level performance and system-level performance. There are a large number of performance studies reported in the literature [7,19,57]. In the following paragraphs, we briefly summarize the findings in Chapter 10 of Reference 57.

At the link level, the major performance metric used is the *block error rate* (BLER). The higher the BLER, the more retransmissions are needed to deliver user data on average. The retransmissions induced a double penalty, namely, a higher packet delay and a higher interference level for a given goodput.[8] On the other hand, a higher BLER also entails a lower required power level. Thus, an important question of performance evaluation is to find the optimal BLER level that leads to the lowest energy per correctly received bit under a certain acceptable level of retransmissions. The crux is to investigate the relationship between the BLER and the *effective* E_b/N_0 as defined below:

$$\left(\frac{E_b}{N_0}\right)_{\text{effective}} = \frac{E_b/N_0}{1-\text{BLER}} \tag{9.1}$$

[8] Equivalently, a link with too many retransmissions will result in a lower effective goodput at a given power budget.

Assuming a mobile speed of 3 km/h and a downlink data rate of 64 kbps, the optimal BLER is found [57] to be 10%. Specifically, if the BLER is higher than 10%, there are unacceptably too many retransmissions, leading to a low goodput and a detrimental level of interference. If the BLER is lower than 10%, the transmitted power is wasted because of the overconservative physical layer.

Wigard et al. also presented a system-level performance study with the following parameters:

- Network size: 18 cells (users are uniformly distributed)
- BLER target: 5%
- Packet scheduling period: 200 ms
- Minimum data rate (initial data rate): 32 kbps
- Maximum data rate: 384 kbps
- Maximum power: 2 W
- Inactivity timer: 2–5 s
- User data arrival: Poisson
- Packet sizes: 500 bytes (minimum), 30 kbytes (median), and 1 Mbytes (maximum)

The study indicated that the cell throughput increases initially with increasing number of users because of the underutilized capacity. The capacity reaches a maximum at 800 kbps per cell at around 300 active users in the network. For a lightly loaded network, almost 60% of the time the users can get the highest data rate. At a high load level (e.g., 250 users), only around 10% of the users can get 384 kbps, while approximately 40% of the users can get only 32 kbps.

9.4 PACKET SCHEDULING IN HSDPA (Rel 5)

The HSDPA concept can be seen as a continue evolution of the DSCH time-division scheduling and a new transport channel, namely, the *high speed DSCH* (HS-DSCH), is defined for more efficient scheduling of high-bit-rate packet data users. In addition, fast scheduling is made possible by introducing a new MAC layer called the *MAC-hs* residing at node B. The fast scheduling (over 2-ms intervals) together with the HS-DSCH transport channel allow the HSDPA system to exploit multiuser selection diversity over the microscopic fading channels and therefore achieve higher peak data rate and spectral efficiency as well as QoS control for bursty and downlink asymmetric packet data users. In the following text, we shall give an overview of HSDPA.

9.4.1 Key Enabling Technologies in HSDPA

There are two key enabling features in HSDPA, namely, the adaptive modulation and channel coding, hybrid ARQ in the physical layer (HS-DSCH), and

TABLE 9.1. Various Data Rates Available in DS-DSCH of HSDPA Using Different Modulation, Channel Coding, and Multicode Allocations

TFRC	Data Rate (1 Code)	Data Rate (5 Codes)	Data Rate (15 Codes)
QPSK, rate $\frac{1}{4}$	120 kbps	600 kbps	1.8 Mbps
QPSK, rate $\frac{1}{2}$	240 kbps	1.2 Mbps	3.6 Mbps
QPSK, rate $\frac{3}{4}$	360 kbps	1.8 Mbps	5.3 Mbps
16QAM, rate $\frac{1}{2}$	477 kbps	2.4 Mbps	7.2 Mbps
16QAM, rate $\frac{3}{4}$	712 kbps	3.6 Mbps	10.8 Mbps

the fast scheduling at the MAC-hs layer in Node B. They are elaborated as follows.

1. *Adaptive Modulation, Channel Coding, and Multicode Transmissions.* On the HS-DSCH channel, two fundamental CDMA features—namely the variable spreading factor and fast power control[9]—have been deactivated and replaced by *short packet size, adaptive modulation and coding* (AMC), *multicode operation*, and *fast hybrid ARQ*. Although more complicated, the replacement of fast power control with AMC yields a power efficiency gain due to the elimination of the inherit power control overheads. In the HS-DSCH, the spreading factor is fixed at 16, which gives good data rate resolution with reasonable complexity. In order to increase the efficiency of AMC to exploit microscopic channel fading, the packet duration is reduced to 2 ms. The means of adaptation in HDSPA are the channel coding rate, the modulation level, and the number of codes employed, as well as the transmit power allocated per code. To be able to exploit the microscopic fading, the AMC and the multiple codes of the HSDPA must be able to cover a wide dynamic range. In the modulation design, the HSDPA incorporates 16QAM in addition to QPSK modulation in order to increase the peak data rate. The combination of rate-$\frac{3}{4}$ encoding and 16QAM gives a peak bit rate of 712 kbps per code channel (SF = 16). On the other hand, higher robustness is available with rate-$\frac{1}{4}$ and QPSK modulation, giving a peak bit rate of 119 kbps per code channel. Given very good channel conditions, a single user can simultaneously receive up to 16 code channels with an aggregate bit rate of 10.8 Mbps. Table 9.1 illustrates various combinations of channel encoding and modulation level as well as the corresponding data rate in HDSPA.

The dynamic range of the AMC for single-code and multicode systems is illustrated in Figure 9.24. The curve includes the gain from fast hybrid ARQ based on chase combining that significantly improves the throughput at low

[9]Moreover, due to the existence of buffer for the bursty data sources and a relaxed delay constraint relative to circuit-switched voice users, the optimal transmission strategy should be *water filling* instead of *equalizing* the fading channel. Hence, the removal of fast power control does not incur any fundamental performance penalty in information theoretical sense.

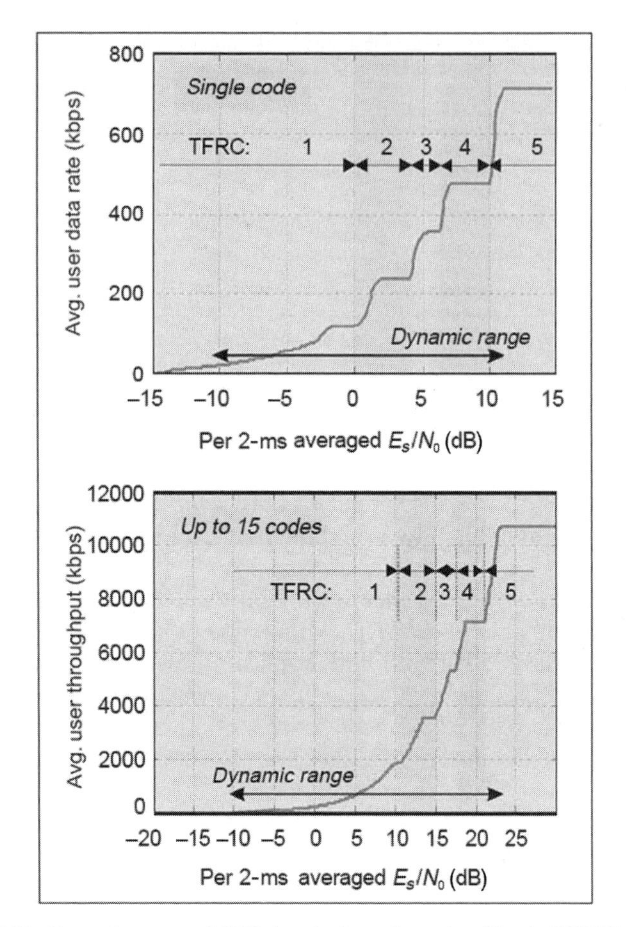

Figure 9.24. Dynamic range of AMC for single-code and multicode HSDPA systems.

E_s/η_0. The dynamic range of E_s/η_0 variation is usually over 20 dB, due to the Rayleigh fading. Observe that multicode operation allows a smaller granularity of the AMC and results in a smoother throughput curve. Furthermore, multicode operation allows the instantaneous throughput to vary over a wider dynamic range of E_s/η_0 (around 32 dB).

2. *Fast Hybrid ARQ*. The hybrid ARQ (HARQ) protocol adopted by the HSDPA system is a stop-and-wait protocol. The transmitter persistently transmits the current packet until it has been successfully received by the UE [indicated through the acknowledgment (ACK)]. In order to avoid resource idling (and therefore a waste) while waiting for the ACK, a maximum of eight parallel HARQ processes may be set up for the UE so that different HARQ processes transmit in different TTI (2 ms each). The control of HARQ is located at node B over the new MAC-hs layer so that storage of the unacknowledged data packets and the subsequent scheduling of retransmissions

does not involve the RNC. Hence, Iub signaling (between node B and RNC) is avoided and the resulting retransmission delay in HSDPA is much lower than the conventional retransmission in the RNC (RLC layer). Typical retransmission delay in HSDPA is around 8–12 ms. The HSDPA concept supports two types of retransmission strategy: *chase combining* (CC) and *incremental redundancy* (IR). The basic idea of CC is to transmit an identical version of the erroneously detected data packet and then for the decoder to combine the soft information of the received copies weighted by the SNR prior to decoding. In the IR scheme, additional redundancy information is incrementally transmitted if the decoding fails in the first attempt. In general, the IR scheme requires a lower SNR relative to the CC scheme at the same target BLER, but the disadvantagc of the IR scheme is a higher memory requirement in the UE. Hence, the possibility of using IR scheme is defined by the UE capability class.

3. *Fast Scheduling*. While the AMC and multicode transmission in HS-DSCH allow a smooth link-level adaptation over a large dynamic range of SNR variation, fast scheduling is introduced in the MAC-hs residing in node B to facilitate multiuser selection diversity over the Rayleigh fading channels. The scheduling decisions as well as the packet buffers are located at node B, and no Iub signaling (between node B and RNC) is needed in HSDPA. Hence, the HS-DSCH can be assigned to one packet data user in 2-ms TTI resolution. This is shorter than the coherence time of most pedestrian users in the system; therefore, the microscopic fading remains quasistatic within a packet transmission.

In HSDPA, three new channels are introduced in the physical layer as outlined below:

- *High-Speed Downlink Shared Channel (HS-DSCH).* This is the channel carrying user data over the downlink and thus is the major resource for the scheduling algorithm to manipulate.
- *High-Speed Shared Control Channel (HS-SCCH).* This is the channel carrying the physical layer control information over the downlink for several key purposes, including decoding, combining, and retransmission.
- *High-Speed Dedicated Physical Control Channel (HS-DPCCH).* This is an uplink channel carrying important control information such as ARQ acknowledgments (ACK) and the downlink channel quality indicator (CQI).

Similar to the DSCH in UMTS Rel 99, there is always a low-data-rate downlink DCH associated with the HS-DSCH. While the HS-DSCH is shared by a number of UEs, the associated downlink DCH is dedicated per UE. Besides the downlink DCH, there is also an uplink HS-DPCCH (per UE) associated with the HS-DSCH in HSDPA. Figure 9.25 illustrates the multiuser fast scheduling concept in HSDPA. Node B tracks the channel quality of each active packet data users in the downlink direction by monitoring the transmit power

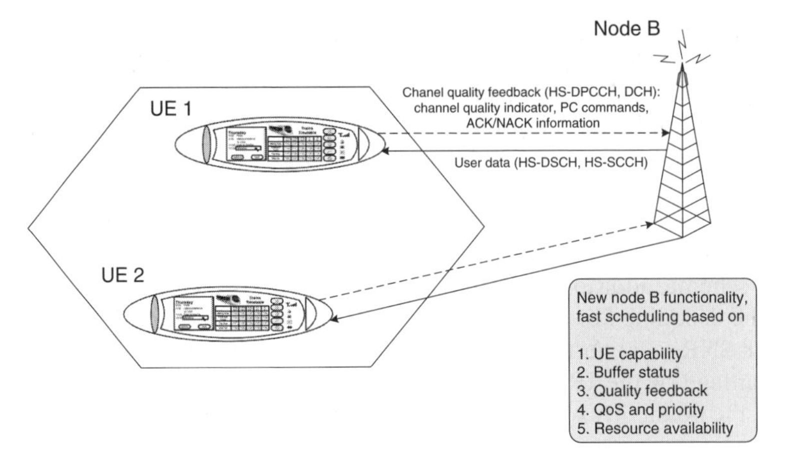

Figure 9.25. A schematic diagram illustrating the scheduling mechanism (HSDPA) in UMTS Rel 5.

of the downlink DCH associated with the HS-DSCH. In addition, the UE can also be requested to report a CQI periodically over the uplink HS-DPCCH. The CQI is an indication of the adaptation mode and number of codes currently supported by the UE according to the current channel condition. The feedback cycle of the CQI can be set as a network parameter adjustable in 2-ms steps. To facilitate HARQ, the UE is also required to send an ACK/NAK response on the HS-DPCCH.

In node B, estimation of the CSI of a UE is based on the power level of the downlink DCH and the ACK/NACK ratio as well as the CQI. Depending on the scheduling algorithm, node B then schedules data transmission to UE on the HS-DSCH. Prior to sending data to the HS-DSCH, node B sends a detailed message to the active packet data users via the associated HS-SCCH. This message describes the employed AMC mode and the multicode set as well as the HARQ process of the HS-DSCH transmission, and it is transmitted two slots in advance of the HS-DSCH as illustrated in Figure 9.26.

9.4.2 Scheduling Algorithms and Performance

The scheduling algorithms to be used in the HSDPA system are not defined in the UMTS standard. Thus, different implementations of scheduling policies are possible. In the literature, several major types of scheduling policies are considered:

- *Round Robin.* This is the simplest type of scheduling policy that does not make use of the available CSI from the UEs to perform optimization. Advantages include the ease of implementation and inherent fairness. Disadvantages are low bandwidth efficiency and possible failures in meeting UE QoS requirements.

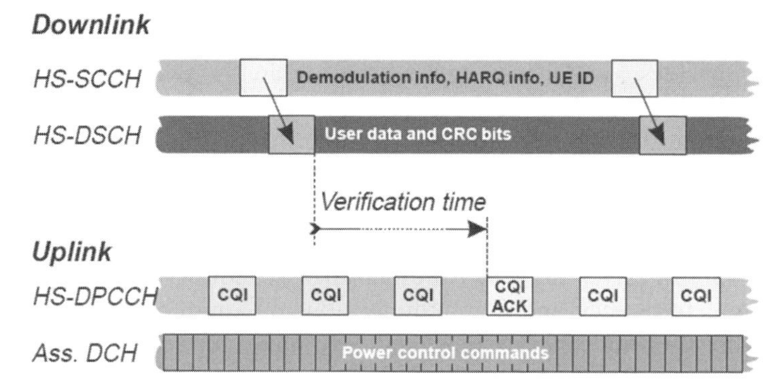

Figure 9.26. Interchannel operation of HSDPA scheduling.

- *Maximum Throughput.* This is the most aggressive form of "opportunistic" scheduling in the sense that it tries to allocate the resources to the user with the best channel quality in a greedy manner as we have illustrated in Chapter 6. Advantages are the ease of implementation and information-theoretic optimal system throughput. Indeed, as repeatedly shown by various studies in the literature, a maximum throughput scheduling algorithm can lead to a 100% increase in system throughput compared with the Rel 99 RNC scheduling mechanism. Disadvantages include inherent unfairness and possible failures in meeting the QoS requirements of the "unlucky" UEs (e.g., on the cell edge).
- *Proportional Fair (PF).* As discussed in Chapter 7, a PF scheduling algorithm tries to balance the issues of fairness and system throughput. PF scheduler can achieve a 60–70% improvement in system throughput over the simple round-robin scheduler (i.e., no adaptation). It offers some notions of fairness because all packet data users have an equal probability of becoming active (allocated with timeslot in HS-DSCH) even though they may experience different average channel qualities.
- *QoS Dependence.* Some more recently proposed scheduling algorithms for HSPDA [18,73] explicitly take into account the UE's QoS requirements, which manifest as traffic class (an application defined parameter), buffer occupancy of pending data, deadline, and delay. Contrary to the PF scheduling algorithm, such QoS-aware scheduling algorithms can meet QoS requirements although they do not exhibit theoretically proven bandwidth efficiency.

Definitely more research is needed in the area of scheduling algorithm design for HSDPA. How to combine the spectral efficiency and application QoS requirement in the scheduling algorithm is still a challenging problem. We shall elaborate more on this topic in Chapter 11.

9.4.3 Continued Evolution

HSDPA provides a significant cell capacity gain for packet data traffic in UMTS and is thus an important part of the continuous 3G evolution. Since HSDPA systems offer improved code efficiency and dynamic range in the user data rate, they can utilize the improvements in the detector performance such as channel equalizations, multiuser and multicode interference cancellations, and advanced MIMO techniques. In addition, the introduction of fast scheduling in HSDPA paves the way for cross-layer scheduling design incorporating the time, frequency, and spatial dimensions on a multiuser perspective.

Naturally, the "high speed" idea can be extended to the uplink also, specifically, to the so-called high-speed uplink packet access (HSUPA). The idea is that except for scheduling, all the actions are reversed in HSUPA compared with HSDPA; that is, scheduling is still under the control of node B. But the measurements and combining actions are also moved to node B from the UEs. However, there is one major obstacle in realizing HSUPA—the power control problem. Over the downlink, the power source is centralized at node B. Thus, the power level can be dynamically adjusted solely at node B's discretion. By contrast, over the uplink, there are numerous heterogeneous power sources—the UEs. Thus, the interference induced is not easily controllable at the receiver located at node B. Indeed, uplink power control cannot be abandoned as in the downlink because of the near–far problem. Much more research is needed in this area.

9.5 SUMMARY

In this chapter, we have elaborated the application packet scheduling using UMTS as an example. We have reviewed a general architecture of UMTS. A UMTS network can be partitioned into a UTRAN and a CN. The UTRAN handles radio-dependent functions (such as radio resource control), whereas CN handles radio-independent functions (such as switching of connections and packets, mobility management, and session management). Inside the UTRAN, we have the RNC acting as a base station controller controlling a number of node Bs (base stations).

In the UTRAN, channels are partitioned into three layers: the logical channels (or radio bearer), the transport channels, and the physical channels. The logical channels form a common interface for both the user plane and control plane protocols. Various logical channels are mapped to different combinations of transport channels in the MAC layer. Various transport channels are further multiplexed into different combinations of physical channels in the PHY layer.

In UMTS Rel 99, packet scheduling resides at the S-RNC (RRC layer) and therefore fails to exploit the microscopic channel variation in the fading channels due to the inherit delay in the Iub signaling (over 500 ms). Packet data users can be served on DCH based on code scheduling or DSCH based on time scheduling. In the former case, the physical radio resource assigned to the active UE are dynamically adjusted according to radio bearer reconfiguration as triggered by traffic volume changes in the RRC layer. It suffers from a larger channel setup overhead and is suitable only for a large volume of bursty data. In the latter case, a number of packet data users are timesharing the DSCH, which incurs a lower setup overhead and delay. In both cases, the peak data rate of the packet user is around 2 Mbps.

In UMTS Rel 5, HSDPA is introduced to significantly improve the spectral efficiency of packet data users. There are three enabling technologies behind HSDPA: AMC, fast HARQ, and fast scheduling. Three new channels (HS-DSCH, HS-SCCH, HS-DPCCH) are created to facilitate an adaptive link with wide dynamic range over the supported bit rates. In addition, a new MAC layer (MAC-hs) is introduced at node B, and fast scheduling can be done at node B without going through Iub signaling. This substantially reduces the scheduling latency, and the scheduling is based on time division over the shared HS-DSCH. The scheduling period is 2 ms, which is sufficient to exploit the microscopic fading for pesdestrian users. As a result of the fast scheduling and the adaptive physical layer, multiuser selection diversity can be effectively exploited and the HSDPA can deliver a much higher spectral efficiency relative to UMTS Rel 99.

EXERCISES

1. To upgrade from GSM to GPRS, the operator does not need to change the base station hardware. [T/F]

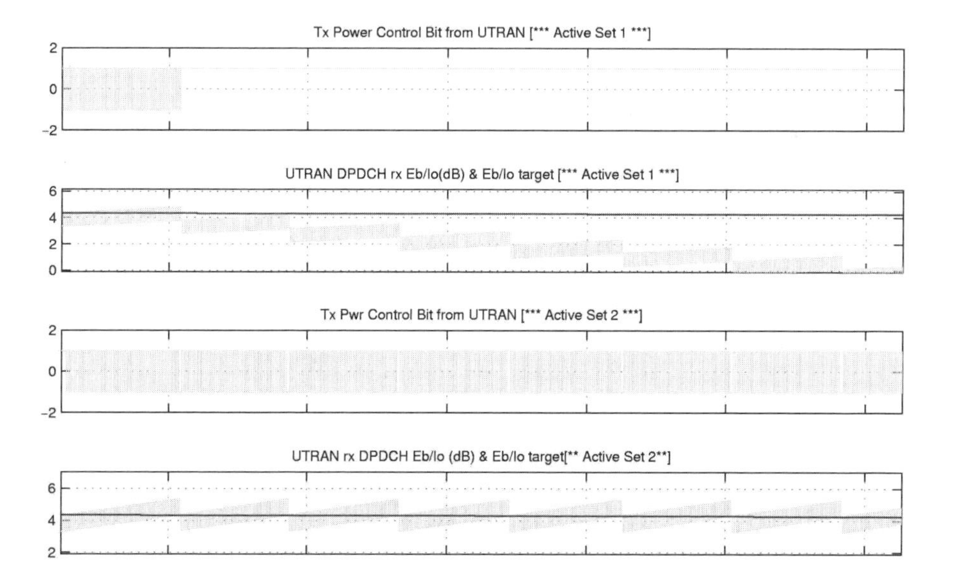

2. UMTS systems allow QoS differentiations in the air interface. [T/F]
3. In UMTS Rel 99, the system capacity of high bit rate packet data users mixed with regular voice users will be limited by the number of orthogonal codes. [T/F]
4. The outerloop power control is done at the Node-B of UMTS systems. [T/F]
5. In UMTS, one way to increase the physical bit rate is to assign multiple codes to the same user. [T/F]
6. In UMTS Rel-99, the RNC is responsible for dynamically controlling the physical resources allocated to the data bearer. [T/F]
7. In UMTS Rel-99, downlink channels of different data rates are not orthogonal in flat fading channels because of the different spreading factors used to spread the downlink channels. [T/F]
8. In the uplink of UMTS mobiles, channels from the same mobiles are spread using orthogonal channelization codes. [T/F]
9. In HSDPA, there is one high speed shared channels setup to be shared dynamically among all voice users. [T/F]
10. A UMTS mobile is in soft handoff with 2 Node Bs (1 and 2). The figure below illustrates the traces of received signal power (Eb/Io) as well as the power control bits transmitted to the UMTS mobile versus time at the 2 BTSs involved in the soft handoff process. Explain why the received power at Node-B (1) is dropping despite the Node-B (1) consistently transmit power control bit of "1" to the mobile. [T/F]

PART 3

ADVANCED TOPICS

10

CROSS-LAYER SCHEDULING FOR WIDEBAND SYSTEMS

10.1 OVERVIEW

In Chapters 1–6, we focused on the point-to-point and multiuser adaptation design for MIMO channels utilizing either the full or limited CSIT. Narrowband transmission and flat fading MIMO channels are assumed in these chapters. However, in practice, as the transmission bandwidth increases, it is increasingly possible to experience frequency-selective fading. Hence, it is the focus of this chapter to extend the discussion on the design of cross-layer jointly adaptive physical layer and MAC layer to cover wideband MIMO transmissions.

Recall from Chapter 1 that the number of resolvable multipaths as experienced by the transmitted signal is given by $L_p = \lceil W_{tx}/B_c \rceil$, where W_{tx} is the signal transmission bandwidth and B_c is the channel coherence bandwidth. Hence, when $W_{tx} \gg B_c$, the receiver will get a number of echos at distinct delay positions from a single signal transmission. These echos at different delay positions will cause intersymbol interference (ISI) and result in an irreducible error floor in the BER–vSNR curves. To combat the effect of ISI, equalization is needed at the receiver, and the optimal equalization algorithm is shown to have a complexity [114] in exponential of the number of resolvable multipaths L_p. Hence, unless the number of paths L_p is small, the complexity of the optimal equalization is usually beyond implementation limitations. In practical systems such as 802.11a/b/g (WiFi), 802.16 (WiMax), and B3G systems, the transmission bandwidth is about 20 MHz. Hence, there are quite a lot of resolvable multipaths L_p for these systems, especially when they operate in outdoor environments, and we cannot rely on the receiver to perform equalization

Channel-Adaptive Technologies and Cross-Layer Designs for Wireless Systems with Multiple Antennas: Theory and Applications. By V. K. N. Lau and Y.-K. R. Kwok
ISBN 0-471-64865-5

alone. Specifically, we need to introduce some special structures in the transmitted signal so as to simplify the task of equalization at the receiver for wideband transmissions.

Two popular technologies are used to tackle the frequency-selective fading channels in wideband transmissions: *direct-sequence code-division multiple access* (DS-CDMA) and *orthogonal frequency-division multiplexing* (OFDM). DS-CDMA is a radio access technology that employs signature codes to partition the system resource into *code channels*. In other words, all the K user signals share the entire bandwidth and the entire time duration but differentiate among themselves with respect to different *code channels*. This extra code dimension is introduced through the *spreading process*. At the receiver, to recover the original information symbol, the receiver has to apply the same code sequence (synchronized in time) and perform the *despreading process*. As a result, the DS-CDMA system has a unique feature, namely, the *interference suppression* capability. On average,[1] a DS-CDMA system can suppress the interference power by a factor called the *processing gain* PG, which is defined as

$$\mathrm{PG} = \frac{T_b}{T_c} \tag{10.1}$$

where T_b is the information bit duration and T_c is the chip duration. With respect to the frequency-selective fading and the effect of ISI, DS-CDMA systems can effectively suppress the ISI by PG times, and therefore, using a simple RAKE finger structure at the receiver, the *equalization* can be done in linear complexity with respect to L_p. Hence, DS-CDMA is the primary technology chosen for 3G cellular systems because of its robustness and low-complexity implementation with respect to frequency-selective fading channels.

On the other hand, OFDM technique splits the information data into parallel streams over a large number of subcarriers. The subcarriers are spaced apart at precise frequencies such that the modulated data remain orthogonal. Unlike traditional frequency-division multiplexing in which the subcarriers are completely separated in frequency domain, subcarriers in OFDM system overlap. However, with appropriate processing at the receiver, the modulated data on these subcarriers appear to be orthogonal to each other. Hence, the transmission bandwidth in each subcarrier is given by $f_s = \frac{W_{tx}}{n_f}$, where n_f is the number of subcarriers. Since f_s is much smaller than the transmission bandwidth W_{tx}, flat fading will be experienced by each subcarrier and there is no need for ISI equalization at the receiver if sufficient guard period is inserted

[1] The actual interference suppression capability depends on specific code design.

between OFDM symbols. One effective way to implement the OFDM idea is to perform IFFT at the transmitter and FFT at the receiver because FFT-based operation is computationally efficient.

In point-to-point communications, the DS-CDMA and OFDM systems are comparable in terms of the effectiveness in combatting the ISI due to frequency-selective fading without expensive equalization at the receiver. However, the two systems behave differently from a multiuser perspective. For DS-CDMA, the signal from a user is spread across the entire transmission bandwidth. Different users are multiplexed through different code channels. From the cross-layer scheduling perspective, the scheduling can exploit the *multiuser diversity* on the spatial (due to multiple antennas) and code dimensions only. However, because of the frequency diversity as a result of spreading per user, the fluctuations (or variance) of the signal quality between users are suppressed, and this degrades the efficiency of multiuser diversity. In addition, the near–far effect may affect the performance of the DS-CDMA system significantly; therefore, accurate and fast power control schemes are required. On the other hand, for OFDM-based system, multiple users can be assigned to use a subset of n_f subcarriers, resulting in OFDMA (OFDM–multiple access). Hence, this is advantageous to the cross-layer scheduling algorithm because the multiuser diversity can be exploited over both the spatial and frequency dimensions. As we have illustrated in the previous chapters, multiuser diversity is very critical to multiuser system performance in fading channels.

In this chapter, we shall focus on the design and performance analysis of the cross-layer scheduling algorithm for wideband transmissions over frequency-selective MIMO fading channels, with emphasis on the OFDMA/MISO system because of the abovementioned advantage in performance. The chapter is organized as follows. In Section 10.2, we will briefly review the backgrounds of DS-CDMA and OFDM systems and outline the physical models of the multiuser DS-CDMA/MISO and OFDMA/MISO systems on the basis of the information-theoretic approach. In Section 10.3, we formulate the cross-layer scheduling problem for OFDMA/MISO and DS-CDMA/MISO physical layers. In Section 10.4, we compare the performance of the two systems. In Section 10.5, we discuss some implementation issues of OFDMA systems in terms of the synchronization requirements. Finally, in Section 10.6, we summarize the main points discussed in this chapter.

10.2 OVERVIEW OF DS-CDMA/MISO AND OFDM/MISO SYSTEMS

In this section, we briefly review the basics of OFDM and DS-CDMA systems. Specifically, we will elaborate the multiuser physical layer model on the basis of the information-theoretic approach for the DS-CDMA/MISO and OFDM/MISO systems. In both cases, we assume slow fading channels where the channel fading remains quasistatic within the entire encoding frame. However, we assume wideband transmissions and therefore consider frequency-selective fading channels in the analysis.

10.2.1 Multiuser Physical Layer Model of DS-CDMA/MISO Systems

DS-CDMA is a radio access technology that employs signature codes to partition the system resource into *code channels*. In other words, all the K user signals share the entire bandwidth and the entire time duration but differentiate among themselves with respect to different *code channels*. This extra code dimension is introduced through the *spreading process* as illustrated in Figure 10.1a.

The individual information bit (or modulation symbol) for user $k(s_k)$ is *spread* with a code sequence $c_k(t) \in \{+1, -1\}$ to form the transmit signal. The code sequence is a time sequence with the smallest time unit of transition

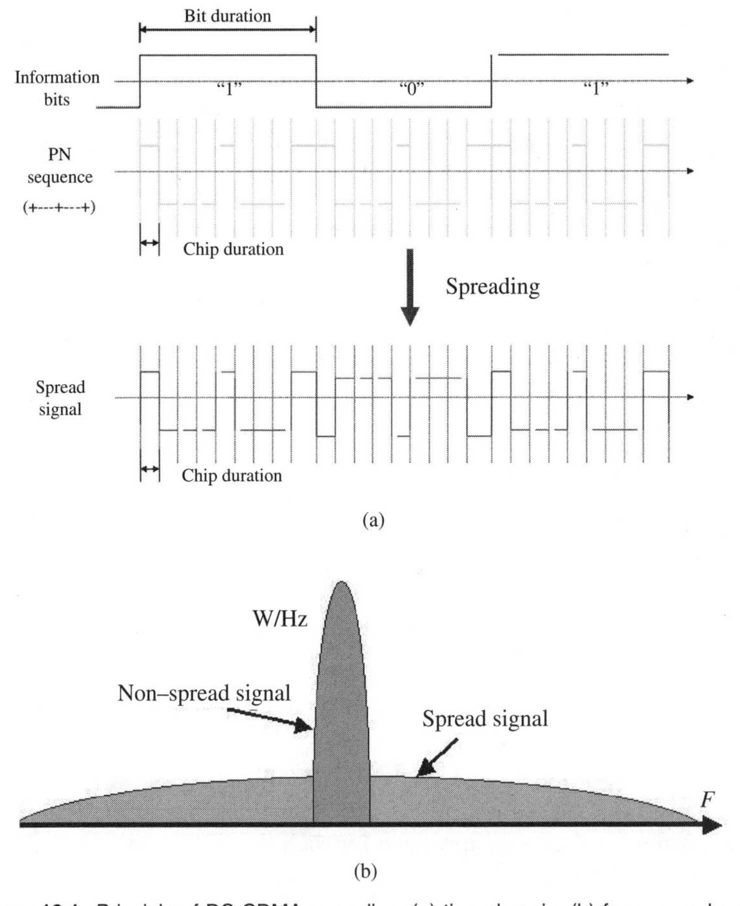

Figure 10.1. Principle of DS-CDMA spreading: (a) time domain; (b) frequency domain.

called a *chip*. The time duration of a chip is called *chip duration* T_c. Hence, a modulation symbol is *chopped* into many tiny pieces by the code sequence. The ratio between the modulation symbol duration T_s and the chip duration T_c is called the *spreading factor* (SF), given by

$$\mathrm{N}_{\mathrm{SF}} = \frac{T_s}{T_c} \tag{10.2}$$

As a result of the spreading process, the transmit bandwidth is increased by N_{SF} times

$$W_{tx} = W_s \times N_{\mathrm{SF}} \tag{10.3}$$

where W_s is the signal bandwidth without spreading. Figure 10.1b illustrates the spreading process in frequency domain.

We consider a base station with n_T transmit antennas and K mobile users (each with a single receive antenna). Figure 10.2 illustrates the multiuser base station transmit architecture. The input datastream from each of the K users is carried in both the spatial and code dimensions. Specifically, input data from users are spacetime-encoded onto the n_T transmit antennas and spread by a $N_{\mathrm{SF}} \times n_T$ complex spreading matrix $\mathbf{C}_k$, where the columns represent the spreading sequence corresponding to the n_T transmit antennas. In other words, the signals from all the active users are differentiated by different spreading matrices. The transmit power gain applied to each of the n_T transmit antennas is determined by a *power control matrix* $\mathbf{\Lambda}_k$ given by

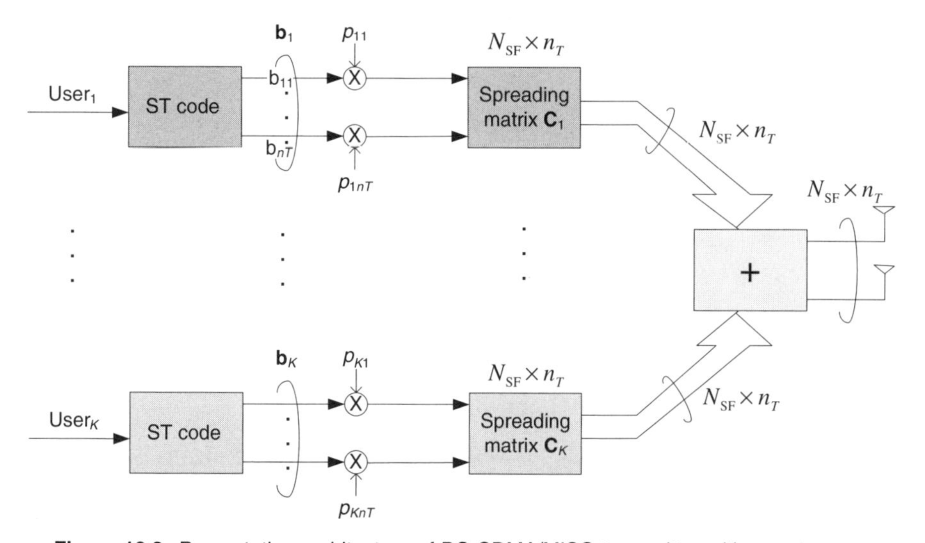

Figure 10.2. Base station architecture of DS-CDMA/MISO transmitter with n_T antennas.

$$\mathbf{\Lambda}_k = \begin{bmatrix} \sqrt{p_{k,1}} & \cdots & 0 \\ \vdots & \ddots & \vdots \\ 0 & \cdots & \sqrt{p_{k,nT}} \end{bmatrix} \tag{10.4}$$

where $\mathrm{tr}(\mathbf{\Lambda}_k\mathbf{\Lambda}_k^*) = p_k$ is the total power allocated to user k. We assume that the total transmit power in the base station is constrained to P_0

$$\sum_{k \in \mathcal{A}} \mathrm{tr}(\mathbf{\Lambda}_k \mathbf{\Lambda}_k^*) = P_0 \tag{10.5}$$

where $\mathcal{A} = \{k \in [1, K] : p_k > 0\}$ is the active user set.

Consider frequency-selective fading channels with L_p resolvable multipaths. The impulse response between the mth transmit antenna and the kth mobile is given by

$$h_{k,m}(t) = \sum_{p=1}^{L_p} h_{k,m}^{(p)} \delta\left(t - \tau_{k,m}^{(p)}\right)$$

where $\tau_{k,m}^{(p)}$ denotes the delay of the pth path for the kth user from the mth transmit antenna. We assume $N_{\mathrm{SF}} T_c \gg \max_{k,m,p} \tau_{k,m}^{(p)}$. In addition, for the sake of simplicity, we assume that all paths are synchronous, although in practice, these paths will arrive at slightly different delays [39, 97]. Hence, the $N_{\mathrm{SF}} \times 1$ received signal at user k is given by

$$\mathbf{Y}_k = \sum_{k \in \mathcal{A}} \mathbf{S}_j \mathbf{B}_j + \mathbf{Z}_k \tag{10.6}$$

where $\mathbf{B}_j$ is the $n_T \times 1$ transmitted vector of user j, $\mathbf{Z}_k$ is the $N_{\mathrm{SF}} \times 1$ complex Gaussian noise vector with covariance matrix $\sigma_z^2 \mathbf{I}_{N_{\mathrm{SF}}}$

$$\mathbf{S}_j = \sum_{p=1}^{L_p} \mathbf{C}_j^{(p)} \mathbf{H}_j^{(p)} \mathbf{\Lambda}_j, \tag{10.7}$$

where $\mathbf{H}_k^{(p)}$ is the $n_T \times n_T$ diagonal matrix consisting of the complex fading gains from the n_T transmit antennas along path p

$$\mathbf{H}_k^{(p)} = \begin{bmatrix} h_{k,1}^{(p)} & \cdots & 0 \\ \vdots & \ddots & \vdots \\ 0 & \cdots & h_{k,nT}^{(p)} \end{bmatrix} \tag{10.8}$$

and $\mathbf{C}_j^{(p)}$ is the $N_{\mathrm{SF}} \times n_T$ spreading sequence matrix (assigned to user j) along the pth path.

We assume that the channel gains across different paths, different antennas, and different users are i.i.d. complex Gaussian random processes with zero mean and variance $\sigma_h^2(p)$ set according to the *power-delay profile* such that $\Sigma_{p=1}^{L_p} \sigma_h^2(p) = 1$. To decouple the physical layer performance from specific spreading sequence design, we assume random sequence. The spreading sequence matrix $\mathbf{C}_j^{(p)}$ consists of i.i.d. entries with zero mean and variance $1/N_{\text{SF}}$. In practice, sequences across different paths are related through shifts. However, we assume small autocorrelation of the sequence and therefore, the sequence matrices $\mathbf{C}_j^{(p)}$ are independent for $p = 1, \ldots, L_p$. This shall greatly simplify the analysis, and it is shown [39] to be quite accurate through simulations.

Without loss of generality, we consider the receive processing for user k. The received signal can be written as follows:

$$\mathbf{Y}_k = \underbrace{\mathbf{S}_k\mathbf{B}_k}_{\text{Desired signal}} + \underbrace{\sum_{j \neq k} \mathbf{S}_j\mathbf{B}_j}_{\text{Interference}} + \underbrace{\mathbf{Z}_k}_{\text{Noise}} \tag{10.9}$$

For low-complexity implementations, we assume linear processing. Specifically, the received vector $\mathbf{Y}_k$ is premultiplied by a linear vector $\mathbf{F}_k^*$:

$$V_k = \mathbf{F}_k^*\mathbf{Y}_k = \mathbf{F}_k^*\mathbf{S}_k\mathbf{B}_k + \mathbf{F}_k^* \sum_{j \neq k} \mathbf{S}_j\mathbf{B}_j + \mathbf{F}_k^*\mathbf{Z}_k \tag{10.10}$$

In particular, we shall focus on the conventional *matched-filter* approach where $\mathbf{F}_k = \mathbf{S}_k$. In this case, the matched-filter (MF) output is given by

$$V_k(\text{MF}) = \mathbf{S}_k^*\mathbf{S}_k\mathbf{B}_k + \mathbf{S}_k^* \sum_{j \neq k} \mathbf{S}_j\mathbf{B}_j + \mathbf{S}_k^*\mathbf{Z}_k \tag{10.11}$$

Note that

$$V_k(\text{MF}) = \mathbf{S}_k^*\mathbf{Y}_k = \sum_{p=1}^{L_p} \mathbf{\Lambda}_k^* \mathbf{H}_k^{(p)*} \mathbf{C}_j^{(p)*} \mathbf{Y}_k$$

indicates that the matched-filter processing is the same as the standard RAKE receiver processing as illustrated in Figure 10.3.

In much the same way as we did in previous chapters, we shall model the performance of the multiuser physical layer using the information-theoretic approach. The maximum data rate of user k (bits per channel use) based on matched-filter processing and perfect CSIR is given by

$$\begin{aligned} C_k &= \max_{p(\mathbf{B}_k)} I(\mathbf{B}_k; V_k | \mathbf{S}) \\ &= \log \left| \mathbf{I}_{n_T} + \left(\mathbf{S}_k^* \mathbf{Q} \mathbf{S}_k \right)^{-1} \mathbf{S}_k^* \mathbf{P} \mathbf{S}_k \right| \end{aligned} \tag{10.12}$$

where $\mathbf{P} = \mathbf{S}_k\mathbf{S}_k^*$, $\mathbf{Q}$ is given by

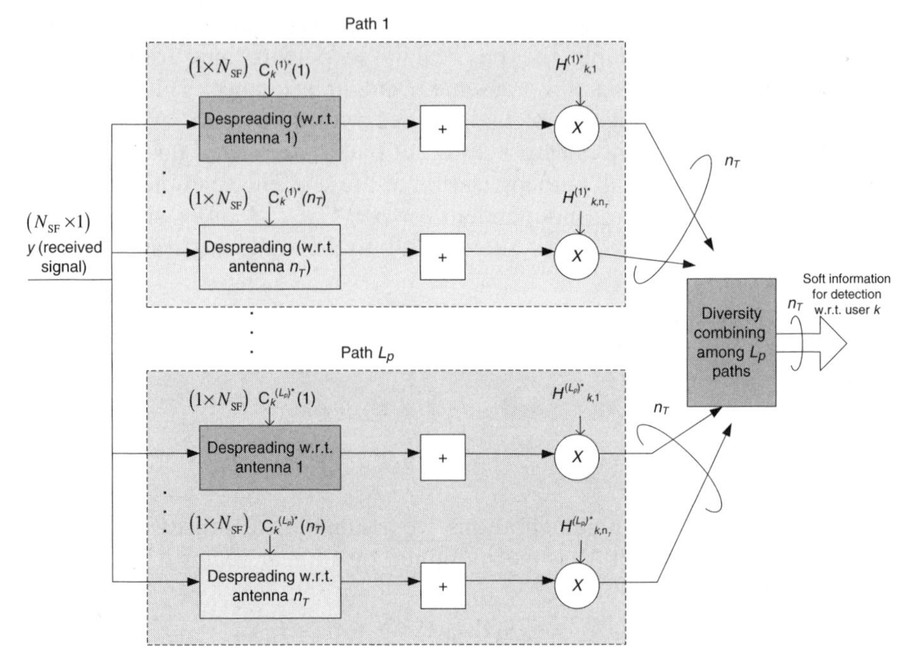

Figure 10.3. RAKE receiver processing of multiuser DS-CDMA/MISO systems.

$$\mathbf{Q} = \sum_{j\neq k} \mathbf{S}_j \mathbf{S}_j^* + \sigma_z^2 \mathbf{I}_{N_{SF}}$$

To further simplify the capacity expression, we consider the asymptotic situation where $N_{SF} \to \infty$. In addition, we assume that we have a wideband transmission where the transmitted bandwidth is much larger than the coherence bandwidth of the frequency-selective fading channels. The results are summarized in the following lemma.

Lemma 10.1 (Asymptotic Capacity) At sufficiently large spreading factor N_{SF} and sufficiently large number of multipaths L_p, the channel capacity of user k converge to a limit that is independent of the sequence matrix $\mathbf{C}_k$. The capacity limit in bits per channel use is given by

$$C_k\left(\mathbf{H}_k^{(p)}\right) = \overline{C}_k = n_T \log_2\left(1 + \frac{p_k}{n_T(\sigma_z^2) + \dfrac{P_0}{N_{SF}}}\right) \tag{10.13}$$

and is achieved by uniformly allocating the transmit power across the n_T antennas: $\mathbf{\Lambda}_k = \sqrt{\dfrac{p_k}{n_T}}\mathbf{I}_{n_T}$.

Proof Consider the first term $\mathbf{S}_k^*\mathbf{S}_k$. By the law of large numbers, we have

$$\begin{aligned}\mathbf{S}_k^*\mathbf{S}_k &= \sum_{p=1}^{L_p}\sum_{p'=1}^{L_p} \mathbf{\Lambda}_k^*\mathbf{H}_k^{(p)*}\mathbf{C}_k^{(p)*}\mathbf{C}_k^{(p')}\mathbf{H}_k^{(p')}\mathbf{\Lambda}_k \\ &\underbrace{\rightarrow}_{a.s}\sum_p \mathbf{\Lambda}_k^*\mathbf{H}_k^{(p)*}\mathbf{H}_k^{(p)}\mathbf{\Lambda}_k \\ &= \mathbf{\Lambda}_k^*\left(\sum_p \mathbf{H}_k^{(p)*}\mathbf{H}_k^{(p)}\right)\mathbf{\Lambda}_k \\ &\rightarrow \mathbf{\Lambda}_k^*\mathbf{\Lambda}_k \end{aligned} \tag{10.14}$$

On the other hand, consider the term $\mathbf{S}_j^*\mathbf{Q}\mathbf{S}_j$. We have

$$\begin{aligned}\mathbf{S}_k^*\mathrm{Q}\mathbf{S}_k &= \sigma_z^2\mathbf{S}_k^*\mathbf{S}_k + \sum_{j\neq k}\mathbf{S}_k^*\mathbf{S}_j\mathbf{S}_j^*\mathbf{S}_k \\ &\rightarrow \sigma_z^2\mathbf{S}_k^*\mathbf{S}_k + \sum_{j\neq k}\mathbf{S}_k^*\varepsilon\left[\mathbf{S}_j\mathbf{S}_j^*\right]\mathbf{S}_k \end{aligned} \tag{10.15}$$

where we have fixed $\mathbf{S}_k$ and apply the law of large numbers elementwise. The term $\varepsilon[\mathbf{S}_j\mathbf{S}_j^*]$ in (10.15) can be further simplified as

$$\begin{aligned}\varepsilon\left[\mathbf{S}_j\mathbf{S}_k^*\right] &= \varepsilon\left[\sum_p\sum_{p'}\mathbf{C}_j^{(p)}\mathbf{H}_k^{(p)}\Lambda_j\Lambda_j^*\mathbf{H}_k^*\mathbf{C}_j^*\right] \\ &\rightarrow \frac{1}{N_{\mathrm{SF}}}\sum_{m=1}^{n_T} p_{j,m}\left(\sum_{p=1}^{L_p}\left|H_{k,m}^{(p)}\right|^2\right)\mathbf{I}_{n_T} \\ &\rightarrow \frac{1}{N_{\mathrm{SF}}}p_j\mathbf{I}_{n_T} \end{aligned} \tag{10.16}$$

Hence, substituting (10.16) into (10.15), we have

$$\begin{aligned}\mathbf{S}_k^*\mathbf{Q}\mathbf{S}_k &\rightarrow \left(\sigma_z^2 + \sum_{j\neq k}\frac{p_j}{N_{\mathrm{SF}}}\right)\mathbf{S}_k^*\mathbf{S}_k \\ &\simeq \left(\sigma_z^2 + \frac{P_0}{N_{\mathrm{SF}}}\right)\mathbf{S}_k^*\mathbf{S}_k \end{aligned} \tag{10.17}$$

where $\Sigma_{j\in\mathcal{A}}p_j = P_0$ is the total downlink transmit power constraint.

Hence, substituting Equations (10.17) and (10.14) into the capacity equation in (10.12), we obtain the results.

This lemma illustrates the effectiveness of frequency diversity as a result of wideband transmission over the frequency-selective fading channels. Hence, the capacity of user k approaches a constant value independent of the channel fading variations as indicated in Lemma 10.1. In addition, the spreading matrix transforms the MIMO channels into n_T parallel scalar channels. Since the channel fadings are i.i.d. across different users, we have $p_1 = p_2 = \ldots = p_K = \frac{P_0}{K}$ and $c_1 = c_2 = \ldots = \bar{c}(P_0/K)$.

10.2.2 Physical Layer Model of OFDMA/MISO Systems

10.2.2.1 Overview of OFDM Systems. Another popular approach in the physical layer to combat frequency-selective fading is to split the transmission bandwidth into a large number of subcarriers. Each subcarrier carries part of the user data. Since the bandwidth of each subcarrier is narrow, the subcarrier will experience flat fading instead of frequency-selective fading, therefore avoiding expensive equalization at the receiver. In conventional multicarrier systems, the whole spectrum is divided into many narrow and nonoverlapping subchannels. While this method can avoid interchannel interference (ICI), the spectral efficiency is low. On the other hand, OFDM systems allow orthogonality between subcarriers while allowing overlapping between them in the frequency domain as illustrated in Figure 10.4. Orthogonality is maintained

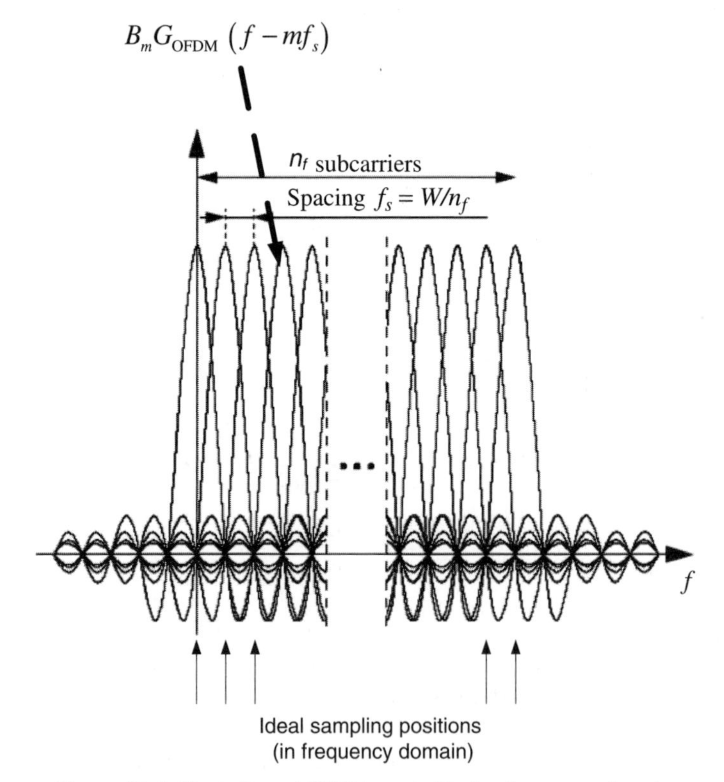

Figure 10.4. Illustration of OFDM symbol in the frequency domain.

because in each subcarrier the zero point aligns with the center of other subcarriers. These concepts are elaborated quantitatively below.

In the conventional *time-domain modulation schemes*, the modulation symbols are carried in the transmitted pulses in the time domain as

$$x_{\text{time-mod}}(t) = \sum_m b_m g_{\text{time-mod}}(t - mT_s) \tag{10.18}$$

where $g_{\text{time-mod}}(t)$ is the modulation pulse satisfying the Nyquist criteria (with the zero crossing points at $t = nT_s$) and T_s is the modulation symbol duration. At the receiver, if we have ideal sampling instants at $t = 0, T_s, 2T_s, \ldots$, there will be no ISI between the modulation symbols b_m and b_j. On the other hand, in OFDM systems, we modulate the information bits in *frequency domain*. Hence, analogous to the time-domain modulation, the information bits are modulated on frequency domain in OFDM systems as below (Figure 10.4)

$$X_{\text{OFDM}}(f) = \frac{1}{n_f} \sum_{m=1}^{n_f - 1} B_m G_{\text{OFDM}}(f - mf_s) \tag{10.19}$$

where $G_{\text{OFDM}}(f)$ is the *modulation pulse* in frequency domain satisfying the Nyquist criteria (with zero crossing points at $f = nf_s$), f_s is the dual of the modulation symbol duration T_s, and n_f is the total number of *subcarriers*. Hence, at the receiver, we shall sample the modulated symbols $\{B_m\}$ in frequency domain. If there is no frequency offset at the receiver to enable us to sample (in frequency domain) at the ideal frequencies $f = 0, f_s, 2f_s, \ldots$, there will be no *interchannel interference* (ICI) between B_m and B_j. Hence, by designing the *shape* of the frequency-domain pulse $G_{\text{OFDM}}(f)$, we can determine the robustness of the OFDM systems with respect to frequency offset at the receiver (similar to the way in which pulseshaping in time domain can reduce the ISI due to timing error in time-domain modulation systems). Considering the time-domain version of the OFDM symbol, we have

$$x_{\text{OFDM}}(t) = \frac{1}{n_f} \sum_{m=0}^{n_f - 1} B_m \exp(j2\pi m f_s t) g_{\text{OFDM}}(t) \tag{10.20}$$

where $g_{\text{OFDM}}(t)$ is the *time-domain* version of the frequency-domain modulation carrier $G_{\text{OFDM}}(f)$ as illustrated in Figure 10.5a. Note that $g_{\text{OFDM}}(t)$ is a time-limited pulse with duration T_s given by

$$T_s = \frac{1}{f} \tag{10.21}$$

Figure 10.5b compares the power spectral density of OFDM signals between a *raised-cosine* window function and a regular rectangular window. We can see that with the raised-cosine time-domain window, the spectrum falls off rapidly compared to the rectangular window. Yet, for simplicity in the analysis, we shall

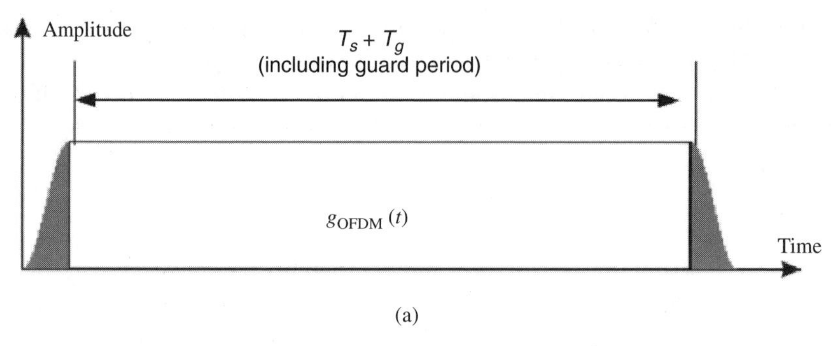

(a)

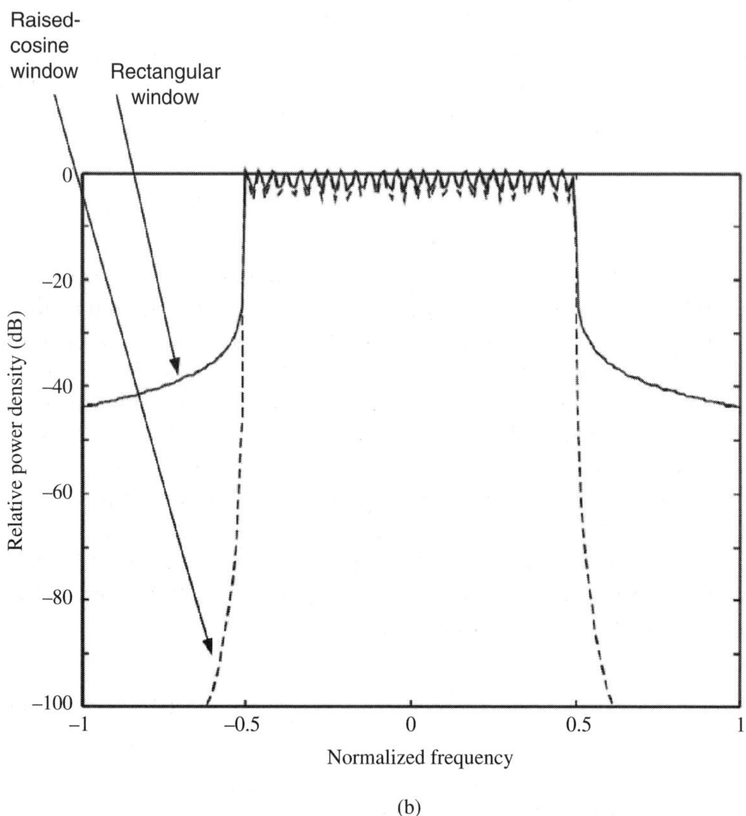

(b)

Figure 10.5. (a) Time-domain windowing $g_{OFDM}(t)$ and its effect on the (b) power spectral density of OFDM signals.

assume $g_{OFDM}(t) = 1$ for all $t \in [0, T_s]$ and zero otherwise. Hence, $G_{OFDM}(f) = \text{sinc}(f/f_s)$.

Consider the discrete-time processing of OFDM systems. The transmission bandwidth of $X_{OFDM}(f)$ (bandpass) is given by

$$W = n_f f_s \tag{10.22}$$

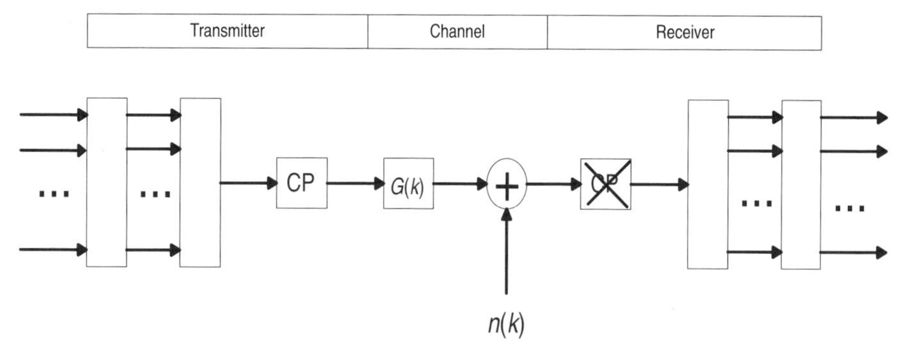

Figure 10.6. Architecture of OFDM systems based on IDFT and DFT.

where n_f is the number of subcarriers used in the OFDM systems. On the basis of the Nyquist sampling criteria, the time-domain signal $x_{\text{OFDM}}(t)$ can be sampled at the sampling frequency $f_0 \geq W = n_f f_s$. The time-domain samples (sampled at $f_0 = n_f f_s$) are given by

$$x_{\text{OFDM}}(n) = \frac{1}{n_f} \sum_{m=0}^{n_f-1} B_m \exp\left(j2\pi \frac{mn}{n_f} \right) \tag{10.23}$$

for $n = 0, 1, \ldots, n_f - 1$, and the sampling period T_0 is given by $T_0 = F_s/n_f$. However, observe that Equation (10.23) is exactly the same as the IDFT equation with respect to the n_f *frequency-domain samples* $\{B_0, \ldots, B_{n_f-1}\}$. Hence, the discrete-time-domain samples of the OFDM signal $x_{\text{OFDM}}(t)$ can be expressed as

$$\mathbf{x} = [x_0, \ldots, x_{n_f-1}] = \text{IDFT}(B_0, \ldots, B_{n_f-1}) \tag{10.24}$$

This is illustrated in Figure 10.6. The transmit data in frequency domain $\mathbf{B} = [B_0, \ldots, B_{n_f-1}]$ are fed to the IDFT unit followed by a parallel-to-serial convertor to get a transmit data block $\mathbf{x} = [x_0, \ldots, x_{n_f-1}]$. Before it is transmitted, a cyclic prefix of duration N_{cyclic} is added in the front of this block. The cyclic prefix is a replica of the last N_{cyclic} samples of the transmitted data block. Hence, the transmitted data block after the addition of cyclic prefix is given by

$$\tilde{\mathbf{x}} = [x_{nf} - N_{\text{cyclic}}, \ldots, x_{n_f-1}, x_0, \ldots, x_{n_f-1}]$$

There are two reasons to add the cyclic prefix:

1. Because of the multipath propagation of the frequency-selective fading channels, the received data block on one path will be superimposed with the received data blocks from the other paths at different delays, and this will cause intersymbol interference (ISI) between adjacent blocks.

2. Because of the multipaths of the frequency-selective fading channels, samples within a block will interfere with each other in frequency domain, resulting in intercarrier interference (ICI).

Hence, cyclic prefix is added to combat these two effects and this is elaborated below quantitatively. In general, we shall have $N_{\text{cyclic}}T_s/n_f > \sigma_\tau$, specifically, duration of the cyclic prefix longer than the delay spread of the channels.

Consider a frequency-selective fading channel with impulse response given by

$$h(t) = \sum_{p=1}^{L_p} h_p \delta(t - pT_0) \tag{10.25}$$

The received samples after the frequency-selective fading channel are expressed as

$$\tilde{\mathbf{y}} = \tilde{\mathbf{x}} * \mathbf{h} + \mathbf{z} \tag{10.26}$$

where $\tilde{\mathbf{y}} = [y_0, \ldots]$ denotes the block of the $n_f + L_p - 1$ received samples, $\mathbf{h} = [h_1, \ldots, h_{L_p}]$ denotes the FIR response of the channel model, $\tilde{\mathbf{x}}$ denotes the block of transmitted OFDM samples with cyclic prefix added, $\mathbf{z} = [z_0, \ldots]$ denotes the block of i.i.d. channel noise, and $*$ denotes linear convolution.

At the receiver, the first N_{cyclic} elements of received data, $[y_0, \ldots, y_{N_{\text{cyclic}}-1}]$, are discarded. Let $\mathbf{y} = [y_{N_{\text{cyclic}}}, \ldots, y_{N_{\text{cyclic}}+n_f-1}]$ be the block of n_f received samples after removal of the cyclic prefix. Applying DFT to $\mathbf{y}$, the block of n_f frequency-domain received samples $\mathbf{Y}$ is given by

$$\begin{aligned}
\mathbf{Y} &= \text{DFT}(\mathbf{y}) \\
&= \text{DFT}(\tilde{\mathbf{x}} * \mathbf{h} + \mathbf{z}) \\
&= \text{DFT}(\mathbf{x} \otimes \mathbf{h} + \mathbf{z}) \\
&= \text{DFT}(\text{IDFT}(\mathbf{B}) \otimes \mathbf{h} + \mathbf{z}) \\
&= \mathbf{B} \odot \text{DFT}(\mathbf{h}) + \text{DFT}(\mathbf{z})
\end{aligned} \tag{10.27}$$

where $\otimes$ denotes cyclic convolution, DFT(*) denotes the DFT transform on the elements whose indexes are $0, 1, \ldots, n_f - 1$, and $\odot$ denotes element-by-element multiplication between two vectors. We can see that because of the cyclic prefix, the linear convolution becomes cyclic convolution. Hence, the DFT transform of cyclic convolution of two sequences is equal to the multiplication of cyclic convolution of these two sequences. Hence, the $n_f \times 1$ received vector $\mathbf{Y}$ can be expressed as

$$\mathbf{Y} = \mathbf{H}\mathbf{B} + \mathbf{Z} \tag{10.28}$$

where $\mathbf{B}$ is the $n_f \times 1$ transmitted vector (in frequency domain), $\mathbf{H}$ is the $n_f \times n_f$ diagonal channel matrix with the mth diagonal element given by the mth output of DFT($\mathbf{h}$), and $\mathbf{Z}$ is the $n_f \times 1$ noise vector, which has the same covariance matrix as $\mathbf{z}$ given by $\sigma_z^2/n_f\mathbf{I}_{n_f}$. In other words, from Equation (10.28), the frequency-selective fading channel is transformed into n_f parallel flat fading channels with fading coefficient given by H_m.

10.2.2.2 Multiuser OFDMA/MISO Systems. In this subsection, we shall outline the multiuser physical layer model of a base station utilizing OFDM and MIMO. While OFDM is a point-to-point technology to combat frequency-selective fading channels, it can be extended to multiuser situations. The *orthogonal frequency-division multiple-access* (OFDMA) system is the multiple-access extension of OFDM systems. On the basis of IDFT and DFT kernels, the frequency-selective fading channel is decomposed into n_f parallel and independent flat fading channels as illustrated in Equation (10.28). Hence, at the base station, different users can share the n_f parallel channels by assigning different subsets of subcarriers to each user.

We consider a system with a base station (n_T transmit antennas) and K mobile users (each of which has single receive antenna). Figure 10.7 illustrates the transmitter architecture of the multiuser base station design (based on OFDMA) with n_T transmit antennas. A user can be assigned resources in *spatial domain* as well as *frequency domain*. Considering the mth subcarrier, information bits from the K users are first encoded to produce the encoded symbols $\{U_{m,1}, \ldots, U_{m,K}\}$. The encoded symbols are assumed to have unit vari-

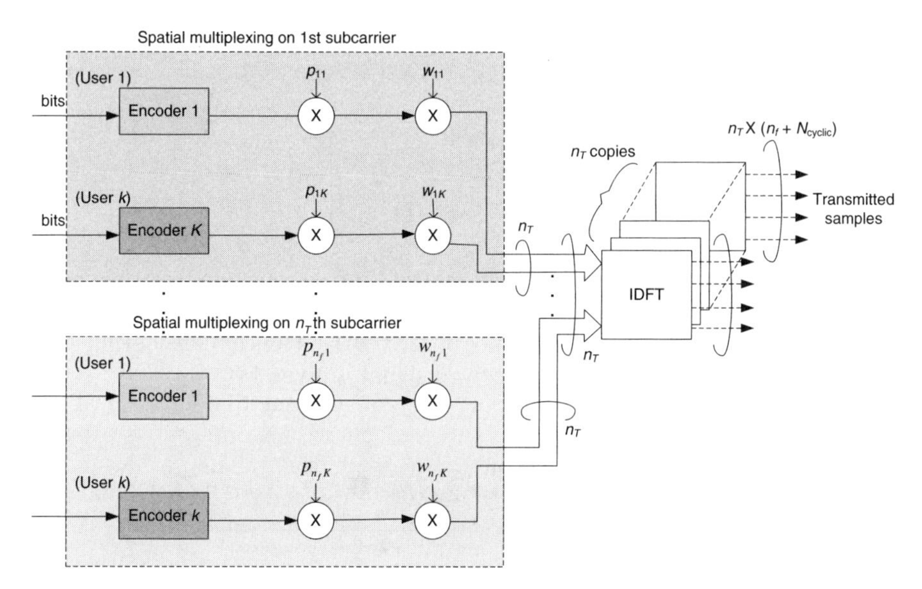

Figure 10.7. Base station architecture of OFDMA/MISO systems with n_T transmit antennas.

ance (for normalization) and are amplified by the respective power gains $\{\sqrt{p_{m,1}}, \ldots, \sqrt{p_{m,K}}\}$ ($p_{m,k} \geq 0$) and carried by the $n_T \times 1$ *spatial multiplexing weight vectors* $\{\mathbf{w}_{m,1}, \ldots, \mathbf{w}_{m,K}\}$, respectively. The processed signals are fed to the mth inputs in the bank of n_T IDFT processors. The $n_T \times n_f$ outputs of the IDFT are first added with cyclic prefix. The resulting $n_T \times (n_f + N_{\text{cyclic}})$ blocks of time-domain samples are transmitted over the n_T transmit antennas column by column. At the receiver of each of the K mobiles, there is a corresponding DFT block. Hence, we can analyze the physical layer performance from the frequency domain. The $n_T \times 1$ transmitted signal, $\mathbf{B}_m$ at the mth input of the IDFT block is given by

$$\mathbf{B}_m = \sum_{k=1}^{K} \sqrt{p_{m,k}}\, \mathbf{w}_{m,k} U_{m,k} \tag{10.29}$$

Without loss of generality, the $n_f \times 1$ frequency-domain received signal vector of the kth mobile user is given by

$$\mathbf{Y}_k = \mathbf{H}_k \mathbf{B} + \mathbf{Z}_k \tag{10.30}$$

where $\mathbf{B}$ is the $n_T n_f \times 1$ aggregate transmit vector (at the input of IDFT), given by

$$\mathbf{B} = \begin{bmatrix} \mathbf{B}_0 \\ \vdots \\ \mathbf{B}_{n_f-1} \end{bmatrix}$$

$\mathbf{H}_k$ is the $n_f \times n_f n_T$ block diagonal channel matrix (aggregate over the n_f subcarriers) of user k, given by

$$\mathbf{H}_k = \begin{bmatrix} \mathbf{H}_{0,k} & 0 & 0 \\ 0 & \ddots & 0 \\ 0 & 0 & \mathbf{H}_{n_f-1,k} \end{bmatrix}$$

where $\mathbf{H}_{m,k}$ denotes the $1 \times n_T$ channel matrix (in frequency domain) between the n_T transmit antennas and the kth mobile and $\mathbf{Z}_k$ is the $n_f \times 1$ zero-mean white Gaussian noise with covariance matrix $\sigma_z^2 \mathbf{I}_{n_f}$. Hence, considering mth subcarrier at the kth mobile, the received signal is given by

$$\begin{aligned} Y_{m,k} &= \mathbf{H}_{m,k} \mathbf{B}_m + Z_{m,k} \\ &= \underbrace{\sqrt{p_{m,k}}\, H_{m,k} \mathbf{w}_{m,k} U_{m,k}}_{\text{Desired signal}} + \underbrace{\sum_{j \neq k} \sqrt{p_{m,j}}\, \mathbf{H}_{m,j} \mathbf{w}_{m,j} U_{m,j} + Z_{m,k}}_{\text{Multiuser intereference}} \end{aligned} \tag{10.31}$$

Similarly, we shall adopt the zero-forcing (ZF) approach in choosing the spatial multiplexing weights $\{\mathbf{w}_{m,k}\}$ as in Chapter 6. Define $\mathcal{A}_m = \{k \in [1, K]:$

$p_{m,k} > 0\}$ as the set of active users corresponding to the mth subcarrier. The *spatial multiplexing weight* $\mathbf{w}_{m,k}$ is chosen according to the following optimization problem.

Problem 10.1 (ZF Weight Determination) Given the active user sets $\{\mathcal{A}_0, \ldots, \mathcal{A}_{n_f-1}\}$. For each subcarrier m, given $k \in \mathcal{A}_m$, choose $\mathbf{w}_{m,k}$ to maximize $|\mathbf{H}_{m,k}\mathbf{w}_{m,k}|^2$ with following the constraints:

$$\mathbf{H}_{m,j}\mathbf{w}_{m,k} = 0$$
$$\mathbf{w}_m^* \mathbf{w}_{m,k} = 1$$

Since there are n_T complex degrees of freedom in $\mathbf{w}_{m,k}$ and there are $2|\mathcal{A}m| - 1$ equations as constraints, we have

$$|\mathcal{A}_m| \le n_T \tag{10.32}$$

for all $m \in [0, n_f - 1]$. The solution of the ZF spatial multiplexing weight is outlined in Chapter 6. Hence, the received signal is given by

$$Y_{m,k} = \sqrt{p_{m,k}}\,\mathbf{H}_{m,k}\mathbf{w}_{m,k}U_{m,k} + Z_{m,k} \tag{10.33}$$

Given the sets of active users across the n_f subcarriers $\mathcal{A}_0, \ldots, \mathcal{A}_{n_f-1}$, the maximum data rate of user k at the mth subcarrier is given by the maximum of the mutual information between $U_{m,k}$ and $Y_{m,k}$. The total data rate of user k (in bits per channel use) is given by

$$C_k(\mathbf{H}_k) = \begin{cases} \sum_{m:k \in \mathcal{A}_m} \log_2\left(1 + \frac{p_{m,k}|\mathbf{H}_{m,k}\mathbf{w}_{m,k}|^2}{\sigma_z^2 / n_f}\right), & \text{if } k \in \mathcal{A}_m \text{ for some } m \\ 0, & \text{otherwise} \end{cases} \tag{10.34}$$

Assuming that the total transmit power from the basc station is constrained at P_0, we have

$$\sum_{m=0}^{n_f-1} \sum_{k \in \mathcal{A}_m} p_{m,k} \le P_0 \tag{10.35}$$

10.3 CROSS-LAYER SCHEDULING DESIGN FOR DS-CDMA/MISO AND OFDMA/MISO SYSTEMS

In this section, we shall consider the cross-layer scheduling design for both the DS-CDMA/MISO and the OFDMA/MISO physical layers for frequency-selective fading channels. We consider a system with a base station (n_T transmit antennas) and K mobile users (each has single receive antenna). For simplicity, we assume that the transmitter and the K receivers have perfect knowledge of CSIT and CSIR, respectively. Furthermore, we assume that the system performance is given by the *network capacity* utility function:

$$U_{\text{cap}}(\overline{R}_1, \ldots, \overline{R}_K) = \sum_{k=1}^{K} \mathrm{E}[r_k] \tag{10.36}$$

10.3.1 Cross-Layer Design for DS-CDMA/MISO Systems

For DS-CDMA/MISO systems, the information from a user is spread across the entire transmit bandwidth W. Assuming wideband transmission such that $W \gg B_c$ (where B_c is the coherence bandwidth of the channels), the frequency diversity order as a result of spreading is given by $L_p = \left\lceil \frac{W}{B_c} \right\rceil$, which is large. Hence, the fluctuation of user data rate among the K users at any fading slot is very small. In fact, as indicated by the Lemma 10.1, the instantaneous capacity of user k approaches a constant independent of the current channel fading as a result of the frequency diversity. Hence, there is no need for any sophisticated scheduling as far as the network capacity is concerned. The optimal network capacity of the K users is simply given by

$$U_{\text{cap}}(\overline{R}_1, \ldots, \overline{R}_K) \approx K n_T \log_2\left(1 + \frac{P_0}{K n_T\left(\sigma_z^2 + \frac{P_0}{N_{\text{SF}}}\right)}\right) \tag{10.37}$$

10.3.2 Cross-Layer Design for OFDMA/MISO Systems

For OFDMA/MISO systems, a user can be assigned a spatial channel and/or multiple subcarriers depending on the instantaneous CSI. Hence, the cross-layer scheduling algorithm can be cast into the following optimization problem.

Problem 10.2 (Cross-Layer Optimization for OFDMA/MISO) Given a CSI realization $\mathbf{H}_1, \ldots, \mathbf{H}_K$, choose the active set allocation $\mathcal{A} = \{\mathcal{A}_0, \ldots, \mathcal{A}_{n_f-1}\}$ and the power allocation $\mathcal{P} = \{p_{m,k}: m \in [0, n_f - 1], k \in \mathcal{A}_m\}$ such that

$$\sum_{m=0}^{n_f-1} \sum_{k \in \mathcal{A}_m} p_{m,k} = P_0$$

and

$$|\mathcal{A}_m| \leq n_T$$

for all $m \in [0, n_f - 1]$ and the instantaneous network capacity $\Sigma_{k=1}^{K} c_k(\mathbf{H})$ is maximized where c_k is given by Equation (10.34).

There are two sets of optimization variables in Problem 10.2: the power allocation $\mathcal{P}$ and the active user allocation $\mathcal{A}$. The former is a continuous variable; the latter is a discrete variable. Hence, the optimization problem can be solved in the following steps.

Step 1: Initialization. For each subcarrier $m \in [0, n_f - 1]$, initialize the active user sets $\mathcal{A}_m$ such that $|\mathcal{A}_m| \le n_T$.

Step 2: Optimization with respect to $\mathcal{P}$. Given the active user sets $\mathcal{A}_m$, the optimization problem can be written into the following Lagrangian function

$$L(p_{m,k}, \lambda) = \sum_{m=0}^{n_f-1} \sum_{k \in \mathcal{A}_m} \left[\log_2\left(1 + \frac{p_{m,k}|\mathbf{H}_{m,k}\mathbf{w}_{m,k}|^2}{\sigma_z^2/n_f}\right) - \lambda p_{m,k} \right] \tag{10.38}$$

where λ is the Lagrange multiplier chosen to satisfy the total transmit power constraint P_0. Using standard optimization techniques and the KKT condition, the optimal power allocation is given by

$$p_{m,k}^* = \left(\frac{1}{(\ln 2)\lambda} - \frac{\sigma_z^2/n_f}{|\mathbf{H}_{m,k}\mathbf{w}_{m,k}|^2} \right)^+ \tag{10.39}$$

where $(x)^+ = \max(x, 0)$.

Step 3: Combinatorial search over $\mathcal{A}$. Since the active user sets $\mathcal{A}_m$ are discrete, one can search over all possible combinations on $\mathcal{A}_m$ satisfying the cardinality constraint and repeat step 2. However, the search for $\mathcal{A}$ might not be straight forward. For instance, due to the total transmit power constraint, the search of the active user sets across the n_F subcarriers may be coupled[2]. The exhaustive search complexity is given by $\left(\sum_{i=1}^{n_T}\binom{K}{i}\right)^{n_F}$ which is extremely complicated for moderate n_F. However, we have the following lemma that establishes the fact that the search of the optimal $\mathcal{A}$ can be decoupled from the n_F subcarriers. Hence, the complexity of the exhaustive search is reduced to $n_f\binom{K}{n_T}$. This complexity is still quite large for moderate K and n_f. Similar to Chapter 6, we shall employ some sub-optimal search algorithms to further reduce the search complexity.

Lemma 10.2 For sufficiently large P_0, the optimal admitted user set $\{\mathcal{A}_n\}$ has full cardinality $|\mathcal{A}_n| = n_T$ and the search for the optimal $\mathcal{A}_n$ can be decoupled from the n_F subcarriers. Specifically, the optimal $\mathcal{A}_n$ is given by:

[2]Due to the transmit power constraint in Problem 20, the Lagrandge multipler λ in (10.38) is dependent on the active user sets $\mathcal{A}$ across the n_F subcarriers. Hence, a change in $\mathcal{A}_m$ might distort the λ and therefore, change the choice for $\mathcal{A}_n$ as well.

$$A_n^* = \arg\max_{\mathcal{A}_n} \Pi_{k\in\mathcal{A}_n} |\mathbf{H}_{n,k}\mathbf{w}_{n,k}|^2$$

Since the search is independent of λ, the search can be decoupled from the n_F subcarriers.

Proof Please refer to Appendix A.

10.4 PERFORMANCE COMPARISONS OF DS-CDMA- AND OFDM-BASED SYSTEMS

In this section, we shall investigate the performance of wideband schedulers in frequency-selective fading channels. We shall consider the overall network capacity gain as L_p and n_T increase. Furthermore, we shall compare the multi-user performance levels of the OFDMA/MISO and DS-CDMA/MISO systems.

Recall that both DS-CDMA and OFDMA introduce additional dimensions into the signal model through different means. For DS-CDMA systems, the extra dimension is introduced through time-domain spreading N_{SF}. For OFDMA systems, the extra dimension is introduced through frequency-domain partitioning n_f. It is important to normalize the capacity in terms of a common unit (bps/Hz) for fair comparisons. For DS-CDMA systems, a vector symbol (per channel use) consists of N_{SF} chips. Hence, the normalized bit rate (bps/Hz) is given by

$$\tilde{C}_{\text{CDMA}} = \frac{1}{N_{\text{SF}}} C_{\text{CDMA}} \tag{10.40}$$

where C_{CDMA} is the capacity *in bits per channel use* and $\tilde{C}_{\text{CDMA}}$ is the capacity in bps/Hz. Using Lemma 10.1, the total system capacity (bps/Hz) is given by

$$\begin{aligned}\tilde{C}_{\text{CDMA}} &= \frac{1}{N_{\text{SF}}} \sum_{k=1}^{K} n_T \log_2\left(1 + \frac{P_0/K}{n_T\left(\sigma_z^2 + (P_0/K)\left(\frac{K}{N_{\text{SF}}}\right)\right)}\right) \\ &= \alpha n_T \log_2\left(1 + \frac{\gamma/n_T}{(1+\alpha\gamma)}\right)\end{aligned} \tag{10.41}$$

where $\gamma = \frac{P_0}{K\sigma_z^2}$ is the per-user SNR and α is the *CDMA loading factor* given by

$$\alpha = \frac{K}{N_{\text{SF}}} \tag{10.42}$$

For the OFDMA systems, a vector symbol per "channel use" consists of $n_f + N_{\text{cyclic}}$ samples. Hence, the normalized bit rate is given by

$$\tilde{C}_{\text{OFDMA}} = \frac{1}{n_f + N_{\text{cyclic}}} C_{\text{OFDMA}} \tag{10.43}$$

where C_{OFDMA} is the capacity in bits per channel use and $\tilde{C}_{\text{OFDMA}}$ is the capacity in bps/Hz.

10.4.1 Performance of DS-CDMA/MISO Systems

Figure 10.8a illustrates the normalized network capacity (bps/Hz) (or network spectral efficiency) of the DS-CDMA/MISO system versus loading factor α at high SNR and low SNR. The number of the resolvable paths in the channel, L_p, is assumed to be 16. It can be deduced from the figures that for a given SNR, the normalized network capacity (bps/Hz) or equivalently the network spectral efficiency increases as α increases. This is because of the factor of $1/N_{\text{SF}}$ in Equation (10.41). For a fixed number of users K, the largest α is K when the spreading factor $N_{\text{SF}} = 1$. This implies that DS-CDMA is not spectrally effective if capacity-achieving codes are employed at the transmitter. However, in practice when the error correction codes are not sufficiently powerful, we would like to use a larger spreading factor because this would enhance the SINR and therefore reduce the error probability of the packet transmission due to channel noise and interference. Thus, the choice of N_{SF} involves a tradeoff between SINR and the overall spectral efficiency. From the figure, the rate of increase in the network spectral efficiency saturates at moderate α (e.g., $\alpha = 2$ at high SNR).

On the other hand, Figure 10.8b illustrates the normalized network capacity (bps/Hz) versus the number of transmit antennas n_T. As n_T increases, the network capacity increases as a result of spatial multiplexing. Yet, there is a diminishing return on the network capacity due to the power splitting per spatial channel.

10.4.2 Performance of OFDMA/MISO Systems

Figure 10.9 illustrates the normalized bit rate of the OFDMA/MISO system versus SNR and the number of subcarriers n_f. The number of transmit antennas is assumed to be 3, the number of the resolvable paths L_p is assumed to be 16, and the number of users K is assumed to be 10. We use the genetic algorithm to search the optimizing admitted user set $\mathcal{A}_n$ because the optimal algorithm has very high computational complexity. Figure 10.9 shows the normalized network capacity (bps/Hz) versus (a) SNR and (b) n_f. We observe that increasing n_f will lead to smaller overheads due to cyclic prefix and hence, an increase in the normalized network capacity (or network spectral efficiency). As we increase the subcarriers, there are two factors which can affect the network spectral efficiency, namely the *guard time overhead* and the *power splitting between subcarriers*. The power splitting between subcarriers is compensated by the noise spitting effect as well because the noise power per sub-carrier is σ_z^2/n_f. Hence, the overhead SNR per subcarrier remains unchanged. On the other hand, when n_f is large, the overhead of the cyclic prefix is smaller and therefore this factor favors large n_f. However, there is diminishing returns on the spectral efficiency gain as we increase n_f. To reduce the implementation complexity (such

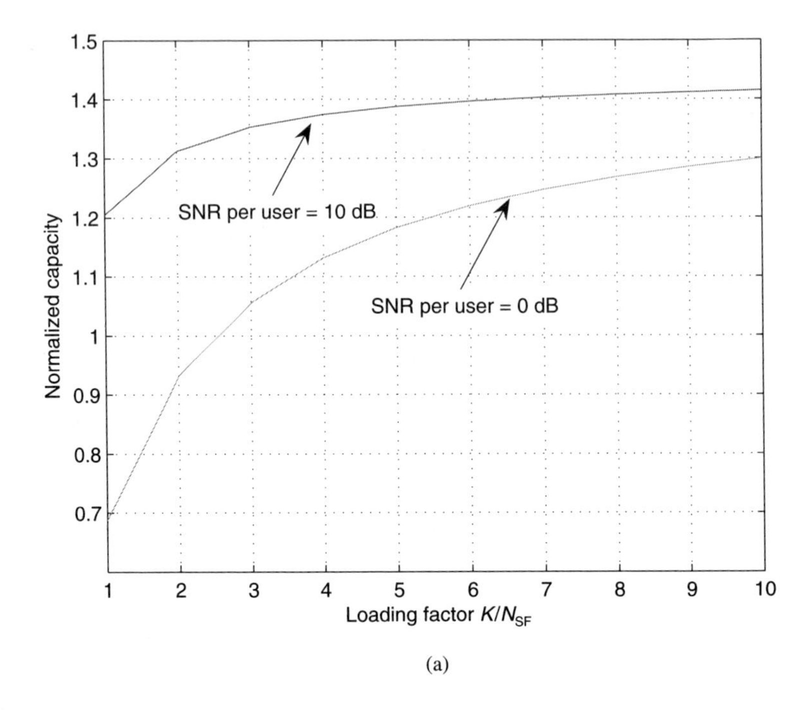

(a)

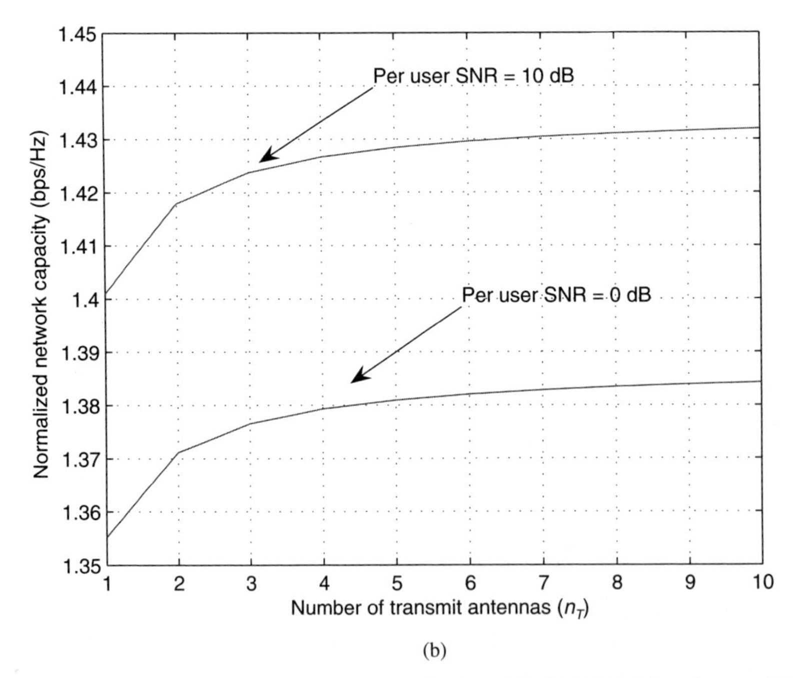

(b)

Figure 10.8. Normalized network capacity (bps/Hz) of DS-CDMA/MISO systems: (a) DS-CDMA/MISO network capacity versus loading factor; (b) DS-CDMA/MISO network capacity versus n_T.

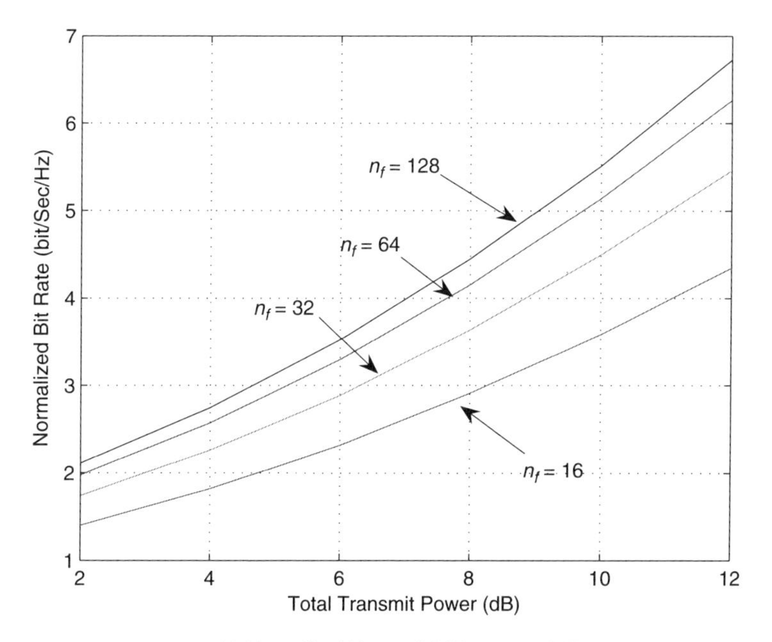

(a) Normalized bit rate (b/s/Hz) versus SNR.

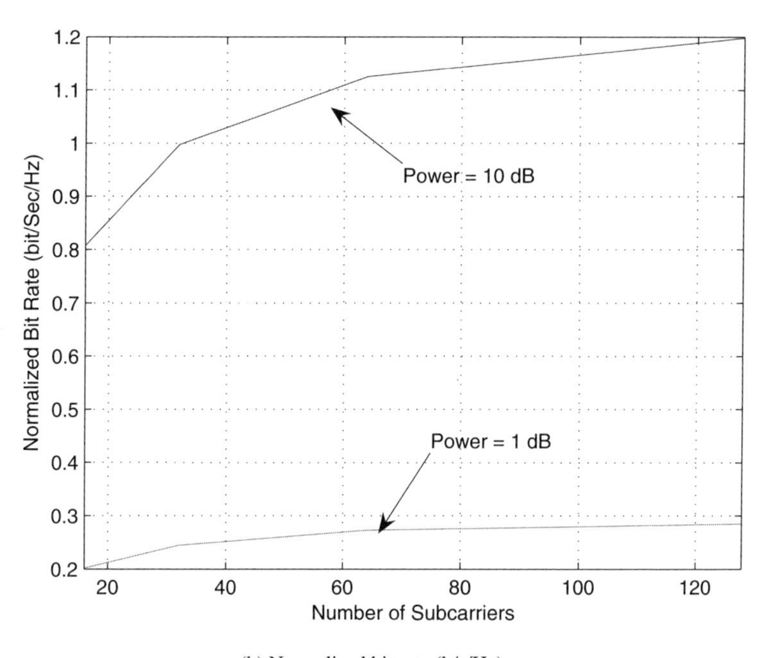

(b) Normalized bit rate (b/s/Hz) versus n_f.

Figure 10.9. Normalized bit rate (kbps/Hz) versus SNR (a) and n_f (b), with $n_T = 3$, $K = 10$, and $L_p = 16$.

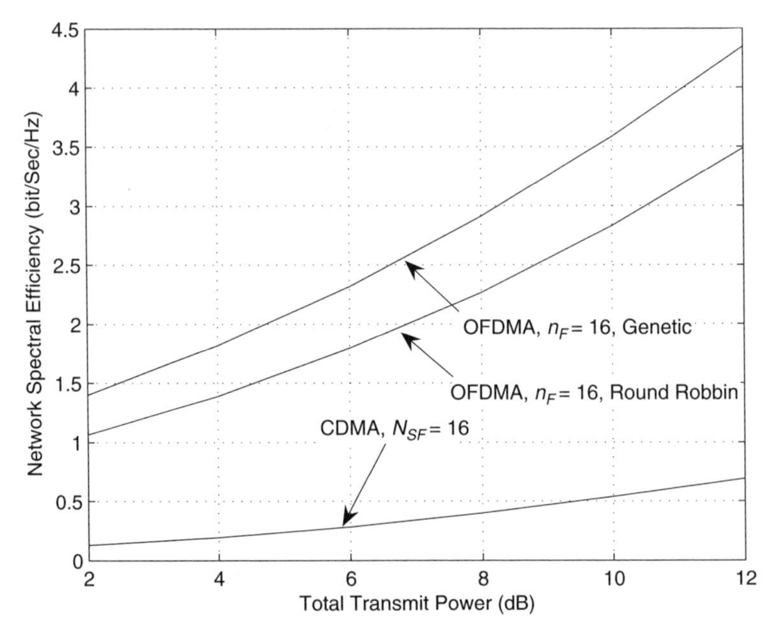

Figure 10.10. Performance comparison of OFDMA/MISO and DS-CDMA/MISO for K = 32, L_p = 16, and n_T = 3.

as the FFT size), we shall pick the smallest possible n_f, which is L_p because otherwise, each subcarrier does not see flat fading channels.

10.4.3 Comparisons of DS-CDMA/MISO and OFDMA/MISO Systems

Figure 10.10 compares the network spectral efficiency (bps/Hz) with the per-user SNR (γ) of the DS-CDMA/MISO and OFDMA/MISO systems. The number of the resolvable paths L_p is assumed to be 16, the number of users $K = 32$, and the number of transmit antennas $n_T = 3$. For the DS-CDMA/MISO system, the spreading factor is $N_{SF} = 16$ and the corresponding loading factor $\alpha = 2$, which corresponds to the high-spectral-efficiency region as illustrated in Figure 10.8a. It is illustrated that there is a significant SNR gain of ~5.5 dB in OFDMA/MISO compared with DS-CDMA/MISO. The significant gain is contributed by two factors. Firstly, the multi-user selection diversity in OFDMA/MISO system is more effective because the best users are selected to transmit on each of the n_f subcarriers only. For DS-CDMA/MISO system, signals from all the users are spread over the entire spectrum and therefore, they benefit from link level diversity but suffer from the lack of multi-user diversity. Secondly, there is multi-user interference in the DS-CDMA/MISO system due to the frequency selective fading channels. This factor is highlighted by comparing the performance of the OFDMA/MISO with round robin scheduling (no multi-user selection diversity) with respect to the DS-CDMA/MISO system performance.

10.5 IMPLEMENTATION ISSUES OF OFDMA SYSTEM

In the previous sections, we have focused on the cross-layer scheduling formulation and the performance comparisons of DS-CDMA/MISO and OFDMA/MISO systems. We found that the multiuser performance of OFDMA-based systems is significantly better than that of the DS-CDMA-based systems because of the multiuser diversity over the frequency dimension. However, the analysis is based on ideal assumptions. While OFDMA-based systems offer significant performance gains, proper frequency and timing synchronization is necessary to maintain user orthogonality among the active users. For instance, Doppler shifts and oscillator instability will contribute to ICI between subcarriers, and timing errors may result in ISI between consecutive OFDM symbols. We shall elaborate these implementation issues in this section [100]. Without loss of generality, we consider a single-antenna system $n_T = 1$ to simplify the analysis.

10.5.1 Downlink versus Uplink

Frequency and timing synchronizations for the single-user OFDM link have received much attention since the late 1990s both 1997. Several solutions are available in the literature [65,119]. These techniques can be applied in the multiuser downlink direction as well where the base station *broadcasts* the transmitted signals to the active users. Since the transmitted signals to all the active users originate from a single point (the base station), the frequency offsets at the receiver will be uniform across all the subcarriers and therefore, conventional techniques used to tackle frequency offsets for point-to-point OFDM systems can be used to correct for frequency and timing offsets in the multiuser downlink direction.

On the other hand, the issue of frequency and timing synchronization becomes more complicated in the uplink direction (from the mobiles to the base station). Consider an OFDMA system with one base station and K active users. The available spectrum is divided into n_f subchannels, and these subchannels are allocated to the K active users by some scheduling strategy as discussed in Section 10.3.1. In general, signals from different mobiles will experience different frequency and timing offsets. Hence, at the base station (receiver), there will be nonuniform frequency and timing offsets across the n_f subcarriers and the K active users, respectively. Furthermore, correction of the frequency and timing offsets for any particular user cannot be performed at the base station because this will misalign the other users. Therefore, the base station needs to perform estimations on the frequency offsets and the timing offsets from all the K active users and instruct the K active users to compensate for the frequency and timing offsets independently at the transmitter side. In fact, the frequency offsets are usually contributed by the instability in the transmitter oscillator, the receiver oscillator, and the Doppler spread. However, in most cases, the Doppler spread[3]

[3] Besides, it is not possible to correct for the effect of Doppler shift because it does not simply shift the frequency but also distorts the frequency spectrum.

is insignificant compared to the offsets contributed by the local oscillators. Since the offsets contributed by the local oscillators are symmetric in both uplink and downlink, the mobiles can make use of the downlink pilot to estimate the frequency offset and correct for the offset locally at the mobile transmitter. In any case, because of the more challenging nature in the uplink synchronization problem, we shall focus our attention on the uplink direction of the multiuser OFDMA systems.

10.5.2 Signal Model

In this subsection, we shall elaborate on the uplink signal model of the OFDMA systems incorporating the effect of frequency and timing offsets from the K active users. The transmitter (mobile) and receiver (base station) block diagrams are shown in Figures 10.11 and 10.12, respectively.

Each mobile transmitter transmits on the assigned subcarriers. Let $\mathcal{B}_k$ denote the set of assigned subcarriers for user k. In the kth mobile transmitter, the datastream $\{c_i^{(k)}\}$ is serial-to-parallel-converted and partitioned into a block of data $\mathbf{c}_m^{(k)} = [c_m^{(k)}(0), \ldots, c_m^{(k)}(L_m - 1)]^T$ of length $L_m = |\mathcal{B}_k|$, where $|\mathcal{B}_k|$ denotes the cardinality of set $\mathcal{B}_k$ and $(\cdot)^T$ denotes transpose. The vector $\mathbf{c}_m^{(k)}$ is extended with the insertion of $n_f - L_m$ zeros to produce the IDFT inputs $\mathbf{B}_m^{(k)} = [B_m^{(k)}(0), \ldots, B_m^{(k)}(n_f - 1)]^T$, where the component of $\mathbf{B}_m^{(k)}$ is defined as

$$B_m^{(k)}(n) = \begin{cases} c_m^{(k)}(l), & \text{if } n = \mathcal{B}_k(l) \\ 0, & \text{otherwise} \end{cases} \tag{10.44}$$

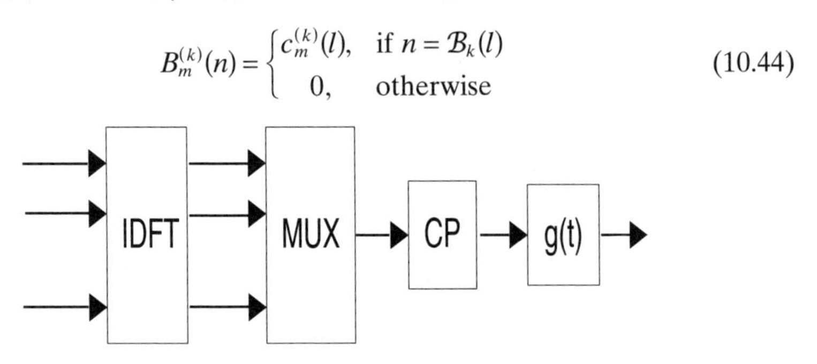

Figure 10.11. Mobile transmitter of the OFDMA system.

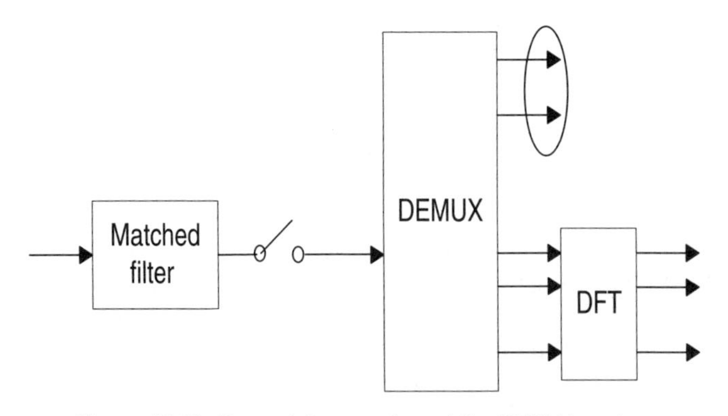

Figure 10.12. Base station receiver of the OFDMA system.

where $\mathcal{B}_k(l)$ denotes the lth element of the set $\mathcal{B}_k$. For example, if $n_f = 64$ and $\mathcal{B}_k = \{1, 5, 64\}$ during the mth block, then there are only three nonzero 64-IDFT inputs of the kth mobile transmitter: $c_m^{(k)}(1)$, $c_m^{(k)}(2)$, $c_m^{(k)}(3)$ mapping to subcarrier 1, subcarrier 5, and subcarrier 64, respectively. The other subcarrier inputs are all zero.

The continuous time-domain signal samples during the mth OFDM symbol after IDFT, cyclic prefix of length N_{cyclic} and D/A conversion is given by

$$\tilde{x}_m^{(k)}(t) = \begin{cases} x_m^{(k)}(t)(l), & \text{if } t \in [0, n_f T_0) \\ x_m^{(k)}(t + n_f T_0), & \text{if } t \in [-N_{\text{cyclic}} T_0, 0) \end{cases} \tag{10.45}$$

where T_0 is the OFDM sample duration given by $1/W$ and $x_m^{(k)}(t)$ is the time-domain signal before cyclic prefix given by

$$x_m^{(k)}(t) = \frac{1}{n_f} \sum_{n=0}^{n_f-1} B_m^{(k)} \exp\left(\frac{j2\pi nWt}{n_f}\right) \tag{10.46}$$

Hence, the transmitted signal from the kth mobile after compensation for the frequency and timing offsets is given by

$$x^{(k)}(t) = \exp(-j2\pi\hat{\Delta}_k t) \sum_{m=-\infty}^{\infty} \tilde{x}_m^{(k)}(t - \hat{\tau}_k) g(t - mT - \hat{\tau}_k) \tag{10.47}$$

where $T = (n_f + N_{\text{cyclic}})T_0$ is the OFDM symbol duration; $\hat{\Delta}_k$ and $\hat{\tau}_k$ are the compensation for frequency offset and timing offset of user k, respectively; and $g(t)$ is the windowing pulse for subcarrier spectral shaping, which is nonzero only for $t \in [-N_{\text{cyclic}}T_0, n_f T_0)$. For simplicity, we assume rectangular pulse for $g(t)$.

At the receiver, the signal from the kth mobile suffers from a frequency offset of Δ_k and a timing offset of τ_k. The frequency offset is contributed by the Doppler shift and the local frequency offsets at the transmitter and the receiver. While the receiver frequency offset at the base station is common for all users, the Doppler shift and the transmitter frequency offset are different for each user. Similarly, the timing offset is contributed by the propagation delay and the timing errors at the mobile transmitter D/A and the base station receiver A/D sampling. While the base station receiver A/D sampling timing offset is common for all users, the first two components are different for each user. Let $\nu_k = \Delta_k - \hat{\Delta}_k$ and $\varepsilon_k = \tau_k - \hat{\tau}_k$ be the residual frequency offset and the residual timing offset of user k respectively. In addition, decompose the residual timing offset ε_k into an integral part and a fractional part as

$$\varepsilon_k = (\mu_k + \delta_k)T_0 \tag{10.48}$$

where $\mu_k = [\varepsilon_k/T_0]$ is the integral part of the residual timing offset and $\delta_k \in [0, 1)$ is the fractional part of the residual timing offset. The received signal after A/D (sampling at $t = nT_0$) is given by

$$y(n) = \sum_{k=1}^{K} \exp\left(\frac{j2\pi\nu_k n}{n_f}\right) y_k(n - \mu_k) + z(n) \tag{10.49}$$

where $z(n)$ is the thermal noise, and

$$\begin{aligned} y_k(n) &= \sum_{m=-\infty}^{\infty} \sum_{i=-N_{\text{cyclic}}}^{n_f-1} \tilde{x}_m^{(k)}(i) \exp\left(\frac{j2\pi nW\delta_k}{n_f}\right) h_k(n - i - mN_T) \\ &= \sum_{m=-\infty}^{\infty} \sum_{i=-N_{\text{cyclic}}}^{n_f-1} \tilde{x}_m^{(k)}(iT_0) \tilde{h}_k(n - i - mN_T), \end{aligned} \tag{10.50}$$

$\tilde{x}_m^{(k)}(i) = \tilde{x}_m^{(k)}(iT_0)$ is the ith IDFT sample given by Equation (10.45), $N_T = n_f + N_{\text{cyclic}}$ is the length of the extended OFDM symbol, $h_k(n)$ is the sample of channel impulse response $h_k(t)$ sampled at $t = nT_0 - \delta_k$, and $\tilde{h}_k(n) = h_k(n)\exp\left(\frac{j2\pi nW\delta_k}{n_f}\right)$. Note that the effect of fractional timing offset becomes a random phase term and can be incorporated into the random channel fading $h_k(n)$ to become $\tilde{h}_k(n)$ as long as δ_k is shorter than $N_{\text{cyclic}} - L_p$, where L_p is the number of significant multipaths in $h_k(t)$. Hence, the fractional timing offset appears as a linear phase shift at the output of the DFT and can be corrected by the channel equalizer, which does not distinguish the random phase introduced by the channel fading from those deriving from the timing offsets.

Without loss of generality, consider the receive processing on user 1. To simplify the analysis, consider a hypothetical case where $\nu_k = 0$ and $\varepsilon_k = 0$ for all $k \neq = 1$. Hence, we are interested to first look at the situation where all the remaining users have been perfectly synchronized except for the first user and try to understand the effect of frequency and timing offsets with respect to user 1. Collecting a block of n_f received samples for DFT processing of the mth OFDM symbol, we have

$$\begin{aligned} \mathbf{y}_m &= \exp\left(\frac{j2\pi\nu_1 mN_T}{n_f}\right) \mathbf{\Gamma}(\nu_1) \mathbf{D}_m^{(1)}(\mu_1) \tilde{\mathbf{h}}_1 \\ &\quad + \sum_{k \neq 1} \exp\left(\frac{j2\pi\nu_k mN_T}{n_f}\right) \mathbf{\Gamma}(\nu_k) \mathbf{D}_m^{(k)}(\mu_k) \tilde{\mathbf{h}}_k + \mathbf{z}_m \end{aligned} \tag{10.51}$$

where $\mathbf{y}_m$ is the $n_f \times 1$ received vector at the input of the DFT processing given by $[y(mN_T), \ldots, y(mN_T + n_f - 1)]T$, $\tilde{\mathbf{h}}_k$ is the $L_p \times 1$ channel vector given by $[\tilde{h}_k(0), \ldots, \tilde{h}_k(L_p - 1)]^T$, and $\mathbf{\Gamma}(\nu_k)$ is a $n_f \times n_f$ diagonal matrix given by

$$[\mathbf{\Gamma}(\nu_k)]_{n,n} = \exp\left(\frac{j2\pi\nu_k n}{n_f}\right) \tag{10.52}$$

while $\mathbf{D}_m^{(k)}(\mu_k)$ is a $n_f \times L_p$ matrix with entries as

$$[\mathbf{D}_m^{(k)}(\mu_k)]_{n,l} = \begin{cases} B_m^{(k)}(n-l-\mu_k), & -N_{\text{cyclic}} + \mu_k \le n-1 \le n_f - 1 \\ B_{m-1}^{(k)}(n-l-\mu_k+N_T), & -N_{\text{cyclic}} \le n-1 \le -N_{\text{cyclic}} - 1 + \mu_k \end{cases} \tag{10.53}$$

where $n \in [0, n_f - 1]$ and $l \in [0, L_p - 1]$. To separate the user of interest (user 1), we compute DFT with respect to the received vector $\mathbf{y}_m$ and select those outputs that correspond to the subcarrier of user 1. The DFT output corresponding to the user 1 is a $|\mathcal{B}_1| \times 1$ vector $\mathbf{Y}_m^{(1)}$given by

$$\mathbf{Y}_m^{(1)} = \mathbf{F}_1 \mathbf{y}_m \tag{10.54}$$

where $\mathbf{F}_1$ is a $|\mathcal{B}_1| \times n_f$ kernel matrix with entries

$$[\mathbf{F}_1]_{i,n} = \exp\left(\frac{-j2\pi n \mathcal{B}_1(i)}{n_f}\right)$$

for $i \in [0, |\mathcal{B}_1| - 1]$ and $n \in [0, n_f - 1]$.

10.5.3 Effect of Timing Offsets

From Equation (10.53), the effect of timing offset can result in ISI because the matrix $\mathbf{D}_m(\mu_1)$ may contain contributions from both the mth and the $(m-1)$th OFDM symbols. In order to eliminate the ISI for all the K users, we require that $\mathbf{D}_m^{(k)}(\mu_k)$ contain contributions from the mth OFDM symbols only. This is equivalent to

$$\max_k \bmod(\mu_k, N_T) \le N_{\text{cyclic}} - L_p \quad \forall k \in [1, K] \tag{10.55}$$

Figure 10.13a illustrates one scenario where the residual timing offsets from user 1 and user 2 exceed the *margin* of the cyclic prefix guard period $(N_{\text{cyclic}} - L_p)T_0$. Hence, depending on the position of the DFT window of size n_f, there will be ISI for at least one of the two users. Figure 10.13b illustrates another scenario where the residual timing offsets from all the users are within the cyclic prefix guard period as depicted in Equation (10.55). In this case, there exist a DFT window position of size n_f such that all the n_f samples for DFT correspond to a single OFDM symbol among all the active users. In this case, the residual timing offset ε_k contributes only to linear phase shift in the DFT outputs. This phase shift can be incorporated into the channel fading $\tilde{h}_k(n)$ and corrected by a regular channel equalizer.

Comparing with the DS-CDMA-based system, the OFDMA-based system offers one additional implementation advantage with respect to the timing synchronization requirement. For instance, in order to reduce the multiuser interference in the uplink direction of DS-CDMA-based systems, *quasi-orthogonal* spreading sequence is utilized. However, for good autocorrelation properties, the synchronization requirement of the DS-CDMA-based system is quite strict and must be within one to two chip durations for quasiorthogo-

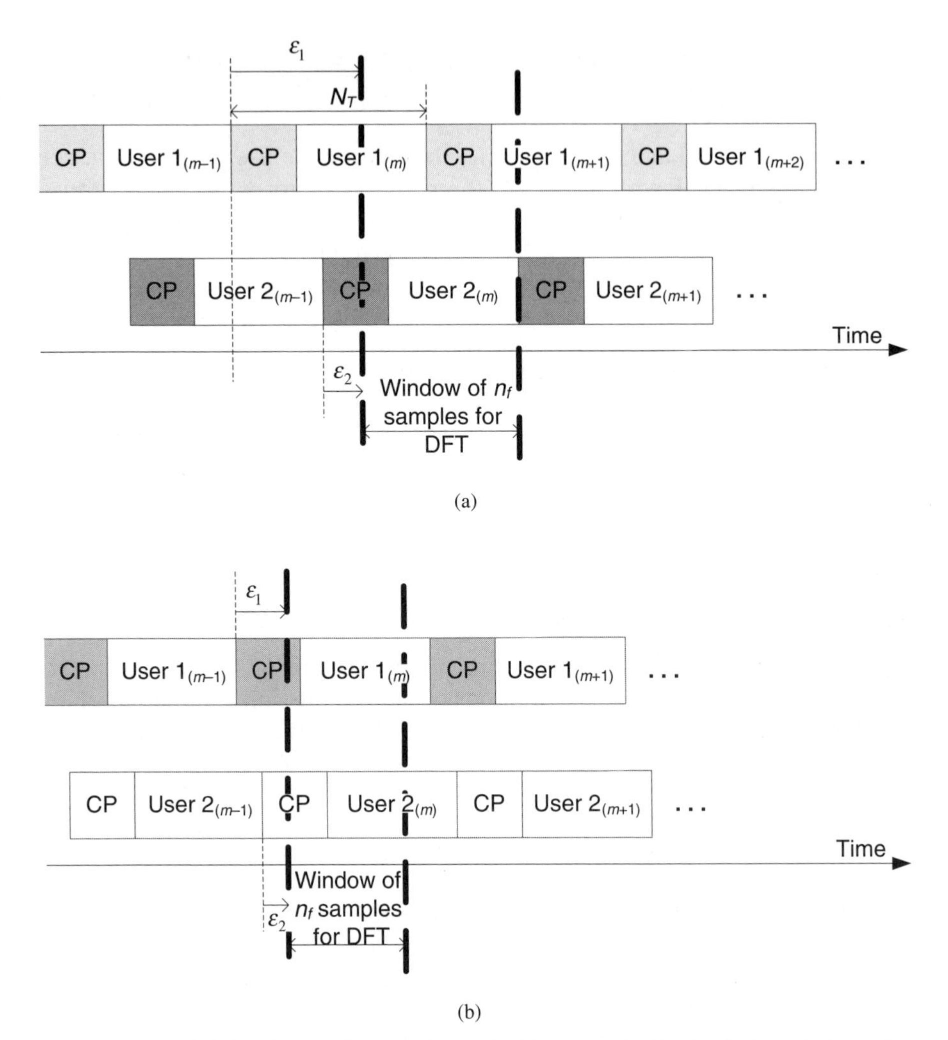

Figure 10.13. OFDMA timing offset requirement with respect to ISI: (a) timing offset causing ISI ($\max(\mu_1, \mu_2) > N_{\text{cyclic}} - L_p$); (b) timing offsets with no ISI ($\max(\mu_1, \mu_2) \leq N_{\text{cyclic}} - L_p$).

nal spreading sequence to ensure good performance. On the other hand, for OFDMA-based systems, there is a flexible tradeoff between the timing synchronization requirement of the K uplink users and the guard period allocated per OFDM symbols N_{cyclic}. Hence, we can always relax the timing synchronization requirement by increasing N_{cyclic} at the expense of higher overhead.

10.5.4 Effect of Frequency Offsets

To consider the effect of the residual frequency offsets on system performance, we expand Equation (10.54) as follows:

$$\mathbf{Y}_m^{(1)} = \exp\left(\frac{j2\pi\nu_1 m N_T}{n_f}\right)\mathbf{F}_1\mathbf{\Gamma}(\nu_1)\mathbf{D}_m^{(1)}(\mu_1)\tilde{\mathbf{h}}_1 + \sum_{k\neq 1}\exp\left(\frac{j2\pi\nu_k m N_T}{n_f}\right)\mathbf{F}_1\mathbf{\Gamma}(\nu_k)\mathbf{D}_m^{(k)}(\mu_k)\tilde{\mathbf{h}}_k + \mathbf{F}_1\mathbf{w}_m \tag{10.56}$$

Hence, we can easily see that the frequency offsets from the K users $\nu_1, \ldots, \nu_K$ will contribute to the multiuser interference in the DFT outputs. In the special case when $\nu_2 = \nu_3 = \ldots = \nu_K = 0$, we have $\Gamma(0) = \mathbf{I}$ and $\mathbf{F}_1\mathbf{D}_m^{(k)}(\mu_k) = \mathbf{0}_{|\mathcal{B}_1|\times L_p}$. Hence, the DFT outputs $\mathbf{Y}_m^{(1)}$ for user 1 is free from the multiuser interference terms and can be expressed as

$$\mathbf{Y}_m^{(1)} = \exp\left(\frac{j2\pi\nu_1 m N_T}{n_f}\right)\mathbf{A}(\mu_1, \nu_1)\tilde{\mathbf{h}}_1 + \mathbf{F}_1\mathbf{w}_m \tag{10.57}$$

where

$$\mathbf{A}(\mu_1, \nu_1) = \mathbf{F}_1\mathbf{\Gamma}(\nu_1)\mathbf{D}_m^{(1)}(\mu_1)$$

In practice, the major contribution of the frequency offset is due to the local oscillators at the mobile transmitters and the base station receiver. Since the uplink and downlink offsets of the local oscillators are the same,[4] the uplink frequency offset synchronization problem can be simplified by utilizing the downlink pilot. All the K mobiles estimate the combined frequency offsets with reference to a common pilot channel from the base station. Each mobile compensates for the frequency offsets in the uplink transmission on the basis of the downlink estimate. Typical residual frequency offset (normalized with respect to the subcarrier spacing f_s) after compensation is of the order of 10^{-3}. Hence, the frequency offset effect can be mitigated. Besides, from Equation (10.56), we can deduce that assigning a group of subcarriers to a user or introducing some empty subcarriers between *band groups* can help to enhance the robustness of the system with respect to multiuser interference due to frequency offsets. For example, the subcarrier allocation for user 1 and user 2 can be $\mathcal{B}_1 = \{1, 2, \ldots, 32\}$ and $\mathcal{B}_2 = \{34, 35, \ldots, 40\}$.

10.6 SUMMARY

In this chapter,we have elaborated on the cross-layer design for wideband transmissions over frequency-selective fading channels. When the transmit bandwidth of the signal is much larger than the channel coherence bandwidth, there will be multiple distinct echos (frequency—selective fading) in the receiver input, and this will result in ISI.We discuss two common and effective physical layer designs—the DS-CDMA/MISO and OFDMA/MISO systems—to combat frequency-selective fading channels. We consider a base

[4] In typical transceiver design, the transmit and receive paths are clocked by the same oscillator.

station with n_T transmit antennas and K mobiles (each with a single receive antenna). For DS-CDMA/MISO systems, information from the users is spread across both the temporal and spatial domains by a spreading matrix. As a result of the spreading action, the signals from a user occupies the entire wide transmission bandwidth and therefore achieves high-frequency diversity order (given by the number of resolvable multipaths L_p). It is shown that for sufficiently large spreading factor N_{SF} and sufficiently large transmission bandwidth, the maximum data rate (in terms of Shannon's capacity) can be decomposed into n_T independent scalar channels, and the capacity value approaches the ergodic capacity. In other words, the frequency diversity suppresses fluctuations of user data rate and therefore does not allow multiuser diversity.

On the other hand, OFDMA is the multiaccess extension of the regular OFDM systems where a user can be assigned a subset of the subcarriers depending on the instantaneous CSI. Together with the n_T transmit antennas, there are n_T spatial channels and n_f frequency channels as system resources that can be assigned to some of the K mobile users. In other words, together with the cross-layer scheduling algorithm, both the n_T spatial channels and the n_f frequency channels contribute to the multiuser selection diversity gain. The optimal cross-layer scheduling algorithm for OFDMA/MISO is formulated, and the optimal solution is obtained. Because of the high computational complexity involved, we use some suboptimal search algorithms (such as the greedy algorithm and the genetic algorithm used in Chapter 6) to obtain the multiuser performance. The performance is compared with the DS-CDMA/MISO systems, and it is found that significant gains of the OFDMA/MISO systems can be obtained because of the expanded dimension of multiuser diversity. From this, we conclude that combatting the fading from a point-to-point level is much less effective than that based on multiuser selection. In the context of multiuser performance, the physical layer should not suppress the fluctuation of user data rate.

Finally, we elaborate on the practical implementation issues of OFDMA systems with respect to the time and frequency synchronization requirements. Timing and frequency offsets of OFDMA-based systems in the uplink direction can be quite complicated because different users may have different residual timing and frequency offsets due to the variation in the propagation delay, transmitter frequency offset and mobile speed. The timing offsets among users may contribute to ISI between consecutive OFDM symbols if the maximum residual timing offsets among the K active users exceed the guard period $N_{\mathrm{cyclic}}T_0$. On the other hand, if the maximum timing offset is within the guard period, the ISI can be completely eliminated because the timing offset is equivalent to a linear phase shift at the DFT outputs. On the other hand, frequency offsets among the K users will contribute to multiuser interference. Since different users may have different timing offsets and frequency offsets, it is not possible to apply correction at the base station receiver. Instead, the base station needs to estimate the frequency offsets and timing offsets for all the K users and instruct the individual mobile transmitters to compensate for these offsets.

EXERCISES

1. *DS-CDMA/MISO* DS-CDMA makes uses of spreading sequence to mitigate interference. In a multi-user CDMA/MISO system, there are two sources of interferences, namely the multi-user interference and the interference due to spatial channels in one user. Using the same transmit architecture as in figure 10.2 and (10.6), comment on the effect on the system performance if the columns of the spreading matrix $\mathbf{C}_k$ are identical. Is it possible to use spreading sequences to mitigate multi-user interference only and apply zero-forcing weights at the transmitter (similar to the OTBF approach in Chapter 6) to separate the interference due to spatial channels per one code channel?

2. *DS-CDMA system with RAKE Receiver and Matched Filter Processing* Consider a CDMA/MISO system as depicted in figure 10.2. Denote B_k to be the transmit symbol from user k, S_k to be the product of power allocation matrix, channel matrix and the spreading sequence matrix (as in (10.7)), Z_k is the Gaussian noise vector with variance $\sigma_z^2 I$. The received signal Y_k is

$$Y_k = \sum_{j \in A} S_j B_j + Z_k$$

and the matched filter output V_k is

$$V_k = S_k^* Y_k$$

Assuming perfect CSIR,

(a) prove that the conditional entropy $H(Y|S)$ is upper bounded by $\log_2 (\pi e S_k^* P S_k + S_k^* Q S_k)$

(b) prove that the conditional entropy $H(Y|B_k, S)$ is given by $\log_2 (\pi e S_k^* Q S_K)$.

(c) prove that the capacity is equal to $\log_2 (I + (S_k^* Q S_k)^{-1} S_k^* P S_k)$

where $P = S_k S_k^*$ and $Q = \Sigma_{j \neq k} S_k S_j^* + \sigma_z^2 I$.

3. *OFDM and Circulant Matrix* If the discrete Fourier transform (DFT) and IDFT are defined as:

$$X_n = \sum_{k=0}^{N=1} x_k \exp(-j2\pi kn/N)$$

and

$$x_k = \frac{1}{N} \sum_{n=0}^{N-1} x_n \exp(j2\pi kn/N).$$

(a) Show that a concise representation of the DFT and IDFT operations are given by $\mathbf{X} = \mathbf{Fx}$ and $x = \frac{1}{N} \mathbf{F}^* \mathbf{X}$.

(b) A circulant matrix is a matrix where row $(n + 1)$ is a right rotation of the row n. For example,

$$\mathbf{B} = \begin{bmatrix} 1 & 2 & 3 \\ 3 & 2 & 1 \end{bmatrix}$$

is a circulant matrix. Prove that after discarding the cyclic prefix (assuming that the cyclic prefix is longer than the channel delay spread), the received signal vector $\mathbf{y}$ with respect to the OFDM transmission in time domain $\mathbf{x}$ (without cyclic prefix) can be expressed as:

$$\mathbf{y} = \mathbf{Hx} + \mathbf{z}$$

where $\mathbf{H}$ is a circulant matrix and $\mathbf{z}$ is a channel noise vector. Relate the eigenvectors of any square circulant matrix $\mathbf{H}$ to the Fourier transform matrix $\mathbf{F}$ in (a).

4. *ISI in OFDM* A clever student finds out that avoiding the ISI in OFDM should be possible even without cyclic prefix, as long as the OFDM symbols are separated in time by a guard interval longer than the channel delay spread. In other words, we transmit:

$$s(t) = \sum_{k,n} a_n(k) g_n(t - k(T + T_g))$$

where

$$g_n(t) = \exp(j2\pi f_n t)[u(t - T_g) - u(t - (T + T_g))]$$

and $T_g > T_m$ (delay spread) is the guard interval between OFDM symbols, k is an index for the OFDM symbol. i.e. OFDM symbol k consists of N PAM modulation symbols and n is the sub-carrier index within an OFDM symbol.

(a) Show that this method of transmission results in no ISI between OFDM symbols as long as $T_g > T_m$.

(b) Show that orthogonality between sub-carriers is not guaranteed when using a bank of filters matched to $g_n(t)$ at the receiver even if $f_n - f_{n-1} = 1/T$.

(c) Suggest a possible algorithm at the receiver to avoid ISI between OFDM sub-carriers.

5. *Cross Layer Design for OFDMA/MISO Systems* Assuming perfect CSIT knowledge at the base station $(H_1, \ldots, H_K)$, the cross layer scheduler has to determine the optimal admitted user sets $A = \{A_0, \ldots, A_{n_f-1}\}$, optimal power allocation $P = \{p_{m,k}: m \in [0, n_f - 1], k \in A_m\}$ every time slot such that

$$\sum_{m=0}^{n_f-1} \sum_{k \in A_m} p_{m,k} = P_0$$

and

$$|A_m| \leq n_T$$

for all $m \in [0, n_f - 1]$ and the instantaneous network capacity $\Sigma_{k=1}^{K} c_k(H)$ is maximized where c_k is given by

$$c_k(H_k) = \sum_{m:k \in A_m} \log_2\left(1 + \frac{p_{m,k}|H_{m,k}w_{m,k}|^2}{\sigma_z^2}\right)$$

if $k \in A_m$ for some m and zero otherwise.

(a) express the above optimization problem (with the contraints) into a Lagrangian $L(p_{m,k}, \lambda)$.

(b) prove that the optimization problem is convex in $\{p_{m,k}\}$.

(c) prove that the optimal power allocation for subcarrier m and user k is $p_{m,k}^*\left(\frac{1}{\lambda} - \frac{\sigma_z^2}{|H_{m,k}w_{m,k}|^2}\right) + \text{where } (x)^+ = \max(x, 0)$.

APPENDIX A: PROOF OF LEMMA 10.2

The objective function is given by:

$$f(\{p_{k,n}\}, \{\mathcal{A}_n\}) = \sum_{n=1}^{n_F} \sum_{k \in \mathcal{A}_n} r_{k,n} = \sum_{n=1}^{n_F} \sum_{k \in \mathcal{A}_n} \log_2\left(1 + \frac{p_{k,n}|\mathbf{H}_{n,k}\mathbf{w}_{n,k}|^2}{\sigma_z^2/n_F}\right)$$

Substituting the optimal power $p_{k,n}^*$ from (10.39) into the above objective function and for sufficiently large P_0, we have $f^*(\{\mathcal{A}_n\})$ given by:

$$f^*(\{\mathcal{A}_n\}) = \sum_{n=1}^{n_F} \sum_{k \in \mathcal{A}_n} \log_2\left(\frac{|\mathbf{H}_{n,k}\mathbf{w}_{n,k}|^2}{\lambda\sigma_z^2/n_F}\right) = \sum_{n=1}^{n_F} \log_2\left(\frac{\Pi_{k \in \mathcal{A}}|\mathbf{H}_{n,k}\mathbf{w}_{n,k}|^2}{(\lambda\sigma_z^2/n_F)|\mathcal{A}_n|}\right)$$

The Lagrandge multiplier λ is chosen to satisfy the transmit power constraint in Problem 20. Assuming large enough P_0, we have

$$\frac{\sum_n |\mathcal{A}_n|}{\lambda} = \left[P_0 + \frac{\sigma_z^2}{n_F} \sum_{n,k} |\mathbf{H}_{n,k}\mathbf{w}_{n,k}|^{-2}\right]$$

Hence, the objective function can be expressed as:

$$f^*(\{\mathcal{A}_n\}) = \sum_{n=1}^{n_F} \log_2 \left(\frac{\Pi_{k \in \mathcal{A}} \gamma_{n,k} \left[P_0 \, n_F / \sigma_z^2 + \sum_{n,k} \gamma_{n,k}^{-1} \right]^{|\mathcal{A}_n|}}{|\mathcal{A}|} \right)$$

where $\gamma_{n,k} = |\mathbf{H}_{n,k}\mathbf{w}_{n,k}|^2$ and $|\mathcal{A}| = \Sigma_n |\mathcal{A}_n|$. For large P_0 and using Lemma 6.3, we have $|\mathcal{A}| \approx n_F n_T$. For sufficiently large P_0, the term $\gamma_{n,k}^{-1} \ll P_0$ and $f^*(\{\mathcal{A}_n\})$ is approximated by:

$$f^*(\{\mathcal{A}_n\}) = \sum_{n=1}^{n_F} \log_2 \left(\frac{\Pi_{k \in \mathcal{A}} \gamma_{n,k} \left[P_0 \, n_F / \sigma_z^2 \right]^{n_T}}{n_F n_T} \right)$$

Hence, the optimization of f w.r.t. $\mathcal{A}$ can be decoupled into $\mathcal{A}_n^* = \arg\max_{\mathcal{A}_n} \Pi_{k \in \mathcal{A}} |\mathbf{H}_{n,k}\mathbf{w}_{n,k}|^2$. These n_F searches are decoupled and hence, the search for $\mathcal{A}_n$ is independent from that of $\mathcal{A}_m$.

11

CROSS-LAYER SCHEDULING DESIGN BASED ON QUEUEING THEORY AND INFORMATION THEORY

11.1 OVERVIEW

This chapter motivates the study of jointly adaptive MAC layer design from a cross-layer perspective. In particular, combining information theory and queuing theory analysis would be justified as one of the most favorable approaches. On different channel models, optimization formulations for throughput maximization and performance analysis based on utility function from a queuing theory perspective would be provided. The concept of a stability region will also be discussed. Finally, different stable scheduling algorithms are suggested to optimize the multiuser system performance on the basis of utility functions introduced.

A fundamental problem in MAC layer design for wireless systems is to provide efficient resource allocation among the users in the systems. On one hand, the scheduling algorithm has to prioritize the requests according to the user quality-of-service (QoS) requirements. On the other hand, the wireless channels offer *physical layer resource* that can be in frequency domain, time domain, or spatial domain as well as combinations of thereof. More recently, the existing literature on communication system design has tended to separate the treatments on the MAC layer and the physical layer [17]. However, unlike the fixed wire network where the channel is time-invariant, the wireless channels are very dynamic and time-varying. Hence, the conventional isolated design approach of optimizing the physical layer and the MAC layer

Channel-Adaptive Technologies and Cross-Layer Designs for Wireless Systems with Multiple Antennas: Theory and Applications. By V. K. N. Lau and Y.-K. R. Kwok
ISBN 0-471-64865-5

separately fails to exploit the dynamic nature of the physical layer and is suboptimal in multiuser wireless channels. This motivates the *cross-layer approach* where the MAC layer and the physical layer are jointly optimized. In the cross-layer optimization design, there are two major challenges. The MAC layer scheduling algorithm has to be adaptive to the dynamics of the systems, which are contributed by the physical layer dynamics and the source statistics of the user applications. For instance, the MAC layer has to assign priority based on the channel conditions (the physical layer condition) and buffer occupancy (the source statistics) in the user queues to ensure that the user QoS requirements are satisfied and the resource utilization is maximized.

There are two common approaches of MAC layer scheduling design: queueing theory and information theory. In the *queueing theory approach*, the physical layer is modeled as a simple abstract pipeline to carry information with a fixed level of reliability. The analysis is focused on the *abstract physical layer* resource allocation among stochastic traffic arrivals to each user and the delay analysis from the queuing perspective. In other words, the MAC layer scheduling adapts only to the *dynamics of the source* without exploiting the *dynamics of the physical layer*. Hence, this approach addresses only half of the problem. On the other hand, in the information-theoretic approach, we assume that all the users have infinite queue size and are delay-insensitive. Hence, when allocated a resource, a user always has sufficient source packets in the queue to transmit. The scheduling algorithm design is therefore decoupled from the source statistics and the MAC layer is designed to adapt to the dynamics of the physical layer only. Again, this approach addresses only the other half of the problem. To enable the design of a jointly adaptive MAC layer that can also provide a complete analysis of delay and throughput, we certainly require a cross-layer design framework that can adapt to both *source dynamics* and *physical layer dynamics*. Thus, it requires a design framework that can combine both queueing theory and information theory.

In this chapter, we shall focus on providing a hierarchical and systematic survey [37,43] on the approach of the cross-layer MAC design[1] that combines both queueing theory and information theory. The efficiency of the cross-layer MAC design usually depends on how well we consider the following four specific issues: channel state information, buffer occupancy, source statistics, and the packet arrival process. There are four noteworthy motivations in the literature that involve elements of information theory and queuing theory in the cross-layer MAC design, which are elaborated below.

1. *Multiaccess Issues.* When we consider multiple access, there are two particular interesting research areas that developed rapidly [43] (Gallager perspectives). One is the extensive study on multiaccess information theory, which

[1]Analysis techniques for the cross-layer MAC scheduler can be categorized into combining information theory and queuing theory approaches, stochastic control approach, and the Markov decision problem formulation approach. In this chapter, we shall focus on the first approach.

combines noise and interference in the modeling but leaves the issue of arrival process nature to be handled by appropriate source coding; the other is from the MAC layer perspective (with consideration such as collision resolution protocol design), which considers resource allocation and the scheduling problem from the queuing-theoretic viewpoint but trivializes the channel model and therefore underestimates the capability of the physical layer issues of noise and interference. Neither approach on its own would be appropriate for multiple-access design because of the lack of the overall picture including both the dynamics of the source statistics and the dynamics of the physical layer. From an information-theoretic point of view, Gallager [43] presents the following argument for the necessity of the cross-layer perspective in multi-access design:

> From point to point channel, one normally assumes an infinite reservoir of data to be transmitted. The reason for this is that it is a minor practical detail to inform the receiver when there is no data to send: furthermore, there is no other use for the channel, so potential lack of data might as well be left out of the model. For multi-access channels, on the other hand, most transmitters have nothing to send most of the time and only a few are busy. The problem is then to share the channel between the busy users, and this is often the central technical problem in multi-access communication.

The key point in Gallager's argument [43] is that in the MAC layer, we are scheduling resources between busy users in the MAC layer in which some users who have no data to send may be idle at this time. It clearly served as a contradiction to the typical assumption in information theory analysis; it always has an infinite bit pool ready for transmission for all users. Thus, there is a need to bridge this gap by considering also the random message arrivals together with the queuing analysis in information theory. Later in this chapter, we will also see how the arrival characteristic of the data source affects the MAC scheduling algorithm design.

2. *Coding Issues.* As Gallager [43] also mentioned, if the burstiness of arrival process is to be smoothed out by the source coding, it would take a much longer time than the usual coding delay for smoothing out the channel noise. Since such source coding delay requirement makes no practical coding achievable, the queuing analysis approach of modeling the arrival characteristic would be preferred to the source coding abstraction approach. In particular, if the traffic is delay-sensitive (i.e., we have a delay bound that needs to be satisfied), the capacity could be zero [52] in the Shannon capacity sense in some fading situations (e.g., channel outage). In order to deal with such an impediment provided by the delay bound, the concept of delay-limited capacity was proposed in Reference 51, which suggested the use of statistical multiplexing to average the fading over users instead of averaging over time. Such a new capacity definition could also be considered as a preparation for the coming integration of cross-layer analysis using both information theory and queuing theory.

3. *Quality of Service Modeling Issues.* Many more recent MAC layer schedulers [13,36] designed from a purely queuing-theoretic perspective can handle the quality-of-service (QoS) requirements well but lack a good channel model. They suffer from the same problem as in multiaccess issues because of their usual simplification of channel model in the ON/OFF channel situation for ease of analysis. Thus, information theory may be a possible choice for extension of the channel model to a more realistic situation.

4. *Scheduling and Control Issues.* Queuing theory is closely related to control theory, in particular stochastic control with the application on call admission control and flow control on the network [37]. Network parameters such as QoS requirements, call blocking probability, layer 2 retransmission probability, leaky-bucket rate, and link reliability can all participate as control parameters in the control systems. However, the control space is usually constrained to unrealistic channel mode. Thus, the parameters controlling the link reliability are particularly worth reinvestigation from the information-theoretic perspective. Besides, stability analysis provided by stochastic queuing control (which will be discussed in a later section) would possibly be modified if the control space were provided from information theory.

There are also many other approaches in the cross-layer scheduler design [11,40] based on very simple model of the physical layer, particularly for the QoS modeling, scheduling, and control issues. Nevertheless, no matter which specific design focuses we are interested in, information theory provides significant extensions to queuing analysis in the existing literature.

11.1.1 Cross-Layer Scheduling Model

Figure 11.1 illustrates the cross-layer scheduling model, where the scheduling algorithm at the MAC layer is modeled as an optimization problem with respect to some *physical layer constraints* as well as *application QoS constraints*. At every timeslot, the scheduling algorithm has to produce the *rate allocation* $\mathbf{r} = (r_1, \ldots, r_K)$ as well as the *power allocation* $\mathbf{p} = (p_1, \ldots, p_K)$ for all the K users based on the observation of the current *channel state information* (CSI) from the physical layer and the *queue state information* (QSI) from the application layer. The rate allocation and power allocation are selected so as to optimize some system objectives, summarized below.

Problem 11.1 (General Cross-Layer Scheduling Problem) A multiuser wireless network can be modeled by a dynamic system with states evolution given by the following:

- *Channel State Information (CSI).* The channel fading states $\tilde{\mathbf{H}}(t) = [\mathbf{H}_1(t), \ldots, \mathbf{H}_K(t)]$, where $\mathbf{H}_k(t)$ is the channel fading process between the base station and the kth user.

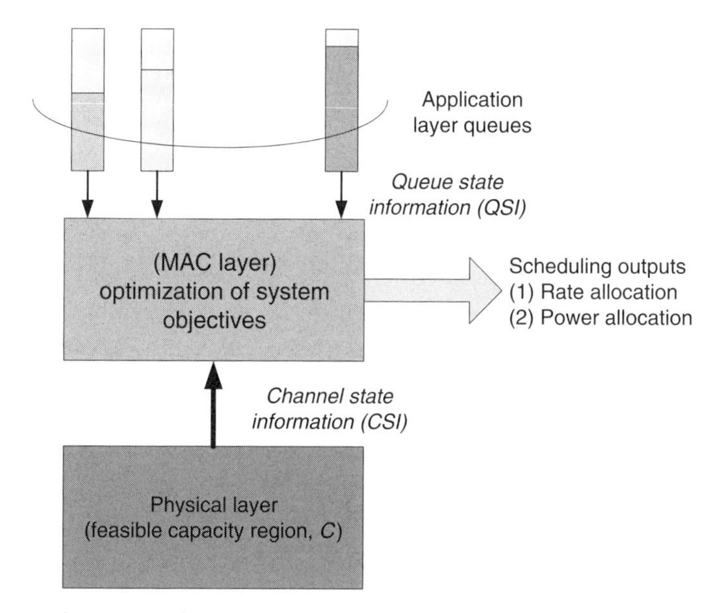

Figure 11.1. Generic optimization model of cross-layer scheduling.

- *Queue State Information (QSI).* The queue length $\mathbf{Q}(t) = [Q_1(t), \ldots, Q_K(t)]$, where $Q_k(t)$ denotes the number of untransmitted packets in buffer of user k at time t.

A *scheduler* is a controller that produces a rate allocation $\mathbf{r} = (r_1, \ldots, r_K)$ as well as power allocation $\mathbf{p} = (p_1, \ldots, p_K)$ at every timeslot t based on the observation on the state of the system $(\tilde{\mathbf{H}}(t), \mathbf{Q}(t))$ at time t. The set of rate allocations over t is called the *rate allocation policy* $\mathcal{R}$ and the power allocation over t is called the *power allocation policy* $\mathcal{P}$. The cross-layer scheduling problem is to find the optimal scheduling policies $\mathcal{R}$ and $\mathcal{P}$ so as to maximize (or minimize) the system utility function:

$$U(\overline{R}_1, \ldots, \overline{R}_K; \overline{Q}_1, \ldots, \overline{Q}_K)$$

where $\overline{R}_k = \varepsilon[r_k]$ is the average data rate of user k. At the same time, the optimization problem may have the following constraints.

1. *Physical Layer Constraint.* At any fading instant, the instantaneous data rate vector has to be in the *feasible region of the physical layer*, namely, $\mathbf{r} = (r_1, \ldots, r_K), \in C(\tilde{\mathbf{H}})$, where $C(\tilde{\mathbf{H}})$ is the instantaneous feasible capacity region. The power allocation vector has to satisfy some power constraints such as the average power constraint or short-term power constraint. These will be elaborated in later sections.

2. *Application Layer QoS Constraint.* There can be many different QoS constraints depending on the specific applications. Depending on the application nature, individual users may require the following form of QoS guarantee [13].
   a. *Delay-Bound.* For realtime applications with different degrees of delay sensitivity, the following form of QoS is usually required

$$\Pr(Q_k > Q_{\max}) \le \sigma_k \tag{11.1}$$

   where Q_k is proportional to the steady-state delay of the kth user, $Q_{\max}$ is proportional to the maximum target delay, and σ_k is the delay violation probability of user k.
   b. *Minimum Throughput Guarantee.* This form of QoS notation is well known in the definition of different classes of service in ATM. For wireless scheduling, the minimum throughput guarantee is defined in statistical sense: for each user i, $\overline{R}_i \ge \rho_i$, where $\overline{R}_i$ is the average throughput of user i and ρ_i is the minimum average throughput requirement for that user.

In general, the choice of the scheduling objective function $U(\overline{R}_1, \ldots, \overline{R}_K; \overline{Q}_1, \ldots \overline{Q}_K)$ in the cross-layer optimization problem will have great impact on system performance. For instance, when the application is delay-insensitive and the buffer size is relatively large, selection of the scheduling objective can be based on the following concerns [40]:

1. *Maximizing the Network Utilization Efficiency.* From the operator's perspective, wireless resources are scarce and it is very important to ensure high resource utilization. One possible system utility function with respect to the network utilization objective is given by $U(\overline{R}_1, \ldots, \overline{R}_K) = \Sigma_k \overline{R}_k$. This objective function has been introduced in Chapter 6.
2. *Fairness.* While maximizing network utilization is an important objective from the operator's perspective, it may result in a bias against certain users whose channels are statistically inferior to the others (e.g., users on the cell edge). Fairness is another important objective from a user's perspective. For example, the system utility function that achieves proportional fairness is given by $U(\overline{R}_1, \ldots, \overline{R}_K) = \Sigma_k \log(\overline{R}_k)$. Other possibilities include log-proportional fairness as well as some other combinations [91].

These system objective functions have been widely adopted in the cross-layer scheduling optimization based on CSIT. In other words, the cross-layer scheduler is adaptive with respect to the channel state $\tilde{\mathbf{H}}(t)$ only. These scheduling objectives do not consider the delay performance of the applications as well as system stability in terms of buffer length. However, when the application is delay-sensitive and when the system buffer size is limited, delay performance and system stability are important considerations. In this case, the cross-layer optimization has to be adaptive with respect to both the channel state $\tilde{\mathbf{H}}(t)$ and the queue status $\mathbf{Q}(t)$. The system objective functions can be chosen according to the following concerns:

1. *System Stability*. When the buffer size of the system is limited, system stability is an important concern where the scheduling objectives have to be properly chosen in order to ensure that all the queues in the system do not blow up. This point will be elaborated in Section 11.4.
2. *Delay Minimization*. In some applications where delay of data users is important, the scheduling objective may focus on the overall average delay rather than the average throughput. For example, one possible objective function that minimizes the overall network delay is given by $U(\overline{Q}_1, \ldots, \overline{Q}_K) = \Sigma_k \overline{Q}_k$.

In any case, the cross-layer scheduling optimization can be decoupled from the detail implementations of the multiuser physical layers. Specifically, the multiuser physical layer and the scheduling algorithm in the MAC layer interface through the *feasible capacity region* C. The scheduling problem formulation will be the same irrespective of the specific multiuser physical layer because different physical layers will present a different feasible region C to the scheduler. This chapter is organized as follows. In Section 11.2, we review a number of commonly used physical layer models in the cross-layer design. In Section 11.3, we elaborate on the motivation behind the cross-layer scheduling taking into account of both the physical channel state and the queue status. In Section 11.4, we focus on the concept of *stability region* and the corresponding scheduling design to achieve stability. In Section 11.5, we discuss another cross-layer optimization design with respect to delay minimization for homogeneous users. Finally, we conclude with a brief summary in Section 11.6.

11.2 MULTIUSER PHYSICAL LAYER MODEL

Different models were proposed to model the performance of the multiuser physical layer. In this section, we elaborate on three important models: the graph model, the ON/OFF channel model, and the information-theoretic model. The choice of modeling depends on the tradeoff between simplicity and accuracy in the particular problem. Before we discuss these different physical layer models, we need to introduce a generic model of the multiuser physical layer—the *feasible region* of the physical layer. This is illustrated in Figure 11.2. Suppose that a system has K users and user i transmits at data rate r_i; then the transmission data rates of the K users can be represented by the rate tuple $\mathbf{r} = (r_1, \ldots, r_K)$. A rate tuple $\mathbf{r}$ is called *feasible* if the physical layer can deliver

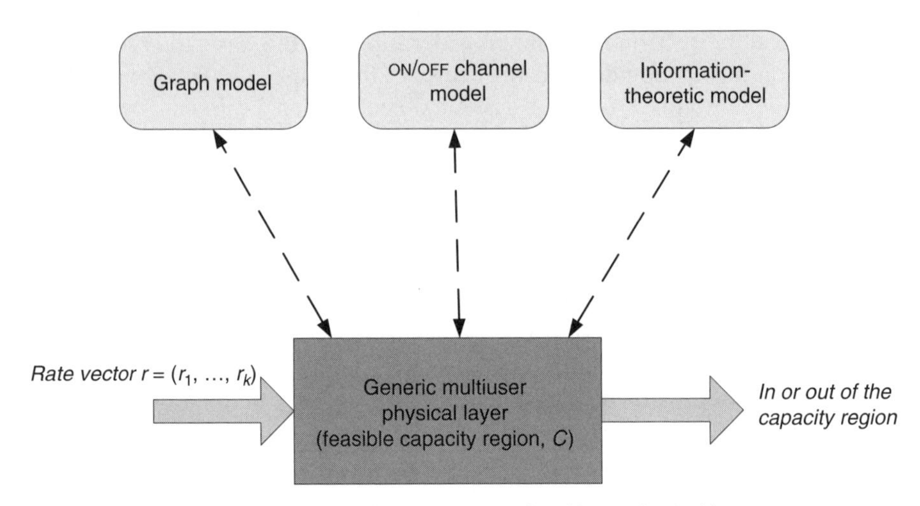

Figure 11.2. Generic feasible region of multiuser physical layer.

the data rates *reliably*[2] to the K users, respectively. The set of all feasible rate tuples **r** is called the *feasible region* of the physical layer. In this way, we decouple the cross-layer optimization problem (to be described in later sections) from the details of the multiuser physical layer designs. This is because different implementations and designs of the multiuser physical layer are all characterized by a generic *feasible region* and the MAC layer simply optimizes the scheduling objective with respect to the feasible region given by the underlying physical layer.

11.2.1 Graph Model

In the graph model, the wireless link is assumed to be static and offers a constant capacity. However, there is a mutual relationship between the links as a result of mutual interference. The interference is usually modeled as a path gain matrix [64,148]. While the graph model is an over simplified physical layer model because it fails to capture the time-varying dynamics of wireless channels, it serves as a good first-order approximation to define a multiuser physical layer's *feasible capacity region* based on the mutual interference relationship between wireless links. For example, the graph model is applicable to model the interference from users in the cochannel cells. It can also be used to model the multiuser interference within a single-cell scenario with power control in the base station such as CDMA systems with nonorthogonal user code.

[2]The term *reliable transmission* is a generic concept that can imply different physical meanings depending on the specific physical layer model.

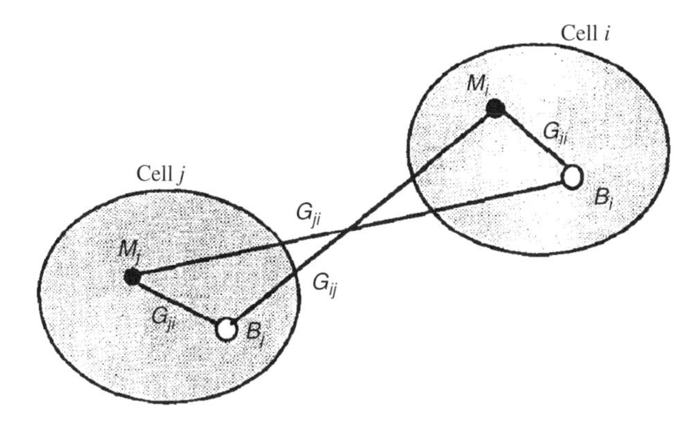

Figure 11.3. Illustration of graph model quoted from Reference 148.

To better illustrate the idea, we apply the graph modeling techniques to model the interference relationship in a cellular radio system (using the model illustrated by Zander [148]) and illustrate the corresponding feasible capacity region of the graph model. Consider a large cellular system with K users operating on the K cochannel cells so that each cell consists of one user only. Consider an active user in cell i. Both the mobile and the base station will experience interference from the other $K - 1$ users in the $K - 1$ cochannel cells. The above mentioned interference relationship in multicell systems can be modeled by a graph as illustrated in Figure 11.3. Suppose that the path gain between the mobile transmitter in cell i and the base station receiver in cell j is denoted by G_{ij}. The links between the base station in cell i and its *intracell users* is called *link i* with a path gain denoted by G_{ii}.

Let $p_i \geq 0$ be the transmit power in link i. The total interference power received by a receiver in cell i is $\Sigma_{j \neq i} G_{ji} p_j$, and the desired signal power received by a receiver in cell i from the transmitter of cell i is $G_{ii} p_i$. Hence, the received SINR, τ_i, of users in cell i is given by

$$\tau_i = \frac{G_{ii} p_i}{\sum_{j \neq 1} G_{ij} p_j + \sigma_z^2} \tag{11.2}$$

for $i \in [1, K]$, where σ_z^2 is the noise power. Let γ_i be the required SINR of link i, where $i \in [1, K]$. For simplicity, assume that the transmit power of link i, p_i, is given by

$$p_i = \begin{cases} P_i, & \text{if link is on} \\ 0, & \text{if link is off} \end{cases}$$

The corresponding data rate of link i, r_i, is given by

$$r_i = \begin{cases} R_i, & \text{if } p_i = P_i \\ 0, & \text{if } p_i = 0 \end{cases}$$

In other words, we assume that the link can be turned on or off only by assigning the transmit power p_i to be P_i or 0. The feasible capacity region in the graph model is characterized by the set of rate tuples $\mathbf{r} = (r_1, \ldots, r_K)$ such that the SIRs of all users, $(\tau_1, \ldots, \tau_K)$, are higher than the required SIR. This is summarized in the definition below.

Definition 11.1 (Feasible Capacity Region in Graph Model) A rate tuple $\mathbf{r} = (r_1, \ldots, r_K)$ (where $r_i \in \{0, R_i\}$) is feasible if there exists a power allocation $\mathbf{p} = (p_1, \ldots, p_K)$ (where $p_i \in \{0, P_i\}$) such that

$$\frac{G_{ii} p_i}{\sum_{j \neq 1} G_{ij} p_j + \sigma_z^2} \leq \gamma_i \tag{11.3}$$

for all $i \in [1, K]$.

We consider the following numerical example to illustrate the concepts of feasible capacity region.

Example 11.1 (Graph Model of Cellular System) Suppose that there are two cochannel cells ($K = 2$) in the system. The SINR requirement for both cells is 10 dB ($\gamma_1 = \gamma_2 = 4$). The path gain matrix $[G_{i,j}]$ in dB between cell 1 and cell 2 is given by

$$\mathbf{G}_{dB} = \begin{bmatrix} -3 & -10 \\ -10 & -3 \end{bmatrix}$$

The noise power σ_z^2 is normalized to 1. Let the transmit power in cell 1 and cell 2 be 10 W ($P_1 = P_2 = 10$ W) and the corresponding data rates be both 1 bps/Hz ($R_1 = R_2 = 1$). Using the feasible rate region definition above, the rate tuple $\mathbf{r} = (1, 1)$ is not feasible because such rate tuple requires power allocation $\mathbf{p} = (10, 10)$. The SIR of link 1 or link 2, given by $\tau_1 = \tau_2 = (10/2)/(10/10 + 1)$, is less than the required SINR of 10 dB. On the other hand, the rate tuple $\mathbf{r} = (1, 0)$ is feasible because this requires a power allocation of $\mathbf{p} = (10, 0)$ and the corresponding SINR of link 1, given by $\tau_1 = (10/2)/1$, is larger than the required SINR. Similarly, the rate tuple r = (0, 1) is also feasible.

The graph model is useful for expressing the interference relationship between users in a simple manner in multiuser networks. However, there are several limitations. For instance, the time-varying fading of the wireless channels is completely ignored. It will be too complicated if the time-varying dynamics of the data rates and transmit powers in the multiuser networks have to be taken into account.

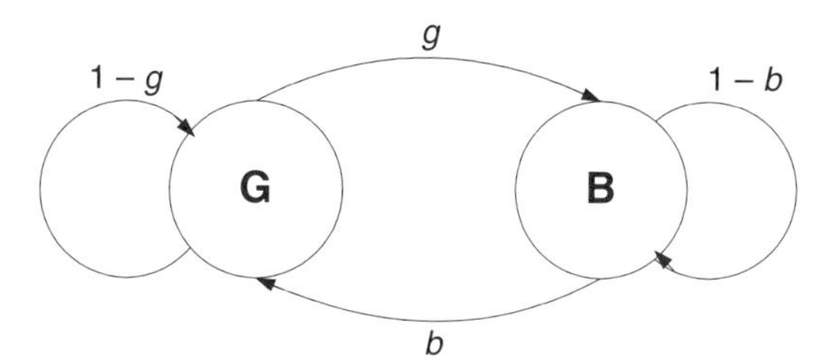

Figure 11.4. The ON/OFF channel model.

11.2.2 ON/OFF Channel Model

As a step in modeling the time-varying dynamics of wireless fading channels, the *ON/OFF* channel model has been proposed [27,69]. In this model, the channel could be in either *good* or *bad* state depending on the realistic user-perceived SNR. If the instantaneous SNR is above a threshold SNR_0, then the channel state S_n is regarded as in good state. On the other hand, if the instantaneous SNR is below SNR_0, the channel state S_n is regarded as in bad state. The sequence of channel states $\{S_n\}$ would typically be modeled as a Markov process [27], or assumed to be i.i.d. [69]. Note that for the ON/OFF channel with Markov process, there is another formal name for it in the literature: the "Gilbert–Elliott channel" model (illustrated in Figure 11.4). We can see from the figure that this model is characterized by two parameters, b and g, which are based on the relationship that $p_{\mathrm{GB}} = P(S_{n+1} = G|S_n = B) = b$ and $p_{\mathrm{BG}} = P(S_{n+1} = B|S_n = G) = g$.

We denote the probability that the channel is in good or bad state at the kth instant by $P^k(G)$ and $P^k(B)$, respectively, and also in matrix form $\mathbf{P}^k = [P^k(G)P^k(B)]^*$. Note that the ON/OFF channel can be characterized by the 2×2 transition matrix $\mathbf{T}$, given by

$$\mathbf{T} = \begin{bmatrix} 1-g & b \\ g & 1-b \end{bmatrix} \tag{11.4}$$

where $[\mathbf{T}]_{i,j}$ denotes the probability that the channel changes from the current state j to the next state i. Hence, the state probability vector at the $(k + 1)$th instant is given by

$$\mathbf{P}^{k+1} = \mathbf{T}\mathbf{P}^k \tag{11.5}$$

The ON/OFF channel model is useful in capturing the memory in wireless fading channels. For instance, if $b = g = 0$, the ON/OFF channel resembles the slow fading situation where the channel state remains quasistatic within a long time. On the other hand, when $b = g = 0.5$, the ON/OFF channel resembles the i.i.d. fading situation. From these equations, the stationary distribution of good state and bad state, denoted as $\mathbf{P}^{\infty} = [P^{\infty}(G), P^{\infty}(B)]^*$, is given by

$$\mathbf{P}^{\infty} = \left[\frac{g}{b+g}, \frac{b}{b+g}\right] \tag{11.6}$$

We consider the following example [144] to illustrate the application of the ON/OFF channel model on Rayleigh fading channels.

Example 11.2 (Rayleigh Fading Channels) Suppose that the amplitude α of the received signal is Rayleigh-distributed. The SNR distribution $\gamma = |\alpha|^2$ is a negative exponential given by

$$f(\gamma) = \frac{1}{\bar{\gamma}} e^{-\gamma/\bar{\gamma}}, \quad \gamma \geq 0 \tag{11.7}$$

If the SNR threshold between good and bad states is γ_t, then

$$P^{\infty}(B) = \int_0^{\gamma_t} \frac{1}{\bar{\gamma}} \exp^{-\gamma/\bar{\gamma}} \, d\gamma = 1 - e^{-\gamma_t/\bar{\gamma}} = 1 - e^{-\rho^2}$$

$$P^{\infty}(G) = e^{-\rho^2}$$

where $\rho^2 = -\gamma_t/\bar{\gamma}$.

In order to match the dynamics of the ON/OFF channel [144] with respect to the dynamics of the actual fading channels, we consider the average time during which the fading amplitude stays below a certain level γ_t and match it to the average time the ON/OFF channel model remaining in the bad state once it has entered it. The two parameters (b, g) in the transmission matrix are given by

$$g = \frac{\rho f_D T_s \sqrt{2}\pi}{e^{-\rho^2} - 1} \tag{11.8}$$

$$b = \rho f_D T_s \sqrt{2}\pi \tag{11.9}$$

where $f_D = v/\lambda$ is the Doppler frequency and T_s is the symbol duration.

The numerical results on different parameter values with $\bar{\gamma} = 15\,\text{dB}$ and $f_D T_s = 0.003$ are given in Table 11.1 [144].

In the multiuser system, the connection between the base station and the K mobiles are modeled as K independent *ON/OFF channels*. For the kth link, the instantaneous data rate the physical layer can support is given by

TABLE 11.1. Numerical Results for ON/OFF Channel Modeling on Underlying Rayleigh Fading Channel with $\bar{\gamma} = 15\,\text{dB}$ and $f_D T_s = 0.003$

γ_t (dB)	$P^{\infty}(G)$	$P^{\infty}(B)$	$Pe(G)$	$Pe(B)$	g	b	P_{ow}
0	0.969	0.031	8.11×10^{-3}	0.266	0.0416	1.34×10^{-3}	3.83×10^{-3}
1	0.961	0.039	6.93×10^{-3}	0.243	0.0369	1.50×10^{-3}	4.38×10^{-3}
2	0.951	0.049	5.70×10^{-3}	0.219	0.0328	1.68×10^{-3}	4.92×10^{-3}
3	0.939	0.061	4.49×10^{-3}	0.195	0.0290	1.89×10^{-3}	5.40×10^{-3}
4	0.924	0.076	3.35×10^{-3}	0.171	0.0256	2.12×10^{-3}	5.76×10^{-3}
5	0.905	0.095	2.33×10^{-3}	0.147	0.0226	2.15×10^{-3}	5.93×10^{-3}
6	0.882	0.118	1.49×10^{-3}	0.125	0.0199	2.67×10^{-3}	5.89×10^{-3}
7	0.853	0.147	8.63×10^{-4}	0.105	0.0174	2.99×10^{-3}	5.66×10^{-3}
8	0.819	0.181	4.37×10^{-4}	0.087	0.0152	3.36×10^{-3}	5.27×10^{-3}

$$r_k = \begin{cases} r_k(G), & \text{if } S_k = G \\ r_k(B), & \text{if } S_k = B \end{cases}$$

where S_k denotes the current channel state of the kth user and $r_k(G)$ and $r_k(B)$ are the corresponding maximum data rates when the channel is in *good state* and *bad state*, respectively. Hence, the capacity region of the multiuser system modeled by the ON/OFF model is a dynamic region depending on the current channel states of the K users $(S_1, \ldots, S_K)$:

$$\mathcal{R}(S_1, \ldots, S_K) = \{(d_1, \ldots, d_K) : d_k \leq r_k(S_k) \forall k \in [1, K]\}$$

Modeling the physical layer as an ON/OFF channel has the advantage of simplifying the queuing analysis while still capturing the time-varying nature of the channel with memory. The channel model can be extended to a multiple-state model by setting more than one SNR threshold [69]. However, as the number of states increases, the complexity of the physical layer increases, and this defeats the purpose of the simplicity in the ON/OFF model. In addition, the extension of the ON/OFF model to MIMO wireless fading channels is not trivial because the channel matrix involves $n_R \times n_T$ complex quantities.

11.2.3 Information-Theoretic Model

In this approach of physical layer modeling, the performance of the physical layer is captured by information theory. For example, the data rate of a user is characterized by the Shannon capacity, which is the maximum data rate at which information can be transmitted with arbitrarily low error probability. While the Shannon capacity is simply a theoretical bound, this model offers an important advantage of decoupling the performance from the specific implementation of coding and modulation. Besides, for sufficiently large coding block length and powerful error correction coding such as turbo codes and LDPC codes, the Shannon capacity limit can be realized to within 0.02 dB.

In the multiuser situation, the physical layer model based on the information-theoretic approach can be characterized by a *capacity region*, which is defined as the *closure* of the set of achievable rate tuples $(r_1, \ldots, r_K)$. In the downlink direction (one-to-many), the multiuser capacity region is characterized by the *broadcast region*. There are two classes of broadcast channels: the *degraded broadcast channels* and the *non-degraded broadcast channels*. For example, the received signal of mobile user k from a base station with n_T transmit antennas is given by

$$Y_k = \mathbf{H}_k \mathbf{X} + Z_k \tag{11.10}$$

for $k \in [1, K]$, where $\mathbf{H}_k$ is the $1 \times n_T$ channel fading coefficient between the base station and the mobile user k, $\mathbf{X}$ is the $n_T \times 1$ transmit vector, and Z_k is the channel noise with variance σ_z^2. When $n_T = 1$, the broadcast channel in Equation (11.10) belongs to the degraded broadcast channels, and the instantaneous capacity region $C_{n_T=1}(H_1, \ldots, H_K)$ conditioned on the channel realizations $(H_1, \ldots, H_K)$ is given by

$$r_k \le \log_2\left(1 + \frac{|H_k|^2 p_k}{\sum_j^K p_j |H_k|^2 \mathbf{1}(|H_k| < |H_j|) + \sigma_z^2}\right) \quad \forall k \in [1, K] \tag{11.11}$$

where $\Sigma_{j=1}^K p_j = P_0$ and P_0 is the total transmit power constraint from the base station. The capacity region of the degraded broadcast channel in general can be achieved by *superposition coding*. On the other hand, when $n_T > 1$, the broadcast channel in Equation (11.10) belongs to *nondegraded broadcast channels*, and the capacity region is no longer given by (11.11). It has been shown that the capacity region can be achieved by *dirty-paper coding* [10].

On the other hand, in the uplink direction (many-to-one), the multiuser capacity region is characterized by the *multiple-access region*. For example, the $n_R \times 1$ received signal at the base station receiver is given by

$$\mathbf{Y} = \sum_{k=1}^K \mathbf{H}_k X_k + \mathbf{Z} \tag{11.12}$$

where $\mathbf{H}_k$ is the $n_R \times 1$ channel fading coefficient between the kth mobile and the base station, X_k is the transmitted symbol from the kth mobile user, and $\mathbf{Z}$ is the $n_R \times 1$ Gaussian channel noise with covariance matrix $\sigma_z^2 \mathbf{I}_{n_R}$. The capacity region of the multiple-access channel in (11.12) conditioned on the channel fading $(\mathbf{H}_1, \ldots, \mathbf{H}_K)$, $C_{n_T=1}(H_1, \ldots, H_K)$, is given by

$$\sum_{k \in S} r_k \le \log_2 \left| \mathbf{I}_{n_R} + \frac{\sum_{k \in S} p_k \mathbf{H}_k \mathbf{H}_k^*}{\sigma_z^2} \right| \tag{11.13}$$

for all $S \subset [1, K]$. The capacity region can be achieved by successive interference cancellation at the base station.

The use of the broadcast region and multiple access region to model the physical layer performance for cross-layer MAC design has been elaborated in Chapter 6. However, as mentioned in the introduction, the cross-layer design approach introduced in Chapter 6 only exploits the dynamics of the fading physical layer, while the dynamics of the source statistics are ignored in the design. In the next sections, we shall extend the cross-layer design framework to take into account the dynamics of the source statistics and delay requirements in addition to the dynamics of the physical layer as well.

11.3 MOTIVATIONS FOR CROSS-LAYER SCHEDULING IN MULTIUSER WIRELESS NETWORKS

In this section, we shall highlight the importance of cross-layer scheduling with combined adaptation with respect to the time-varying physical layer as well as the application source statistics. In general, there are two essential components in cross-layer scheduler design: the *channel state information* (CSI) and the *queue length* status of the user buffer. The following examples are used to illustrate the importance of cross-layer scheduling designs adaptive to both components.

Example 11.3 (Channel State Information–Aware Scheduler Gives Better Throughput Performance than Does Regular Scheduler that Does Not Exploit Knowledge of Channel Conditions) Andrews *et al.* [13], show that the user with a better channel condition should be given some priority to transmit. For instance, consider the following system:

System Description. Consider a system with two users in a time-varying channel where the channel characteristics for both users are independent. User 1 (who may be far away from the base station) can transmit at either 76.8 or 153.6 kbps with equal probability. User 2 (who may be near the base station) perceives a better channel, and so can transmit at either 153.6 or 307.2 kbps with equal probability. Assume that both users have infinite data to send.

Transmission Strategy 1 (Round-Robin). Since in round-robin transmission strategy, every one takes a turn and gets an equal chance to transmit their data, users are served at the average rates of (user 1) $R_1 = 0.5 \times (0.5 \times 76.8 + 0.5 \times 153.6) = 57.6$ kbps and (user 2) $R_2 = 0.5 \times (0.5 \times 153.6 + 0.5 \times 307.2) = 115.2$ kbps, respectively.

Transmission Strategy 2 (Proportional Fair). As in proportional fair strategy, users with the highest relative rate, $r_k/\overline{R}_k$, will be selected, where r_k is the instantaneous data rate and $\overline{R}_k$ is the average data rate of user k.

For example, user 1 will be selected to transmit when the instantaneous data rate is 153.6 kbps, and user 2 will be selected to transmit when the instantaneous data rate is 307.2 kbps. When a "draw" occurs (i.e., both users are relatively better or relatively worse), the user selected to serve is chosen randomly with no bias (i.e., each with equal probability). The average rates of the two users in this transmission strategy are given by $R_1 = 0.5 \times (0.25 \times 76.8 + 0.75 \times 153.6) = 67.2$ kbps and $R_2 = 0.5 \times (0.25 \times 153.6 + 0.75 \times 307.2) = 134.4$ kbps, which is 16% higher for each user than with the round-robin strategy.

From this example, we see that the scheduler that makes efficient use of channel condition information would outperform scheduler without any channel condition consideration. The fundamental reason for the performance gain is the "multiuser selection diversity," and this has been elaborated in Chapter 6.

Example 11.4 ("Buffer Condition"–Aware Scheduler Gives Better Long-Term Average Delay Performance than Does Non– "Buffer Condition"–Aware Scheduler.) In this example, we will illustrate the importance of "buffer condition"–aware strategy on the average delay minimization.

Consider a system with two users. At $t = 0$, two packets arrive at the buffers of the two users. The first user (if selected) can transmit at a data rate of $\frac{1}{100}$ packets per second, while the second user (if selected) can transmit at a data rate of $\frac{1}{2}$ packets per second. If we use the first-come first-serve (FCFS) scheduling discipline, then the packet in the first user would be consumed at $t = 100$ s and the packet in the second user would be consumed at $t = 102$ s. The average delay of system is $(100 + 102)/2 = 101$ s. On the other hand, if we schedule the second user to transmit before the first user, the packet in the second user would be transmitted at $t = 2$ s, while the packet in the first user would be transmitted at $t = 102$ s. The average delay is given by $(2 + 102)/2 = 52$ s. Hence, there is a great reduction in the average delay if the scheduling strategy can consider the buffer information (i.e., the size of the packets in the buffer to be scheduled). The latter strategy in this example is called "shortest remaining transmit first" (SRTF) scheduling. In SRTF, the scheduler gives priority for the users with the smallest remaining service requirement so that a short packet would not suffer from a large delay due to transmission of a long packet.

Example 11.5 (Information-Theoretic Water Filling Is Not Optimal) In Example 11.3, we have illustrated that channel state information would be useful in scheduling to exploit multiuser diversity. In Example 11.4, we have illustrated the importance of "buffer condition"–aware scheduling strategy, especially if our objective is to minimize the average packet delay. In this example, we shall illustrate that it is also important to have a cross-layer scheduling strategy that adapts to both the channel state information and the buffer conditions.

From the information-theoretic perspective, when there is perfect channel state information (CSI) available at both transmitter and receiver in the single-input single-output (SISO) situation, it is well known that the optimal scheduler design that can give maximum capacity would be temporal water filling [30]. However, as mentioned in the introduction of the chapter, the validity of information-theoretic argument is based on the assumption that there is an infinite pool of data bits to be transferred in the user buffer. Kittipiyakul and Javidi [69] give a counterexample to the optimality argument of water filling, which is summarized as follows.

In this example, a multiuser physical layer based on OFDMA is considered. To simplify the analysis, each OFDM subcarrier is modeled as an ON/OFF channel model. A particular user is either "connected" or in the ON state (able to transmit data through that carrier) or "disconnected" or in the OFF state (unable to transmit data through that carrier). Figure 11.5 illustrates the multiuser physical layer, where the white box stands for the ON state and the blue box stands for the OFF state. Note that each OFDM subcarrier is able to transmit only one packet at a time, where the tick in Figure 11.5 stands for the assigned subcarriers for a particular user in the current timeslot. The total number of packets being scheduled (counting the number of ticks from timeslots 1 to 4) in four time-slots by water-filling strategy (illustrated in Figure 11.5d) is less than that obtained by the MTLB (maximal throughput and load balancing) scheduling policy (illustrated in Figure 11.5c). In addition, Figure 11.5e shows that water filling would cause the queue length to eventually become unbounded.

From this example, we can interpret that pure water-filling in Figure 11.5d is "blind" about the queue status of user buffer. It may transmit all the packets available in a single buffer (say, user 1's buffer) when the channel condition is good in one timeslot (say, at $t = 0\,\mathrm{s}$), while in the next timeslot (at $t = 1\,\mathrm{s}$), there is already no packet in the buffer for that user to transmit even when the channel condition is still good. So it may be suggested that even when the channel condition is good for a particular user at a given time, it should leave the chance for another user transmission (depending on all buffer status) in order to achieve an overall system throughput gain.

From these three examples, we can see that it is very important for the cross-layer scheduling design to exploit the channel state information from the physical layer as well as the queue status from the application layers. This is critical if the performance objectives include maximum throughput, minimum delay, and the queuing system stability. In later sections, we shall elaborate on these specific optimization problems, including the stability of the system as well as the delay minimization and throughput maximization.

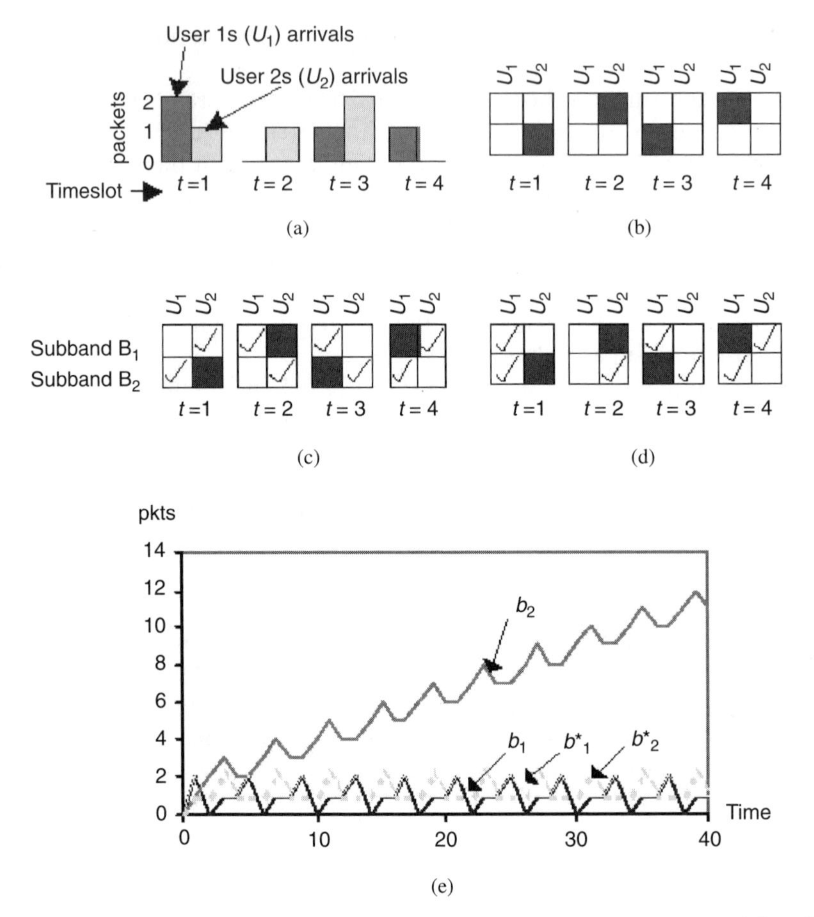

Figure 11.5. A counterexample showing that water filling neither guarantees stability of the queue nor gives maximum throughput. The packet arrival process and the channel connectivity process have periodic structures with a period of four timeslots. (a) Packet arrivals for users 1 and 2; (b) channel connectivity profile (white square indicates ON state); (c) MTLB policy; (d) (pure water filling) maximal instantaneous throughput policy π; (e) queue occupancy over time—queue lengths for users U_1 and U_2 under (b_1, b_2) and under (b_1^*, b_2^*). Under policy π, queue of user U_2 (b_2) is building up over time. (Example extracted from Reference 69.)

11.4 STABILITY REGION

In Section 11.2, we discussed an important attribute of the physical layer, namely, the feasible rate region (capacity region). In this section, we discuss an equivalently important attribute, namely, the *stability region*, in the MAC layer that is used to determine the operating region where a system remains stable. When a system is stable, the length of all queues in that system would be bounded. Hence, the buffer size is sufficient for the queuing system and there should no packet loss due to buffer overflow. This concept is very important in the sense that not all scheduling objectives will make the system stable, and it is ignored in the cross-layer design optimization based on information theory only because we have assumed infinite buffer size to decouple the queue dynamics from the scheduling operations. However, in practice, all buffers have finite length and it is very important to ensure stability in the system. The concept of stability region provides a guideline in the choice of system optimization objectives. The stability region can be explicitly defined as follows.

Definition 11.2 (Stability Region) For a queuing system with K users, the *stability region* refers to the set of average arrival rates $\lambda = [\lambda_1, \lambda_2, \ldots, \lambda_K]$ of the K users for which there exists a scheduling rule such that the buffer lengths in the system are bounded.

Conceptually, this is the same as saying that the stability region is the set of admissible arrival rates for which no queue *blows up*. Study of the stability region enables systematic design of the admission control over the system of data users. Before we discuss the scheduler design to achieve stability, we consider the following examples regarding stability region of several multiuser physical layers.

11.4.1 Stability Region of Time-Invariant Physical Layer

Consider a multiuser physical layer modeled by the time-invariant graph model [64] as described in Section 11.2.1. Let $\mathbf{G} = [G_{i,j}]$ denote the path gain matrix, where $G_{i,j}$ is the path gain from the transmitter of link i to the receiver of link j. At a given timestep, let p_i be the power level of transmitter of link i and $p_i > 0$ if and only if link i is active. Suppose that there are K links in the system and the required SIRs of the K links are specified by the $K \times K$ diagonal matrix $\Gamma = \text{diag}(\gamma_1, \ldots, \gamma_K)$. Define $\mathbf{G}' = [G'_{ij}]$ as the normalized interference matrix where $G'_{i,j} = G_{ji}/G_{ii}$ if $i \neq j$ and $G'_{i,i} = 0$. A set of links S is active (or feasible) at any given time if there exists a power vector $\mathbf{p} = \begin{bmatrix} p_1 \\ \vdots \\ p_K \end{bmatrix}$ whose support is S (i.e., $p_i > 0$ if and only if $i \in S$) such that

$$\Gamma \mathbf{G}' \mathbf{p} + \nu \leq \mathbf{p}$$

where ν is the noise power vector and the comparison is elementwise. It is assumed that all the links in the active set $\mathcal{A}$ have the same data rate r_0. For example, if $\mathcal{A} = \{1, 3, 4\}$ and $K = 6$, then the data rates of all the links are $\mathbf{r} = [r_0, 0, r_0, r_0, 0, \ldots, 0]$. The capacity region of the m links is a set of all *feasible rate vectors* of the links:

$$C = \{\mathbf{r} : \mathbf{r} \in \mathcal{A}_n \text{ for some active sets } \mathcal{A}_n\}$$

Each link i has a queue associated with it. For simplicity, we assume that both the arrival and the departure processes of the queue are specified in bit level and that a source bit arrives at link i according to the Possion arrivals at a rate of λ_i [bits per second (bps)]. When the source bit arrives at link i, it is appended to the tail of the queue. When a set of links $\mathcal{A}$ is activated at a timestep, the head of the queue of each link in $\mathcal{A}$ is removed. We have the following necessary and sufficient condition on the stability of the system.

Theorem 11.1 (Condition for Stability for the Poisson Arrivals) If there is a scheduling algorithm for which the system is stable, then the rate vector of the Poisson arrival (bps) $\lambda = [\lambda_1, \ldots, \lambda_K]$ is strictly dominated by a convex combination of the rate vectors $\mathbf{r} = [r_1, \ldots, r_K]$ in the capacity region C. Conversely, if λ is strictly dominated by a convex combination of the rate vectors in the capacity region C, then there exists a scheduling algorithm for which the system is stable.

Proof Please refer to Appendix 11A.

Theorem 11.1 provides a necessary and sufficient condition for the stability of systems with Poisson arrivals. For instance, the stability region of the system S can be written as

$$S = \left\{ \lambda = (\lambda_1, \ldots, \lambda_K) : \lambda \le \sum_n \alpha_n \mathbf{r}_n \text{ for some } \mathbf{r}_n \in C \text{ and } \alpha_n \ge 0 \right.$$
$$\left. \text{such that } \sum_n \alpha_n = 1 \right\} \tag{11.14}$$

where the operator $\le$ refers to componentwise comparison. Since $\Sigma_n \alpha_n \mathbf{r}_n$ represents a convex combination of all the feasible rate vectors $\mathbf{r}_n$ in the capacity region of the time-invariant physical layer, the stability region can be written as

$$S = \text{int}(\overline{C}) \tag{11.15}$$

where $\overline{C}$ denotes the convex hull[3] of the capacity region in the physical layer and int(.) denotes the interior of a set. Figure 11.6 illustrates an example of the capacity region of a two-user time-invariant channel as well as the corresponding stability region. Any Poisson arrival rate vector $\lambda \in S$ will make the system stable with an appropriately designed scheduler. On the other hand, for any Possion arrival rate vector λ outside the stability region S, no scheduling algorithm that can make the system stable exists. The stability region is useful in the admission control of the data traffic in the cellular system. The

[3]The *convex hull* of a set of vectors is the smallest convex set that contains all the vectors.

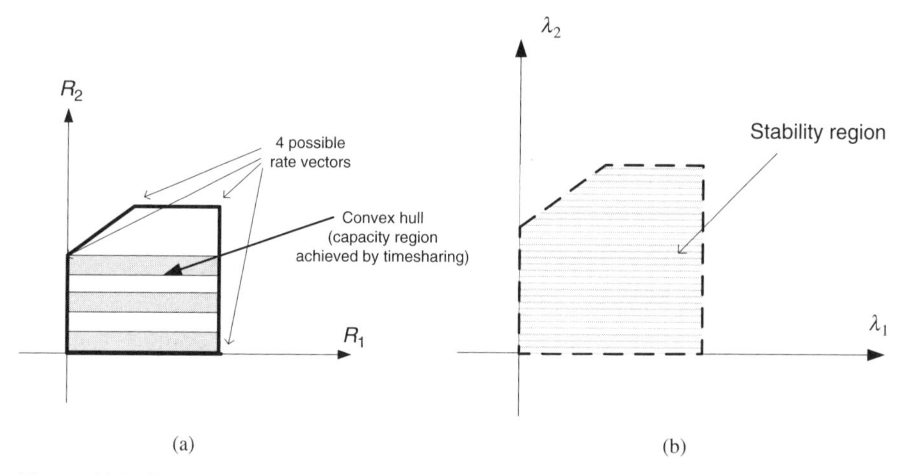

Figure 11.6. Illustration of capacity region and stability region for a two-user time-invariant channel: (a) capacity region C of a two-user channel; (b) stability region $S = \text{int}(\overline{C})$.

system should keep the the arrival rates within the stability region as otherwise the buffers in the system will overflow.

11.4.2 Stability Region of Stochastic Physical Layer

In Section 11.4.1, we characterized the stability region for the deterministic (or time-invariant) physical layer. The stability region is strictly dominated by the convex combination of the rate vectors in the capacity region of the underlying physical layer. In the following, we shall characterize the stability region when the underlying physical layer is time-varying in a random manner. In other words, we shall focus on the stability region of the system with the stochastic multiuser physical layer. This is very important because in wireless communications, channels are randomly varying with time because different and random interference levels are observed by different users, as well as fast fading of a signal received by a user.

We consider a general multiuser physical layer shared by K users where the capacity region is time-varying and is governed by an underlying random process $\mathrm{m}(t) \in \mathcal{M}$, which is referred to as the state of the wireless channel. The multiuser channel is characterized by a discrete channel state $m \in \mathcal{M} = [1, \ldots, M]$. The capacity region when the multiuser physical channel is at state $m(t)$ is denoted by $C(m)$. When the physical layer is at state $m(t)$ and when a feasible rate vector $\mathbf{r}^m = [r_1^m, \ldots, r_K^m] \in C(m)$ is selected by the scheduler, the data rate of user i is given by r_i^m (bps). Here, we allow more than one user to be selected (i.e., there exists $r_i^m > 0$ and $r_j^m > 0$ for some $i \neq j$). For example, if $\mathbf{r}^m = [1, 2, 0, 0]$ is selected by the scheduler, then the instantaneous scheduled rate of users 1 and 2 are 1 and 2 bps, respectively. The scheduled data rates of users 3 and 4 are both zero. Without loss of generality, assume that the set of feasible rate vectors $\mathbf{r}^m$ in the capacity region $C(m)$ [when the channel is in state $m(t)$] is indexed by a parameter j from some index set that (with some abuse of notation) is also denoted by $C(m)$.

Note that this multiuser physical layer model captures the practical scenario of downlink scheduling of the 3G CDMA high-data-rate (HDR) system. In HDR, multiple mobile stations in a cell share the same CDMA wireless channel. On the downlink, time is divided into fixed-size timeslots (1.67 ms). This slot size is short enough to ensure that the channel fading remains approximately constant within one or a few consecutive timeslots. In each timeslot, data can be transmitted to only one user. This multiuser physical layer model also captures a more general scenario where the base station has multiple transmit antennas and more than one user can be selected at any particular time (by the scheduler) and transmit over the same time and bandwidth through spatial multiplexing. In both cases, because of the action of channel fading between users, the channel state $m(t)$ is given by the joint state of the fading coefficients among the K users.

Evolution of the channel state of the multiuser physical layer model, $m(t)$, $t = 0, 1, 2, \ldots$, is assumed to be based on an irreducible discrete-time Markov chain with finite state space $\mathcal{M}$. The unique stationary distribution of this Markov chain is denoted by $\pi = (\pi_1, \ldots, \pi_M)$. Due to irreducibility, we have $\pi_m > 0$ for all m. In addition, we make a nondegeneracy assumption that for each user i, there is at least one channel state m such that $r_i^m > 0$. Otherwise, we shall have some users that can never be served. Let $A_i(t)$ be the number of bit arrivals of user i at time t. Assume that the arrival process of user i, $A_i(t)$, is an irreducible positive recurrent Markov chain with countable state space and the input processes among the K users are mutually independent. Let λ_i be the mean arrival rate of user i (bps). The vector of mean arrival rates is denoted by $\lambda = (\lambda_1, \ldots, \lambda_K)$. We have the following theorem regarding the necessary and sufficient condition for stability of the system.

Theorem 11.2 (Necessary and Sufficient Conditions for Stability of Stochastic Physical Layer) Let $\phi = [\phi_{m,j}]$, $m \in \mathcal{M}$, $j = 1, \ldots, C(m)$ be a stochastic matrix (i.e., $\phi_{m,j} \geq 0$ for all m and j and $\Sigma_j\, \phi_{m,j} = 1$ for every m). Consider the vector $\mathbf{v}(\phi) = (v_1(\phi), \ldots, v_K(\phi))$ where

$$\mathbf{v}(\phi) = \sum_m \pi_m \sum_{j \in C(m)} \phi_{mj} \mathbf{r}_j^m \tag{11.16}$$

If

$$\lambda < \mathbf{v}(\phi) \tag{11.17}$$

for some stochastic matrix ϕ, then there exists a scheduling rule for which the system is stable. Conversely, if there exists a scheduling rule for which the system is stable, then

$$\lambda \leq \mathbf{v}(\phi) \tag{11.18}$$

for some stochastic matrix ϕ.

Note that the operators $\leq$ and $<$ refer to componentwise comparison. For example, $\mathbf{x} \leq \mathbf{y}$ means that $x_i \leq y_i$ for all i. On the other hand, $\mathbf{x} < \mathbf{y}$ means that $x_i \leq y_i$ for all i and the strict inequality $x_i < y_i$ holds for at least one i. The set of all arrival vectors λ satisfying Equation (11.17) is called the *stability region* of the system.

Proof Please refer to Appendix 11B.

This theorem can be considered as an extension of Theorem 11.1 for the time-invariant channels. In fact, Equation (11.16) can be expressed as

$$\mathbf{v}(\phi) = \varepsilon\left[\sum_{j \in C(m)} \phi_{mj} \mathbf{r}_j^m\right] \tag{11.19}$$

for some stochastic matrices ϕ_{mj} where the expectation is taken with respect to the channel state m. Evaluating the region $\lambda > \mathbf{v}(\phi)$ over all possible stochastic matrix ϕ, the stability region S can be characterized by

$$\lambda < \varepsilon[\overline{C}(m)] \tag{11.20}$$

where $\overline{C}(m)$ denotes the *convex hull* of the capacity region $\overline{C}(m)$.

To illustrate the stability region concept, consider the following example.

Example 11.6 (Stability Region of Multiantenna Base Station with ZF Processing) Consider the wireless system with a multiantenna base station (n_T transmit antennas) and K single-antenna client users. Assume that the base station has full knowledge of the channel states. Using the "zero-forcing approach" as in Chapter 6, the received signal of user k is given by

$$Y_k = \sqrt{p_k}\, \mathbf{h}_k \mathbf{w}_k U_k + Z_k$$

where p_k is the transmit power allocated to the spatial channel for user k satisfying the total transmit power constraint $\Sigma p_R \leq P_0$, $\mathbf{w}_k$ is the $n_T \times 1$ zero-forcing weight at the transmitter, $\mathbf{h}_k$ is the $n_R \times 1$ channel fading between the base station and the kth user, U_k is the encoded symbol for user k, and Z_k is the AWGN channel noise with variance σ_z^2. Let $\mathcal{A}$ be the "set of admitted users" at any timeslot t. The physical layer capacity region conditioned on the channel states $\tilde{\mathbf{h}} = (\mathbf{h}_1, \ldots, \mathbf{h}_K)$, $C(\tilde{\mathbf{h}})$, is given by

$$C(\tilde{\mathbf{h}}) = \overline{\bigcup_{\mathcal{A}:|\mathcal{A}| \leq n_T} C(\tilde{\mathbf{h}}, \mathcal{A})}$$

where

$$C(\tilde{\mathbf{h}}, \mathcal{A}) = \left\{ \mathbf{r} : r_k \le W \log_2 \left(1 + \frac{p_k |\mathbf{h}_k \mathbf{w}_k|^2}{\sigma_z^2} \right) \quad \forall k \in \mathcal{A}; r_k = 0 \; \forall k \ne \mathcal{A}; \varepsilon_{p_k} \le P_0 \right\}$$

and W is the channel bandwidth.

Assume that the channel state sequence $\tilde{\mathbf{h}}(t)$ is i.i.d. at $t = 0, 1, 2, \ldots$. According to Theorem 11.1 and Equation (11.20), the stability region is characterized by

$$\lambda < \varepsilon\left[\overline{C(\tilde{\mathbf{h}})}\right] \tag{11.21}$$

where the expectation is taken with respect to the channel state $\tilde{\mathbf{h}}$. Figure 11.7 illustrates the unions of the instantaneous capacity region $C(\tilde{\mathbf{h}}, \mathcal{A})$ and the corresponding convex hull $\overline{C(\tilde{\mathbf{h}})}$ for $n_T = 2$ and $K = 2$. As long as the arrival rate vector λ is within the "interior" of the average of the "convex hull" int $(\varepsilon[\overline{C(\tilde{\mathbf{h}})}])$, there exists a scheduler that can achieve system stability.

Note that while Theorem 11.2 gives the necessary and sufficient conditions for the system stability, the scheduling policy that can achieve system stability may not be unique. In Section 11.4.3, we shall elaborate various specific scheduling designs that achieve system stability.

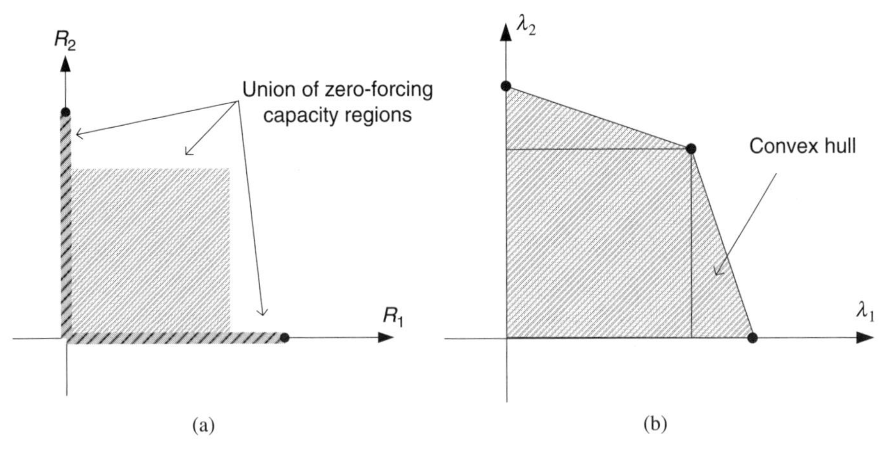

Figure 11.7. Illustration of capacity region for multiantenna base station $n_T = 2$ with $K = 2$ using zero-forcing processing at the transmitter (note that the average of the convex hulls denotes the stability region): (a) union of the instantaneous capacity regions $C(\tilde{\mathbf{h}}, \mathcal{A} = \{1\})$, $C(\tilde{\mathbf{h}}, \mathcal{A} = \{2\})$, and $C(\tilde{\mathbf{h}}, \mathcal{A} = \{1,2\})$; (b) instantaneous capacity region $\overline{C(\tilde{\mathbf{h}})}$.

11.4.3 Scheduling Design for Stability

In the previous subsections, we have focused on characterization of the stability region for the system. Given an arrival rate vector λ within the stability region, we know that there exists a scheduler that can achieve stability. In other words, not all schedulers can achieve stability in the system. In the following paragraphs, we focus on the connection between the scheduling objectives and system stability.

Definition 11.3 (Throughput Optimal Scheduler) A scheduler is "throughput-optimal" if it can maintain the stability of the system without knowing the arrival rate vector λ as long as λ is within the system's stability region S.

In fact, there is quite a wide class of scheduling rules that are *throughput-optimal* (i.e., they guarantee that all the queues remain stable in the system as long as the vector of arrival rates is within the system stability region). For example, the *max-weight scheduler* (MW), the *modified max-weight scheduler* (MMW), and the *exponential scheduler* (EXP) are all *throughput-optimal*. We shall elaborate these schedulers below.

Max-Weight Scheduler (MW). The MW scheduling rule is summarized in the following definition.

Definition 11.4 [Max-Weight (MW) Scheduler] At any time t, the MW scheduler selects the user(s) with the largest $\mathbf{Q}(t) \cdot \mathbf{r}(t)$, where $\mathbf{Q}(t) = [Q_1(t), \ldots, Q_K(t)]$ is the vector of queue lengths (number of untransmitted bits at time t) of K users and $\mathbf{r}(t) = [r_1(t), \ldots, r_K(t)]$ is the data rate vector within the instantaneous feasible capacity region $C(t)$ in the underlying multiuser physical layer

$$\max_{\mathbf{r}(t) \in C(t)} \mathbf{Q}(t) \cdot \mathbf{r}(t)$$

where $\cdot$ denotes vector dot product.

We have the following theorem regarding the stability property of the MW scheduler.

Theorem 11.3 (Throughput Optimiality of MW Scheduler) Suppose that at any timeslot t, the channel state is $m(t)$. The MW scheduler choosing the rate vector $\mathbf{r} \in C(m)$ that optimizes $\mathbf{Q}(t) \cdot \mathbf{r}(t)$ achieves stability for all vector of arrival rates λ strictly dominated by (11.16) for some ϕ.

In other words, the MW scheduler picks the most favorable rate allocation $\mathbf{r}$ within the capacity region of the physical layer such that it optimizes the weighted sum of the queue length and data rate. Users with longer queue and/or higher instantaneous data rate will be given priority. Such a scheduling policy can ensure stability of the system if the arrival rate vector is within the stability region of the system. The proof [138] is elaborated in Appendix 11C.

The objective function in Theorem 11.3 is chosen to make sure the queues in the system will remain stable. It is clear that this rule tries to equalize the queue sizes for different users, and hence it is useful for delay-sensitive applications. The feasible set of rate vectors in the capacity region depends of the choice of coding, modulation, and beamforming strategies employed. We shall consider an example of the MW scheduling rule over a multiantenna base station.

Example 11.7 (MW Scheduling over Multiantenna Base Station) Consider the same system as in Example 11.6. The MW scheduling rule can be cast into the following optimization problem (Problem 11.2).

Problem 11.2 (MW Scheduling) Select the admitted user set $\mathcal{A}$ and power allocation $\mathbf{p} = (p_1, \ldots, p_K)$ such that

$$\sum_{k \in \mathcal{A}} r_k Q_k$$

is maximized with a transmit power constraint $\mathcal{E}[\Sigma_{k \in \mathcal{A}} p_k] \leq P_0$ where Q_k denotes the queue length of user k.

The Lagrangian function of this optimization problem is given by

$$L(\mathbf{p}, \mathcal{A}, \mu) = \sum_{k \in \mathcal{A}} r_k Q_k - \mu \sum_{k \in \mathcal{A}} p_k \tag{11.22}$$

The optimizing variables consist of the real vector $\mathbf{p}$ as well as the discrete set $\mathcal{A}$. Hence, it can be solved by a mixed convex optimization (with respect to $\mathbf{p}$) and a combinatorial search (with respect to $\mathcal{A}$). Given a particular set $\mathcal{A}$, the optimizing power allocation $\mathbf{p}$ is given by $\partial L / \partial p_k = 0$. Using standard optimization techniques, the optimizing power is given by

$$p_k^* = \left(\frac{Q_k}{\mu \ln 2} - \frac{\sigma_z^2}{|\mathbf{h}_k \mathbf{w}_k|^2} \right)^+$$

where the Lagrange multiplier μ is chosen to satisfy the average power constraint P_0. The optimizing admitted user set $\mathcal{A}$ can be obtained by combinatorial search similar to the algorithms elaborated in Chapter 6.

Figure 11.8 illustrates the delay–throughput performance of the multiantenna system based on the MMW, MW, and EXP scheduling rules. We assume that all K users have a common average packet arrival rate λ. Note that the delay blows up when the average arrival rate exceeds a threshold ($\lambda \approx 0.8$ bps for $n_T = 3$), illustrating that the system becomes unstable when

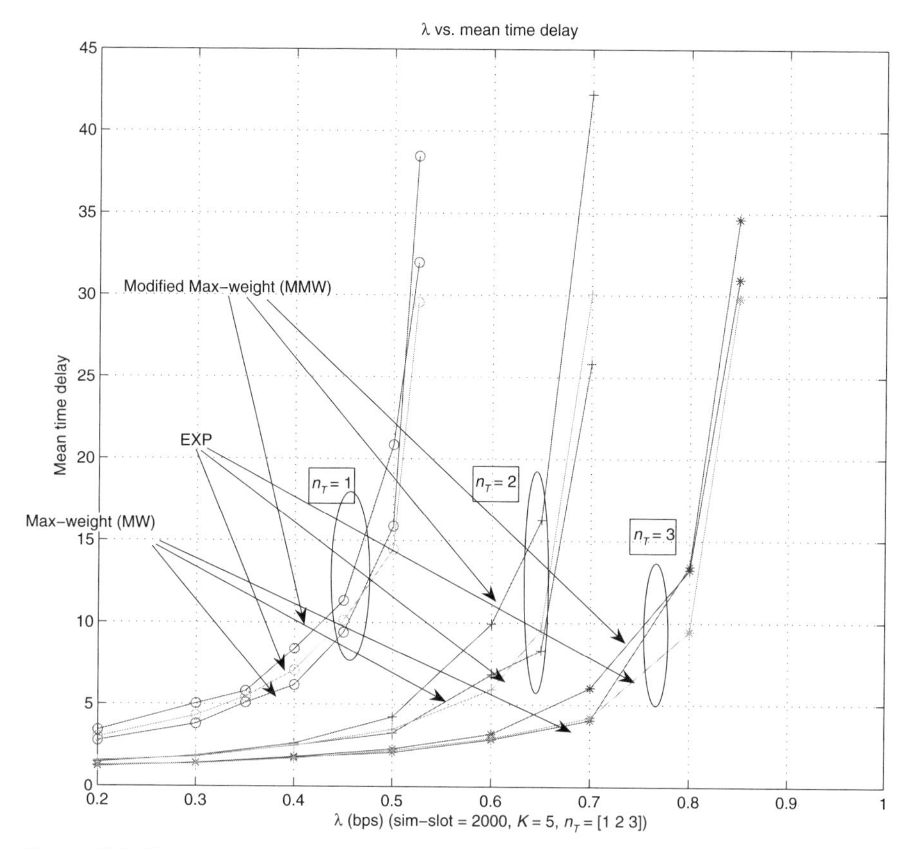

Figure 11.8. The average delay versus average arrival rate of multiantenna system with n_T = 1,2,3, SNR = 10 dB, and $K = 5$.

the arrival rate is outside the stability region. When the arrival rate is within the stability region, all the three schedulers can guarantee system stability (with a finite delay). Observe that as n_T increases, the stability threshold increases because of the capacity gain of spatial multiplexing.

Modified Max-Weight (MMW). In fact, we can generalize the MW scheduling to a "modified max-weight" scheduling rule defined as follows.

Definition 11.5 (Modified MW Scheduler) At anytime t, the MMV scheduler selects the rate vector $\mathbf{r}(t)$ with the largest $\Sigma_{k=1}^{K} \alpha_k Q_k(t) r_k(t)$, where $Q_k(t)$ is the queue lengths (number of untransmitted bits at time t) of user k, $\{a_k\}$ are any positive constants, and $\mathbf{r}(t) = [r_1(t), \ldots, r_k(t)]$ is the data rate vector within the instantaneous feasible capacity region $C(t)$ in the underlying multiuser physical layer:

$$\max_{\mathbf{r}(t)\in C(t)} \sum_{k=1}^{K} \alpha_k Q_k(t) r_k(t)$$

Similarly, we have the following theorem on the "throughput optimality" of MMW scheduling.

Theorem 11.4 (Throughput Optimality of MMW) The MMW scheduling policy as defined in Definition 11.5 is throughput-optimal; specifically, if there exists any scheduling rule that makes all user queues stable, then the MMW rule also makes all user queues stable.

Proof Please see Reference 77.

Exponential Scheduler (EXP). Besides the MW and MMW scheduling, we shall illustrate that the exponential scheduling rule is also throughput-optimal.

Definition 11.6 (Exponential Scheduler) For fixed $\alpha_k = \Theta(\in)$,[4] let $\varphi(\mathbf{Q}(t)) = [e^{\alpha_1 Q_1(t)}, \ldots, e^{\alpha_K Q_K(t)}]$ and $\varphi(Qi(t)) = e^{\alpha_i Q_i(t)}$. At any time t, the "exponential scheduler" selects the user(s) with the largest $\varphi(\mathbf{Q}(t)) \cdot \mathbf{r}(t)$, where $\mathbf{Q}(t) = [Q_1(t), \ldots, Q_K(t)]$ is the vector of queue lengths (number of untransmitted bits at time t) of K users and $\mathbf{r}(t) = [r_1(t), \ldots, r_K(t)]$ is the vector of data rate within the instantaneous feasible capacity region C(t) in the underlying multiuser physical layer

$$\max_{\mathbf{r}(t)\in C(t)} \varphi(\mathbf{Q}(\mathbf{t})) \cdot \mathbf{r}(t)$$

where · denotes vector dot product.

Theorem 11.5 (Throughput Optimality of EXP Scheduler) At any timeslot t, the EXP scheduler choosing the rate vector r ∈ $C(t)$ that optimizes $\varphi(\mathbf{Q}(t)) \cdot \mathbf{r}$ is "throughput-optimal."

Proof Please Refer to Appendix 11D.

11.5 CROSS-LAYER SCHEDULER DESIGN FOR DELAY MINIMIZATION

In the last section, we focused on the characterization and cross-layer scheduler design for system stability. While system stability is important for systems with finite buffer size, we would also like to minimize the average length of

[4]The symbol denotes Θ *asymptotically equal*; specifically, $f(x) = \Theta(g(x))$ means $f(x) = cg(x)$ for some constant c.

the queues (delay minimization). In the following text, we shall focus on the cross-layer scheduling design to minimize the average delay of the users. The system delay is an important design objective for delay-sensitive data applications. We shall first cast the cross-layer design as an optimization problem based on a combined framework of information theory and queuing theory. The optimal solution is derived, and the physical interpretations of the optimal scheduling are discussed. For illustration, we shall focus on the uplink multiaccess capacity region in the following text.

11.5.1 Problem Formulation

Consider the uplink of a wireless system with a base station and K mobiles. The received signal at the base station is given by

$$Y = \sum_k H_k X_k + Z$$

where X_k and H_k are the transmitted symbol and the channel fading of user k and Z is the AWGN channel noise with variance σ_z^2. In addition, we assume that the transmitted power of user k, namely, p_k, satisfies the average power constraint $\varepsilon[p_k] = P_k$ and the peak power constraint $p_k \le \hat{P}_k$. The multiaccess channel is characterized by the instantaneous channel state (CSI) $\tilde{\mathbf{H}}(t) = (H_1(t), \ldots, H_K(t)) \in \mathcal{H}^K$, where $\mathcal{H}$ is the state space of the CSI and $|\hat{H}_k| = \sup|\mathcal{H}| < \infty$ for all k. At any timeslot t, the feasible rate vector (in bps) $\mathbf{r} = (r_1, \ldots, r_K)$ must be within the physical layer capacity region $C(\tilde{\mathbf{H}})$ given by

$$\sum_{k \in B} r_k \le W \log_2 \left(1 + \frac{\sum_{k \in B} p_k |h_k|^2}{\sigma_z^2} \right) \tag{11.23}$$

for all $B \subseteq [1, 2, \ldots, K]$, where W is the channel bandwidth.

For simplicity, we assume that the input process and the output process of the queues in the system are operating at the bit level. Let the packet arrival (bit arrival) of the kth user be Poisson arrival[5] with average arrival rate λ_k (bps). Let $\mathbf{Q}(t) = (Q_1(t), \ldots, Q_K(t))$ be the vector of queue length (the number of untransmitted bits) at time t for K users. The cross-layer scheduling problem is equivalent to the stochastic control problem as illustrated in Figure 11.9.

At any timeslot t, the dynamics of the system can be specified by a state $\mathbf{S}(t) = (\tilde{\mathbf{H}}(t), \mathbf{Q}(t))$ constituted by the channel state (CSI) and the queue state. The scheduler is characterized by two mappings, namely, the *rate allocation policy* and the *power allocation policy* from the system state $\mathbf{S}(t)$. The rate

[5] It can be shown that the Poisson assumption can be relaxed to the Markov modulated case as well.

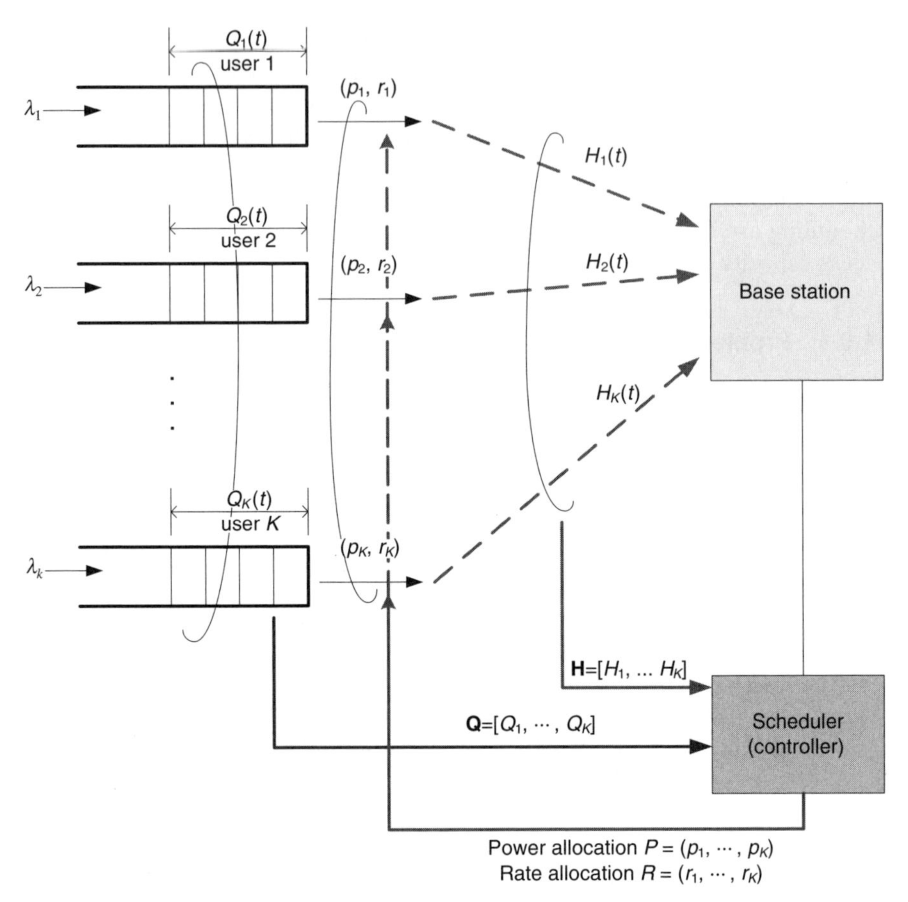

Figure 11.9. Equivalence of the cross-layer scheduling design and the stochastic control problem.

allocation policy $\mathcal{R}$ is a mapping from the system state $\mathbf{S}(t)$ to the rate vector $\mathbf{r}(t) \in C(t)$. The power allocation policy $\mathcal{P}$ is a mapping from the system state $\mathbf{S}(t)$ to the power allocation vector $\mathbf{p}(t)$ such that $\mathcal{E}[\mathbf{p}] \leq \mathbf{P} = (P_1, \ldots, P_K)$ (average power constraint) and $\mathbf{p} \leq \hat{\mathbf{P}}$ (peak power constraint). The *delay-optimal* cross-layer scheduling design can be cast into the following optimization problem [145].

Problem 11.3 (Delay Minimization Problem) Find the rate allocation policy $\mathcal{R}$ and the power allocation policy $\mathcal{P}$ such that the total average delay of the system $\lim_{t\to\infty}\mathcal{E}[\Sigma_k Q_k(t)]$ is minimized for $\lambda \in S$ where S is the stability region of the system.

In other words, the optimization problem is targeted to find a scheduling policy to minimize the average total queue length. By Little's law, minimizing the average queue length is equivalent to minimizing the average packet delay in steady state. In fact, the cost function can be modified to a more general objective $\lim_{t\to\infty}\mathcal{E}[\varphi(\mathbf{Q}(t))]$ for some Schur convex and increasing functions [146] $\varphi: \mathcal{R}^K \mapsto \mathcal{R}$.

11.5.2 Optimal Solution

In this section, we focus on the optimal solution to the scheduling problem (Problem 11.3). Before we discuss the delay-optimal solution, we consider the following theorem regarding the *throughput optimality* [146] of a simple rate allocation and power allocation policy, namely, the *Tse–Hanly* (TH) policy [132], in the fading multiaccess channel in Equation (11.23).

Theorem 11.6 (TH Policy Is Throughput-Optimal) For any given realization of the channel state $\tilde{\mathbf{h}}$, let $\mathcal{P}_{\text{TH}}(\tilde{\mathbf{h}}, \mu)$ and $\mathcal{R}_{\text{TH}}(\tilde{\mathbf{h}}, \mu)$ be the optimizing solution of the following optimization problem

$$(\mathcal{P}_{\text{TH}}, \mathcal{R}_{\text{TH}}) = \arg\max_{\mathbf{p},\mathbf{r}} \sum_{k=1}^{K} \mu_k r_k - \sum_{k=1}^{K} \beta_k p_k \tag{11.24}$$

where $\mathbf{r} \in C(\tilde{\mathbf{h}}, \mathbf{p})$, $\mathbf{p} \le \hat{\mathbf{P}}$ (peak power constraint) and β are the Lagrange multipliers chosen to satisfy the average power constraint $\mathcal{E}[\mathbf{p}] \le \mathbf{P}$. Then a throughput-optimal scheduling policy is given by $\mathcal{P}_{\text{TH}}(\tilde{\mathbf{h}}, \alpha * \mathbf{q})$ and $C_{\text{TH}}(\tilde{\mathbf{h}}, \alpha * \mathbf{q})$, where α is any positive vector, $\mathbf{q}$ is the current queue length, and $\alpha * \mathbf{q}$ is the vector whose component is given by $\alpha_i q_i$.

Proof Please refer to Appendix 11F.

Using the proof of Theorem 11.6 in Appendix 11F and Theorem 11.8 in Appendix 11E, we can have an average delay bound for the TH policy as summarized in the following corollary.

Corollory 11.1 (System Delay) For any arrival rate vector, $\lambda \in S$, under the TH policy $(\mathcal{P}_{\text{TH}}(\tilde{\mathbf{h}}, \mathbf{v}), C_{\text{TH}}(\tilde{\mathbf{h}}, \mathbf{v}))$ in Equation (11.24), the steady-state queue length has a finite mean $\overline{Q}_k$ for all k and

$$\sum_{k=1}^{K} \alpha_k \overline{Q}_k \le \frac{\theta}{2T\delta} \tag{11.25}$$

for some constants θ and δ where T is the time duration of timeslots. By Little's law, the average bit delay $\overline{D}_k$ satisfies

$$\sum_{k=1}^{K} \alpha_k \lambda_k \overline{D}_k \le \frac{\theta}{2T\delta} \tag{11.26}$$

Proof Please refer to Appendix 11F.

Let $\mathbf{v} = \alpha^* \mathbf{q}$ be the vector of weighted queue sizes. In the case of one user $K = 1$, it can be shown [132] that $(\mathcal{P}_{\mathrm{TH}}(\tilde{\mathbf{h}}, \mathbf{v}), C_{\mathrm{TH}}(\tilde{\mathbf{h}}, \mathbf{v}))$ reduces to the well-known water filling scheme whereby more power is allocated to favorable channel states and less or no power is allocated to unfavorable channel states. Water filling is known to maximize throughput in a single-user channel with an infinitely backlogged transmitter. Theorem 11.6 shows that it is also throughput-optimal for a communication system with random packet arrivals. In the general case of multiple users $K > 1$ and unequal weighted queue sizes, several users typically transmit, and little can be said about the optimal power allocation policy. The optimal rate allocation policy, on the other hand, satisfies a general principle known as the *longest weighted queue–highest possible rate* (LWQHPR). Given any power allocation policy $\mathcal{P}$, the LWQHPR is as defined below

Definition 11.7 [Longest Weighted Queue–Highest Possible Rate (LWQHPR)] For any feasible power allocation policy $\mathcal{P}$ (satisfying both the peak and the average power constraints), the rate allocation policy (LWQHPR) is given by

$$\mathcal{R}^*(\tilde{\mathbf{H}}, \mathbf{Q}) = \arg \max_{r \in C(\tilde{\mathbf{H}}, \mathcal{P}(\tilde{\mathbf{H}}, \mathbf{Q}))} \sum_{k=1}^{K} \alpha_k Q_k r_k \tag{11.27}$$

for some $\alpha_k > 0$.

In fact, following the same proof as inTheorem 11.6, for any given feasible power allocation policy $\mathcal{P}$, the LWQHPR rate allocation policy is *throughput-optimal* and the delay bound in Corollary 11.1 holds. In fact, the LWQHPR policy is equivalent to the MMW scheduling policy introduced in Definition 11.5. The LWQHPR becomes the regular MW scheduler if $\alpha_k = 1$ for all $k \in [1, \ldots, K]$, and it is also known as the *longest queue–highest possible rate* (LQHPR) policy [146].

Besides throughput optimality, we shall illustrate that for systems with symmetric fading, the LQHPR rate allocation policy is also *strongly delay-optimal* given any symmetric power allocation policy [146]. We shall first introduce the following definitions.

Definition 11.8 (Symmetric Fading Process $\tilde{\mathbf{H}}(t)$) A fading process $\tilde{\mathbf{H}}(t)$ is called "symmetric" if for any $\tilde{\mathbf{h}} = (h_1, \ldots, h_K)$, we have

$$\Pr[H_1(t) = h_1, \ldots, H_K(t) = h_K] = \Pr[H_1(t) = h_{\pi(1)}, \ldots, H_K(t) = h_{\pi(K)}] \tag{11.28}$$

for any permutation π. For example, at every t, $\mathbf{H}_1(t), \ldots, \mathbf{H}_K(t)$, being i.i.d. is one example of a symmetric fading process.

Definition 11.9 (Symmetric Power Control $\mathcal{P}$) A power control policy $\mathcal{P}$ is "symmetric" if for any $\tilde{\mathbf{h}} = (h_1, \ldots, h_K)$, we have

$$\mathcal{P}_i(\tilde{\mathbf{H}}(t) = (h_1, \ldots, h_K)) = \mathcal{P}_{\pi^{-1}(i)}(\tilde{\mathbf{H}}(t) = (h_{\pi(1)}, \ldots, h_{\pi(K)})) \qquad (11.29)$$

for any permutation π where $\mathcal{P}_i$ denotes the allocated power for the ith user. For example, if $K = 2$ and $\mathcal{P}_1(h_1, h_2) = \mathcal{P}_2(h_2, h_1)$, then the power control policy is symmetric.

In the following paragraphs, we shall define the concept of *majorization* and *stochastic coupling*. For any vector $\mathbf{x} = (x_1, \ldots, x_K) \in \mathcal{R}^K$, let $x_{[1]} \geq \ldots \geq x_{[M]}$ denote the components of $\mathbf{x}$ in decreasing order.

Definition 11.10 (Majorization) For vectors $\mathbf{x}$, $\mathbf{y} \in \mathcal{R}^K$, $\mathbf{x}$ is "weakly majorized" by $\mathbf{y}$ (notation $\mathbf{x} \prec_w \mathbf{y}$) if

$$\sum_{i=1}^{k} x_{[i]} \leq \sum_{i=1}^{k} y_{[i]}, \quad k = 1, \ldots, K \qquad (11.30)$$

Note that $\mathbf{x}$ is said to be majorized by $\mathbf{y}$ (notation $\mathbf{x} \prec \mathbf{y}$) if equality holds for $k = K$.

For example, $(1, 1) \prec_w (3, 0)$ and $(1, 1) \prec (2, 0)$.

Definition 11.11 (Stochastic Majorization) Let $\mathbf{X}$ and $\mathbf{Y}$ be two random vectors in $\mathcal{R}^K$. $\mathbf{X}$ is "stochastically weak-majorized" by $\mathbf{Y}$ (denoted by $\mathbf{X} \prec_w^{st} \mathbf{Y}$) if there exist random vectors $\tilde{\mathbf{X}}$ and $\tilde{\mathbf{Y}}$ also in $\mathcal{R}^K$, with the Following properties:

1. $\mathbf{X}$ and $\tilde{\mathbf{X}}$ are identically distributed
2. $\mathbf{Y}$ and $\tilde{\mathbf{Y}}$ are identically distributed
3. $\tilde{\mathbf{X}} \prec_w \tilde{\mathbf{Y}}$ *a.s.* (i.e. weakly majorization relationship is only in "almost sure" probability sense).

It can be shown that the LQHPR rate allocation policy is delay-optimal under symmetric fading channels and symmetric power allocation policy. This is outlined in the theorem below.

Theorem 11.7 (LQHPR Policy Is Delay-Optimal) Let $\mathcal{P}: \tilde{\mathbf{H}} \mapsto \mathcal{R}_+^K$ be any given symmetric power allocation policy as per Definition 11.9. Let $\mathbf{Q}(t)$ be the queue length evolution under the LQHPR policy [given by Equation (11.27) with $\alpha_i = 1$] and $\mathbf{Q}'(t)$ be the queue length evolution under any other feasible rate policy. If the channel fading is symmetric (as per Definition 11.8), then

$$\mathbf{Q}(t) \prec_w^{st} \mathbf{Q}'(t) \text{ for all } t \geq 0$$

Proof Please see Reference 145.

While Theorem 11.7 establishes the stochastic majorization of the LQHPR policy, we still have to establish the connection between the stochastic majorization and the average delay of the system. We shall first introduce the following definitions and lemmas [98].

Definition 11.12 ($<_w$-Persevering Function) Let $\mathcal{A} \subset \mathcal{R}$. A function $\phi: \mathcal{A} \mapsto \mathcal{R}^K$ is said to be $<_w$-persevering if $\mathbf{x} <_w \mathbf{y} \Rightarrow \phi(\mathbf{x}) \leq \phi(\mathbf{y})$ for $\mathbf{x}, \mathbf{y} \in \mathcal{A}$.

The following lemma [98] draws the links between the cost function and stochastic weak majorization.

Lemma 11.1 (StochasticWeak Majorization) Let $\mathbf{X} = (X_1, \ldots, X_K)$ and $\mathbf{Y} = (Y_1, \ldots, Y_K)$ be random vectors taking values in $\mathcal{R}^K$. We have $\mathbf{X} <_w^{st} \mathbf{Y}$ if and only if $\varepsilon[\phi(\mathbf{X})] \leq \varepsilon[\phi(\mathbf{Y})]$ for all $<_w$-perserving functions $\phi: \mathcal{R}^K \mapsto \mathcal{R}$ for which the expectation exists.

In fact, it turns out that a real-valued function ϕ defined on $\mathcal{A} \subset \mathcal{R}^K$ is $<_w$-perserving if and only if it is increasing and *Schur convex* [98]. This includes all symmetric,[6] convex, and increasing real functions on $\mathcal{R}^K$. For example, $\phi(\mathbf{x}) = \max_{i=1,\ldots,K}|x_i|$ and $\phi(\mathbf{x}) = \Sigma_{k=1}^K |x_i|^r$ for $r \geq 1$ or $r < 0$, $\phi(\mathbf{x}) = (\Sigma_{i=1}^K |x_i|^r)^{1/r}$ for $r \geq 1$ are both $<_w$-perserving functions.

From Lemma 11.1, we have the following corollary to establish the linkage between stochastic majorization of the LQHPR policy (in Theorem 11.7) and the average packet delay in the system.

Corollory 11.2 (Delay and Majorization) Given any symmetric power allocation policy $\mathcal{P}$, let $\mathbf{Q}(t)$ be the vector of queue length under the LQHPR policy and $\mathbf{Q}'(t)$ be the vector of queue length under any other rate allocation policy. If the channel fading process $\tilde{\mathbf{H}}(t)$ is symmetric, we have

$$\varepsilon[\phi(\mathbf{Q}(t))] \leq \varepsilon[\phi(\mathbf{Q}'(t))], \quad ;\forall t \geq 0 \tag{11.31}$$

for all $<_w$-preserving functions $\phi: \mathcal{R}^K \mapsto \mathcal{R}$ for which the expectation exists.

Since $Q_i(t) \geq 0$, the function $\phi(\mathbf{Q}) = \sum_{i=1}^{K} Q_i(t)$ is $<_w$-preserving and the LQHPR policy minimizes the average total delay in the system for any given symmetric power allocation policy and symmetric fading process.

[6] A function ϕ is symmetric on $\mathcal{A} \subset \mathcal{R}^K$ if $\phi(\mathbf{x}) = \phi(\mathbf{xP})$ for any $\mathbf{x} \in \mathcal{A}$ and any $K \times K$ permutation matrix $\mathbf{P}$.

11.5.3 Description of LQHPR Policy

Since the capacity region in the multiaccess fading channels $C(\tilde{\mathbf{H}},\mathcal{P}(\tilde{\mathbf{H}},\mathbf{Q}))$ is a polymatroid, the solution of the LQHPR rate allocation policy corresponds to the rates achievable in a successive decoding scheme. LQHPR is a greedy rate allocation procedure where the longest queue receives the highest possible rate. This is elaborated as follows:

- The LQHPR policy can be viewed as an adaptive version of the Tse–Hanly problem [132]. The Tse–Hanly problem is given by

$$\max_{\mathbf{p},\mathbf{r}} \mu \cdot \mathbf{r} - \beta \cdot \mathbf{p} \text{ subject to } \mathbf{r} \in C(\tilde{\mathbf{h}},\mathbf{p}) \text{ and } \varepsilon[\mathbf{p}] \le \mathbf{P}$$

  where μ is a vector of constant weights and β is a Lagrange multiplier vector. Given μ and β, this problem can be solved by using a greedy algorithm [132] similar to that elaborated in Chapter 6.

- If we order the queue lengths as $q_{[1]} \ge q_{[2]} \ge \cdots \ge q_{[K]}$, the LQHPR solution has the rate allocation given by

$$r_{[i]}^{*} = \log_2\left(1 + \frac{|h_{[i]}|^2 \mathcal{P}_{[i]}(\tilde{\mathbf{h}},\mathbf{q})}{\sum_{j<i} |h_{[j]}|^2 \mathcal{P}_{[j]}(\tilde{\mathbf{h}},\mathbf{q}) + \sigma_z^2}\right)$$

  where $\tilde{\mathbf{h}}$ is a channel fading realization and $\mathbf{q}$ is a queue length realization of the corresponding random process. The solution is given by *adaptive successive decoding* at the base station. For instance, the user with queue $q_{[K]}$ is decoded first and the user with queue $q_{[1]}$ is decoded last. Figure 11.10 illustrates an example of the two-user Gaussian MAC channel with two extreme points $\mathbf{r}_A = (r_{A,1}, r_{A,2})$ and $\mathbf{r}_B = (r_{B,1}, r_{B,2})$. From the figure, we have $r_{A,1} < r_{B,1}$ and $r_{A,2} > r_{B,2}$. The LQHPR policy assigns the

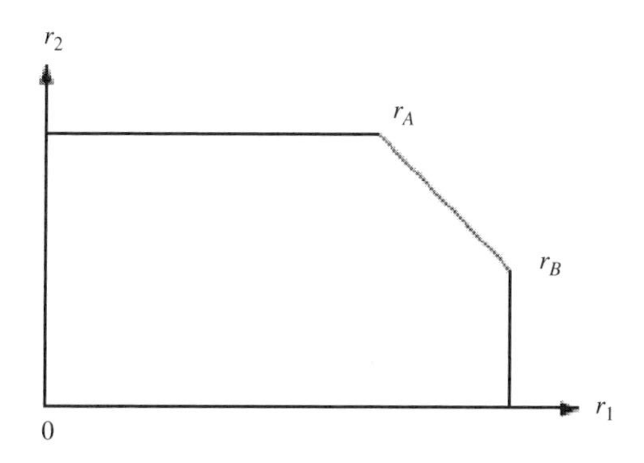

Figure 11.10. Two-user Gaussian MAC capacity region C.

rate vector $\mathbf{r}_A$ whenever $Q_1(t) \leq Q_2(t)$ and assigns the rate vector $\mathbf{r}_B$ whenever $Q_1(t) \geq Q_2(t)$.

- The LQHPR policy has the advantage of *load balancing* or *equalization of queue lengths*. This also echos with respect to Example 11.5 in Section 11.3. From the physical layer perspective, we would like to assign a rate vector on the dominant face of the capacity region because that represents the highest possible physical data rate assignment. However, in practical scenarios, we may not have not enough packets to be transmitted on the assigned rate vector because of empty queues. In this case, the conventional water-filling solution (non-work-conserving scheduling policy) may not be optimal because of the empty queues. In the LQHPR policy, user priority will decrease if the queue length is short, avoiding the situation of assigning resource to empty queues. On the other hand, when the queue length increases, the priority of that user also increases. Hence, the LQHPR policy is a cross-layer scheduling design that is adaptive to both the channel state $\tilde{\mathbf{H}}(t)$ and the queue length $\mathbf{Q}(t)$. It maximizes the rate at which overall unfinished work is processed at any given time, thus minimizing the average total queue lengths in the system.

11.5.4 Performance of an Example System

Consider an example multiaccess system where we have $K = 2$ (two users) and each user sees an i.i.d. fading process $H_i(t)$. Suppose $|H_i|^2(t) = \{10(\text{good}), 1/10(\text{bad})\}$ with probability 0.5. The average transmit powers of users 1 and 2 are given by $P_1 = P_2 = 1$. The packet arrival processes for the two users are independent Poisson processes with the same arrival rate λ. We compare the performance of the following scheduling policies.

LQHPR Policy with Power Adaptation. The LQHPR scheduling policy with power adaptation is specified by the following optimization problem:

$$\max_{\mathbf{p},\mathbf{r}}(Q_1 r_1 + Q_2 r_2) - (\beta_1 p_1 + \beta_2 p_2)$$

where $\mathbf{r} = (r_1, r_2) \in C(H_1, H_2; p_1, p_2)$ and (β_1, β_2) are Lagrange multipliers chosen to satisfy the average power constraint $\varepsilon[p_i] = 1$. The solution is given by the greedy procedure [132].

LQHPR Policy with Constant Power $p_1 = p_2 = 1$. The LQHPR scheduling policy with constant power allocation is given by the optimization problem

$$\max_{\mathbf{r}}(Q_1 r_1 + Q_2 r_2)$$

where $\mathbf{p} = (p_1, p_2) = (1, 1)$. The solution is a special case of the greedy procedure in Reference 132 without power adaptation.

Single-User Scheduling with Constant Power. In single-user scheduling, we select the user with the best channel fading at any given time. The transmit power (for fair comparison) is set to $p_1 = p_2 = 2$.

Knopp–Humblet Scheduling. In Knopp–Humblet [72] scheduling, queue size is ignored and the scheduling algorithm is adaptive with respect to channel fading states only. The optimization problem is given by

$$\max_{\mathbf{p},\mathbf{r}} (r_1 + r_2) - (\beta_1 p_1 + \beta_2 p_2)$$

where $\mathbf{r} = (r_1, r_2) \in C(H_1, H_2; p_1, p_2)$ and (β_1, β_2) are Lagrange multipliers chosen to satisfy the average power constraint $\varepsilon[p_i] = 1$.

Constant Power Best Channel Highest Possible Rates (BCHPR). In the BCHPR scheduling strategy, the transmit power is constant ($p_1 = p_2 = 1$) and the user with the worst fading will be decoded first irrespective of queue size.

Figure 11.11 illustrates the performance comparison (the average packet delay versus arrival rate) of the scheduling rules listed above. We observe that

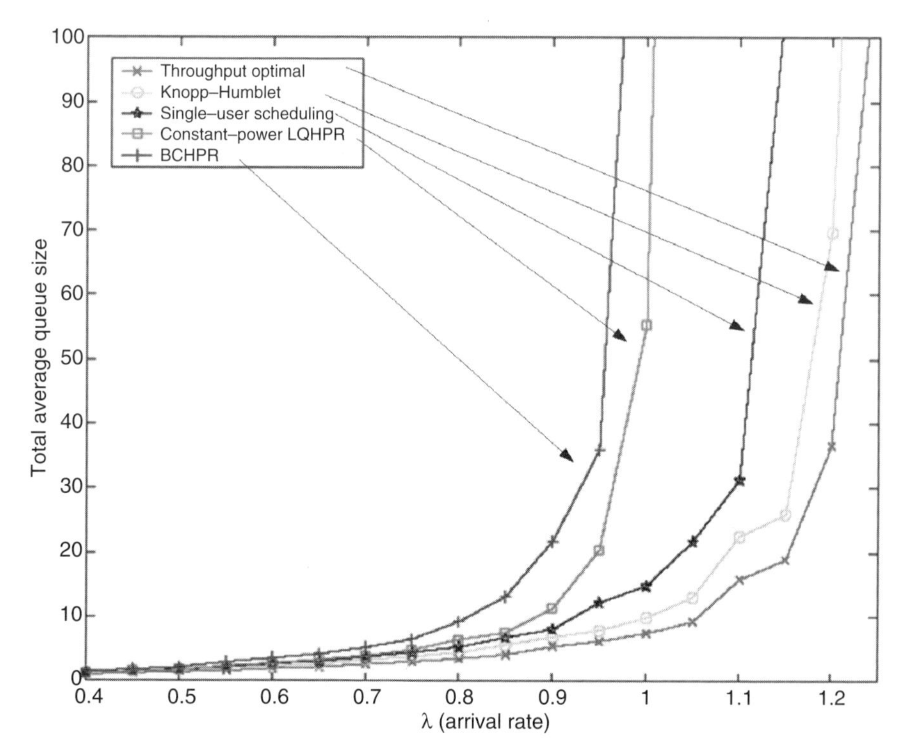

Figure 11.11. Comparison between average packet delay and average arrival rate of various scheduling algorithms in a two-user MAC channel [145].

the LHQPR policy with power adaptation is throughput-optimal in the sense that it ensures system stability as long as the packet arrival rate is within the stability region. At the same time, the LHQPR policy with power adaptation achieves the smallest average packet delay at a given λ. The Knopp–Humblet scheduler (ignoring the queue length) is shown to be inferior in the delay sense, and the worst algorithm is the BCHPR strategy.

11.6 SUMMARY

In this chapter, we have discussed various aspects of the adaptive cross-layer MAC layer scheduler design based on a combined framework of information theory and queuing theory. The multiuser wireless system can be modeled as a stochastic process with the system state determined by the channel state (CSI) $\tilde{\mathbf{H}}(t)$ as well as the queue status $\mathbf{Q}(t)$ among all the users. When the channel state (CSI) is at state $\tilde{\mathbf{H}}(t)$, the capacity region of the underlying multiuser physical layer is given by $C(\tilde{\mathbf{H}}(t))$. The cross-layer scheduler design can be cast as an optimization problem where the scheduler computes a rate allocation vector $\mathbf{r} = (r_1, \ldots, r_K) \in C(\tilde{\mathbf{H}}(t))$ at every timeslot t based on the system state $S(t) = (\tilde{\mathbf{H}}(t), \mathbf{Q}(t))$. Various optimization objectives are specified by the objective functions $U(\overline{R}_1, \ldots, \overline{R}_K; \overline{Q}_1, \ldots, \overline{Q}_K)$, where $\overline{R}_k$ and $\overline{Q}_k$ denote the average data rate and the average queue length of user k. When the application is delay-insensitive and the system buffer size is large, the cross-layer optimization objective and the scheduler are functions of the CSI $\tilde{\mathbf{H}}(t)$ only. This reduces to the conventional cross-layer scheduling design as elaborated in Chapter 6. On the other hand, when the application is delay-sensitive or the system buffer size is limited, the cross-layer optimization objective as well as the scheduler are in general a function of the entire system state $S(t) = (\tilde{\mathbf{H}}(t), \mathbf{Q}(t))$. In other words, the cross-layer scheduler should be adaptive to both the channel condition and buffer status. In any case, the cross-layer optimization design is decoupled from the specific implementation of the coding, modulation, and multiple-access schemes.

We have elaborated the common multiuser physical layer modeling, namely, the graph model, the ON/OFF channel model, and the information-theoretic model. The graph model is useful to capture the high-level interference relationship between multiple links. However, it fails to capture the time-varying dynamics of individual links. The ON/OFF channel model allows certain modeling on the time-varying dynamics of the fading channels but is not general enough to cover multiuser interference. The information-theoretic model offers the most general and flexible multiuser modeling. In any case, the multiuser physical layer can be modeled as a capacity region that is a set of all feasible rate vectors $\mathbf{r}$.

After the physical layer modeling, we have elaborated on the importance of the cross-layer scheduling design adapting to both the physical channel state and the queue state. We consider three examples of cross-layer scheduling. In the first example, we have illustrated that channel state information would be useful in scheduling to exploit multiuser diversity. System capacity can be substantially increased by giving users with good CSI higher priorities. However, if the scheduler adapts to the CSI only, there is no guarantee on the delay performance. In the second example, we have illustrated the importance of *buffer-condition*-aware scheduling strategy, especially if our objective is to minimize the average packet delay. In the final example, we have illustrated that the conventional power water-filling solution may not be truly optimal when buffer status is taken into account. Hence, it is important to have a cross-layer scheduling strategy that adapts to both the channel state information and the buffer conditions. This requires a combined framework based on information theory (for the physical layer) and queuing theory (for the queue dynamics).

In Section 11.4.2, we consider one important design objective of the cross-layer scheduling adapting to both the CSI and queue state, namely, the stability of the system. We define the *stability region* as the set of packet arrival rate vectors λ (Poisson arrivals) where the length of the buffers in the system remains bounded (using an appropriately chosen scheduler). We have characterized the stability region of multiuser systems with a time-invariant physical layer as well as a stochastic physical layer. For instance, the stability region is given by the *convex hull* of the capacity region of the underlying physical layer. For example, the stability regions of the time-invariant physical layer and the stochastic physical layer are given by $\mathrm{int}(\bar{C})$ and $\mathrm{int}(\overline{\varepsilon[C(\tilde{\mathbf{H}})]})$, respectively. Finally, we elaborate on the specific scheduler designs that can maintain system stability.

Finally, in Section 11.5, we consider another important cross-layer design objective to minimize the packet delay. The delay-optimal scheduler design is cast into an optimal control problem, and the solution is given by the LQHPR policy, which schedules packets according to load balancing rules.

APPENDIX 11A: PROOF OF THEOREM 11.1 ON STABILITY REGION OF DETERMINISTIC PHYSICAL LAYER

Necessary Condition. Let $Q_i(t)$ be the queue size of user i at time t, and let $\mathbf{Q}(t) = (Q_i(t))$ be the vector of queue sizes. Let $\mathbf{r}(t) \in C$ be the scheduled rate vector within the time invariant capacity region and $\mathbf{A}(t)$ be the vector of bits arrival for the K users at time t. We have

$$\mathbf{Q}(t) = \mathbf{A}(0) + \mathbf{A}(1) + \cdots + \mathbf{A}(t-1) - t\mathbf{v} \tag{11.32}$$

where $\mathbf{v} = 1/t \times (\mathbf{r}(0) + \mathbf{r}(1) + \ldots + \mathbf{r}(t-1))$. Since $\mathbf{v}$ belongs to the set S of the convex convex combinations of the feasible rate vectors $\mathbf{r}(t) \in C$, its expected value belongs to the convex combination. Thus, it follows from (11.32) that $\lambda - E[\mathbf{q}(t)]/t \in S$. By taking the limit as t goes to infinity, we conclude that $\lambda \in S$, since S is a closed set.

We now show that λ is strictly dominated by a vector in S. The total bit arrival of user i during the interval $[0, t - 1]$ has a Poisson distribution with mean $\lambda_i t$. Let B_i be the event that it exceeds $\lambda_i t + \sqrt{\lambda_i t}$. The probability of B_i is at least an absolute constant c. It follows from (11.32) that

$$\varepsilon[\mathbf{Q}(t)|B_1, B_2, \ldots, B_K] \geq t\lambda + \sqrt{t\lambda} - t\varepsilon[\mathbf{v}|B_1, \ldots, B_K] \tag{11.33}$$

The left side of (11.33) is upper-bounded by $E[\mathbf{Q}(t)]/c^K$, which is bounded by a constant independent of t. By choosing t sufficiently large, we conclude the $\mathbf{r}$ is strictly dominated by the vector $\varepsilon[\mathbf{v}|B_1, \ldots, B_K]$, which belongs to S.

Sufficiency Condition. We show that if $\lambda = \varepsilon[\mathbf{A}]$ is strictly dominated by a vector in S, then there is a scheduling algorithm for which the system is ergodic. Assume that $(1 + \varepsilon)\lambda \leq \Sigma_{n=1}^{h} \alpha_n r_n$ with $\alpha \geq 0$ and $\Sigma_{n=1}^{h} \alpha_n = 1$, where $\mathbf{r}_n$ is the nth rate vector in the feasible rate region C. At each timestep, schedule rate $\mathbf{r}_n$ with probability α_n. Since the arrival rate of user i is λ_i and the departure rate, when queue i is nonempty, is at least $(1 + \varepsilon)\lambda_i$, the system is ergodic and the expected queue size of each link is $\mathcal{O}(1/\varepsilon)$. (see the paper by Kahale and Wright [64] for details).

APPENDIX 11B: PROOF OF THEOREM 11.2 ON STABILITY REGION OF STOCHASTIC PHYSICAL LAYER

Necessary Condition. The random process describing the behavior of the entire system is given by $S(t)$

$$S(t) = \{(U_{i1}(t), \ldots, U_{iQi(t)}(t)), m(t), i = 1, \ldots, K\}$$

where $Q_i(t)$ is the queue length of user i at time t and $U_{ik}(t)$ is the current delay of the kth packet in the queue of user i at time t. Let $D_i(t) = \min(Q_i(t), \mu_i^{m(t)})$ be the number of packets for user i served at time t. Hence, we have

$$Q_i(t+1) = Q_i(t) - D_i(t) + A_i(t+1)\ \forall i$$

By stability of the system, we mean that the Markov chain $S(t)$ is stable.[7] Consider a scheduling policy "static service split" (SSS) parameterized by a $\alpha_j(S)$. When the system is in state S, the SSS rule chooses for rate vector $\mathbf{r}_j^m$ with probability $\alpha_j(S)$. Assume that the stability of the system is achievable. For any fixed timeslot t, we have

[7] A Markov chain $S(t)$ is stable if the set of positive recurrent states is nonempty and contains a finite subset that is reached with probability 1 from any initial state [71]. In this context the term *stability* implies the existence of a stationary probability distribution. In addition, if all positive recurrent states are connected, the stationary distribution is unique.

$$
\begin{aligned}
\lambda &= E[\mathbf{A}(t)] = E[\mathbf{D}(t)] \\
&= \sum_m \pi_m E(\mathbf{D}(t)|m(t) = m) \\
&\leq \sum_m \pi_m \sum_j \left[\sum_s P(S(t) = s|m(t) = m)\alpha_j(s) \right] \mathbf{r}_j^m \\
&= \sum_m \pi_m \sum_j \phi_{mj} \mathbf{r}_j^m
\end{aligned} \tag{11.34}
$$

where $\phi_{mj} = \Sigma_s P(S(t) = s|m(t) = m)\alpha_j(s)$. Since $\Sigma_j \phi_{mj} = 1$ for each m, $[\phi_{mj}]$ is a stochastic matrix and therefore, $\Sigma_m \pi_m \Sigma_j \phi_{mj} \mathbf{r}_j^m = \mathbf{v}(\phi)$ is as defined in Equation (11.16). Hence, this implies the necessary condition in Equation (11.18).

Sufficiency Condition. On the other hand, the proof of sufficient condition could be shown through the standard drift criteria in the Lyapunov function $\max_i Q_i(t)$ (see the article by Andrews et al. [12] for details).

APPENDIX 11C: PROOF OF THEOREM 11.3 ON THROUGHPUT OPTIMALITY OF MW SCHEDULING

At any time t, let the arrival rate vector (in bps) be $\lambda = (\lambda_1, \ldots, \lambda_K) \in S$ (where S is the stability region) and the data rate vector (bps) be $\mathbf{r}(t) = (r_1(t), \ldots, r_K(t))$. The dynamics of the queue length in the system is given by

$$\mathbf{Q}(t+1) = \mathbf{Q}(t) + \lambda - \mathbf{r}(t)$$

where $\mathbf{Q}(t) = [Q_1(t), \ldots, Q_K(t)]$ is the queue length vector in the system at time t. We have

$$\|\mathbf{Q}(t+1)\|^2 = \|\mathbf{Q}(t) + \lambda\|^2 + \|\mathbf{r}(t)\|^2 - 2(\mathbf{Q}(t) + \lambda) \cdot \mathbf{r}(t)$$

where $\|\mathbf{x}\|^2$ denotes the vector norm. Note that

$$\|\mathbf{Q}(t+1)\|^2 \leq \|\mathbf{Q}(t)\|^2 + \|\lambda\|^2 + \|\mathbf{r}(t)\|^2 + 2\mathbf{Q}(t) \cdot (\lambda - \mathbf{r}(t))$$

On the other hand, by the definition of the max-weight scheduler, $\mathbf{Q}(t) \cdot \mathbf{r}(t)$ would be greater than any dot product between $\mathbf{Q}(t)$ and the convex combination of the rate vector from the feasible capacity region. This implies

$$\mathbf{Q}(t) \cdot \mathbf{r}(t) \geq \sum_j \phi_{mj} \mathbf{Q}(t) \cdot \mathbf{r}_j^m$$

for any ϕ_{mj}. Thus

$$\|\mathbf{Q}(t+1)\|^2 \le \|\mathbf{Q}(t)\|^2 + \|\lambda\|^2 + \|\mathbf{r}(t)\|^2 + 2\mathbf{Q}(t)\cdot\left(\lambda - \sum_j \phi_{mj}\mathbf{r}_j^m\right)$$

Taking expectation on both sides of this equation under the assumption that future arrivals are independent of the current queue size, we obtain

$$\varepsilon\left[\|\mathbf{Q}(t+1)\|^2\right] \le \varepsilon\left[\|\mathbf{Q}(t)\|^2\right] + \varepsilon\left[\|\lambda\|^2\right] + \varepsilon\left[\|\mathbf{r}(t)\|^2\right] + 2\varepsilon[\mathbf{Q}(t)]\cdot\left(\lambda - \sum_m \pi_m \sum_j \phi_{mj}\mathbf{r}_j^m\right)$$

where λ is the average arrival rate vector and ϕ_{mj} is some stochastic matrix. This can be further expressed as

$$2\left(\lambda - \sum_m \pi_m \sum_j \phi_{mj}\mathbf{r}_j^m\right)\cdot \varepsilon[\mathbf{Q}(t) + \mathbf{Q}(t-1) + \cdots + \mathbf{Q}(0)] \le (t+1)\left(\varepsilon\left[\|\lambda\|^2\right] + \alpha\right)$$

where $\alpha = \max_{\mathbf{r}\in C(m)\forall m}\varepsilon[\|\mathbf{r}\|^2]$. Since $\lambda < \Sigma_m \pi_m \Sigma_j \phi_{mj}\mathbf{r}_j^m$, the expected value of $(\mathbf{Q}(t) + \cdots + \mathbf{Q}(0))/(t+1)$ is bounded and therefore the system is stable.

APPENDIX 11D: PROOF OF THEOREM 11.5 ON THROUGHPUT OPTIMALITY OF EXP SCHEDULING

The proof is modified from Reference 64 as shown below. At any time t, let the arrival rate vector (bps) be $\lambda = (\lambda_1, \ldots, \lambda_K)$ and the data rate vector (bps) be $\mathbf{r}(t) = (r_1(t), \ldots, r_K(t))$. Let $\mathbf{A}(t)$ be the vector of number of bits arrived for the K users.

If $\mathbf{Q}(t)$ and $\mathbf{A}(t)$ are independent, then

$$\varepsilon[\varphi(Q_i(t) + A_i(t))] = \varepsilon[\varphi(Q_i(t))]\varepsilon[\varphi(A_i(t))]$$

where

$$\varepsilon[\varphi(A_i(t))] = \varepsilon\left[e^{\alpha_i A_i(t)}\right]$$

By Taylor series, each component of $\varepsilon[\omega(\mathbf{A}(t))]$ is given by

$$\varepsilon\left[e^{\alpha_i A_i(t)}\right] = \varepsilon\left[1 + \alpha_i A_i(t) + \frac{1}{2}(\alpha_i A_i(t))^2 + \cdots\right]$$

and for Poisson arrival, $\varepsilon[A_i(t)] = \lambda_i$, $\varepsilon[A_i^2(t)] = \text{var}(A_i^2(t)) + (\varepsilon[A_i(t)])^2 = \lambda_i + \lambda_i^2$ and so on. Thus

$$\varepsilon\left[e^{\alpha_i a_i(t)}\right] = 1 + \lambda_i\left(\alpha_i + \frac{1}{2}\alpha_i^2 + \cdots\right) + \frac{1}{2}\lambda_i(a_i^2 + \cdots) + \cdots$$

On the other hand

$$e^{\lambda_i(e^{\alpha_i}-1)} = 1+\lambda_i(e^{\alpha_i}-1)+\frac{1}{2}(\lambda_i(e^{\alpha_i}-1))^2+\cdots$$
$$= 1+\lambda_i\left(\alpha_i+\frac{1}{2}\alpha_i^2+\cdots\right)+\frac{1}{2}\left(\lambda_i\left(\alpha_i+\frac{1}{2}\alpha_i^2+\cdots\right)\right)^2+\cdots \quad (11.35)$$

Thus, we have

$$\varepsilon[\varphi(Q_i(t)-A_i(t))] = \varepsilon[\varphi(Q_i(t))]e^{\lambda_i(e^{\alpha_i-1})}$$

for all $i \in [1, K]$.

Let p be the probability that $\mathbf{r} \in C(t) > \mathbf{0}$ is chosen. The system queue dynamics is given by

$$\mathbf{Q}(t+1) = \mathbf{Q}(t)+\lambda-\mathbf{r}(t)$$

We have

$$\varepsilon[\varphi(Q_i(t+1))] = (1-p)\varepsilon[\varphi(Q_i(t)+A_i(t))]+p\varepsilon[\varphi(Q_i(t)+A_i(t)-r_i(t))]$$
$$\leq (1-p+pe^{-r_i\alpha_i})\varepsilon[\varphi(Q_i(t)+A_i(t))]$$
$$= (1-p+pe^{-r_i\alpha_i})e^{\lambda_i(e^{\alpha_i}-1)} \times \varepsilon[\varphi(Q_i(t))] \quad (11.36)$$

By Taylor expansion of this equation, we obtain

$$\varepsilon[\varphi(Q_i(t+1))] \leq \left(1-p+p\left(1-r_i\alpha_i+(r_i\alpha_i)^2\right)+\ldots\right)(1+\lambda_i(e^{\alpha_i}-1)+\ldots)\varepsilon[\varphi(Q_i(t))]$$
$$= \left(1-p+p\left(1-r_i\alpha_i+(r_i\alpha_i)^2\right)+\ldots\right)(1+\lambda_i(\alpha_i+\alpha_i^2+\ldots)+\ldots)$$
$$\varepsilon[\varphi(Q_i(t))]$$
$$= (1-pr_i\alpha_i+\lambda_i\alpha_i+\mathcal{O}(\alpha_i^2))\varepsilon[\varphi(Q_i(t))]$$
$$\leq (1+(\lambda_i-pr_i)\alpha_i)\varepsilon[\varphi(Q_i(t))] \quad (11.37)$$

On the other hand, by the definition of the EXP scheduler, $\varphi(\mathbf{Q}(t))\cdot\mathbf{r}(t)$ would be greater than any dot product between $\varphi(\mathbf{Q}(t))$ and the convex combination of rate vector from the feasible capacity region. This implies

$$\varphi(\mathbf{Q}(t))\cdot\mathbf{r}(t) \geq \sum_j \phi_{mj}\varphi(\mathbf{Q}(t))\cdot\mathbf{r}_j^m = \varphi(\mathbf{Q}(t))\cdot\mathbf{r}'(t)$$

for any ϕ_{mj} where $\mathbf{r}'(t) = \Sigma_j\phi_{mj}\mathbf{r}_j^m$. Thus

$$\varepsilon[\varphi(Q_i(t+1))] \leq (1+(\lambda_i-pr_i')\alpha_i)\varepsilon[\varphi(Q_i(t))]$$

Since $\mathbf{r}' \in \mathcal{E}[C(m)]$ and $\lambda < \mathcal{E}[\bar{C}(m)]$ (which is the criterion for stability), we have $r'_i = (1 + \epsilon_i)\lambda_i$ for some $\epsilon_i > 0$. Thus, $\lambda_i - pr'_i = -\Theta(\lambda_i \epsilon_i)$. Let $\lambda_{min} = \min_i \lambda_i$. Since $\alpha_i = \Theta(\epsilon_i)$, if the scheduler follows the "exponential rule" [i.e., the rate vector with the largest $\varphi(\mathbf{Q}(t)) \cdot \mathbf{r}$ will be scheduled] and λ is within the stability region, we have

$$\mathcal{E}[\varphi(Q_i(t+1))] \leq (1 - \Theta(\lambda_{\min} \varepsilon_i^2)) \mathcal{E}[\varphi(Q_i(t))] \tag{11.38}$$

It follows from Equation (11.38) that

$$\mathcal{E}[\varphi(Q_i(t))] = \mathcal{O}\left(\frac{1}{\lambda_{\min} \varepsilon_i^2}\right)$$

By the Markov bound, we have

$$\Pr[\varphi(Q_i(t)) \geq \varphi(B)] \leq \frac{\mathcal{E}[\varphi(Q_i(t))]}{e^{\alpha_i B}}$$

Thus

$$\Pr[Q_i(t) \geq B] \leq \frac{1}{\lambda_{\min} \varepsilon_i^2} e^{-\alpha_i B} \tag{11.39}$$

Thus, each queue under the exponential scheduling not only is bounded but also has tail distribution exponentially decreasing with a decay rate.

APPENDIX 11E: CONDITIONS FOR STABILITY OF A STOCHASTIC PROCESS

We quote a theorem from the literature [88,104] concerning the sufficiency condition for the stability of a stochastic process.

Theorem 11.8 (Conditions for Stability) Let $V(\mathbf{q})$ be a lower-bounded Lyapunov function defined on the unfinished work $\mathbf{q} \in \mathcal{R}^K$. If there exists a compact region $\Lambda \in \mathcal{R}^K$ and $\varepsilon < 0$ such that

1. $\mathcal{E}[V(\mathbf{Q}(t + T))|\mathbf{Q}(t) = \mathbf{q}] < \infty$ for all $\mathbf{q} \in \mathcal{R}^K$.
2. $\mathcal{E}[V(\mathbf{Q}(t + T)) - V(\mathbf{Q}(t))|\mathbf{Q}(t) = \mathbf{q}] < -\varepsilon$ for all $\mathbf{q} \notin \Lambda$.
3. Given $\mathbf{Q}(t + T) = \mathbf{q} \in \Lambda$, there is a nonzero probability that $\mathbf{Q}(t + mT) = \mathbf{0}$ for some finite integer m, where T is the time duration of a timeslot, then there exists a steady-state distribution for $\mathbf{Q}(t)$. Moreover, if condition 3 hold and there exist positive values θ and $\{\gamma_k\}$ such that

$$\varepsilon[V(\mathbf{Q}(t+T))-V(\mathbf{Q}(t))|\mathbf{Q}(t)=\mathbf{q}]\leq\theta-\sum_{k=1}^{K}\gamma_k q_k \tag{11.40}$$

then there exists a steady-state distribution for $\mathbf{Q}(t)$ with bounded first moments $\overline{\mathbf{Q}}$ such that $\Sigma_k\gamma_k\overline{\mathbf{Q}}\leq\theta$.

Please see References 104 and 88 for the proof.

APPENDIX 11F: PROOF OF THEOREM 11.6 ON THROUGHPUT OPTIMALITY OF TSE–HANLY POLICY

We choose the Lyapunov function $V(\mathbf{q})=\Sigma_k\alpha_k q_k^2$. Conditions 1 and 3 (From Appendix 11E) are straightforward from the assumptions of finite expected arrivals and nonzero probability of no arrivals to all the queues. Hence, we shall focus on condition 2.

For condition 2, we first show that

$$\begin{aligned}&\varepsilon[V(\mathbf{Q}(t+T))-V(\mathbf{Q}(t))|\mathbf{Q}(t)=\mathbf{q}]\\&\leq-2T\sum_{k=1}^{K}\alpha_k q_k(\varepsilon[r_k(t)|\mathbf{Q}(t)=\mathbf{q}]-\lambda_k)+\theta\end{aligned}$$

where $\theta=\Sigma_{k=1}^{K}\alpha_k\left(\varepsilon\left[A_k^2(t)+T^2\hat{C}_k^2\right]\right)<\infty$, $A_K(t)$ is the number of bit arrivals to queue k in the tth timeslot, and $\hat{C}_k=W\log_2\left(1+\frac{|\hat{H}_k|^2\hat{P}_k}{\sigma_z^2}\right)$. This follows directly from the definition of the Lyapunov function since $Q_k(t+T)=(Q_k(t)+A_k(t)-r_k(t)T)^+$ and $[(x)^+]^2\leq x^2$ as well as by the peak power constraint $p_k(t)\leq\hat{P}_k$ and the peak fading constraint $|H_k(t)|^2\leq|\hat{H}_k|^2$.

Let $\mathcal{P}_{\mathrm{TH}}(\tilde{\mathbf{h}},\mu)$ and $\mathcal{R}_{\mathrm{TH}}(\tilde{\mathbf{h}},\mu)$ be the power control and rate control policies given by Equation (11.24) for the weight μ, respectively. We have $r'_k=\varepsilon_{\tilde{\mathbf{h}}}[\mathcal{R}_{\mathrm{TH}}(\tilde{\mathbf{h}},\mu)]$ satisfies $\Sigma_k\mu_k r'_k\geq\Sigma_k\mu_k r_k$ for all $\mathbf{r}\in\varepsilon[C]$, where $\varepsilon[C]$ is the average capacity region of theMAC channel over ergodic fading realizations. Define $\mu_k=\alpha_k q_k$ for some positive constants α_k, $\mathcal{P}^*(\tilde{\mathbf{h}},\mathbf{q})=\mathcal{P}_{\mathrm{TH}}(\tilde{\mathbf{h}},\mu)$, and $\mathcal{R}^*(\tilde{\mathbf{h}},\mathbf{q})=\mathcal{R}_{\mathrm{TH}}(\tilde{\mathbf{h}},\mu)$. Since $\mathcal{P}_{\mathrm{TH}}(\tilde{\mathbf{h}},\mu)$ satisfies the power constraints $\overline{\mathbf{P}}$ and $\hat{\mathbf{P}}$ in every queue state $\mathbf{q}$, we see that $\mathcal{P}^*(\tilde{\mathbf{h}},\mathbf{q})$ also satisfies the average and peak power constraints and hence is feasible. Under $\mathcal{P}^*$ and $\mathcal{R}^*$ policies, we have

$$\sum_k\alpha_k q_k\varepsilon[r_k(t)|\mathbf{Q}(t)=\mathbf{q}]\geq\sum_k\alpha_k q_k r_k$$

for all $\mathbf{r}\in\overline{C}$. For any $\lambda\in S=\mathrm{int}\,\overline{C}$, there exists $\delta=(\delta,\ldots,\delta)$ such that $\lambda+\delta\in\overline{C}$. Thus we have

$$\sum_k \alpha_k q_k \varepsilon[r_k(t)|\mathbf{Q}(t)=\mathbf{q}] \geq \sum_k \alpha_k q_k(\lambda_k+\delta)$$

and

$$\varepsilon[V(\mathbf{Q}(t+T))-V(\mathbf{Q}(t))|\mathbf{Q}(t)=\mathbf{q}] \leq \theta - 2T\sum_k \alpha_k q_k \delta \tag{11.41}$$

Choose any $\varepsilon > 0$ and define the compact region as

$$\Lambda = \left\{\mathbf{q} \in \mathcal{R}_+^K : \sum_k \alpha_k q_k \leq \frac{\theta+\varepsilon}{2T\delta}\right\}$$

Then for any $\mathbf{q} \notin \Lambda$, the right side of Equation (11.41) is strictly less than $-\varepsilon$. Therefore, by Theorem 11.8 in Appendix 11E, there exists a steady-state distribution for the queue length $\mathbf{Q}(t)$. Furthermore, the inequality from Equation (11.41) can be combined with the last part of Theorem 11.8 to provide the delay guarantee as in Corollary 11.1.

EXERCISES

1. **[Optimality of Power Water-Filling]**
   Consider a point to point fast fading channel with channel outputs given by: $Y = HX + Z$ where H is the complex Gaussian channel fading coefficient, X is the complex lowpass equivalent transmit signal and Z is the complex white Gaussian noise. Assume that an encoding frame spans across an ergodic realization of the channel fading H(t) and the transmitter has an average power constraint $E\left[|X|^2\right] = P_0$.
   (a) Derive the optimal power control policy for delay insensitive user to maximize the ergodic capacity.
   (b) From (a), explain whether the power control algorithm is to compensate the effect of fading or not? In commercial systems like IS95 (CDMA), the power control is to compensate the effect of channel fading (increase power when fading is poor and decrease power when the fading is good). Is this consistent with the power adaptation policy in (a)? Explain the reason if there is a difference.
   (c) Derive the optimal power control policy for a point to point fading channel for minimizing the user's delay for delay sensitive user.

2. **[Importance of stable scheduler]**
   Assume all users have data in the buffer at all times. Consider a system with 2 users. Arrival rate of user A is 94 bits per time slot; while arrival

rate of user B is 49 bits per time slot. In all time slots, the maximum transmission rate (denoted by DRC_A) of user A is 100 bits per slot. In time slot 10n (for integer n), the DRC_B for user B is 1000 bits per slot. In slot $10n+i$, $1 \le i \le 9$, the DRC for user B is 100 bits per slot. Suppose the proportional fair scheduling rule of selecting the user i (i is either A or B) in each time slot is $i = \arg\max_i \frac{DRC_i(t)}{R_i(t)}$, where the moving window average of service rate $R_i(t)$ is governed by the following rule:

$$R_i(t+1) = (1-\lambda)R_i(t) + \lambda x_i(t) DRC_i(t) \quad \text{where } \lambda \text{ is } 1/1000.$$

$$x_i(t) = \begin{cases} 0, \text{if user } i \text{ is not selected} \\ 1, \text{if user } i \text{ is selected} \end{cases}$$

(a) Suppose there is a time slot T_0 such that $89 < R_A(T_0) < 90$ and $100 < R_B(T_0) < 101$, show that $89 < R_A(T_0 + j) < 90$, $100 < R_B(T_0 + j) < 101$, for all $j = 1, 2, \ldots$

(b) Deduced that user A is only assigned 90% of the time slots.

(c) Explain why proportional fair scheduler is unstable.

(d) Design of a scheduler so that scheduler for user B is stable (Hint: 95% of time slot should be dedicated to user A and remaining 5% to user B)

(e) Assume the buffers are initially empty. Design the minimum buffer size of each user such that there is no packet overflow in scheduler designed in Problem 2 d.

(f) Is it possible to design a queuing buffer for user A such that there is no packet overflow if proportional fair scheduler is used? If so, explain how to do so; if not possible, on average, how many bits will be lost if the size of the buffer of each user is C bits)?

(g) Assume that non-work conserving situation is allowed (i.e. the number of bits transmitted in one time slot can be larger than the number of bits in the buffer); compute the average throughput (DRC only) of the two users using proportional fair scheduler and the scheduler designed in section d.

(h) Using the results in (e) and (f) or otherwise, explain the importance of stability of the scheduler. Try to justify the use of proportional fair scheduler using the results in (g).

3. **[Not all schedulers can achieve any points in the stability region]**

(a) Express the capacity region for 2-user MAC capacity region in AWGN channels.

(b) Derive the stability region for 2-user MAC capacity region in AWGN channels.

(c) Prove that MAX-Weight (described in section 11.4.3) can achieve any points in the stability region (where the scheduling does not need to know about the source arrival rates of the 2 users).

(d) Show that using a TDMA strategy (selecting a single user with the highest instantaneous throughput), we cannot achieve some points in stability region in (b).

4. **[Modified Max Weight is QoS-supporting]**
Consider a fading multi-access channel with 2 users and long-term average power constraint of 1 W per user (averaged over many time slots) where the fading remains quasi-static within a packet duration. The fading process has only 2 states, namely |H_k(t)^2| = {Good: 4, Bad: 0.1}. Assume that the base station has perfect estimation of the CSI from all the 2 users. Suggest a throughput optimal scheduler that could guarantee that the delay of user 1 is 2 times that of delay of user 2. What would be the scheduling policy if the transmit power constraint becomes short-term average (averaged over one packet)?

12

CHANNEL-ADAPTIVE AD HOC ROUTING

12.1 OVERVIEW

Ad hoc routing is important for mobile devices due to its robustness in hostile radio propagation environment. While traditional table-based or on-demand routing protocols can be used, it is much more efficient to use a routing protocol that is channel-adaptive—judiciously selecting links that can transmit at higher data rates to form a route. However, devising channel-adaptive routing protocols is still largely unexplored. In this chapter, we describe a reactive ad hoc routing algorithm, the RICA (receiver-initiated channel-adaptive) protocol, to intelligently utilize the multirate services (based on different modulation schemes). Our NS-2 simulation results show that the RICA protocol is highly effective compared with non-adaptive designs.

12.2 BACKGROUND

Wireless LANs, based on the IEEE 802.11b standard, are becoming ubiquitous because of the almost seamless integration with wireline Ethernet LANs. Nevertheless, it is widely envisioned that a huge potential of wireless networking is yet to be realized until we can deploy a robust and large-scale ad hoc mobile computing network. In an ad hoc network, devices join and leave the network at will and in a totally asynchronous manner. Thus, such a wireless network can provide truly "anytime anywhere" computation because of its robustness and inherent fault tolerance. In addition, many peer-to-peer applications such as location-based services can be supported in such a

Channel-Adaptive Technologies and Cross-Layer Designs for Wireless Systems with Multiple Antennas: Theory and Applications. By V. K. N. Lau and Y.-K. R. Kwok
ISBN 0-471-64865-5

network [34,128,129]. However, to make such an ad hoc mobile computing network feasible, we have to meet a number of challenges related to wireless infrastructure problems. Most notably, for example, when the devices in an IEEE 802.11b wireless LAN are operating in a peer-to-peer manner (i.e., invoking the distributed coordination function) to form an ad hoc mobile computing system [29,121,152], we need to tackle the problem of finding optimized ad hoc routes to enable point-to-point communications between two devices that are possibly out of each other's range.

There are two major classes of ad hoc routing protocols, namely, on-demand and table-based [34,90]. As many researchers have pointed out [21,34,61,86], table-based algorithms are notoriously inefficient in that they require periodic update of the routing information stored in the routing tables even when there is no data traffic. On the other hand, as compared with on-demand algorithms, its merit is that the setup delay for a data transfer is expected to be shorter because a route is presumably stored in the table for use. However, when the actual data transfer is to take place, such a route may no longer exist or may be unusable for at least two reasons: (1) because of the mobility of the mobile devices in the network, their geographic locations may have changed when a data transfer is required, rendering a previously setup route useless; or (2) the quality of the channels among the mobile devices may be inevitably time-varying (due to shadowing and fast fading [109]) and thus, the links in a route may no longer be usable even if the geographic locations do not change much. In our study, we consider mainly on-demand routing algorithms for ad hoc networks but also examine the effectiveness of one table-based protocol. In particular, we are interested in studying the behavior and performance of routing protocols when the time-varying nature of wireless channels is taken into account. Indeed, because the IEEE 802.11b standard [4] also provides multi-rate services with different rates supported by different modulation schemes, it is useful to dynamically change routes by selecting links that can use higher-bandwidth modulation schemes.

In this chapter, we describe a new ad hoc routing algorithm for an IEEE 802.11b–based wireless LAN operating in the ad hoc mode (i.e., using the distributed coordination function without any centralized access point). Our algorithm, called *receiver-initiated channel-adaptive* (RICA) routing, works by proactively changing routes through judicious selection of links that can support higher data rates. After reviewing some state-of-the-art ad hoc routing techniques in Section 12.3, the features and design considerations of the RICA protocol are detailed in Section 12.4. To illustrate the effectiveness of the RICA protocol, we extensively tested it in a simulated IEEE 802.11b wireless LAN environment using the well-known NS-2 platform [105] and compared it with three well-known protocols—AODV (ad hoc on-demand distance vector) [112], DSDV (destination-sequenced distance vector) [110], and DSR (dynamic source routing) [62] protocols—that are not channel-adaptive. The results are described in Section 12.5. Section 12.6 concludes this chapter.

12.3 OVERVIEW OF AD HOC ROUTING PROTOCOLS

12.3.1 AODV Protocol

The *ad hoc on-demand distance vector* (AODV) [112] is a kind of pure reactive routing protocol. In this protocol, each device does not need to keep a view of the whole network or a route to every other device. Nor does it need to periodically exchange route information with neighboring devices. In addition, it needs to discover and maintain a route to a destination device only when a mobile device has packets to send to that destination. As it is considered by the IETF as a standard ad hoc routing protocol [113], the major distinctive characteristics of the ADOV protocol are

- Routes are discovered or maintained only when necessary (or on an as-needed basis).
- It is loop-free [112] at all times; this is accomplished through the use of a device sequence number that increases monotonically (this technique also ensures that the most current route is always used when a route is discovered).
- It is a bandwidth-efficient routing algorithm, which greatly reduces the use of limited bandwidth (the number of route broadcasts is minimized on an as-needed basis).
- It responds very quickly to the topology changes of the network and can recover a broken route in a timely manner.
- It has a low storage requirement on the device because it needs to maintain only the active neighbors' information instead of a full route to the destination.

In AODV, each device contains a route table for a destination. The route table stores the following information: destination address and its sequence number, active neighbors for the route, hop count to the destination, and expiration time for the table. The expiration time is updated each time this route is used. If this route has not been used for a specified period of time, it is discarded.

12.3.2 DSDV Protocol

The *destination-sequenced distance vector* (DSDV) routing protocol [110] is an improvement of the original distributed Bellman–Ford (DBF) routing protocol because it uses a destination sequence number for each device. DSDV is a table-driven routing protocol. Each device in DSDV maintains a routing table containing entries for all the devices in the network. Each entry contains the following information: the destination's address, the distance, and the next device to the destination as well as the sequence number as stamped by the destination.

In order to keep the routing table completely updated at all times, each device periodically broadcasts routing messages to its neighboring devices. In each routing message, a monotonically increasing even sequence number for the device is included. Thus the sequence number is disseminated in the network via update messages. This sequence number is used to address the "freshness" of a route. The largest sequence number is always adopted to update a routing entry. If the sequence number is the same, the one with the smaller distance metric is used. By using the sequence number concept, DSDV avoids the longlived routing loop and counting-to-infinity problems. For the proof of this concept and a detailed description of DSDV, interested readers can refer to the paper by Perkins and Bragwat [110].

As to the route maintenance, when a device—say, A—finds that its route to a destination—say, B—has broken, it advertises the route to B with an infinite distance metric and a sequence number (odd) one greater than its original sequence number for the route that has broken. This causes any other device C routing packets through A to the corresponding destination to update its routing entry by rendering the distance metric infinite until device A gets a new route to B with a higher sequence number [110].

12.3.3 DSR Protocol

The *dynamic source routing* (DSR) protocol was developed by researchers at Carnegie Mellon University [62]. In DSR, the data packet carries the whole hop-to-hop route from the source to the destination in its header. One merit of DSR is that, when receiving a data packet, the intermediate device needs to route the packet only to the next hop as listed in the source list in this packet, instead of bothering to find the next hop from its routing table as in AODV.

DSR is also an on-demand routing protocol. The device does not keep a route to every other potential device all the time unless it has a packet to send to that device. Each device in DSR has a route cache that keeps one or several routes to a destination. When a source device has a packet to send to its destination, it first checks whether it has a route to that destination. If it does, it attaches that route to the data packets and begins the session; otherwise, it should initiate a route discovery process by flooding a RREQ (route request) to find such a route.

When the RREQ is received by a destination device or an intermediate device having a route in its cache, it sends a RREP (route reply) along the route that the RREQ has traversed. If an intermediate device has no route in its cache and the RREQ is not a duplicate one, it puts its identifier number (ID) in the device list in the RREQ and rebroadcasts it out. If the intermediate device receives further duplicate RREQ, it simply discards it. The source might at last receives several RREPs, and it records these routes in its route cache for future use. Note that the RREQ and RREP are both source-routed and carry a device list that they have passed by or to pass by.

In case a link breakage occurs, a RRER (route error) is sent back to the source by the upstream device. Each device receiving this RRER removes the routes including this broken link from its route cache. The source device might initiate another RREQ flooding if the session is not finished yet.

Several mechanisms have been proposed to optimize the performance of DSR [34]. These mechanisms include nonpropagation route request, gratuitous route replies, salvaging, and snooping. For the further details on DSR, interested readers can refer to the text by Johnson and Maltz [62].

12.3.4 ABR Protocol

Associativity-based routing (ABR) [118,127] is a source-initiated on-demand routing protocol. In other words, a mobile device in the network does not need to keep a route to every other devices. The major distinctive characteristic of ABR is that the route is not chosen on the shortest-path basis as in other protocols, but on a longlived basis. Selecting a longlived route between the source and the destination device has many merits, such as (1) the chosen route is more robust (not easy to break due to mobility), (2) the maintenance of route is easier, and (3) the number of route reconstructs (RRC) is reduced (thus the routing overhead is reduced and more bandwidth is saved). In ABR, longevity of a route is given first priority. In other protocols, such as link state and AODV, the primary goal is to find the shortest path. However, in these protocols, data flow transmission interruptions occur more often and more route reconstructions are needed. The essence of ABR is that as a mobile device moves, its associativity with neighboring devices also changes, and this associativity can be quantified by using associativity ticks. Each mobile device in ABR periodically transmits beacons (such as "hello" messages) to signify its existence. When these beacons are received by its neighboring devices, the associativity of this device with its neighboring devices increases. The greater the associativity is, the more stable this device will be. A high associativity of the device means a low mobility of a device. If device A moves out of the transmission range of another device, the associativity record of A in this device is reset.

12.3.5 Link State Routing Protocol

The link state routing protocol was originally designed for wireline networks [34]. In this protocol, each mobile device keeps its own view of the whole network. When a mobile device has a packet to forward, it uses a shortest-path algorithm to determine the next hop to forward that packet to its destination. Every mobile device must keep an up-to-date view of the network. When a mobile device finds a link cost change (due to the change in channel quality) with one of its neighbors, it floods this change throughout the network. Once this flooding packet is received by a mobile device, this mobile device updates its view of the network accordingly. Here, it can be seen that asynchronous updates cause a transient routing loop, but this loop is eliminated eventually as the flooding packet is propagated throughout the network.

12.4 RECEIVER-INITIATED CHANNEL-ADAPTIVE (RICA) ROUTING

The major feature of RICA is to make use of the time-varying property of the wireless channel in that the routing between the source and destination devices is adaptive to the change in *channel state information* (CSI), which corresponds to the SNR (signal-to-noise ratio) of the received signal. Specifically, in the RICA algorithm, the entire route may be changed in response to a change in CSI. Before describing the RICA protocol in detail, we first introduce the channel model below.

12.4.1 Channel Model

The wireless link between two mobile devices is characterized by two components, namely, the *fast fading* component and the *long-term shadowing* component [109] as discussed in Chapter 1. To exploit the time-varying nature of the wireless channel, typically a variable-throughput channel-adaptive physical layer is incorporated in the transceiver of a mobile device in that a variable amount of error protection redundancy as well as a variable modulation level are incorporated to the information packet according to different channel conditions. For example, the IEEE 802.11b working group adopted *complementary code keying* (CCK) [49] in 1998 as the basis for the high-rate physical layer extension to transmit data rates up to 11 Mbps [4]. Specifically, an IEEE 802.11b wireless device can support variable physical bit rates at 1, 2, 5.5, and 11 Mbps [68] based on differential binary phase shift keying (DBPSK), differential quaternary phase shift keying (DQPSK) (for both 2 and 5.5 Mbps), and CCK, respectively. The rate adaptation of 802.11b is done at the MAC layer and is not specified in the standards. In most existing designs, the 802.11b device transmits data at an appropriate data rate using a particular modulation mode based on the estimated signal-to-noise ratio (SNR) as well as the accumulated frame error rate (FER). For details on the IEEE 802.11b standard, the reader is referred to the literature [4,135].

We define a CSI-based "hop" in the following manner. In terms of the CSI that can be estimated from the preambles of the received signal, we can classify the channel quality into four classes, A, B, C, and D, corresponding to data rates of 11, 5.5, 2, and 1 Mbps, respectively, as specified in the IEEE 802.11b standard. Thus, if a link between two mobile devices is with channel quality of class A (i.e., it is able to support the data rate of 11 Mbps), then the distance between these two devices is defined as *one* hop. We then use this "distance" as a baseline as follows. If a link between two mobile devices has a channel quality of class B (with a data rate of 5.5 Mbps), the distance between two devices is *two* hops because now the transmission delay is twice that of a class

A link. In summary, the distance between two devices, with a link having class A (11 Mbps), class B (5.5 Mbps), class C (2 Mbps), or class D (1 Mbps), is 1, 2, 5.5, and 11 hops, respectively.

12.4.2 Route Discovery

The RICA protocol is a reactive and on-demand algorithm in that a source mobile device does not permanently keep a route to any destination. The source device will try to determine a route only when it has packets to send to a particular destination. When the source device has packets to transmit, it generates a *route request* (RREQ) packet that includes the following information: type of the packet, source address, destination address, hop count from the source (initialized to zero), hop distance based on CSI (initialized to zero), broadcast identifier (ID) of the RREQ, and a list of intermediate devices (initialized to an empty list). Whenever the source generates a RREQ, the broadcast ID is increased by one. Thus, the source and destination addresses together with the broadcast ID uniquely identify a RREQ. The source broadcasts the RREQ to all devices within the transmission range. These neighboring devices will relay the RREQ to other farther devices in the following fashion. On receiving the RREQ, an intermediate device first checks whether it has seen this packet before by looking up its RREQ cache. If it is in the cache (indexed by the source and destination addresses, as well as the broadcast ID), the newly received copy is discarded; otherwise, the RREQ is stored in the cache and is forwarded to the neighbors after the following modifications are done:

- The intermediate device inserts its ID number in the list of intermediate devices in the RREQ packet.
- The intermediate device increments the hop count field in the RREQ packet.
- The CSI-based hop distance is also updated as follows. The intermediate device measures the signal quality (i.e., SNR) of the link through which the RREQ is transmitted and computes the CSI-based hop distance from the upstream device. The intermediate device then resets the hop distance to the original hop distance plus this CSI-based hop distance to the upstream device.

This RREQ relaying process continues until the RREQ reaches the destination. To avoid undue overburdening of a particular device, the RICA protocol also requires that a device under a heavy traffic load (e.g., serving as the hotspot pivot devices for two distinct connections) discard only the newly received RREQ. Thus, within a short timeframe in the future no new routes will include such a heavily burdened device.

Figure 12.1a illustrates the broadcast of the RREQ in an ad hoc mobile computing network. As can be seen, eventually the destination device receives

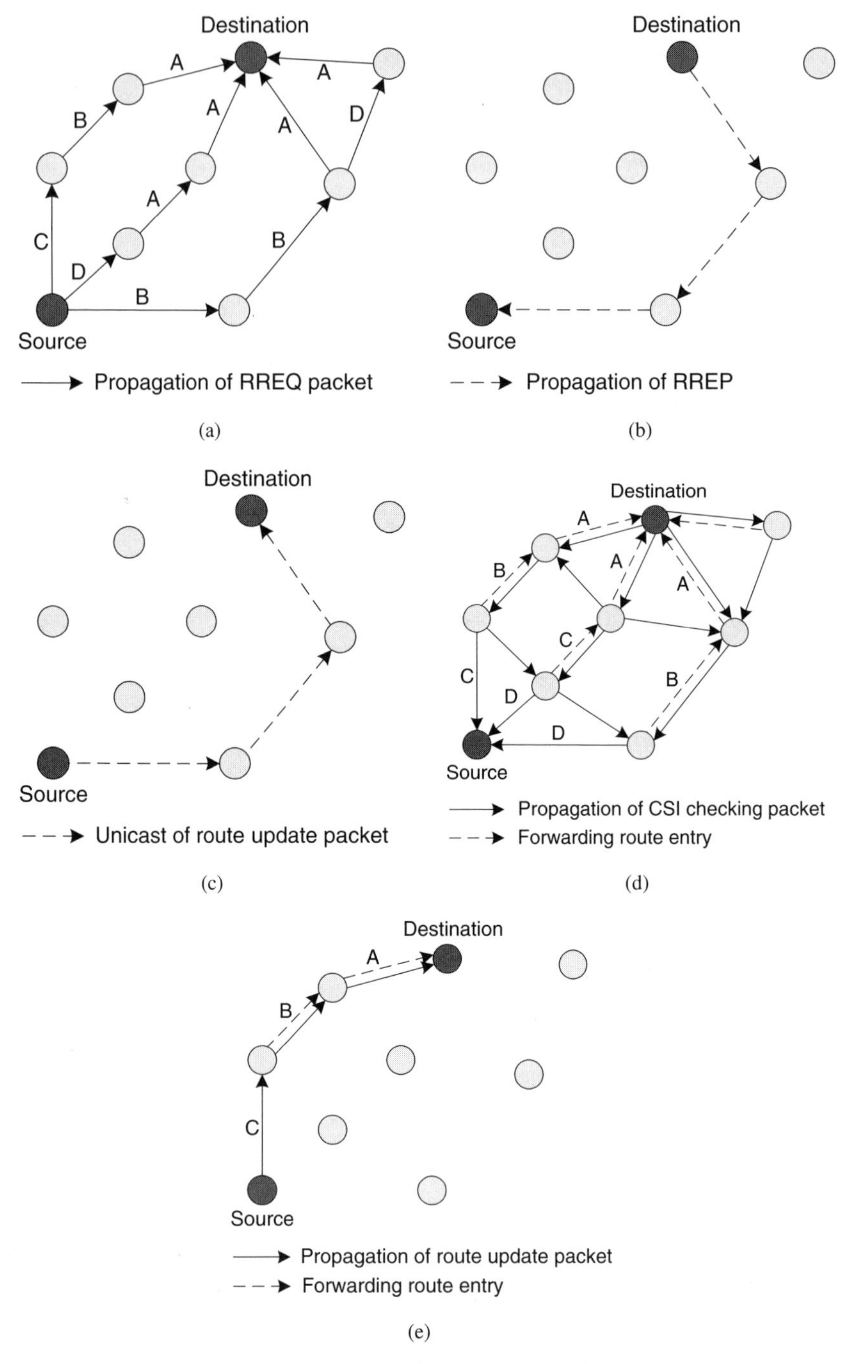

Figure 12.1. Routing mechanisms in the RICA protocol: (a) broadcast of RREQ; (b) propagation of RREP; (c) route update; (d) CSI checking; (e) selection of the best route.

several copies of the RREQ from the same source via all possible routes. The destination can determine its CSI-based hop distance (as defined earlier) from the source on all the routes and thus can choose a route with the minimum distance. As shown in the figure, the RREQ reaches the destination device through four distinct routes (note that the links are labeled with the channel classes) with hop count distances 8.5, 13, 5, and 15, respectively. The destination device then generates a *route reply* (RREP), which includes the following information: type of packet, source address, destination address, route reply ID (corresponding to the broadcast ID of the RREQ), hop distance (CSI-based) and hop count of the route, and list of intermediate devices. The destination device unicasts the RREP along the selected route to the source device[1] as shown in Figure 12.1b. Note that during the transmission of RREP from the destination to the source, the CSI-based hop distance field in RREP has to be recomputed as before. After receiving the RREP, the source device can decide whether to update the route on the basis of the RREP. If the source decides to update the route, it has to construct a route update packet and send it along the selected route to the destination as depicted in Figure 12.1c. This process is elaborated in the following section. Note that route discovery in RICA is an end-to-end process and the intermediate devices are not allowed to send a RREP even if they have the route information for the destination. This is because the route information stored in their route entries may have been obsolete because of the time-varying nature of the wireless channel. This is an important difference between RICA and AODV (or DSR).

12.4.3 Broadcast of CSI Checking Packets

Because the channel quality between two devices is a time-varying function, the throughput of the route to the destination also changes all the time. Thus, the primary goal of the RICA algorithm is to maintain a route between a communicating source–destination pair such that the highest throughput is achieved. Essentially, to attain this goal, a route will have to be updated, possibly frequently, according to the changing channel conditions. Our approach is to let the destination device broadcast a CSI checking packet periodically. Note that the period depends on the coherence time of the shadowing conditions, and typically 3–4 s is acceptable for pesdestrian mobility. The CSI checking packet, acting as a probe, is used for measuring the CSI of every link that it has traversed. Thus, an updated CSI-based hop distance can be obtained. During the lifetime of a communication session, the source could receive several CSI checking packets periodically from the destination and thus, can update the route accordingly. We explain this process in more detail with the help of Figure 12.1d,e.

[1]Note that each device knows to which of its upstream neighbors it should forwarding the RREP, from the intermediate device list in that RREP.

First, the destination device generates a CSI checking packet that includes the type of packet, source address, destination address, hop count and hop distance field (the latter is CSI-based and both are initialized to zero), time-to-live (TTL) field, and checking packet ID. Whenever the destination broadcasts a new CSI checking packet within the same communication session, the checking packet ID is increased by one. The TTL field is used for limiting the broadcast scope of the packet because exhaustive flooding should be avoided to save bandwidth. Specifically, the TTL field is set to the originally known hop count (not based on CSI) of the path plus one. Every time the packet is rebroadcast, the TTL field is decremented. When TTL becomes zero, the checking packet is discarded. The destination first broadcasts the CSI checking packet to its neighboring devices. When a neighbor device receives the checking packet, it updates several fields of the packet as follows. First, it measures the CSI of the link from which the checking packet is transmitted and computes the hop distance based on the measured CSI. Then it resets the hop count to the original value plus one, decrements the TTL, and relays the checking packet to its neighbors. On the basis of the received checking packet, an intermediate device also sets up a forwarding route entry, including source and destination IDs, route entry ID, and downstream device, which is the device from which the intermediate device receives the checking packet. This implies that the forwarding route entry points to the device from which the checking packet is transmitted. The route entry ID is equal to the value of the CSI checking packet ID. The forwarding route entry also has a lifetime that is set to the broadcast period of the CSI checking packet. After the lifetime has elapsed, the forwarding route entry is deleted.

Note that within the same CSI checking process, an intermediate device relays a CSI checking packet only once because further received copies are simply discarded. The checking packet is also discarded if it comes from a link in deep fading (e.g., with only 1 Mbps of effective bandwidth). Moreover, as in the route discovery process, a device under a heavy traffic load does not forward the CSI checking packet but simply discards it. Eventually, the source device receives several checking packets from all possible routes as shown in Figure 12.1d; then the source device selects the shortest path and uses it to replace the original route. For example, in Figure 12.1d, there are three candidate routes with hop distances 8.5, 17.5, and 14, respectively. The source device can then choose the shortest one (the top one that has a hop distance of 8.5) by using a route update packet. The route update packet includes the following fields: type of packet, source address, destination address, hop count, update sequence number, and route entry ID. The updated sequence number is used to identify a new route update from the source to destination and increases monotonically in the course of successive updating. On receiving the route update packet, the downstream device updates its routing table by setting the next device to the same destination as the one in its forwarding route entry (i.e., the downstream device in the entry) and then passes the route update packet to its downstream device. We will further explain the route updating process in subsequent sections. Finally, it should be noted that

- The original route eventually might automatically expire, probably because there is no traffic for a specified timeout period, and be deleted.
- Breaking of the link in the original route has no impact on data transmission in the current route if the link is not in the current route (this is illustrated below).

12.4.4 Route Maintenance

In the RICA protocol, updating of the routing table can be quite frequent, and thus an upstream device has to be sensitive to the status of the connection with its downstream device. The feedback information from the physical layer [78,82] can be used to detect the connectivity of the link. When a device is notified that its downstream device has moved out of its transmission range, the device generates a *route error* (REER) packet, which includes the following information: type of packet, source address, destination address, and last route update sequence number. The device then unicasts the REER to the upstream device. The upstream device first checks whether the device unicasting the REER is its downstream device by looking up its routing table route entry and the related route update sequence number. If either one of these two fields does not match, the device ignores this REER because such an REER comes from a broken route that is obsolete and is useless on the data transmission that is going on in the current route. On the other hand, if both fields match, the upstream device unicasts the REER to its upstream device. This process continues until the REER reaches the source.

For example, as shown in Figure 12.2, mobile device C finds that the link to the destination is broken and thus sends a REER to A. However, mobile device A ignores this REER because it knows that device C is not its downstream device and the REER comes from an old link that is not used by the current route. If the device unicasting the REER is its current downstream device, it unicasts this REER to its upstream device. The process continues, and if the REER reaches the source, then the source can decide whether it should initiate another route discovery process based on the following two considerations:

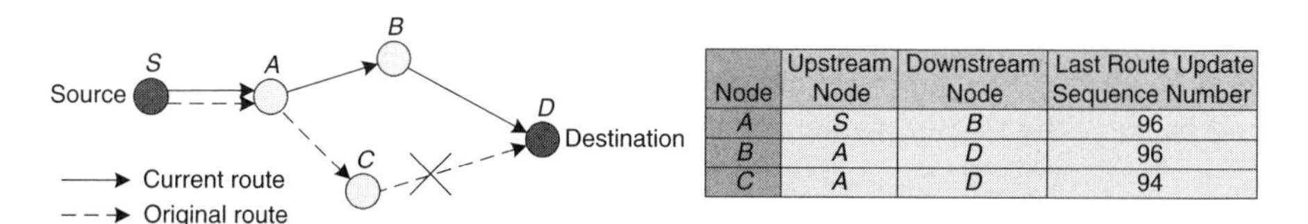

Node	Upstream Node	Downstream Node	Last Route Update Sequence Number
A	S	B	96
B	A	D	96
C	A	D	94

Figure 12.2. Route updating induced by the breaking of a link.

1. If the source device is now receiving CSI checking packets, then the source device ignores the REER and chooses the shortest route based on the packet.
2. If the source device is not receiving CSI checking packets, it broadcasts a RREQ in search of the destination and waits for a RREP. There are three possible scenarios:
   a. If the RREP reaches the source together with the CSI checking packets,[2] the source selects the shortest route based on both the CSI checking packets and RREP packet.
   b. If the CSI checking packets arrive before the RREP, the source decides the route based on these packets. Afterward, if the RREP also arrives and is with a shorter hop distance, the source chooses the route based on RREP (some communications may have already been taken place).
   c. If the RREP arrives before CSI-checking packets, the source chooses the route based on RREP. Afterward, when the CSI-checking packets arrive, the route is decided according to the packets (again some communications may have already been taken place).

12.4.5 Route Updating

As described above, updating of a route might be based on the CSI checking packets or RREP. However, these two updating mechanisms can lead to different results. If the route updating is based on RREP, the route update packet has the following format: type of packet, source address, destination address, hop count, update sequence number, and list of intermediate devices. Because the route update could be based on the CSI checking packet or RREP, routing loops [47] might be formed. To avoid the formation of loops and to differentiate the two cases of route updating based on RREP and the CSI checking packet, an update sequence number is also used. Each source–destination connection pair is related to an update sequence number, which is stamped by the source. Each time the source device decides to have a route update, the route update sequence number is incremented. When the intermediate device receives such a route update packet, it first compares the update sequence number in the packet with the last related route update sequence number that it has seen before. If the update sequence number in the packet is greater, it unicasts the route update packet to the next downstream devices[3] and updates its routing table entry to record its next upstream and downstream devices. If the update sequence number is smaller, the devices can ignore the route update packet. On the other hand, if the route update is based on CSI checking packets, the related route update packet has the following format: type of

[2]The source device waits for a certain time period so that it may receive all the CSI checking packets; during this period, the RREP also reaches the source.
[3]Note that there is a list of intermediate devices in the route update packet.

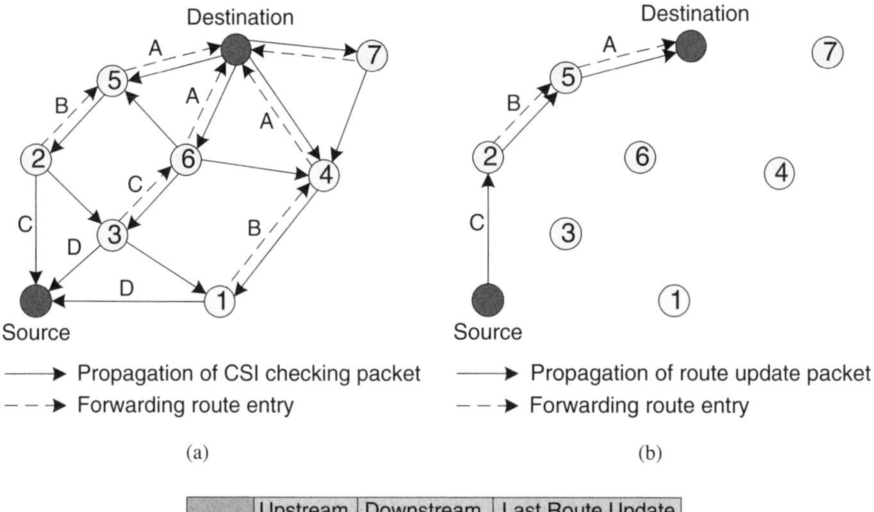

Node	Upstream Node	Downstream Node	Last Route Update Sequence Number
S	—	2	93
1	S	4	91
2	S	5	93
3	S	6	92
4	1	D	91
5	2	D	93
6	3	D	92
7*	—	—	—
D	5	—	93

* The routing table entry has expired.

(c)

Figure 12.3. Illustration of the route updating mechanism based on the CSI checking packets: (a) broadcast of CSI checking packets; (b) selection of the best route; (c) routing table.

packet, source address, destination address, hop count, update sequence number, the intermediate device's (the device sending this packet) address, and route entry ID. If an intermediate device receives such a route update packet, it also compares the update sequence number as mentioned above. If the sequence number in the packet is greater, it further checks the route entry ID in the packet to see whether it matches the one kept in its forwarding route entry in order to ensure the freshness of the route. If the checking is positive, the device unicasts the packet to the downstream device as indicated in its forwarding route entry, and updates its routing table so as to record the next upstream and downstream devices of the route.

The route updating process is illustrated in Figures 12.3 and 12.4. The solid arrows depict the propagation of CSI checking packets, while the dotted arrows represent the setup of forwarding route entries. As can be seen from Figure 12.3, when the route update packet reaches the destination, a new route

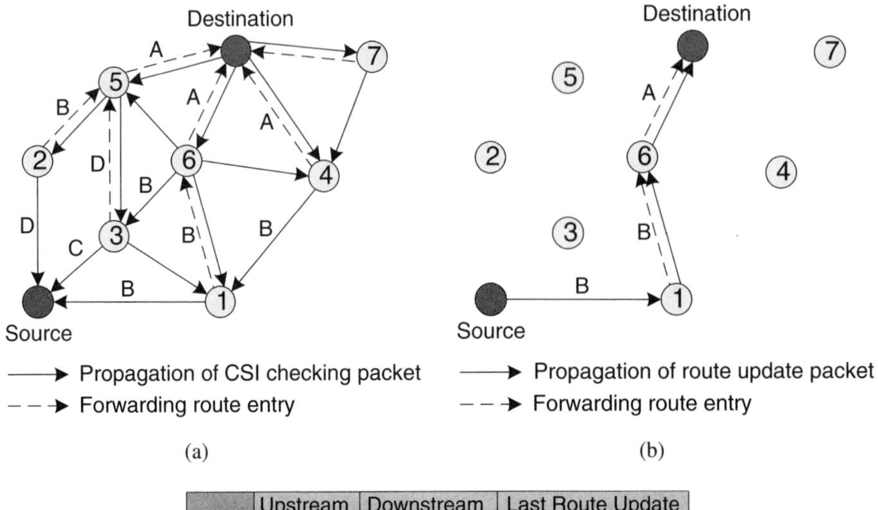

Node	Upstream Node	Downstream Node	Last Route Update Sequence Number
S	—	1	94
1	S	6	94
2	S	5	93
3	S	6	92
4	—	—	—
5	2	D	93
6	1	D	94
7*	—	—	—
D	6	—	94

* The routing table entry has expired.

(c)

Figure 12.4. Illustration of the route updating mechanism based on the CSI checking packets with a slightly different broadcasting sequence: (a) broadcast of CSI checking packets; (b) selection of the best route; (c) routing table.

from the source to destination is set up. The route entry corresponding to a source–destination pair in each device is shown in Figure 12.3c. We can see that the current full route is S-2–5-D and that each device in the route is with the greatest last route update sequence number (i.e., 93) in its route entry. Note that the greater this number is, the fresher the route would be. As time passes, the link quality may change and the destination device may initiate another round of broadcast of CSI checking packets and a new route may be set up as illustrated in Figure 12.4. The updating process is the same as that described above. Now the full route is S-1–6-D, and each device in the route has a greater last route update sequence number (i.e., 94). Note that some old route entries may automatically expire and be deleted because they are obsolete after a specified period of time has elapsed.

TABLE 12.1. Summary of Protocol Characteristics

Feature AODV [112]	DSDV [110]	DSR [62]	ABR [118]		RICA
Power consumption	Low	High	Low	Low	Medium–high
Source routing?	No	No	Yes	No	No
Routing metric	Shortest path	Shortest path	Shortest path	Associativity of device and link load	Shortest path (CSI-based)
Alternative route available?	No	Yes	Yes	No	No
Adaptive to CSI?	No	No	No	No	Yes
Protocol type	Demand-driven	Table-driven	Demand-driven	Demand-driven	Demand-driven

12.4.6 Comparison with Other Protocols

Table 12.1 summarizes the comparisons among key characteristics of the RICA protocol with those of other contemporary protocols.

12.5 PERFORMANCE RESULTS

In this section, we present the results obtained in our extensive simulations comparing the four protocols RICA, AODV, DSDV, and DSR considered in this chapter. These four protocols have been implemented[4] in the standard NS-2 environment [105], which is described below.[5]

12.5.1 Simulation Environment

In our simulation environment, we use an indoor wireless channel model capturing fast fading and long-term shadowing factors. Specifically, we incorporate our channel model as described in Section 12.4.1 into the NS-2 platform. This additional channel model component provides a time-varying transmission environment to all four protocols simulated under the NS-2 system. The maximal transmission ranges for four modulation schemes are specified as

[4]Only RICA is coded by ourselves; the other three already exist in NS-2.

[5]Note that ABR and link state protocols are not included in our comparison because the standard NS-2 platform does not include these two protocols.

follows: 70m for 1Mbps, 60m for 2Mbps, 45m for 5.5Mbps, and 35m for 11Mbps. To model an ad hoc network, we also use the distributed-coordination function (DCF) in the simulated IEEE 802.11b wireless LAN. Using a collision avoidance scheme as well as handshaking with request-to-send/clear-to-send (RTS/CTS) exchanges between the sender and receiver and acknowledgment (ACK) from the receiver, packets can be reliably unicast between any two neighbors within an appropriate range. Through the exchange of RTS/CTS/DATA/ACK, the MAC protocol can detect any data link disconnection with its neighbors and then report this to the network layer. In all the simulations, the broadcast packets (such as RREQ) and control packets (such as RTS/CTS and ACK) are transmitted at the basic data rate of 1Mbps. Other simulation parameters used are as follows:

- *Testing field*—200 × 200m; such a large field can model the environment in a shopping mall or an exhibition center.
- *Mobile speed*—uniformly distributed between 0 and MAXSPEED (which will be elaborated later).
- *Mobility model*—we use the random waypoint model as defined in the movement files in the NS-2 system: when the device reaches its destination, it pauses for 3s and then randomly chooses another destination point within the field with a randomly selected constant velocity.
- *Traffic load*—10 source–destination pairs for the 50-device scenarios and 20 pairs for the 100-device scenarios; in the former test cases, the traffic load varies from 10 to 15 packets per second; in the latter test cases, the traffic loads varies from 5 to 8 packets per second.
- *Simulation time*—600s.

In addition, the data packet size is 512 bytes and the capacity of data buffer size is set to 50. The transmission of packets is a store-and-forward process. When a packet reaches an intermediate device, it waits in the queue for service in a first-come first-served (FCFS) manner. Each packet can remain in the buffer for no more than 30s such that if it has not been transmitted during this period, it will be discarded. Such a relatively short timeout period is chosen because we would like to exert a high pressure on the routing protocols to test their responsiveness in dealing with congested routes (possibly due to poor channel qualities in some links). Finally, the generation of data packets in each source device is based on a constant-bit-rate (CBR) traffic source defined in the standard NS-2 simulator platform. Each simulation scenario is repeated 10 times with a different random seed, and each data point is the average of these 10 trials. The various parameter values of the four protocols are listed in Tables 12.2–12.5. To compare the routing algorithms, we use three major performance metrics as described below.

- *Average End-to-End Delay.* Measured in milliseconds (ms), the end-to-end delay includes the processing time, queueing time, and transmission time of packets in each device in the route.

TABLE 12.2. Parameter Values of AODV

Parameter	Value
Active route timeout	5 s
Reverse route timeout	5 s
Time to keep a broadcast ID	6 s
Broadcast “hello” message?	No
MAC link breakage detection?	Yes
Local route repair?	No
Expanding ring route search?	Yes

TABLE 12.3. Parameter Values of DSDV

Parameter	Value
Periodic route update interval	9 s
Periodic updates missed before link declared broken	3 s
Initial triggered update weighted settling time	6 s
Weighted settling time weighting factor	$\frac{7}{8}$
Route advertisement aggregation time	1 s
Maximum packets buffered per destination device	5

TABLE 12.4. Parameter Values of DSR

Parameter	Value
Time between retransmitted requests	0.5 s
Size of source route header carrying n addresses	$(4n + 4)$ bytes
Timeout for nonpropagation search	30 ms
Time to hold packets awaiting routes	30 s
Maximum rate for sending replies for a route	$1\,s^{-1}$

TABLE 12.5. Parameter Values of RICA

Parameter	Value
Active route timeout	5 s
CSI broadcast interval	4 s
Time to keep a broadcast ID	6 s
Broadcast “hello” message?	No
MAC link breakage detection?	Yes
Time to wait for receiving all RREQ/CSI checking packets	80 ms
Maximum time of unicast retry	3 s

- *Successful Percentage of Packet Delivery*. This is the ratio of packets reaching the destination to total packets generated in the sources. A packet may be dropped either if there is not enough data buffer due to the congestion or if it has stayed in the buffer for more than 30 s.
- *Routing Control Overheads*. This parameter reflects the efficiency of the routing protocol and is measured in bits per second (bps). We count the

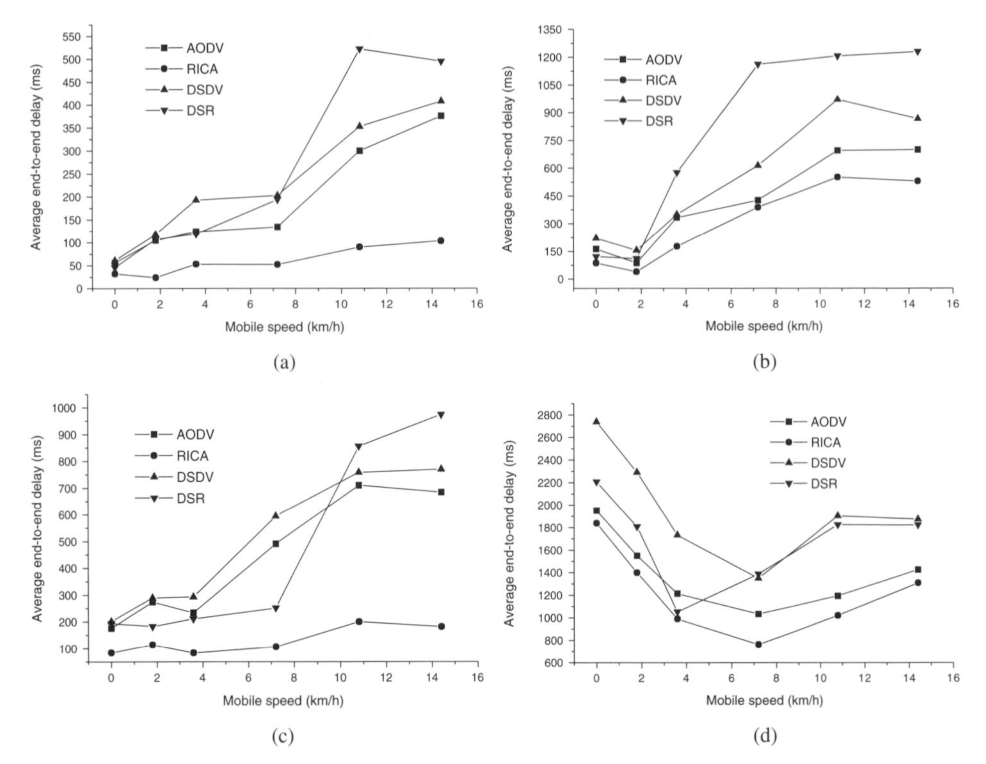

Figure 12.5. Average end-to-end delays of all protocols: (a) 50 devices, 10 packets per second (pps); (b) 50 devices, 15 pps; (c) 100 devices, 5 pps; (d) 100 devices, 8 pps.

total number of routing control packets in each round of simulation. We then average the amount of routing control overheads (in bits) to the whole simulation time.

12.5.2 Average End-to-End Delay

The first set of results is the average packet end-to-end delay against mobile speed with the traffic load varying from 10 to 15 packets per second. The mean mobile speed is varied from 0 to 14.4 km/h and thus, the maximum speed is varied from 0 to 28.8 km/h. This speed range can reasonably model a stationary user, a user moving with pedestrian speed, or a running user. As can be seen in Figure 12.5, taking the CSI into consideration can greatly shorten the end-to-end delay from the source to the destination in the RICA protocol, which outperforms the other three reactive and proactive protocols for the following reasons:

- The source can update the route to the destination frequently and adaptively to the change of the CSI of the links in a route. Indeed, a chosen route is temporarily the shortest one with a much better channel quality than those found by the AODV, DSR, and DSDV protocols. Thus, the transmission of the packet is greatly speeded up.

- The periodic update of the route is adaptive to the geographically sensitive changes of the CSI, which occurs frequently in an indoor environment (e.g., a shopping mall). Thus, packets for the same source–destination pair (i.e., in the same session) can travel through different routes, and load balancing is therefore automatically achieved. Indeed, in the RICA protocol, a busy device will refrain from forwarding RREQ and CSI checking packets, thereby shifting the load to other devices that are having fewer burdens. Such a load balancing effect can help keep the packet queue short and hence reduce the transmission delay.
- Using the CSI checking mechanism, sometimes a full broadcast in search of a route can be avoided. The data queueing delay at the source will be reduced because the source device can swiftly choose a route to the destination. In addition, the CSI checking mechanism can also result in shorter routes, which reflect the current topology changes and thus, the delay incurred at the extra intermediate devices will be reduced.

In all four routing algorithms, when the traffic loads are relatively light, the end-to-end delay increases with the increase in speed of the mobile devices. The reason is that when the mobility is increased, there will be more link breakage. Therefore, the data packets have to be buffered in the source device and wait for the route recovery. This will obviously increase the packet end-to-end delay. However, when we test all four protocols in the heavy load scenario (i.e., 8 packets per second with 20 source–destination pairs) under a denser environment (100 devices in the field), the end-to-end delay is the highest when the devices in the network are in low mobility. This can be explained by the following argument. For this rather high traffic load, more devices are competing for the wireless channel with only a very limited bandwidth; thus, a long queue in each device can very easily form because the route is longlived when the mobility is low. As the mobile speed increases, a long queue is less likely to form since link breakage occurs more often, and this leads to (1) most packets in the broken route being dropped and (2) traffic load spread to a larger number of devices and therefore fewer devices being unduly burdened. These two effects decrease the packet queueing delay at the source but at the same time increase the number of dropped packets as detailed below. The same phenomenon has also been observed in simulations reported by Johansson et al. [61] and Perkins et al. [111], respectively.

12.5.3 Successful Percentage of Packet Delivery

From the simulation results shown in Figure 12.6, we can see that taking CSI into consideration contributes to the reliability of packet delivery. Indeed, we can see that the RICA protocol outperforms the other three routing proto-

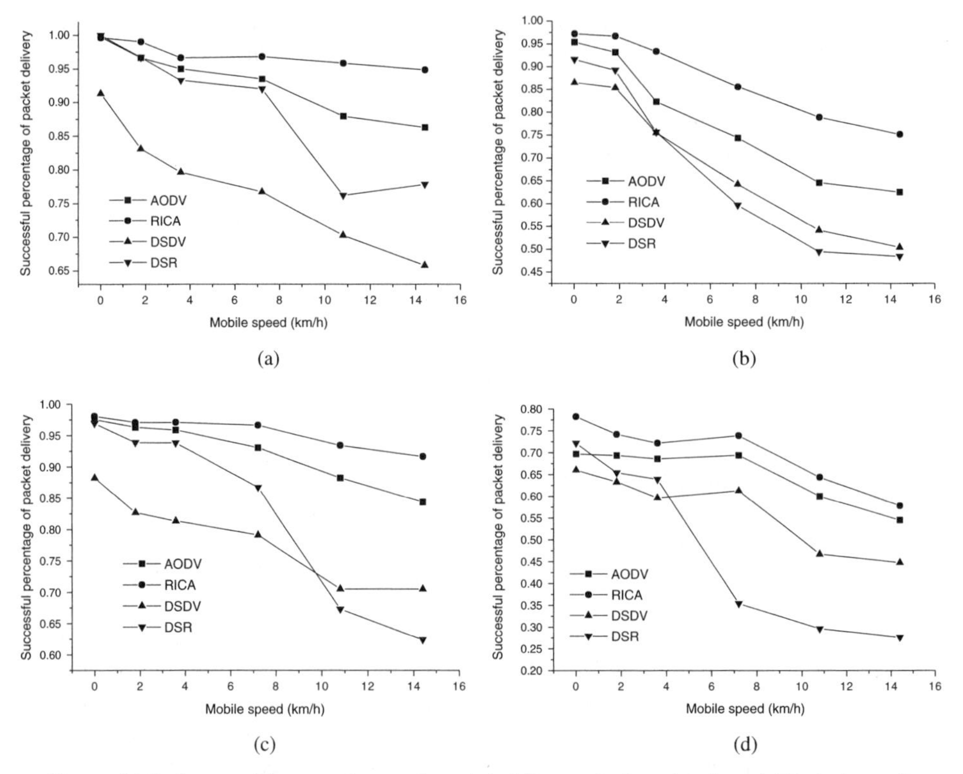

Figure 12.6. Successful percentages of packet delivery of all protocols: (a) 50 devices, 10 packets per second (pps); (b) 50 devices, 15 pps; (c) 100 devices, 5 pps; (d) 100 devices, 8 pps.

cols under this performance metric. On scrutiny of the behaviors of the RICA protocol, we find that

- Usually, the links chosen by RICA are with a higher throughput, and this can help avoid the discarding of packets due to buffer full.
- Frequent and adaptive updates of the route and the load balancing effect can facilitate even distribution of the traffic in the network and therefore, no link is particularly overloaded.

In the other three routing protocols, however, the CSI is not considered. Therefore, they cannot respond timely enough to the change of the link throughput, which fluctuates with time. A long queue is thus more likely to form when the channel quality of some links is not good. Congestion and data loss will also result.

As we have expected, packet delivery rate of all three protocols decreases with the increase of the mobility because the link breakage happens more often. The packet delivery rate also decreases with increasing traffic load because more packets are dropped as a result of the data buffer saturation. The relationship between offered traffic loads and aggregate network throughput is discussed latter.

12.5.4 Routing Control Overhead

To study the efficiency of the routing protocols, we measure the routing control overhead at the MAC layer including all the routing overheads at the IP layer and the ACK, RTS, and CTS MAC control packets that are used to exchange information to unicast data and routing packets. The results are shown in Figure 12.7.

In RICA, we have seen that taking channel quality into consideration when choosing a route can improve the network performance in terms of delay and packet delivery rate. But the tradeoff is that a higher routing control overhead is needed because of (1) broadcast of periodical CSI checking packets to the source and (2) higher traffic delivery rate (thus generating more MAC control overheads).

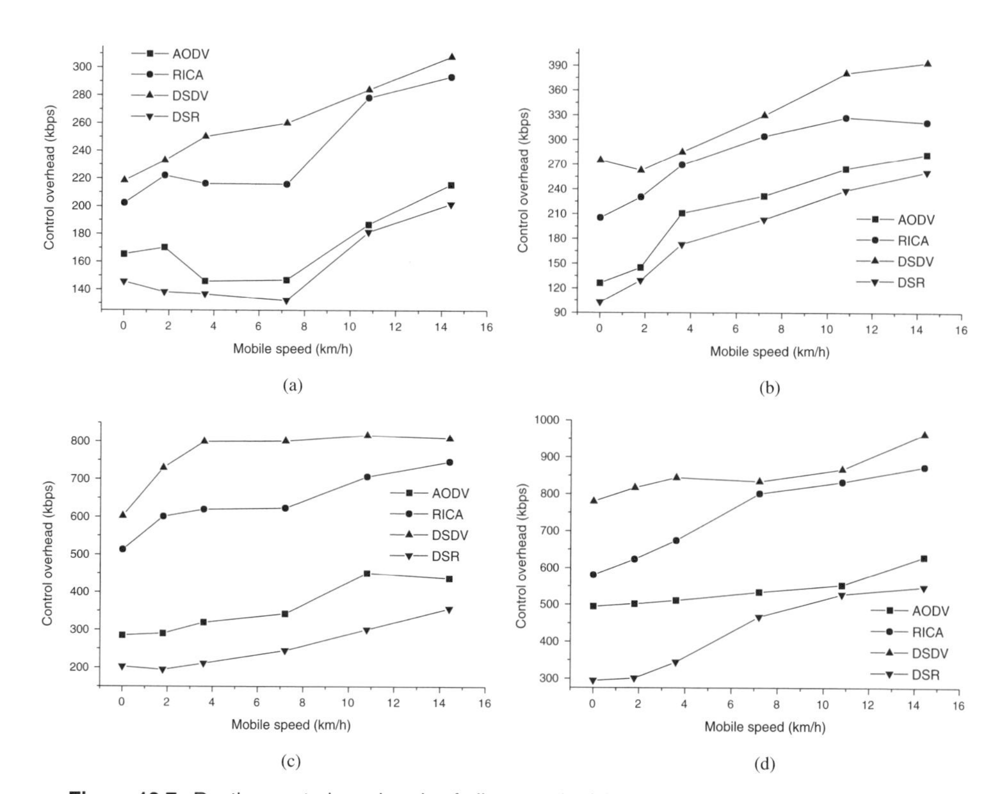

Figure 12.7. Routing control overheads of all protocols: (a) 50 devices, 10 packets per second (pps); (b) 50 devices, 15 pps; (c) 100 devices, 5 pps; (d) 100 devices, 8 pps.

On the other hand, DSR has the least level of routing control overhead due to increased access to route information and aggressive use of route cache. Thus, the RREQ flooding by the source is not as frequent as in AODV and RICA. DSDV, however, generates the greatest amount of routing control overhead. This is a typical drawback of table-driven routing protocols. In order to maintain consistency of the routing table, each device has to periodically broadcast all its routing information to its neighbors. Usually, the whole routing table is huge and much information in the routing table has never been used. Doing so is a waste of limited bandwidth of the wireless channel. Just as expected, routing control overhead in all algorithms increases with mobility because link breakage ismore frequent so that a higher load is needed in route maintenance.

12.5.5 Scalability

To test the scalability of the three protocols, we further increase the number of devices to 100 in a field; thus the population density is twice that of before. The number of source–destination pairs is increased to 20. The traffic load is varied from five to eight packets per second. The average end-to-end delay, packet delivery rate, and routing control overhead are shown in parts (c) and (d) of Figures 12.5, 12.6, and 12.7, respectively. As can be seen, RICA still outperforms the other three routing protocols in that higher data throughput and shorter end-to-end delay are achieved.

12.5.6 Varying Offered Traffic Load

To test the network performance under different levels of offered load, we increase the load until the network becomes saturated. All the protocols are tested in the same two rectangular fields with different populations, and the numbers of source–destination pairs are 10 and 20, respectively. The mobile speed is fixed at 7.2 km/h, which is a reasonable pedestrian speed. The average end-to-end delay, aggregate network throughput, and control overhead are shown in Figures 12.8 and 12.9. Note that *aggregated network throughput* here refers to the total amount of data (in bits) reaching the destination devices per second. Figure 12.8 depicts the performance of protocols with different levels of offered load for 50 devices with 10 source–destination pairs. As can be observed, RICA has much greater aggregate network throughput than do the other three protocols, which are saturated earlier at a relatively light load. We can also see that the network capacity of RICA is about 1.2–1.3 times higher than that of AODV and DSDV and nearly 2 times higher than that of DSR. The DSR protocol has the lowest aggregate throughput because the aggressive use of the route cache and the lack of an effective mechanism to eliminate the aged routes, which might pollute the whole network. The use of an aged route leads to waste of bandwidth and delay in route reconstruction

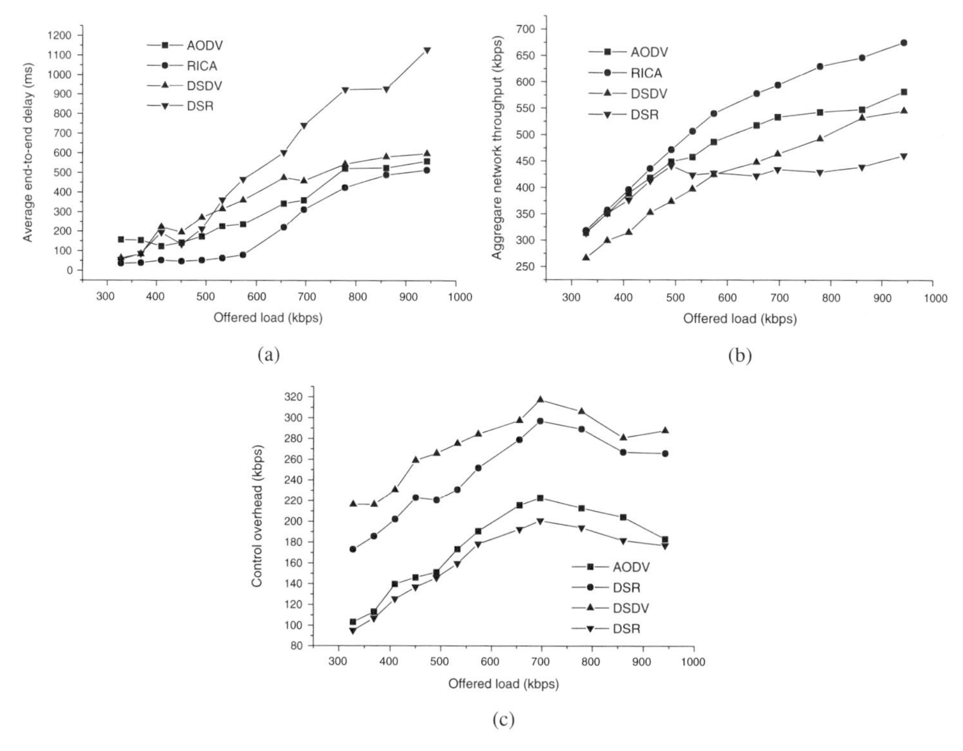

Figure 12.8. Performance of all protocols under various levels of offered load for cases with 50 devices and 10 source–destination pairs: (a) end-to-end delay; (b) throughput; (c) control overhead.

by the source. Thus, this leads to deterioration of network performance. For the end-to-end delay, RICA again outperforms the others because of the higher channel quality. Again, the price to pay is that the routing control overhead is higher than AODV and DSR, while close to DSDV in the case of extremely high load. The results of 100 devices with 20 source–destination pairs show similar trends and are depicted in Figure 12.9.

12.5.7 Quality of Routes

It is also interesting to compare the quality of the routes selected by different algorithms. Figure 12.10 shows the average link throughput, which is defined as the total bandwidth of the link that all packets reaching destinations have traversed, divided by the total number of hops. This parameter reflects the quality of the selected links in each routing algorithm. As can be observed, the average link throughput in other three protocols are very close to each other because channel quality was not considered in route selection.

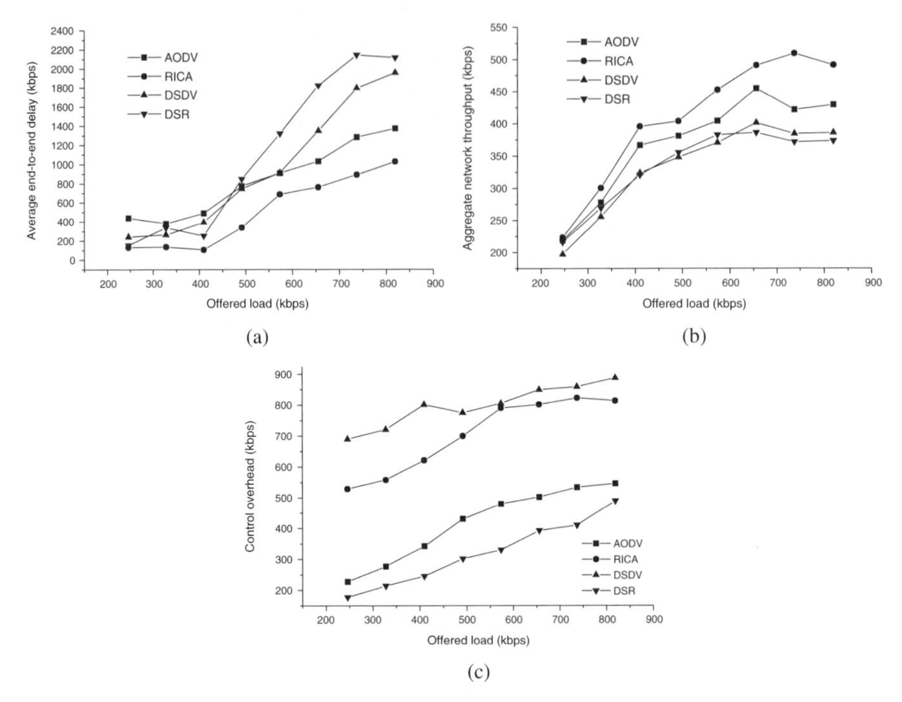

Figure 12.9. Performance of all protocols under various levels of offered load for cases with 100 devices and 20 source–destination pairs: (a) end-to-end delay; (b) throughput; (c) routing control overhead.

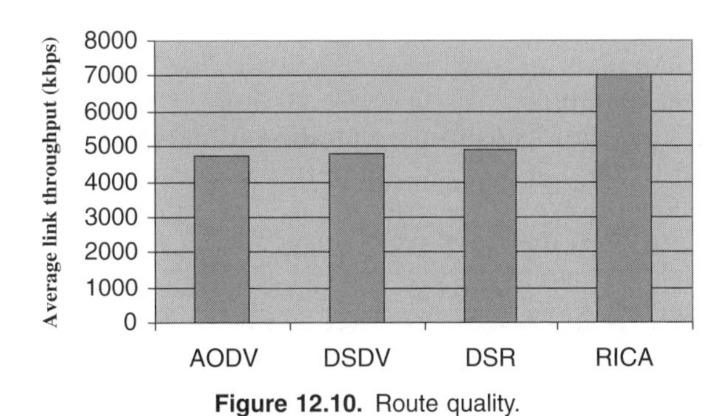

Figure 12.10. Route quality.

12.5.8 Observations in the Simulations

Through the periodic broadcast of CSI checking packets by destinations, RICA chooses routes with high throughputs to render the packet routing adaptive to a complicated CSI fluctuating environment. However, we might ask: How effective is RICA as each destination periodically broadcast CSI-checking packets in such bandwidth limited wireless environment? Could a broadcast storm occur and cause the congestion of the whole network?

To avoid such a broadcast storm, we have taken two measurements: (1) limit the broadcast scope of the CSI checking packet by setting its TTL field as the original hop distance plus one; and (2) if a CSI-checking packet comes from a link in deep fading, the receiving device simply discards this packet instead of rebroadcasting it out.

To answer these questions more comprehensively, we have further studied the relationship between the performance of RICA and the broadcast period together with the channel characteristic. The effectiveness of RICA is constrained by two factors: (1) the "time variation" of the wireless channel and (2) the broadcast period of the CSI checking packet. Here, the time variation of the wireless channel refers to the coherence time of the shadowing component.

Obviously, the time variation of the wireless channel has a direct influence on the routing protocol. If the channel quality changes too rapidly (e.g., less than 1 s), the routing protocols in the network layer cannot react quickly enough to adapt to the time variation inthe physical layer. Thus, we naturally want to know the scope of applicability of RICA. Here, the broadcast period is also another important factor. If the period is too short, a great amount of overhead will be generated, leading to congestion of the network. If the period is too long, the device cannot react timely to the change of the wireless channel. So the question becomes: What is the optimal value?

To obtain quantitative answers to these questions, we have also done another set of extensive simulations so as to determine the relationship between these factors and the scope of applicability of RICA. The results are listed in Figure 12.11. We take AODV as the reference because it is the best protocol among the three other protocols as seen in the abovementioned simulation results. Here, we normalize the results of RICA in that the average end-to-end delay and throughput are divided by those of AODV. For example, when the link's coherence time is 1 s and the CSI checking period is also 1 s,

Normalized performance gain of RICA

	CSI Checking Packet Broadcast Period (seconds)											
Link Change Speed (seconds)	1		2		3		4		5		6	
	Delay	Throughput	Delay	Throughput	Delay	Throughput	Delay	Throughput	Delay	Throughput	Delay	Throughput
1	1.080	0.923	—	—	—	—	—	—	—	—	—	—
2	1.058	1.079	0.845	1.112	—	—	—	—	—	—	—	—
3	0.983	1.067	0.786	1.114	0.641	1.126	—	—	—	—	—	—
4	0.891	1.107	0.394	1.136	0.459	1.137	0.488	1.127	—	—	—	—
5	1.048	1.074	0.665	1.108	0.481	1.116	0.533	1.114	0.591	1.117	—	—
6	0.829	1.107	0.556	1.119	0.262	1.153	0.374	1.133	0.451	1.124	0.426	1.138

Figure 12.11. Normalied performance improvements of RICA with respect to AODV.

the delay achieved by RICA is 1.08 times that of AODV, indicating that the timespan of a stable channel quality is so short (i.e., 1 s) that the channel adaptation performed by RICA (using CSI checking) is rather ineffective. Indeed, we can see that if the link quality changes too fast (e.g., 1 s; a situation that can happen if the background interference and/or mobility is extremely high), RICA is not applicable as the CSI information will soon be outdated. Thus, RICA has to broadcast CSI checking packets more frequently (every second), and this leads to network congestion. When the channel state changes relatively slowly (≥2 s), the RICA becomes more efficient. The CSI broadcast period also has a direct influence on performance, as can be seen in Figure 12.11. Broadcasting too fast or too slowly is not good. We can see from the Figure that the optimal value is around 3 s.

12.5.9 Critiques on the Four Protocols

Four routing protocols have been studied in detail in our simulations. RICA is the best because it is devised for such a CSI fluctuating environment, which is obviously better than the other three routing protocols. It is very effective in small- or moderate-scale networks. However, in larger networks (e.g., 1000 devices) with more source–destination pairs (e.g., 100 pairs) within a small area, it might not perform well. This is because in such a scenario, the destination has to broadcast CSI checking packets periodically, which might be a waste of limited bandwidth and battery power.

DSDV is a table-driven protocol, which exchanges routing information among adjacent devices. Normally, this information is the whole routing table, which might be very band width-consuming. Another drawback of DSDV is that it lacks an effective mechanism to timely recover the broken routes, which might lead to dropping of the data packets.

In contrast, DSR uses source routing to deliver packets to the destination. This routing mechanism lightens the burden of the intermediate devices, but it also has a severe drawback—it does not apply any aging mechanism to the cached routes that have expired. The aged routes can pollute the route caches of other devices because the intermediate devices may have stale routes stored in their route caches and pass the information about these invalidated routes to the source. Using these invalidated routes by the source potentially leads to the drop of a large amount of packets and more delay as the source may need more time to reconstruct a valid route. This will be more obvious when the mobility and traffic load is high. In the worst case, DSR performs even poorer than DSDV. In terms of routing control overhead, DSR in fact is not so conservative. Note that we have not taken the source route in the header of each data packet into account. If we have considered these routing overheads, the amount of overheads by DSR might be comparable to that of AODV. Thus, considering the analysis described above, DSR might not be so bandwidth-saving as it seems to be.

On the contrary, AODV does not have as much access to the routing information as does DSR and therefore, must resort to route recovery more often. AODV is a strong candidate protocol for ad hoc routing (to be standardized by IETF). However, there is still room for improvement. In AODV, route selection is based on the smallest hop count. This may cause an unfair burden on some intermediate devices, which might create a bottleneck of the network. Thus, load balancing cannot be achieved as in RICA.

12.6 SUMMARY

We have presented a channel-adaptive ad hoc routing protocol that works by dynamically updating routes through judiciously selecting links with better channel quality (and hence, higher effective bandwidths). Under a wide range of performance metrics, including delays, packet delivery rates, overheads, scalability, offered load, and route quality, our simulations using the standard NS-2 platform indicate that the RICA protocol outperforms three contemporary well-known protocols. The applicability of the RICA protocol is also analyzed through a careful scrutiny of the relationship between (1) variation time (as governed by various extents of fading effects) and (2) checking period, of the channel quality.

EXERCISES

(1) What is ad hoc networking? Illustrate several application examples of ad hoc network.

(2) Compared to traditional wireless networks such as cellular networks, what are the differences of wireless ad hoc networks? Try to show these differences at the lower three OSI layers.

(3) Compared with wired line communication, the wireless communication is constrained by the limitations of the radio channel. Please elaborate the factors that influence the wireless channel.

(4) From the point of view of route maintenance, the ad hoc routing protocols can be divided into two classes: table-driven and on-demand. What are the advantages and disadvantages of these two kinds of routing protocols respectively?

(5) At the network layer, explain the concept of "channel adaptive" Why we need a routing protocol that is channel adaptive?

(6) In this chapter, we have proposed a routing protocol that is channel adaptive—RICA. Is RICA a table-driven protocol or on-demand protocol?

(7) How can RICA adapt the packet routing to a channel time-varying environment?

(8) Any routing protocol should have its own shortcomings. Raise your viewpoints on the weakness of RICA. Try to suggest how to improve it?

REFERENCES

1. 3GPP, *Physical Channels and Mapping of Transport Channels onto Physical Channels (FDD)*, 3GPP Technical Specifications, TS 25.211, 2004.
2. 3GPP, *Physical Layer Aspects of UTRA High Speed Downlink Packet Access*, 3GPP Technical Specifications, TS 25.848, 2004.
3. 3GPP, *Spreading and Modulation (FDD)*. 3GPP Technical Specifications, TS 25.213, 2004.
4. IEEE Standard 802.11 b/D5.0, *Wireless LAN Medium Access Control (MAC) and Physical Layer (PHY) Specifications: Higher Speed Physical Layer (PHY) Extension in the 2.4 GHz Band*, IEEE, April 1999.
5. I. C. Aboy-Faycal, M. D. Trott, and S. Shamai, The capacity of discrete-time memoryless Rayleigh-fading channels, *IEEE Trans. Inform. Theory* **47**(4):1290–1301 (May 2001).
6. N. Abramson, The throughput of packet broadcasting channels. *IEEE Trans. Commun.* **COM-25**(1):117–128 (Jan. 1977).
7. A. Agnetis, G. Brogi, G. Ciaschetti, P. Detti, and G. Giambene, Optimal packet scheduling in UTRA-TDD, *IEEE Commun. Lett.* **7**(3):112–114 (March 2003).
8. D. Agrawal, T. J. Richardson, and R. L, Urbanke, Multiple-antenna signal constellations for fading channels, *IEEE Trans. Inform. Theory* **47**(6):2618–2626 (Sept. 2001).
9. S. M. Alamouti, A simple transmit diversity technique for wireless communications, *IEEE J. Select. Areas Commun.* **16**(8):1451–1458 (Oct. 1998).
10. A. Amraoui, G. Kramer, and S. Shamai, Coding for the MIMO broadcast channels, *Proc. Int. Symp. Information Theory,* June–July 2003.
11. M. Andrews, K. Kumaran, K. Ramanan, A. Stolyar, R. Vijayakumar, and P. Whiting, *CDMA Data QoS Scheduling on the Forward Link with Variable Channel Conditions*, Bell Laboratories Technical Memorandum, April 2000.

Channel-Adaptive Technologies and Cross-Layer Designs for Wireless Systems with Multiple Antennas: Theory and Applications. By V. K. N. Lau and Y.-K. R. Kwok
ISBN 0-471-64865-5

12. M. Andrews, K. Kumaran, K. Ramanan, A. Stolyar, R. Vijayakumar, and P. Whiting, Scheduling in a queuing system with asynchronously varying service rates, *Probab. Eng. Inform. Sci.* **18**:191–217 (2004).
13. M. Andrews, K. Kumaran, K. Ramanan, A. Stolyar, P. Whiting, and R. Vijayakumar, providing quality of service over a shared wireless link, *IEEE Commu. Mag.* **39**(2):150–154 (Feb. 2001).
14. J. Bannister, P. Mather, and S. Coope, *Convergence Technologies for 3G Networks: IP, UMTS, EGPRS, and ATM*, Wiley, 2004.
15. P. A. Bello. Characterization of randomly time-variant linear channels, *IEEE Trans. Commun. Syst.* **11**(4):360–393 (Dec. 1963).
16. P. Bender, P. Black, M. Grob, R. Padovani, N. Sindhushyana, and S. Viterbi, CDMA/HDR: A bandwidth efficient high speed wireless data service for nomadic users, *IEEE Commun. Mag.* **38**(7):70–77 (July 2000).
17. R. A. Berry and E. M. Yeh, Cross-layer wireless resource allocation—fundamental performance limits for wireless fading channels, *IEEE Signal Process. Mag.* **21**(5):59–68 (Sept. 2004) (special issue on signal processing for networking).
18. M. Boaretto, D. Angelini, and M. Zorzi, Fast scheduling policies using WCDMA high speed downlink packet access, *Proc. VTC*, Oct. 2003, Vol. 4, pp. 2650–2654.
19. F. Borgonovo, A. Capone, M. Cesana, and L. Fratta. Performance evaluation of UMTS packet services, *ST J. Syst. Res.* **1**(1):38–56 (Feb. 2004).
20. M. J. Borran, *Non-coherent and Partially Coherent Space-Time Constellations*, PhD thesis, Rice Univ. June 2003.
21. J. Broch, D. A. Maltz, D. B. Johnson, Y.-C. Hu, and J. Jetcheva, A performance comparison of multi-hop wireless ad hoc network routing protocols, *Proc. MOBICOM*, July 1998, pp. 85–97.
22. G. Caire and S. Shamai, On achievable rates in a multi-antenna broadcast downlink, *Proc. 38th Annual Allerton Conf. Communications, Control and Computing*, Oct. 2000, pp. 1188–1193.
23. G. Caire, G. Taricco, and E. Biglieri, Optimal power control for minimum outage rate in wireless communications, *Proc. ICC*, June 1998, Vol. 1, pp. 58–62.
24. Y. Cao and V. O. K. Li, Scheduling algorithms in broadband wireless networks, *Proc. IEEE* **89**(1):76–87 (Jan. 2001).
25. C. J. Chen and L. C. Wang, Coverage and capacity enhancement in multiuser MIMO systems with scheduling, *Proc. GLOBECOM*, Nov.–Dec. 2004, Vol. 1, pp. 101–105.
26. D. Chizhik, F. Rashid-Farrokhi, J. Ling, and A. Lozano, Effect of antenna separation on the capacity of BLAST in correlated channels, *IEEE Commun. Lett.* **4**(11):337–339 (Nov. 2000).
27. B. E. Collins and R. L. Cruz, Transmission policies for time varying channels with average delay constraints, *Proc. 37th Allerton Annual Conf. Communication, Control and Computing*, 1999, pp. 709–717.
28. J. H. Conway, R. H. Hardin, and N. J. A. Sloane, Packing lines, planes, etc., packings in Grassmannian spaces, *Exp. Math.* **5**(2):139–159 (1996).
29. M. S. Corson, J. P. Macker, and G. H. Cirincione, Internet-based mobile ad hoc networking, *IEEE Internet Comput.* **3**(4):63–70 (July–Aug. 1999).

30. T. M. Cover and J. A. Thomas, *Elements of Information Theory*, 2nd ed., Wiley, 1991.
31. P. A. Dighe, R. K. Mallik, and S. S. Jamuar, Analysis of transmit-receive diversity in Rayleigh fading, *IEEE Trans. Commun.* **51**(4):694–703 (April 2003).
32. V. A. Dubendorf, *Wireless Data Technologies*, Wiley, 2003.
33. D. A. Eckhardt and P. Steenkiste, Effort-limited fair (ELF) scheduling for wireless networks, *Proc. INFOCOM*, March 2000, Vol. 3, pp. 1097–1106.
34. C. E. Perkins, ed., *Ad Hoc Networking*, Addison-Wesley, 2000.
35. L. D. Davis, ed., *Handbook of Genetic Algorithms*, Van Nostrand Reinhold, New York, 1991.
36. N. Ehsan and M. Liu, Properties of optimal power and admission control for a single user in a time varying wireless channel, *Proc. 42nd Annual Allerton Conf. Communication, Control and Computing*, 2004.
37. A. Ephremides and B. Hajek, Information theory and communication networks: An unconsummated union, *IEEE Trans. Inform. Theory* **44**(6):2416–2434 (Oct. 1998).
38. U. Erez and R. Zamir, *Capacity and Coding for Symmetric Channels with Side Information at the Transmitter*, Technical Report, EE-S-97–10, Tel Aviv Univ. Oct. 1997, pp. 1–18.
39. J. Evans and D. N. C. Tse, Large system performance of linear multiuser receivers in multipath fading channels, *IEEE Trans. Inform. Theory* **46**(6):2059–2078 (Sept. 2000).
40. H. Fattah and C. Leung, An overview of scheduling algorithms in wireless multimedia networks, *IEEE Wireless Commun.* **9**(5):76–83 (Oct. 2002).
41. G. J. Foschini and M. J. Gans, On limits of wireless communications in a fading environment using multiple antennas, *Wireless Personal Commun.* **6**(3):315–335 (March 1998).
42. R. Fourer, D. M. Gay, and B. W. Kernighan, *AMPL: A Modeling Language for Mathematical Programming*, Duxbury Press, Wadsworth Publishing Company, 1992.
43. R. Gallager, A perspective on multiaccess channels, *IEEE Trans. Inform. Theory* **31**(2):124–142 (Oct. 1985).
44. R. G. Gallager, *Information Theory and Reliable Communications*, Wiley, 1968.
45. H. E. Gamal and A. R. Hammons Jr., A new approach to layered space-time coding and signal processing, *IEEE Trans. Inform. Theory* **47**(6):2321–2334 (Sept. 2001).
46. A. Ganz, Z. Ganz, and K. Wongthavarawat, *Multimedia Wireless Networks: Technologies, Standards, and QoS*, Prentice-Hall, 2003.
47. J. J. Garcia-Luna-Aceves and S. Murthy, A path-finding algorithm for loop-free routing, *IEEE/ACM Trans. Networking* **5**(1):148–160 (Feb. 1997).
48. D. J. Goodman, R. A. Valenzuela, K. T. Gayliard, and B. Ramamurthi, Packet reservation multiple access for local wireless communications, *IEEE Trans. Commun.* **37**(8):885–890 (Aug. 1989).
49. K. Halford, S. Halford, M. Webster, and C. Andren, Complementary code keying for rake-based indoor wireless communication, *Proc. ISCAS*, May–June 1999, Vol. 4, pp. 427–430.

50. T. Halonen, J. Romero, and J. Melero, eds., *GSM, GPRS and EDGE Performance: Evolution Towards 3G/UMTS*, 2nd ed., Wiley, 2003.
51. S. V. Hanly and D. N. C. Tse, Multiaccess fading channels: Shannon and delay-limited capacities, *Proc. 33rd Annual Allerton Conf. Communication, Control and Computing*, Oct. 1995.
52. S. V. Hanly and D. N. C. Tse, Multiaccess fading channels—Part II: Delay-limited capacity, *IEEE Trans. Inform. Theory* **44**(7):2816–2831 (Nov. 1998).
53. B. Hassibi and B. M. Hochwald, *How Much Training Is Needed in Multiple-Antenna Wireless Links*, Bell Labs Technical Memorandum, 2000.
54. B. M. Hochwald and T. L. Marzetta, Unitary space-time modulation for multiple-antenna communications in Rayleigh flat fading, *IEEE Trans. Inform. Theory* **46**(2):543–564 (March 2000).
55. B. M. Hochwald, T. L. Marzetta, T. J. Richardson, W. Sweldens, and R. Urbanke, Systematic design of unitary space-time constellations, *IEEE Trans. Inform. Theory* **46**(6):1962–1973 (Sept. 2000).
56. J. H. Holland, *Adaptation in Natural and Artificial Systems*, Univ. Michigan Press, 1975.
57. H. Holma and A. Toskala, eds., *WCDMA for UMTS: Radio Access for Third Generation Mobile Communications*, 3rd ed., Wiley, 2004.
58. J. M. Holtzman, Asymptotic analysis of proportional fair algorithm, *Proc. PIMRC*, Sept.–Nov. 2001, Vol. 2, pp. 33–37.
59. R. A. Horn and C. R. Johnson, *Matrix Analysis*, 2nd ed., Cambridge Univ. Press, 1985.
60. A. Jalali, R. Padovani, and R. Pankaj, Data throughput of CDMA-HDR: A high efficiency-high data rate personal communication wireless system, *Proc. VTC Spring*, May 2000, Vol. 3, pp. 1854–1858.
61. P. Johansson, T. Larsson, N. Hedman, B. Mielczarek, and M. Degermark, Scenario-based performance analysis of routing protocols for mobile ad-hoc networks, *Proc. MOBICOM*, July 1999, pp. 195–206.
62. D. B. Johnson and D. A. Maltz, Dynamic source routing in ad hoc wireless networks, in T. Imielinski and H. Korth, eds., *Mobile Computing*, Kluwer Academic Publishers, 1996, Chapter 5.
63. N. Joshi, S. R. Kadaba, S. Patel, and G. S. Sundaram, Downlink scheduling in CDMA data networks, *Proc. MOBICOM*, Aug. 2000, pp. 179–190.
64. N. Kahale and P. E. Wright, Dynamic global packet routing in wireless networks, *Proc. INFOCOM*, April 1997, Vol. 3, pp. 1414–1421.
65. S. Kaiser and K. Fazel, A spread-spectrum multi-carrier multi-access system for mobile communications, *Proc. 1st Int. Workshop on Multicarrier Spread Spectrum*, April 1997, pp. 49–56.
66. F. Kelly, Charging and rate control for elastic traffic, *Eur. Trans. Telecommun.* **8**:33–37 (1997).
67. J. H. Kim and J. K. Lee, Capture defects of wireless CSMA/CA protocols in Rayleigh and shadow fading channels, *IEEE Trans. Vehic. Technol.* **48**(4): 1277–1286 (July 1999).
68. K. I. Kim, *Handbook of CDMA System Design, Engineering, and Optimization*, Prentice-Hall, 2000.

69. S. Kittipiyakul and T. Javidi, *Resource Allocation in OFDMA: HowLoad-Balancing Maximizes Throughput When Water-Filling Fails*, UW Technical Report, UWEETR-2004-0007, 2004.
70. K. Kopsa, H. Artes, G. Matz, and F. Hlawatsch, Space-time algorithms for multi-user channel estimation in the downlink of UMTS/TDD, *Proc. ICC* **4**:2406–2410 (May 2003).
71. L. Kleinrock, *Queueing Systems. Vol. 1: Theory*, Wiley, 1975.
72. R. Knopp and P. Humblet, Information capacity and power control in single cell-multi-user communications, *Proc. ICC* **1**:331–335 (June 1995).
73. T. E. Kolding, K. I. Pedersen, J. Wigard, F. Frederiksen, and P. E. Mogensen, High speed downlink packet access: WCDMA evolution, *IEEE Vehic. Technol. Soc. News* 4–10, (Feb. 2003).
74. J. Korhonen, *Introduction to 3G Mobile Communications*, 2nd ed., Artech House, 2003.
75. S. V. Krisnamurthy, A. S. Acampora, and M. Zorzi, Polling-based media access protocols for use with smart adaptive array antennas, *IEEE/ACM Trans. Networking* **9**(2):148–161 (April 2001).
76. R. Kuehner, T. D. Todd, F. Shad, and V. Kezys, Forward-link capacity in smart antenna base stations with dynamic slot allocation, *IEEE Trans. Vehic. Technol.* **50**(4):1024–1038 (July 2001).
77. S. S. Kullkarni and C. Rosenberg, Opportunistic scheduling for wireless systems with multiple interfaces and multiple constraints, *Proc. MSWiM*, Sept. 2003.
78. Y.-K. Kwok and V. K. N. Lau, A quantitative comparison of multiple access control protocols for wireless ATM, *IEEE Trans. Vehic. Technol.* **50**(3):796–815 (May 2001).
79. Y.-K. Kwok and V. K. N. Lau, System modeling and performance evaluation of rate allocation schemes for packet data services in wideband CDMA systems, *IEEE Trans. Comput.* **52**(6):804–814 (June 2003).
80. A. Lapidoth and I. E. Telatar, The compound channel capacity of a class of finite-state channels. *IEEE Trans. Inform. Theory* **44**(3):973–983 (May 1998).
81. V. K. N. Lau, Variable rate adaptive channel coding for DS-CDMA, *Bell Labs Tech. J.* 38–52 (Nov. 2001).
82. V. K. N. Lau, Performance analysis of variable rate: Symbol-by-symbol adaptive bit interleaved coded modulation for Rayleigh fading channels, *IEEE Trans. Vehic. Technol.* **51**(3):537–550 (May 2002).
83. V. K. N. Lau, Optimal downlink space-time scheduling design with convex utility functions—multiple antenna systems with orthogonal beamforming, *IEEE Trans. Vehic. Technol.* (in press).
84. V. K. N. Lau and Y. K. Kwok, On the synergy between adaptive physical layer and multiple-access control for integrated voice and data services in cellular wireless network, *IEEE Trans. Vehic. Technol.* **51**(6):1338–1351 (Nov. 2002).
85. V. K. N. Lau, Y. J. Liu, and T. A. Chen, On the design of MIMO block-fading channels with feedback-link capacity constraint, *IEEE Trans. Commun.* **52**(1):62–70 (Jan. 2004).

86. S.-J. Lee, M. Gerla, and C.-K. Toh, A simulation study of table-driven and on-demand routing protocols for mobile ad hoc networks, *IEEE Network* **13**(4):48–54 (July–Aug. 1999).
87. M. Lei, X. Tao, and P. Zhang, A novel channel estimation method for combating fast fading in TDD system, *Proc. PIMRC*, Sept. 2002, Vol. 5, pp. 2107–2111.
88. E. Leonardi, M. Mellia, F. Neri, and M. Ajmone Marson, Bounds on average delays and queue size averages and variances in input-queued cell-based switches, *Proc. INFOCOM*, April 2001, Vol 2, pp. 1095–1103.
89. P. Lin, B. Bensaou, Q. L. Ding, and K. C. Chua, CS-WFQ: A wireless fair scheduling algorithm for error-prone wireless channels, *Proc. 9th Int. Conf. Computer Communications and Networks*, Oct. 2000, pp. 276–281.
90. X.-H. Lin, Y.-K. Kwok, and V. K. N. Lau, BGCA: Bandwidth guarded channel adaptive routing for ad hoc networks, *Proc. WCNC*, March 2002, Vol. 1, pp. 433–439.
91. X. Liu, E. K. P. Chong, and N. B. Shroff, Opportunistic transmission scheduling with resource-sharing constraints in wireless networks, *IEEE J. Select. Areas Commun.* **19**(10):2053–2064 (Oct. 2001).
92. Z. Liu, M. J. Karol, M. El Zarki, and K. Y. Eng, Channel access and interference issues in multi-code DS-CDMA wireless packet (ATM) networks, *Wireless Networks* **2**(3):173–193 (1996).
93. D. J. Love, R. W. Heath Jr., and T. Strohmer, Grassmannian beamforming for multiple-input multiple-output wireless systems, *IEEE Trans. Inform. Theory* **49**(10):2735–2747 (Oct. 2003).
94. D. J. Love, R. W. Heath Jr., and T. Strohmer, Limited feedback precoding for spatial multiplexing systems, *Proc. GLOBECOM*, Dec. 2003, Vol. 4, pp. 1857–1861.
95. S. P. Loyld, Least square quantization in PCM, *IEEE Trans. Inform. Theory* **28**(2):129–137 (March 1982).
96. S. Lu, V. Bharghavan, and R. Srikant, Fair scheduling in wireless packet networks, *IEEE/ACM Trans. Network.* **7**(4):473–489 (Aug. 1999).
97. A. Mantravadi and V. V. Veeravalli, Sum capacity of CDMA systems with multiple transmit antennas, *Proc. ISIT*, June 2002, p. 280.
98. A. W. Marshall and I. Olkin, *Inequalities: Theory of Majorization and Its Applications*, Academic Press, New York, 1979.
99. T. L. Marzetta and B. M. Hochwald, Capacity of a mobile multiple-antenna communication link in Rayleigh flat fading, *IEEE Trans. Inform. Theory* **45**(1):139–157 (Jan. 1999).
100. J. L. Massey. Network Information Theory. *Proceedings of DIMACS Workshop on Network Information Theory*, 2003.
101. M. Morelli, Timing and frequency synchronization for the uplink of an OFDMA system, *IEEE Trans. Commun.* **52**(2):296–306 (Feb. 2004).
102. A. Moustakas, H. Baranger, L. Balents, A. Sengupta, and S. Simon, Communication through a diffusive medium: Coherence and capacity, *Science* **287**:287–290 (Jan. 2000).
103. T. Nandagopal, S. Lu, and V. Bharghavan, A unified architecture for the design and evaluation of wireless fair queueing algorithms, *Wireless Network* **8**(2/3): 231–247 (March/May 2002).

104. S. G. Nash and A. Sofer, *Linear and Nonlinear Programming*, McGraw-Hill, 1996.

105. M. J. Neely, E. Modiano, and C. E. Rohrs, Power and server allocation in a multi-beam satellite with time varying channels, *Proc. INFOCOM*, 2002, Vol. 3, pp. 1451–1460.

106. The network simulator (NS-2), http://www.isi.edu/nsnam/ns/, 2005.

107. T. S. E. Ng, I. Stoica, and H. Zhang, Packet fair queueing algorithms for wireless networks with location-dependent errors, *Proc. INFOCOM*, March–April 1998, Vol. 3, pp. 1103–1111.

108. T. Ojanperä and R. Prasad, An overview of air interface multiple access for IMT-2000/UMTS, *IEEE Commun. Mag.* **36**(9):82–86,91–95 (Sept. 1998).

109. A. K. Parekh and R. G. Gallagar, A generalized processor sharing approach to flow control in integrated services networks: The single-node case, *IEEE/ACM Trans. Network.* **1**(3):344–357 (June 1993).

110. J. D. Parsons, *The Mobile Radio Propagation Channel*, 2nd ed., Wiley, 2000.

111. C. E. Perkins and P. Bhagwat, Highly dynamic destination-sequenced distance-vector routing (DSDV) for mobile computers, *Proc. ACM SIGCOMM Conf. Communications Architectures, Protocols, and Applications*, Aug. 1994, pp. 234–244.

112. C. E. Perkins, S. R. Das, E. M. Royer, and M. K. Marina, Performance comparison of two on-demand routing protocols for ad hoc networks, *IEEE Personal Commun.* **8**(1):16–28 (Feb. 2001).

113. C. E. Perkins and E. M. Royer, Ad-hoc on-demand distance vector routing, *Proc. WMCSA*, Feb. 1999, pp. 90–100.

114. C. E. Perkins, E. M. Royer, and S. R. Das, Ad hoc on-demand distance vector (AODV) routing, *IETF Internet Draft*, Feb. 2002, draft-ietf-manet-aodv-10.txt.

115. J. G. Proakis, *Digital Communications*, 3rd ed., McGraw-Hill, 1995.

116. X. Qiu, V. O. K. Li, and J.-H. Ju, A multiple access scheme for multimedia traffic in wireless ATM, *Wireless Networks* **1**(3):259–272 (Dec. 1996).

117. P. Ramanathan and P. Agrawal, Adapting packet fair queueing algorithms to wireless networks, *Proc. MOBICOM*, 1998, pp. 1–9.

118. T. S. Rappaport, *Wireless Communications: Principles and Practice*, Prentice-Hall, 1996.

119. E. M. Royer and C.-K. Toh, A review of current routing protocols for ad hoc mobile wireless networks, *IEEE Personal Commun.* **6**(2):46–55 (April 1999).

120. T. M. Schmidl and D. C. Cox, Robust frequency and timing synchronization for OFDM, *IEEE Trans. Commun.* **45**(12):1613–1621 (Dec. 1997).

121. C. E. Shannon, Channels with side information at the transmitter, *IBM J. Res. Devel.* **2**:289–293 (1958).

122. A. Prasad Sistla, O. Wolfson, and Y. Huang, Minimization of communication cost through caching in mobile environments, *IEEE Trans. Parallel Distrib. Syst.* **9**(4):378–390 (April 1998).

123. D. Slepian and H. O. Pollak, Prolate spheroidal wave functions, fourier analysis and uncertainty—i, *Bell Syst. Techn. J.* **40**(1):43–63 (Jan. 1961).

124. N. J. A. Sloane, Packings in Grassmannian spaces. http://www.research.att.com/ ñjas/grass/index.html, 2003.

125. T. Strohmer and R. W. Heath Jr., Grassmannian frames with applications to coding and communications, *Appl. Comput. Harmonic Anal.* **14**(3):257–275 (May 2003).

126. Z. Sun and T. T. Tjhung, On performance analysis and design criteria for trellis coded unitary space-time modulation, *IEEE Commun. Lett.* **7**(4):156–158 (April 2003).

127. I. E. Telatar, Capacity of multi-antenna Gaussian channels, *Eur. Trans. Telecommun.* **10**(6):585–595 (Nov./Dec. 1999).

128. C. K. Toh, A novel distributed routing protocol to support ad-hoc mobile computing, *Proc. 1996 IEEE 15th Annual Int. Phoenix Conf. Computers and Communications*, March 1996, pp. 480–486.

129. C.-K. Toh, M. Delwar, and D. Allen, Evaluating the communication performance of an ad hoc wireless network, *IEEE Trans. Wireless Commun.* **1**(3):402–414 (July 2002).

130. C. K. Toh, *Ad Hoc Mobile Wireless Networks: Protocols and Systems*, Prentice-Hall, 2002.

131. C. H. Tse, K. W. Yip, and T. S. Ng, Performance tradeoffs between maximum ratio transmission and switched-transmit diversity, *Proc. PIMRC*, Sept. 2000, Vol. 2, pp. 1485–1489.

132. D. N. C. Tse, Optimal power allocation over parallel Gaussian broadcast channels, *IEEE Trans. Inform. Theory* 731–739 (Sept. 2001).

133. D. N. C. Tse and S. V. Hanly, Multiaccess fading channels—Part I: Polymatroid structure, optimal resource allocation and throughput capacities, *IEEE Trans. Inform. Theory* **44**(7):2796–2815 (Nov. 1998).

134. G. Ungerboeck, Channel coding with multilevel/phase signals, *IEEE Trans. Inform. Theory* **IT-28**(1):55–67 (Jan. 1982).

135. G. Ungerboeck, Trellis-coded modulation with redundant signal sets Part II: State of the art, *IEEE Commun. Mag.* **25**(2):12–21 (Feb. 1987).

136. R. van Nee, G. Awater, M. Morikura, H. Takanashi, M. Webster, and K. W. Halford, New high-rate wireless LAN standards, *IEEE Commun. Mag.* **37**(12):82–88 (Dec. 1999).

137. S. Verdu, *Multiuser Detection*, Cambridge Univ. Press, 1998.

138. S. Verdu and T. S. Han, A general formula for channel capacity, *IEEE Trans. Inform. Theory* **40**(4):1147–1157 (July 1994).

139. H. Viswanathan and K. Kumaran, *Rate Scheduling in Multiple Antenna Downlink Wireless Systems*, Bell Labs Technical Memorandum 10009626-010720-01TM, July 2001.

140. H. Viswanathan, S. Venkatesan, and H. Huang, Downlink capacity evaluation of cellular networks with known-interference cancellation, *IEEE J. Select. Areas Commun.* **21**(5):802–811 (June 2003) (special issue on MIMO).

141. B. Vucetic and J. Yuan, *Space-Time Coding*, Wiley, 2003.

142. B. Walke, P. Seidenberg, and M. P. Althoff, UMTS: *The Fundamentals*, Wiley, 2003.

143. L. Wang, Y.-K. Kwok, W. C. Lau, and V. K. N. Lau, Channel adaptive fair queueing for scheduling integrated voice and data services in multicode CDMA systems, *Comput. Commun.* **27**(9):809–820 (June 2004).

144. L. Wang, Y.-K. Kwok, W. C. Lau, and V. K. N. Lau, Efficient packet scheduling using channel adaptive fair queueing in distributed mobile computing systems, *ACM/Kluwer Mobile Networks Appl.* **9**(4):297–309 (Aug. 2004).

145. L. Wilhelmsson and L. B. Milstein, On the effect of imperfect interleaving for the Gilbert-Elliott channel, *IEEE Trans. Commun.* **47**(5):681–688 (May 1999).

146. E. M. Yeh, *Multiaccess and Fading in Communication Networks*, PhD thesis, MIT, Sept. 2001.

147. E. M. Yeh and A. Cohen, Throughput and delay optimal resource allocation in multiaccess fading channels, *Proc. ISIT*, June–July 2003, p. 245.

148. T. Yoo and A. Goldsmith, Capacity of fading mimo channels with channel estimation error, *IEEE Trans. Wireless* (in press).

149. J. Zander, Performance of optimum transmitter power control in cellular radio systems, *IEEE Trans. Vehic. Technol.* **41**(1):57–62 (Feb. 1992).

150. H. Zhang, Service disciplines for guaranteed performance service in packet-switching networks, *Proc. IEEE* **83**(10):1374–1396 (Oct. 1995).

151. W. Zhao, G. Leus, and G. B. Giannakis, Orthogonal design of unitary constellations for uncoded and trellis-coded noncoherent space-time systems, *IEEE Trans. Inform. Theory* **50**(6):1319–1327 (June 2004).

152. L. Zheng and D. N. C. Tse, Diversity and multiplexing: A fundamental tradeoff in multiple antenna channels, *IEEE Trans. Inform. Theory* **49**(5):1073–1096 (May 2003).

153. M. Zorzi and R. R. Rao, Error control and energy consumption in communications for nomadic computing, *IEEE Trans. Comput.* **46**(3):279–289 (March 1997).

154. M. Zorzi, R. R. Rao, and L. B. Milstein, On the accuracy of a first-order Markov model for data transmission on fading channels, *Proc. ICUPC*, Nov. 1995, pp. 211–215.

TOPIC INDEX

Channel-Adaptive Technologies and Cross-Layer Designs for Wireless Systems with Multiple Antennas: Theory and Applications. By V. K. N. Lau and Y.-K. R. Kwok
ISBN 0-471-64865-5